ESO ASTROPHYSICS SYMPOSIA
European Southern Observatory

Series Editor: Bruno Leibundgut

H. Böhringer   G.W. Pratt   A. Finoguenov
P. Schuecker

Heating versus Cooling in Galaxies and Clusters of Galaxies

Proceedings of the MPA/ESO/MPE/USM
Joint Astronomy Conference held
in Garching, Germany, 6-11 August 2006

Springer

Volume Editors

Hans Böhringer
Gabriel W. Pratt
Alexis Finoguenov
Peter Schuecker

Max-Planck-Institut
für extraterrestrische Physik
Giessenbachstraβe
85748 Garching
Germany

Series Editor

Bruno Leibundgut
ESO
Karl-Schwarzschild-Str. 2
85748 Garching
Germany

ISBN 978-3-642-09257-2   e-ISBN 978-3-540-73484-0

Springer is a part of Springer Science+Business Media
springer.com
© Springer-Verlag Berlin Heidelberg 2007
Softcover reprint of the hardcover 1st edition 2007

Cover design: WMXDesign, Heidelberg

Peter Schuecker
1959 - 2006

Our dear friend, colleague and co-editor, Peter Schuecker, died suddenly just as we commenced preparations to edit this volume. It was with great sadness that we realised the loss of an exceptional scientific colleague and friend. Peter was deeply and profoundly devoted to his astrophysical and cosmological studies. He set himself high goals and brought a perfectionist's eye to testing hypotheses and to the derivation of scientific results. His quest for ever-deeper understanding of his and other's work stimulated those working with him, and the persistence and purity of his scientific approach will always remain a model for us. His friendly, insistent discussion of scientific issues was enriching, and we will miss these exchanges very deeply.

Peter's untimely death occurred at a time when his long involvement in many observational cosmology projects appeared to be reaching a culminating point. He was the scientific PI of several important projects aimed at a precise measurement of the large-scale structure, with the potential of providing stringent tests for cosmological models with various forms of 'Dark Energy', using data from KIDS, PanSTARRS, HEDTEX and the eROSITA X-ray survey. It is unfortunate that he could not complete the work that he so diligently prepared himself for. Alongside his scientific work, Peter made time to advise, teach and encourage others.

In recognition and appreciation of his outstanding scientific personality we would like to honour his memory with these proceedings.

Preface

The Universe would be a dull and dark place if the gaseous baryons did not cool, collapse, and form stars and galaxies. However, if this gas is allowed to cool unimpeded at the rate predicted from known atomic physics, in the context of a well-established cosmological model, the gaseous matter would form stars and galaxies with a high efficiency, so that far more than the observed fraction of about 10–15% of the baryonic matter would be found in luminous stellar systems. Therefore, cooling must be damped or regulated by heating processes, and observations show that this 'feedback' is a widespread astrophysical phenomenon.

The place where this cooling and feedback manifests itself most dramatically is in the inner regions of the most massive Dark Matter halos found in our Universe, in the cores of galaxy clusters. The effect of cosmic cooling in these objects was recognized in the mid-1970s, and the term 'cooling flow' was introduced by Andy Fabian and colleagues at Cambridge as a possible explanation of the physical processes obtained in the central regions of clusters. The name refers to the fact that gas cooling in the inner regions is associated with a loss of central pressure, causing an inflow of hot gas from larger radii.

'Cooling flows' soon became a hotly debated astrophysical topic, especially when it was realized that, while the hot intracluster plasma was observed to radiatively cool at high rates in X-rays, no residual matter was found at lower temperatures in longer wavelength observations. Also, no correspondingly strong signatures of star formation were found, as would be expected if the gas would enter a 'cooling catastrophe'. It became clear that complex modelling was required to find a solution to this problem, with many parameters for which it was imperative to use constraints from detailed observational studies. Progress in this field is, therefore, very much observationally driven, requiring data throughout almost all of the electromagnetic spectrum. Much observational and theoretical expertise has to be brought together to solve this fundamental and complex astrophysical problem.

Within this context of combining a wide range of expertise, well-organized conferences, at which the full scope of research on 'cooling flows' can be reviewed, constitute important events. The first such conference was organized by Andy Fabian and his team at Cambridge in 1987. Attended by all the experts in the field, covering the rich 'cooling flow' phenomenolgy at all wavelengths, it was a great success. The following were among the important questions posed at this conference: What is the fate of the cooling gas? Why does the enormous energy input from the central radio AGN not leave more traces in the structure of the thermal intracluster plasma observed in X-rays?

This original conference set the model for a series of conferences to follow. The next conference, organized by Noam Soker in Israel in 1995, reviewed in particular the insights gained with the then newly launched X-ray satellites ROSAT and ASCA. The third meeting, organized by Craig Sarazin and Thomas Reiprich in Charlottesville, Virginia, in 2003, allowed a first discussion of the paradigm change initiated by results from the new X-ray satellites XMM-Newton and Chandra. Insights included the realization that spectral signatures of massive cooling are absent in X-rays while, conversely, clear signatures of AGN interaction with the intracluster medium in several cooling flow clusters are indeed observed, suggesting that AGN may be the source of heating.

In recent years the effort to understand the dense central regions of clusters has grown, both in terms of observation and in terms of detailed numerical hydrodynamical simulations. Furthermore, it has been more commonly appreciated that the suppression of gas cooling in the centre of galaxy clusters may be a model for the effects of feedback in galaxy and structure formation in general. In our meeting we consequently broadened the view to include feedback and self-regulation during galaxy formation.

The conference 'Heating versus Cooling in Galaxies and Clusters of Galaxies' was held in Garching from 6–11 August 2006, as part of the joint astronomy conference series involving the four Munich/Garching astronomy institutes (the Max Planck Institutes for Astrophysik and extraterrestriche Physik, the European Southern Observatory, and the Universitäts Sternwarte München of the Ludwig-Maximillians Universität). The meeting was attended by more than 125 participants from 13 countries all over the world. There was an enthusiastic response from the worldwide scientific community, and the conference was attended by many of the scientists active in this field.

The conference sessions comprised the topics of the observational phenomenology of elliptical galaxy and galaxy cluster cooling cores, of the detailed observations and modelling of the AGN–ICM interaction in cooling cores, observations of the thermal and entropy structure of cooling core clusters, and the efforts of their theoretical understanding, the diagnostic role of metal abundances in clusters, and the significance of other ICM components as cosmic rays and magnetic fields. The results of ambitious numerical simulations of galaxy clusters and their cooling cores were presented, and the effects of cooling cores on the application of clusters for cosmological model testing were discussed. In a final session a connection was made between heating in clusters and the role of feedback in galaxy evolution, and the observations and modelling of heating versus cooling in galaxy formation were reviewed.

Each session was introduced by one or two excellent invited review talks, and followed by contributed talks providing the latest results in the field. A very busy poster session completed the spectrum of presentations, which altogether provided a very comprehensive overview of the state of the art of this research field. We thank all the conference participants whose enthusiasm greatly contributed to the success of this meeting.

We are grateful to ESO for help with Local Organisation Committee logistics and for sponsoring these proceedings and to MPA and MPE for inviting some of the review speakers as institute guests and for essential help with technical issues. We would like to thank the Scientific Organization Committee for help in the planning of the conference, and the Local Organizing Committee for their great commitment to the smooth running of the meeting. We would also like to thank Ms Pamela Bristow from ESO for support in the editing of this book.

Finally, we would like to thank all those who have contributed to this book, and we hope that you will enjoy it both as a record of the scientific content and as a memory of the meeting.

Our conference webpage http://www.mpe.mpg.de/ cool06, which will be maintained for some time, provides further material concerning the conference, including the visual presentations of most of the talks and posters.

Scientific Organizing Committee:
M. Arnaud, M. Begelman, H. Böhringer, M. Donahue, A. Fabian, G. Hasinger, T. Heckman, C. Jones, B. McNamara, T. Ohashi, F. Owen, M. Pettini, T. Reiprich, A. Renzini, P. Rosati, C. Sarazin, N. Soker, R. Sunyaev, S. White.

Local Organizing Committee:
H. Böhringer, B. Boller, F. Braglia, E. Churazov, E. Collmar, M. Depner, R. Fassbender, A. Finoguenov, W. Frankenhuizen, A. Langer, D. Pierini, G.W. Pratt, C. Rickl, P. Rosati, J. Santos, P. Schuecker, A. Simionescu, Y.-Y. Zhang.

Garching,   *Hans Böhringer*
December 2006   *Gabriel W. Pratt*
                                                        Alexis Finoguenov

Contents

Part II AGN-ICM Interaction in Cool Cores

Part VII Optical and Sub-mm Observations of Cool Cores

Part VIII Numerical Simulations and Cosmological Applications

Part IX Suzaku Observations

Part X Heating vs. Cooling in Galaxy Formation

**Powerful Radio Galaxies at z=2–3: Signposts of AGN
Feedback in the Early Universe**
N.P.H. Nesvadba and M.D. Lehnert 416

**2dF–SDSS LRG and QSO (2SLAQ) Survey: Evolution of the
Most Massive Galaxies**
*R. C. Nichol, R. Cannon, I. Roseboom and David Wake, for the
2SLAQ Collaboration* 422

Part XI Conference Summary

Heating vs. Cooling 2006: Conference Summary
R. Bower 429

Part XII Colour Plates

List of Participants

Akahori, Takuya
Tokyo Metropolitan University
akataku@phys.metro-u.ac.jp

Allen, Steven
KIPAC (Stanford University and
SLAC)
swa@stanford.edu

Arnaud, Monique
Saclay
marnaud@discovery.saclay.cea.fr

Asai, Naoki
Chiba University
asai@astro.s.chiba-u.ac.jp

Balestra, Italo
MPE
balestra@mpe.mpg.de

Balogh, Michael
University of Waterloo
mbalogh@uwaterloo.ca

Belsole, Elena
IoA Cambridge
elena@ast.cam.ac.uk

Best, Philip
IfA Edinburgh
pnb@roe.ac.uk

Biernacka Monika
Swietokrzyska Academy
bmonika@pu.kielce.pl

Birzan, Laura
Ohio University
birzan@helios.phy.ohiou.edu

Blanton, Elizabeth
Boston University
eblanton@bu.edu

Böhringer, Hans
Max-Planck-Institut fuer extraterr.
Physik
hxb@mpe.mpg.de

Borgani, Stefano
Dept. of Astronomy - University of
Trieste
borgani@oats.inaf.it

Bower, Richard
ICC, University of Durham
r.g.bower@durham.ac.uk

Braglia, Filiberto
MPE
fbraglia@mpe.mpg.de

Briel, Ulrich
MPE
ugb@mpe.mpg.de

Brüggen, Marcus
IUB, Bremen
m.brueggen@iu-bremen.de

Burns, Jack
University of Colorado
jack.burns@cu.edu

Cavaliere, Alfonso
Università di Roma Tor Vergata
cavaliere@roma2.infn.it

Champavert, Nicolas
CRAL (Centre de Recherche
Astronomique de Lyon)
nicolas.champavert@obs.
univ-lyon1.fr

Chen, Yong
IHEP (Institute of High Energy
Physics)
ychen@mail.ihep.ac.cn

Churazov, Eugene
MPA
echurazov@mpa-garching.mpg.de

Clarke, Tracy
Naval Research Laboratory
tracy.clarke@nrl.navy.mil

Crawford, Carolin
IoA, Cambridge
csc@ast.cam.ac.uk

Croston, Judith
University of Hertfordshire
jcroston@star.herts.ac.uk

Dalla Vecchia, Claudio
Leiden Observatory
caius@strw.leidenuniv.nl

David, Laurence
Harvard-Smithsonian Center for
Astrophysics
david@head.cfa.harvard.edu

De Grandi, Sabrina
INAF - Osservatorio Astronomico di
Brera
sabrina.degrandi@brera.inaf.it

de Plaa, Jelle
SRON Netherlands Institute for
Space Research
j.de.plaa@sron.nl

De Young, David
National Optical Astronomy
Observatory
deyoung@noao.edu

Diehl, Steven
Los Alamos National Laboratory
(LANL)
stevendiehl@gmail.com

Docenko, Dimitrijs
Max-Planck-Institut for Astrophysics
dima@mpa-garching.mpg.de

Dogiel, Vladimir
P. N. Lebedev Institute
dogiel@lpi.ru

Donahue, Megan
Michigan State University
donahue@pa.msu.edu

Dunn, Robert
IoA, Cambridge
rjhd2@ast.cam.ac.uk

Dupke, Renato
University of Michigan
rdupke@umich.edu

Edwards, Louise
Université Laval
louise.edwards.1@ulaval.ca

Eilek, Jean
New Mexico Tech
jeilek@aoc.nrao.edu

Ettori, Stefano
INAF Observatory, Bologna
stefano.ettori@oabo.inaf.it

Fabian, Andrew
IoA, Cambridge
acf@ast.cam.ac.uk

Fassbender, Rene
MPE
rfassben@mpe.mpg.de

Finoguenov, Alexis
MPE
alexis@mpe.mpg.de

Forman, William
Harvard-Smithsonian Center for
Astrophysics
wrf@cfa.harvard.edu

Fujita, Yutaka
Osaka University
fujita@vega.ess.sci.
osaka-u.ac.jp

Gastaldello, Fabio
University of California Irvine
gasta@uci.edu

Ghizzardi, Simona
IASF-MILANO/INAF
simona@iasf-milano.inaf.it

Giacintucci, Simona
Istituto di Radioastronomia
sgiaci_s@ira.inaf.it

Gitti, Myriam
Ohio University
gitti@ohio.edu

Graham, James
IoA, Cambridge
jgraham@ast.cam.ac.uk

Griffiths, Richard
Carnegie Mellon University
griffith@astro.phys.cmu.edu

Hallman, Eric
University of Colorado
hallman@casa.colorado.edu

Hardcastle, Martin
University of Hertfordshire
mjh@star.herts.ac.uk

Hatch, Nina
IoA, Cambridge
nah@ast.cam.ac.uk

Heckman, Timothy
Johns Hopkins University
heckman@pha.jhu.edu

Heinz, Sebastian
MIT
heinzs@space.mit.edu

Henry, J. Patrick
IfA-Honolulu / MPE
henry@IfA.Hawaii.Edu

Hopp, Ulrich
USM
hopp@usm.uni-muenchen.de

Hudson, Daniel
AIfA Univeristät Bonn
dhudson@astro.uni-bonn.de

Iapichino, Luigi
Universität Würzburg
luigi@astro.uni-wuerzburg.de

Johnston-Hollitt, Melanie
University of Tasmania
Melanie.JohnstonHollitt@utas.edu.au

Johnstone, Roderick
IoA, Cambridge
rmj@ast.cam.ac.uk

Jones, Christine
CfA, Cambridge
cjf@head.cfa.harvard.edu

Kaiser, Christian
University of Southampton
crk@soton.ac.uk

Kauffmann, Guinevere
MPA
gkauffmann@mpa-garching.mpg.de

Kirkpatrick, Charles
Ohio University
ck345105@ohio.edu

Komossa, Stefanie
MPE
skomossa@mpe.mpg.de

Kravtsov, Andrey
University of Chicago
andrey@oddjob.uchicago.edu

Lagana, Tatiana
Instituto de Astronomia
ferraz@iap.fr

Leccardi, Alberto
Università di Milano / INAF-IASF
Milano
leccardi@iasf-milano.inaf.it

Lorena, Gazzola
University of Nottingham
ppxlg@nottingham.ac.uk

Mazzotta, Pasquale
Università di Roma Tor Vergata
mazzotta@roma2.infn.it

McCarthy, Ian
ICC, Durham University
i.g.mccarthy@durham.ac.uk

McNamara, Brian
Ohio University
mcnamarb@ohio.edu

Mohamed, Nasser
Kapteyn Insitute
nasser@astro.rug.nl

Molendi, Silvano
IASF Milano/INAF
silvano@iasf-milano.inaf.it

Morandi, Andrea
Università di Bolonga
andrea.morandi@studio.unibo.it

Nagai, Daisuke
California Institute of Technology
daisuke@tapir.caltech.edu

Nakazawa, Kazuhiro
ISAS/JAXA
nakazawa@astro.isas.jaxa.jp

Nenestyan, Oxana
AIfA
oxana@astro.uni-bonn.de

Nesvadba, Nicole
MPE
nicole@mpe.mpg.de

Nulsen, Paul
Harvard-Smithsonian Center for
Astrophysics
pnulsen@cfa.harvard.edu

O'Dea, Christopher
Rochester Institute of Technology
odea@cis.rit.edu

O'Sullivan, Ewan
Harvard-Smithsonian Center for
Astrophysics
ejos@head.cfa.harvard.edu

Ohashi, Takaya
Tokyo Metropolitan University
ohashi@phys.metro-u.ac.jp

Okabe, Nobuhiro
Tohoku University
okabe@astr.tohoku.ac.jp

Ota, Naomi
RIKEN
ota@crab.riken.jp

Overzier, Roderik
Leiden Observatory
overzier@strw.leidenuniv.nl

Owen, Frazer
NRAO
fowen@aoc.nrao.edu

Pavlovski, Georgi
University of Southampton
gbp@phys.soton.ac.uk

Pfrommer, Christoph
CITA
pfrommer@cita.utoronto.ca

Pierini, Daniele
MPE
dpierini@mpe.mpg.de

Pizzolato, Fabio
Technion
fabio@physics.technion.ac.il

Poole, Gregory
University of Victoria
gbpoole@uvic.ca

Pope, Edward
University of Southampton
edpope@soton.ac.uk

Pratt, Gabriel
MPE
gwp@mpe.mpg.de

Rafferty, David
Ohio University
rafferty@helios.phy.ohiou.edu

Rasia, Elena
Università di Padova
rasia@pd.astro.it

Rasmussen, Jesper
University of Birmingham
jesper@star.sr.bham.ac.uk

Rebusco, Paola
MPA
pao@mpa-garching.mpg.de

Reiprich, Thomas
Argelander-Institut fuer Astronomie
thomas@reiprich.net

Revaz, Yves
Lerma / Observatoire de Paris
yves.revaz@obspm.fr

Ricker, Paul
University of Illinois at Urbana-
Champaign
pmricker@uiuc.edu

Rosati, Piero
ESO
prosati@eso.org

Rossetti, Mariachiara
INAF-IASF Milano
rossetti@iasf-milano.inaf.it

Rudd, Douglas
University of Chicago
drudd@oddjob.uchicago.edu

Ruszkowski, Mateusz
JILA
mr@quixote.colorado.edu

Sakelliou, Irini
Max-Planck-Institut for Astronomy
irini.s@mpia.de

Salome, Philippe
IRAM
salome@iram.fr

Sanders, Jeremy
IoA, Cambridge
jss@ast.cam.ac.uk

Santos, Joana
MPE
jsantos@mpe.mpg.de

Sarazin, Craig
University of Virginia
sarazin@virginia.edu

Sato, Kosuke
Tokyo Metropolitan University
ksato@phys.metro-u.ac.jp

Scannapieco, Evan
Kavli Institute for Theoretical
Physics
evan@physics.berkeley.edu

Schmidt, Robert
University of Heidelberg
rschmidt@ari.uni-heidelberg.de

Schuecker, Peter
MPE
peters@mpe.mpg.de

Sijacki, Debora
MPA
deboras@mpa-garching.mpg.de

Simionescu, Aurora
MPE
aurora@mpe.mpg.de

Soker, Noam
Technion
soker@physics.technion.ac.il

Statler, Thomas
Ohio University
statler@ohio.edu

Takei, Yoh
ISAS/JAXA
takei@astro.isas.jaxa.jp

Takizawa, Motokazu
Yamagata University
takizawa@sci.kj.yamagata-u.ac.jp

Thomas, Peter
University of Sussex
p.a.thomas@sussex.ac.uk

Tozzi, Paolo
INAF - Osservatorio Trieste
tozzi@ts.astro.it

Vernaleo, John
University of Maryland
vernaleo@astro.umd.edu

Vikhlinin, Alexey
SAO
avikhlinin@cfa.harvard.edu

Voges, Wolfgang
MPE
wvoges@mpe.mpg.de

Voit, Mark
Michigan State University
Voit@pa.msu.edu

von der Linden, Anja
Max-Planck-Institute for
Astrophysics
anja@mpa-garching.mpg.de

Werner, Norbert
SRON National Institute for Space
Research
n.werner@sron.nl

White, Simon
MPA
swhite@mpa-garching.mpg.de

Wise, Michael
University of Amsterdam
wise@science.uva.nl

Zappacosta, Luca
University of California
lzappaco@uci.edu

Zhang, Yu-Ying
MPE
yyzhang@mpe.mpg.de

Part I

Observations of Cooling Cores

Introduction to Cluster Cooling Cores

C.L. Sarazin

Department of Astronomy, University of Virginia, P.O. Box 400325,
Charlottesville, VA 22904-4325, USA
sarazin@virginia.edu

1 Introduction

Following earlier sounding rocket observations, measurements with the Uhuru
satellite showed that clusters of galaxies were, as a class, luminous extended
X-ray sources [17, 19, 24]. Early observations showed that the radiation was
thermal emission from hot gas, indicating that clusters of galaxies were full
of hot intracluster medium (ICM) at a typical temperature of $T \sim 10^8$ K
and electron density of $n_e \sim 10^{-3}$ cm^{-3} (e.g., [40]). This intracluster plasma
dominates the baryon content of clusters, with a total mass of ~ 5 times
larger than that of all the stars and galaxies.

Based on these early observations, it was found that the cooling times in
the centers of clusters could be comparable to the Hubble time [25]; better
resolution observations have shown that the cooling times are often much
shorter. The cooling is radiative, and due to the same X-rays that we observe.
Based on these observations and on theoretical work, it was suggested that
the ICM in the centers of clusters was strongly affected by radiative cooling,
and formed a "cooling flow" [9, 14, 28].

In this paper, I will briefly summarize the early observational and theoretical results on cooling core clusters, and discuss some of the Chandra and
XMM-Newton results from the first few years of their operation. Of course,
more recent results will be covered extensively in the many other papers in
this volume.

2 Early Results

2.1 Early X-ray Results

Early observations of cooling core clusters established a number of basic X-ray
properties. First, these clusters had very bright central peaks in their X-ray
surface brightness. Second, X-ray spectra indicated that the gas had a positive
radial temperature gradient; the gas got cooler as the radius decreased within
the cooling core. Third, the observed gas densities and temperatures implied
that the radiative cooling time of the gas was less than the Hubble time
within the cooling cores. In the densest regions in the centers of cooling core

clusters, the radiative cooling times typically drop to values of $t_{cool} \sim 2 \times 10^8$ yr. These properties were found in a significant fraction of clusters, generally at least 50%. (For reviews of the early properties of cooling cores, see [12, 40].)

This led to a simple theoretical model for cooling flows. The basic assumption was that the energy in the X-rays which we see comes from the thermal energy content of the ICM. A second assumption was that no heating process balances this radiative cooling; this is the idea which now appears to be wrong in detail. Third, it is assumed that the gas at the centers of clusters is not disrupted or disturbed or mixed into the outer ICM. If this is the case, then the gas will cool radiatively. The weight of the surrounding cluster gas will compress the cooling central gas, and it will flow slowing (subsonically) toward the center of the cluster. This will result in denser, cooling gas at the center of the cluster, in agreement with the observations. If no heating process balances the cooling, the gas should cool down to very low temperature, certainly below the X-ray emitting range. If the gas cools nearly isobarically and if the energy input from inflow in the gravitational potential is small, then the total cooling luminosity would be

$$L_{cool} \approx \frac{5}{2} \frac{kT}{\mu m_p} \dot{M} , \qquad (1)$$

where T is the initial temperature of the cooling gas, $\dot{M}$ is the cooling rate (mass per unit time), and μm_p is the mean mass per particle in the gas. Applying this equation to the X-ray surface brightnesses of cooling cores in clusters led to estimated cooling rates of $\dot{M} \gtrsim 100 \, M_\odot/\text{yr}$ in many clusters.

2.2 Early Results in Other Wavebands

Early optical observations established that cluster cool cores were always centered on a large, cluster-dominant elliptical galaxy. These ellipticals were unusual in having extensive systems of optical emission line filaments from gas at $\sim 10^4$ K (e.g., [15, 10, 20, 11]). The central elliptical galaxies in cooling cores also were found to have very blue colors, which indicated that they contained relatively young stars (e.g., [23, 30, 21]). The emission lines and young stars appeared to be correlated with structures in the X-ray and radio images of the same systems (e.g., [42]).

Many cooling core clusters also have dust, and cooler gas observed through IR emission, 21 cm emission and absorption, and CO emission (e.g., [39]).

2.3 Reservoir for the Cooling Gas?

Initially, it was thought that these cooler forms of gas or star formation might represent the ultimate reservoir of the hot gas which was seen to cool through part of the X-ray band. Unless the gas is reheated in some way, radiative cooling should accelerate as the temperature drops. Thus, one might expect the gas to cool down to low temperatures, and form stars. The discovery of significant amounts of cooler gas and of young stars in cooling core ellipticals appeared to be qualitatively consistent with this expectation. Unfortunately, this result did not work quantitatively. The amounts of cool gas were generally much smaller than the estimates of the cooling rates integrated over a significant fraction of the Hubble time. Similarly, the present star formation rates are much lower than the early estimates of the cooling rates. It appeared that a few percent of the cooling gas could be consumed by star formation or stored as cold gas in the observed forms, but not the full expected amount. This discrepancy became known as the "cooling flow problem": where does most of the cooling gas go?

2.4 Early Radio Results

In a very high proportion ($\gtrsim 70\%$) of cooling core clusters, the central galaxy is a moderately bright radio source (e.g., [6]), and this proportion is much higher than for other elliptical galaxies. These radio galaxies are typically FR I sources, with small, spatially distorted structures. The extended portions often have rather steep radio spectra, which is consistent with their being confined by the gas X-ray emitting gas. Many of the brightest, nearby radio galaxies are at the centers of cooling core clusters (e.g., Virgo A, Perseus A, Hydra A, Cygnus A, etc.)

In general, the radio lobes of these cooling core radio sources are strongly polarized, but have extremely large Faraday rotations (e.g., [46]), indicating that there are very strong magnetic fields in cooling cores [45].

3 XMM-Newton and Chandra Results: High Resolution Spectra

The RGS gratings on XMM-Newton (and, to a lesser extent, the gratings on Chandra) have provided high resolution X-ray spectra of a sample of cooling core clusters. In this regard, the low energy grating spectra are particularly important. In the standard cooling flow model without reheating, all of the gas which cools from higher temperatures should continue to cool out of the X-ray band. As the gas cools below 10^7 K, heavy elements will recombine and produce very strong line emission. One particularly strong set of lines produced at lower temperature come from Fe XVII (or Fe^{+16}); these lines mainly lie below 1 keV in the spectrum. The XMM-Newton spectra showed that Fe XVII and other low ionization lines were missing from the spectra of cluster cooling cores at the level predicted by the standard cooling flow model without reheating (e.g., [35]), and this result was confirmed with Chandra.

In detail, the high resolution spectra showed that most of the gas in cluster cooling cores only cools down to $\sim 1/2$ or $1/3$ of its initial temperature

([35]). If the emission measure is modeled as a power-law function of the temperature, the luminosity emitted in each temperature range is found to vary as $dL_X/dT \propto T^{1-2}$, rather than $dL_X/dT \propto T^0$ as predicted by the standard cooling flow model without reheating (Eqn. 1). Thus, there is much less emission at low temperature than expected in this model. The upper limits on the cooling rate to low temperatures (and a few possible detections of low temperature X-ray emission) are all roughly consistent with only 1–10% of the gas cooling at high temperature continuing to low temperatures. These results appear to resolve at least one version of the standard cooling flow problem; the quantities of gas cooling to low temperatures are now consistent with the amounts of cool gas and rates of star formation observed in these systems.

On the other hand, it is likely that some heating process is required to explain the small amounts of gas which continue to cool to low temperatures. In fact, the heating cannot merely balance cooling, which would cause the gas to accumulate at $\sim 1/3$ of its original temperature; gas which has cooled must be reheated back to nearly the ambient temperature. Although very inhomogeneous abundances [32] or nonradiative cooling (e.g., [13, 44]) may also play a role, it is likely that some heating process accounts for most the deficit of low ionization X-ray lines.

Another important result is that the high resolution spectra do not show evidence for large amounts of local X-ray absorption in cooling cores ([35]), as had been suggested based on previous lower resolution spectra (e.g., [49]).

The existing high resolution spectra of cooling cores mainly give upper limits to the rate of cooling to very low temperatures. Since some cooler gas and star formation are seen in these systems, it is likely that some gas is cooling to very low temperatures, at least part of the time. As noted above, the existing limits are generally consistent with the amounts of cool gas and star formation. It would be very useful to improve the spectral measurements to either give measured values for the cooling rate to low temperatures, or to give upper limits which were inconsistent with the measurements of cooler materials.

4 Chandra and XMM-Newton Results: Cluster Mergers and Cooling Cores

Clusters of galaxies form hierarchically by the merger of smaller systems, which I will refer to as "subclusters." In these mergers, the subclusters collide at speeds of ~ 2000 km s^{-1}. Although the most common events involve the merger of systems with very disparate masses, major mergers also occur in which two already quite massive clusters collide to form a particularly large system. These major mergers are the most energetic events which have occurred in the Universe since the Big Bang. They can release energies of

$\sim 10^{64}$ erg, of which $\sim 10^{63}$ erg will be dissipated in the intracluster medium through shocks.

4.1 Cold Fronts

Early Chandra images showed very sharp X-ray surface brightness discontinuities in merging clusters [26, 48]. These were not merger shocks, as spectral observations showed that the denser gas was also cooler, had lower entropy, and had about the same pressure as the less dense gas. Instead, these features, which are called "cold fronts", appear to be contact discontinuities between gas which originally was the cool core of one of the merging clusters and more diffuse, hot, shocked gas from the other subcluster [26, 48]. In some cases, double cold fronts are seen in merging clusters. Based on simple hydrodynamical arguments, such cold fronts can be used to determine the kinematics of mergers in clusters (e.g., [48, 41]).

In some cases, the cold fronts have long tails of cool gas behind them (e.g., [37]), which are probably due to ram pressure stripping of gas from the cooling core of one of the merging clusters. Cold fronts and these cool tails indicate that cooling cores survive for some period during mergers, but are probably disrupted in the end in major mergers.

4.2 Cold Fronts Due to Sloshing

Many cooling core clusters have weak surface brightness discontinuities near the cooling core, which are probably due to subsonic motions (e.g., [27]). Normally, there is only one cooling core in the region. The lack of a second subcluster center, slower motions, and lack of other evidence of a major merger suggests that these features are not merger cold fronts. Instead, they are probably due to oscillatory motions in the centers of these clusters, due either to the sloshing of the cooling core gas [27] and/or to oscillations of the dark matter potential [47]. These oscillations might be produced by the momentum from AGN jets, or a minor, off-center cluster merger [1].

5 Chandra and XMM-Newton Results: Cooling Cores and Radio Sources

As noted in above, in almost all cases the central galaxies in cooling cores are radio sources. Chandra and XMM-Newton observations have provided dramatic evidence for the interaction of these radio sources with the intracluster gas. For example, Fig. 1 (left) shows the Chandra X-ray image (greyscale) of Abell 2052, with radio contours superposed [4]. In the X-ray, these radio bubbles generally show two holes in the X-ray surface brightness on opposite sides of the galaxy nucleus. In most cases, the holes are surrounded by bright shells of X-ray emission. The X-ray holes correspond to the lobes of the central radio source. For the systems with a relatively simple geometry, deprojection analysis indicates that the X-ray surface brightness in the holes is consistent with foreground and background cluster emission; that is, the holes appear to be empty of X-ray emitting gas. The masses of X-ray gas in the surrounding shells are consistent with the mass which is missing from the holes [3]. All of this is consistent with a picture in which the central radio source has sent out two jets, which have been stopped in the cooling core gas. The jets have inflated two lobes, and the radio plasma has displaced the X-ray gas and compressed it into the two surrounding shells. In a few cases, similar radio bubbles were seen originally with ROSAT [43, 5, 22, 38]. The most spectacular case, in terms of the detail of the observations, is in the Perseus cluster, where there is a nearly million second Chandra exposure [16].

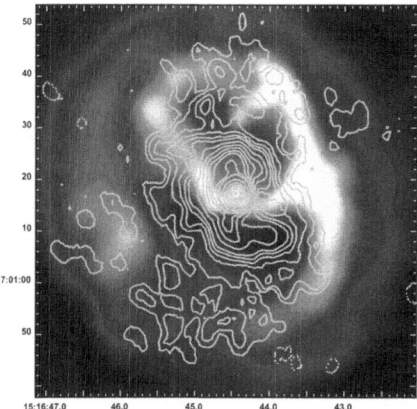 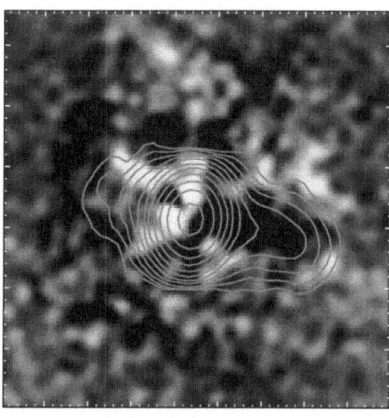

Fig. 1. (Left) Chandra X-ray image (greyscale) of the center of Abell 2052, with radio contours superposed [4, 3]. There are holes in the X-ray image surrounded by bright shells. The radio emission fills the holes. (Right) Greyscale is the Chandra residual image of Abell 2597 after subtraction of an elliptical model, with low frequency (330 MHz) radio contours superposed [8]. The low frequency radio emission extends out into the "ghost bubble" to the west; also, there may be a channel in the X-ray image connecting the ghost bubble with the AGN at the center.

In many cooling core clusters, "ghost bubbles" are also seen at larger radii from the center. These are holes in the X-ray emission without associated high frequency radio emission. Figure 1 (right) shows Abell 2597, an example of a cooling core with ghost bubbles [31]. Recently, low frequency radio images have detected radio emission in many of the ghost bubbles (e.g., [8]). In general, the properties of the ghost bubbles are consistent with older radio bubbles which have risen buoyantly in the cluster atmosphere.

Although the radio bubbles show an anticorrelation between radio and X-ray emission, in a few systems there is evidence for a form of positive correlation. Specifically, columns of cool, dense gas are seen going from the center of the central galaxy out to the radio lobes. Examples include Virgo/M87 [50] and Abell 133 [18] (see Fig. 2 [left]). One suggestion is that these X-ray features are due to cooler gas from the center of the cooling core which has been entrained and uplifted by a buoyant radio lobe.

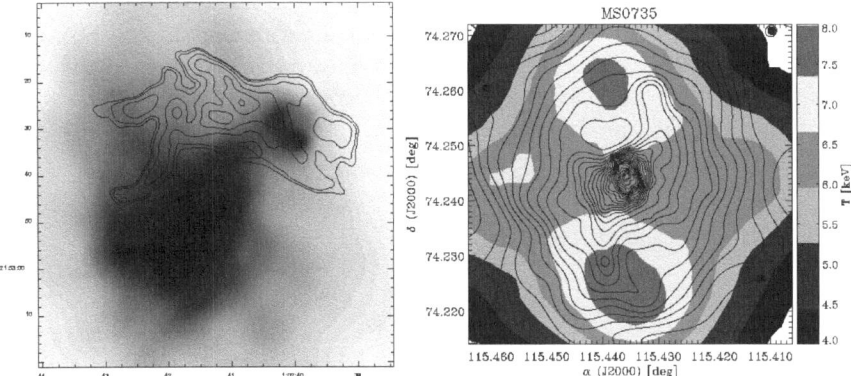

Fig. 2. (Left) Chandra X-ray image (greyscale) of the center of Abell 133, with radio contours superposed [18]. There is a column of cool, dense, X-ray bright gas extending from the nucleus (at the lower left) to the extended radio source (upper right). (Right) Greyscale is the Chandra temperature map of the center of the cluster MS0735.6+7421, while the contours are the X-ray surface brightness [29]. The radio bubbles in this cluster are bounded by hot regions, which indicates that the radio source is driving shocks into the intracluster gas.

X-ray spectral observations show that the X-ray bright shell around radio bubbles are generally cool, and that the pressures in these shells are similar to those of the surrounding hotter gas. This indicates that these shells are not due to shocks, and that the radio sources are not expanding very supersonically. However, recently a few clusters have been found in which the radio lobes are surrounded by shells of hot gas, indicating that the radio sources are driving fairly strong shocks. Examples include MS0735.6+7421 [29], which is shown in Fig. 2 (right), Hydra A [34], and Herc A [33].

For the majority of the bubbles which are expanding subsonically or mildly transonically, the pressure within the bubbles must be comparable to the pressure in the X-ray gas external to the bubbles. For the sources with supersonic expansion, the internal pressures must be even higher. However, when the pressures in the radio lobes are estimated by the standard minimum energy or equipartition arguments, they are found to be ∼20 times smaller than required in most cases (e.g., [3]). This indicates that we have not identified the primary energy content and pressure source within radio sources. It may be that the magnetic fields are larger than given by equipartition. The extra pressure might be due to a large population of low energy relativistic electrons, or to a very large population of relativistic ions. Alternatively, the radio sources may contain diffuse but very hot thermal gas, which provides most of the energy and pressure. Within the radio sources, jet kinetic energy is dissipated by shocks or other frictive processes, so it would not be surprising if most of the energy was thermalized. So far, it has been difficult to detect such hot gas in X-ray spectra of the radio lobes (e.g., [3]). However, it should be detectable with high spatial resolution SZ images [36].

Radio bubbles are very useful systems for determining the total energies supplied by the jets in the radio sources. For the systems which are expanding subsonically, the total energy is the "PdV" work done to displace the X-ray emitting gas, plus the internal energy in the radio bubble [7]. (For systems with shocks, the shock energy needs to be included [29]). This gives a total energy of

$$E_{\mathrm{radio}} = PV + \frac{PV}{\gamma - 1} = \frac{\gamma}{\gamma - 1} PV = (2.5 - 4) PV, \qquad (2)$$

where P is the pressure in the radio lobe, V is its volume, and γ is the mean adiabatic index of the contents of the radio lobe. The two terms in the left side of the equation are the work done by the bubble and its internal energy. The range of values at the right of the equation correspond to the range from non-relativistic gas ($\gamma = 5/3$) to relativistic material ($\gamma = 4/3$). For example, in Abell 2052 the total energy is $E_{\mathrm{radio}} \approx 10^{59}$ ergs [3], which is a typical value. Similar measurements have provided the best direct evidence of the total energy content of radio jets.

In addition, the energy in the radio source can be compared to the cooling X-ray luminosity of the cooling core to see if it is energetically possible for the radio source to inhibit the radiative cooling of the gas. To make this comparison, one needs a time scale for the activity of the radio source; typically, this is given by the buoyancy rise time of the radio bubbles. In most cases, this comparison indicates that energy from the radio source could balance radiative cooling (e.g., [2]), although this is not true in all cases. Radio heating (as well as the common occurrence of radio sources in cooling cores) requires that the radio activity be episodic; it may be that the average radio power is more important than the current value.

6 Conclusions

Recent observations have shown that cluster cooling cores are more complex than was thought a decade ago. They are certainly affected by radiative cooling, but also by AGN heating, and possibly nonradiative cooling, thermal conduction, and cluster hydrodynamics. High resolution X-ray spectra have shown that much less gas is cooling to low temperatures than had been predicted in the past. The limits on the cooling rates are now consistent with the observed rates of star formation and amount of cool gas detected in these systems, thus possibly resolving one version of the "cooling flow problem". However, it would be very useful if low ionization X-ray lines could be observed in a significant number of systems, which would give actual values for the cooling rate, rather than merely limits.

AGN heating is probably the leading candidate for the process which prevents gas from cooling to low temperatures. Although we have abundant evidence for the interaction of the radio sources and X-ray gas, and the energetics appear plausible, we still do not understand in detail how the X-ray gas is heated by the radio sources.

Acknowledgement. Support for this work was provided by the National Aeronautics and Space Administration through Chandra awards GO4-5133X and GO5-6126X, XMM-Newton awards NNG04GO80G, NNG04GO34G, NNG04GP46G, NNG05GA34G, NNG05GO50G, and NNG06GD54G, and Suzaku awards NNX06AI37G and NNX06AI44G.

References

1. Y. Ascasibar, M. Markevitch, 2006, ApJ, 650, 102
2. L. Bîrzan, D. A. Rafferty, B. R. McNamara et al, 2004, ApJ, 607, 800
3. E. L. Blanton, C. L. Sarazin, B. R. McNamara, 2003, ApJ, 585, 227
4. E. L. Blanton, C. L. Sarazin, B. R. McNamara et al., 2001, ApJ, 558, L15
5. H. Böhringer, W. Voges, A. C. Fabian et al., 1993, MNRAS, 264, L25
6. J. O. Burns, 1990, AJ, 99, 14
7. E. Churazov, R. Sunyaev, W. Forman, H. Böhringer, 2002, MNRAS, 332, 729
8. T. E. Clarke, C. L. Sarazin, E. L. Blanton et al, 2005, ApJ, 625, 748
9. L. L. Cowie, J. Binney, 1977, ApJ, 215, 723
10. L. L. Cowie, E. M. Hu, E. B. Jenkins, D. G. York, 1983, ApJ, 272, 29
11. C. S. Crawford, S. W. Allen, H. Ebeling et al., 1999, MNRAS, 306, 857
12. A. C. Fabian, 1994, ARA&A, 32, 277
13. A. C. Fabian, S. W. Allen, C. S. Crawford et al., 2002, MNRAS, 332, L50
14. A. C. Fabian, P. E. J. Nulsen, 1977, MNRAS, 180, 479
15. A. C. Fabian, P. E. J. Nulsen, G. C. Stewart et al., 1981, MNRAS, 196, 35
16. A. C. Fabian, J. S. Sanders, G. B. Taylor et al., 2006, MNRAS, 366, 417
17. W. Forman, E. Kellogg, H. Gursky et al., 1972, ApJ, 178, 309
18. Y. Fujita, C. L. Sarazin, J. C. Kempner et al., 2002, ApJ, 575, 764
19. H. Gursky, A. Solinger, E. M. Kellogg et al., 1972, ApJ, 173, L99
20. T. M. Heckman, S. A. Baum, W. J. M. van Breugel et al., 1989, ApJ, 338, 48
21. A. K. Hicks, R. Mushotzky, 2005, ApJ, 635, L9
22. Z. Huang, C. L. Sarazin, 1998, ApJ, 496, 728
23. R. M. Johnstone, A. C. Fabian, P. E. J. Nulsen, 1987, MNRAS, 224, 75
24. E. Kellogg, H. Gursky, H. Tananbaum et al., 1972, ApJ, 174, L65
25. S. M. Lea, J. Silk, E. Kellogg, S. Murray, 1973, ApJ, 184, L105
26. M. Markevitch, T. J. Ponman, P. E. J. Nulsen et al., 2000, ApJ, 541, 542
27. M. Markevitch, A. Vikhlinin, P. Mazzotta, 2001, ApJ, 562, L153

28. W. G. Mathews, J. N. Bregman, 1978, ApJ, 224, 308
29. B. R. McNamara, P. E. J. Nulsen, M. W. Wise, D. A. Rafferty, C. Carilli, C. L. Sarazin, E. L. Blanton, 2005, Nature, 433, 45
30. B. R. McNamara, R. W. O'Connell, 1989, AJ, 98, 2018
31. B. R. McNamara, M. W. Wise, P. E. J. Nulsen, L. P. David, C. L. Carilli, C. L. Sarazin, C. P. O'Dea, J. Houck, M. Donahue, S. Baum, M. Voit, R. W. O'Connell, A. Koekemoer, 2001, ApJ, 562, L149
32. R. G. Morris, A. C. Fabian, 2003, MNRAS, 338, 824
33. P. E. J. Nulsen, D. C. Hambrick, B. R. McNamara, D. Rafferty, L. Birzan, M. W. Wise, L. P. David, 2005, ApJ, 625, L9
34. P. E. J. Nulsen, B. R. McNamara, M. W. Wise, L. P. David, 2005, ApJ, 628, 629
35. J. R. Peterson, S. M. Kahn, F. B. S. Paerels, J. S. Kaastra, T. Tamura, J. A. M. Bleeker, C. Ferrigno, J. G. Jernigan, 2003, ApJ, 590, 207
36. C. Pfrommer, T. A. Enßlin, C. L. Sarazin, 2005, A&A, 430, 799
37. T. H. Reiprich, C. L. Sarazin, J. C. Kempner, E. Tittley, 2004, ApJ, 608, 179
38. E. Rizza, C. Loken, M. Bliton, K. Roettiger, J. O. Burns, F. N. Owen, 2000, AJ, 119, 21
39. P. Salomé, F. Combes, A. C. Edge et al., 2006, A&A, 454, 437
40. C. L. Sarazin, 1986, Rev. Mod. Phys., 58, 1
41. C. L. Sarazin, The physics of clusters mergers, in: *Merging Processes in Galaxy Clusters*, ed by L. Feretti, I. M. Gioia, G. Giovannini (Kluwer, Dordrecht 2002)
42. C. L. Sarazin, J. O. Burns, K. Roettiger, B. R. McNamara, 1995, ApJ, 447, 559
43. C. L. Sarazin, R. W. O'Connell, B. R. McNamara, 1992, ApJ, 389, L59
44. N. Soker, E. L. Blanton, C. L. Sarazin, 2004, A&A, 422, 445
45. N. Soker, C. L. Sarazin, 1990, ApJ, 348, 73
46. G. B. Taylor, A. C. Fabian, S. W. Allen, 2002, MNRAS, 334, 769
47. E. R. Tittley, M. Henriksen, 2005, ApJ, 618, 227
48. A. Vikhlinin, M. Markevitch, S. S. Murray, 2001, ApJ, 551, 160
49. D. A. White, A. C. Fabian, R. M. Johnstone, R. F. Mushotzky, K. A. Arnaud, 1991, MNRAS, 252, 72
50. A. J. Young, A. S. Wilson, C. G. Mundell, 2002, ApJ, 579, 560

Star Formation and Feedback in Cooling Flows

B. R. McNamara

Dept. of Physics & Astronomy, University of Waterloo, 200 University Avenue
West, Waterloo, Ontario, Canada
mcnamara@uwaterloo.ca

1 Star Formation in cD Galaxies

A small fraction of central dominant galaxies (hereafter referred to as cD
galaxies) harbor star formation proceeding at rates of a few $M_\odot$ yr^{-1}, as in
the disk of the Milky Way, to more than $100\,M_\odot$ yr^{-1} [1,2], rivaling the rates
in burgeoning galaxies at redshifts $z = 2 - 3$. This star formation is fueled
by reservoirs of molecular gas [3] and is often accompanied by spectacular
nebular emission [4, Crawford and Hatch, this volume], as in NGC 1275 of
the Perseus cluster. The incidence of line emission in optically-selected cDs
culled from the Sloan and NOAO Fundamental Plane surveys is $\sim 10\%$,
which is smaller than the line-emitting fraction found in normal galaxies
outside of clusters [5,6]. This fraction rises to $\sim 40\%$ in cDs centered in
dense X-ray atmospheres with short central cooling times ($< 10^9$ yr) [1,5],
and a large fraction of these are forming stars. Thus, several percent of the
general cD galaxy population hosts active star formation, and most if not all is
associated with "cooling flows" ie., bright cusps of X-ray emission associated
with rapidly cooling gas [24]. Their properties have been reviewed previously
in this conference series [7,8], so I will focus here on recent developments
after briefly summarizing the earlier reviews: 1) Star formation occurs in
short $> 10^7$ yr to mid duration $\lesssim 10^9$ yr episodes that are incompatible with
fueling by a long-duration cooling flow. 2) Star formation regions are generally
embedded in X-ray emitting gas with cooling times less than $\sim 3 \times 10^8$ yr.
3) Star formation rates are approaching or agree with the revised (lower)
cooling rates derived from XMM-*Newton*, Chandra, and Fuse, a conclusion
that has strengthened in intervening years. 4) Positive feedback or enhanced
star formation by radio jet interactions is evident in some systems. 5) Disk
star formation rarely occurs in cooling flows.

2 Cooling Flows: A Controversial History

The relationship between cold gas, nuclear radio emission, star formation
and accretion from cooling flows has for decades been a controversial one.
The dispute centered on the gulf between the star formation rates $\dot{s} \lesssim$
$1 - 100\,M_\odot$ yr^{-1} and the vastly larger X-ray cooling rates $\dot{m} \sim 100 -$

$1000 \, M_\odot \, yr^{-1}$ implying the cooling rates were misestimated or the existence of a dark repository for the cooling gas (ie., low-mass stars, Jupiters, a mist of cold gas, etc). Adopting the former interpretation, some argued that star formation and its accompanying gas were produced by the occasional merger or stripping event with a nearby, gas-rich galaxy. This model has problems of its own, the most troublesome being the difficulty explaining the connection between star formation and the existence of $\sim 10^{12} \, M_\odot$ of dense, X-ray emitting gas centered tightly on the cD. Moreover, the relatively low incidence of line emission in cDs and other galaxies in cluster cores suggests that gaseous merger activity is *suppressed*, not enhanced, with respect to typical galactic environments [6]. cDs are usually thought to grow by dissipationless "dry" mergers with very little accompanying star formation.

3 The New View of Cooling Flows

New developments over the past seven years have changed our understanding of cooling flows. Moderate resolution spectra of clusters obtained with XMM-*Newton*'s Reflection Grating Spectrometer show that the dominant soft X-ray cooling channels, such as the 12 Å Fe XVII feature, are suppressed relative to classical cooling flow predictions [9]. The bulk of the gas is apparently cooling down to only $\sim 1/3$ of the ambient gas temperature and only 10% or less is apparently condensing out and fueling star formation. In fact, upper limits on the level of condensation from XMM-*Newton*, *Chandra*, and FUSE are approaching the star formation rates estimated using far-UV, U-band, and far infrared *Spitzer* observations [3,8,10,11,12], and in some cases they formally agree (Fig. 1, left). This significant development side-steps the problem of the "missing repository," but it requires that the radiation lost from the bulk of the cooling gas be replenished, most likely by AGN outbursts.

4 Cooling Regulated by AGN in cD and gE Galaxies

The discoveries of jet-blown cavities in the X-ray atmospheres of galaxies, groups, and clusters [eg., 13] provide a direct and relatively accurate way of measuring AGN jet power. Cavities are probably close to being in pressure balance and thus rise buoyantly in the ICM. As they rise, they apparently deposit their enthalpy into the surrounding gas at a mean instantaneous rate of $P_{jet} \sim 4pV/t_{buoy}$, where t_{buoy} can take several forms but is approximately R_{bub}/v_{bub}, where R_{bub} is the distance of the center of the bubble from the nucleus of the host galaxy and v_{bub} is approximately the buoyancy speed, which itself is a substantial fraction of the local sound speed [13]. *Chandra* observations provide the sizes (volumes) and surrounding pressures, giving a direct and accurate lower limit to jet power independent of the radio properties. Jet power measured in this fashion lies between 10^{42} erg s^{-1} in groups to upward of 10^{46} erg s^{-1} in rich clusters. By comparison, the X-ray cooling corresponding to a classical cooling rate $\dot{m}$ is $L_{\text{cool}} \sim 10^{44} \left(\frac{\dot{m}}{100\,\text{M}_\odot\,\text{yr}^{-1}} \right)$ erg sec^{-1}. Jet power correlates with and is comparable to L_{cool} in roughly half of clusters with prominent cavity systems [11, and see Fig. 1, right]. If the coupling between jet power and the surrounding gas is close to unity, AGN outbursts would substantially reduce or quench cooling entirely. The existence of multiple cavity systems, ripples (sound waves), and detection statistics show AGN outbursts repeat on timescales of 10^7 yr to 10^8 yr. Quiescent or underpowered systems may be in a rapid cooling phase and will experience a future outburst as fuel reaches the nucleus [13]. This imperfect balance between heating and cooling may allow for periods of rapid accretion and star formation in cD galaxies, which eventually shuts down following a suitably energetic outburst. More than half of the systems shown in the right panel of Fig. 1 are experiencing star formation, and the fraction increases with increasing L_{cool} (ie., L_{ICM}). This behavior is consistent with intermittent fueling.

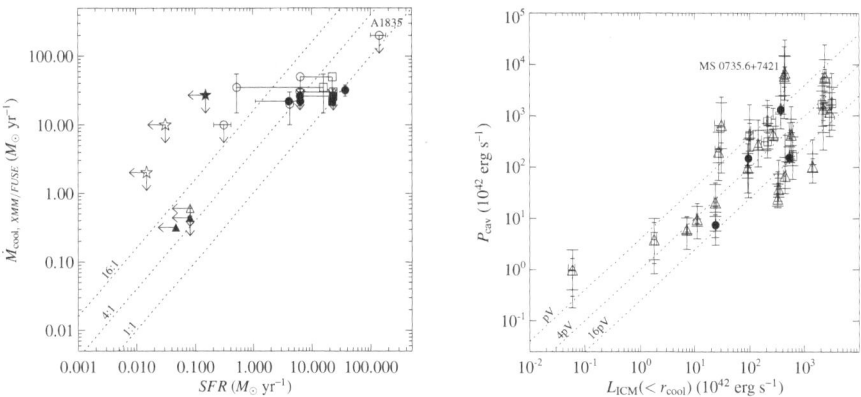

Fig. 1. Left Plot of star formation rate against X-ray cooling rate from [11]. The diagonal lines represent equality (1:1) between star formation and cooling rates, and offsets corresponding to four to one and 16 to one respectively. **Right** Plot of X-ray cooling luminosity against jet power from [11] and Rafferty, this volume. Diagonal lines represent equality between heating and cooling assuming $1pV$, $4pV$, and $16pV$ of enthalpy per cavity.

The situation in giant elliptical (gE) galaxies contrasts markedly with cDs in clusters. Nulsen et al. (this volume) measured jet power in Christine Jones's sample of nearby gE's and plotted it against X-ray luminosity following the procedures of Birzan [13] and Rafferty [11]. Unlike clusters, nearly all gE's lie near or above equality between X-ray cooling and jet heating. Cooling flows in gEs appear to be quenched by AGN heating, which could explain their dearth of star formation and prominent emission nebulae.

5 AGN Feedback in the Context of Galaxy Formation

Three lines of evidence suggest that ICM cooling is regulated primarily by AGN feedback. First, the radiative cooling time of the hot gas in the cores of cooling flows is less than 10^9 yr [14], yet very little gas is actually cooling to low temperatures (ie., $\dot{m} \ll m_{\mathrm{gas}}/t_{\mathrm{cool}}$) [9]. A feedback mechanism coupled to cooling operating on a timescale $\lesssim t_{\mathrm{cool}}$ may be required to maintain the gas in this state. Second, X-ray cooling luminosity and jet power in giant elliptical and cD galaxies are correlated over eight orders of magnitude of power, suggesting that supermassive black holes know about the X-ray atmospheres surrounding them and vice versa [11, 13, Nulsen et al, this volume, and Fig. 1]. Third, the existence of entropy floors in the centers of groups and clusters suggests that the hot gas is being centrally heated by AGN [15]. Apparently, thermostatically controlled AGN outbursts are able to supply enough energy to regulate or quench cooling flows, provided the shock and cavity energy is efficiently coupled to and distributed throughout the cooling atmospheres.

This picture has broader consequences for the general problem of galaxy formation. Standard CDM models predict a power law distribution of dark matter halos, yet the luminosity function of bright galaxies declines exponentially [16]. The most luminous gE and cD galaxies are fainter than expected, implying star formation is suppressed at late times, most likely by feedback during the violent process of galaxy formation. Opinion has shifted away from starburst winds to AGN outflows in massive (bright) halos, based on observation and energetic arguments [17, 18]. This is consistent with the situation in massive starbursts in cooling flows which liberate $\lesssim 10^{42}$ erg s^{-1} , much less than the $\sim 10^{45}$ erg s^{-1} from AGN [2]. AGN feedback may also play a role in maintaining the roughly constant ratio of bulge mass (luminosity and stellar velocity dispersion) to black hole mass in galaxies. There is intriguing new evidence that cDs grow in cooling flows in a way that maintains a constant ratio [3], although their bulges and black holes do not always grow in lock-step [11].

6 Present and Future Prospects

The new and more detailed understanding of cooling in cluster cores that has emerged in recent years calls for a renewed effort to measure star formation rates and star formation histories in central galaxies. X-ray and UV cooling limits and existing star formation rates in a half dozen or so systems are formally consistent with each other (see Fig. 1). If these systems are indeed fueled by cooling gas, the soft X-ray cooling lines, such as the 12Å Fe XVII feature, should be detectable in $200 - 500$ ksec exposures with XMM-*Newton*'s Reflection Grating Spectrometer [3] and in much shorter exposures taken in the future with Constellation-X. Combining this with observations of ultraviolet O VII emission from gas cooling below $\sim 10^5$ K [19] and newer, more accurate star formation rates would critically test the viability of the modern cooling flow framework. High spatial and spectral resolution with broad wavelength coverage are required to measure colors, fluxes, and line indices sensitive to star formation and to correct them for the effects of dust extinction and dilution by bright galactic backgrounds.

Although only a small fraction of cDs are forming stars, the implications are significant. These systems anchor the high end of the galaxy luminosity function and they are perhaps the *only* massive systems where the entire cycle of cooling, accretion, star formation, black hole growth, and feedback—a cycle relevant to the formation of all massive galaxies—can be examined and modeled in detail. Simulations and semianalytic models of galaxy formation should strive to reproduce these systems. I list below several noteworthy properties and constraints taken from an Annual Reviews article being written by Paul Nulsen and myself:

(1) A small fraction ($\lesssim 10\%$) of cDs at modest redshifts are experiencing star formation, and this star formation is embedded in dense $\sim 10^{-1}$ cm^{-3} atmospheres with cooling times $\ll 10^9$ yr. Star formation may be a consequence of the inability of AGN to balance radiative cooling losses at all times in cD clusters with X-ray cooling luminosities above $\sim 10^{43}$ erg s^{-1} [11]. In contrast to cDs, AGN heating is largely able to quench star formation in galaxies and groups with cooling luminosities below $\sim 10^{43}$ erg s^{-1} (Nulsen et al, this volume).

(2) In both giant ellipticals and cDs AGN feedback power scales in proportion to the cooling luminosity as expected in an operational feedback loop [11,13, Fig. 1, Nulsen et al, this volume].

(3) Bondi accretion may power the AGN in lower mass giant ellipticals [20] but is probably unable to do so in more massive cD galaxies [11]. Only the most powerful cluster outbursts experience accretion rates approaching $\sim 1 \, M_\odot$ yr^{-1}; typical rates are $\sim 10^{-2} \, M_\odot$ yr^{-1}, which is a small fraction of the Eddington accretion rate of a $10^8 \, M_\odot$ black hole [11].

(4) The jet model most directly associated with the "Magorrian" mode of [17] and the "radio" feedback mode operating at late times [18] must account for the enormous range in radiative efficiency and large jet kinetic energy observed in present day gE's and cDs. It should also recover the observed temperature and cooling time profiles in the cores of clusters with active AGN activity which scale with radius as $T \propto r^{\sim 0.4}$ and $t_{\mathrm{cool}} \propto r^{\sim 1.3}$, respectively [21]. This constraint is difficult to satisfy in models of gas heated from the center.

(5) The existence of very powerful AGN outbursts and the persistent energy demands of cooling imply substantial black hole growth in cD galaxies at late times. Averaged over long time scales, this growth appears to occur with a roughly constant ratio of bulge to black hole mass [22] when star formation is taken into account [11]. This ratio may differ between normal gEs and cD galaxies [23].

(6) Star formation parameterized with a Schmidt-Kennicutt law is probably a reasonable approximation to these systems [3], although this issue is in need of further study [eg., 12]. Disk formation is rare or short lived. Energy pumped into the hot gas by starburst winds $\lesssim 10^{43}$ erg s^{-1} is negligible on the scale of the cooling flow [3], but may be important near the nucleus of the cD where fueling of the AGN is actively occurring.

(7) Finally, the observed level of chemical enrichment, abundance ratios, and spatial distribution of the chemical elements in the hot gas levy significant constraints on history of star formation in the cores of clusters and the level of mixing generated by AGN outbursts and mergers.

Acknowledgement. This short review benefited greatly from conversations with Laura Bîrzan, David Rafferty, Paul Nulsen, and Michael Wise. I thank Hans Böhringer and his colleagues for agreeing to host this excellent conference and for their patience with this manuscript.

References

1. Crawford, C. S., Allen, S. W., Ebeling, H., Edge, A. C., & Fabian, A. C. 1999, MNRAS, 306, 857
2. McNamara, B. R., et al. 2006, ApJ, 648, 164
3. Edge, A. C. 2001, MNRAS, 328, 762
4. Heckman, T. M., Baum, S. A., van Breugel, W. J. M., & McCarthy, P. 1989, ApJ, 338, 48
5. Edwards, L.O.V, Hudson, M.J., Balogh, M.L., Smith, R.J. 2007, MNRAS, submitted
6. Best, P.N., von der Linden, A., Kauffman, G., Heckman, T.M., Kaiser, C.R. 2006, MNRAS submitted, astro-ph/0611197
7. McNamara, B. R. 1997, ASP Conf. Ser. 115: Galactic Cluster Cooling Flows, 115, 109
8. McNamara, B. R. 2004, The Riddle of Cooling Flows in Galaxies and Clusters of galaxies, 177
9. Peterson, J. R., Kahn, S. M., Paerels, F. B. S., Kaastra, J. S., Tamura, T., Bleeker, J. A. M., Ferrigno, C., & Jernigan, J. G. 2003, ApJ, 590, 207
10. Hicks, A. K., & Mushotzky, R. 2005, ApJ, 635, L9
11. Rafferty, D. A., McNamara, B. R., Nulsen, P. E. J., & Wise, M. W. 2006, ApJ, 652, 216
12. Egami, E., et al. 2006, ApJ, 647, 922
13. Bîrzan, L., Rafferty, D.A., McNamara, B.R., Wise, M.W., Nulsen, P.E.J., 2004, ApJ, 607, 800
14. Bauer, F. E., Fabian, A. C., Sanders, J. S., Allen, S. W., & Johnstone, R. M. 2005, MNRAS, 359, 1481
15. Voit, G. M., Donahue, M. 2005, ApJ, 634, 955
16. Benson, A. J., Bower, R. G., Frenk, C. S., Lacey, C. G., Baugh, C. M., & Cole, S. 2003, ApJ, 599, 38
17. Sijacki, D., & Springel, V. 2006, MNRAS, 366, 397

18. Croton, D. J., et al. 2006, MNRAS, 365, 11
19. Bregman, J. N., Fabian, A. C., Miller, E. D., & Irwin, J. A. 2006, ApJ, 642, 746
20. Allen, S. W., Dunn, R. J. H., Fabian, A. C., Taylor, G. B., Reynolds, C. S. 2006, MNRAS, 372, 21
21. Sanderson, A.J.R., Ponman, T.J., O'Sullivan, E. 2006, MNRAS, 372, 1496
22. Häring, N., Rix, H. 2004, ApJ, 604, L89
23. Lauer, T.R., Faber, S.M., Richstone, D., et al. 2006, astro-ph/0606739
24. Sarazin, this volume

Cold Gas in Cluster Cores

M. Donahue

Michigan State University, Physics & Astronomy Dept., East Lansing, MI US 48824-2320 donahue@pa.msu.edu

Summary. I review the literature's census of the cold gas in clusters of galaxies. Cold gas here is defined as the gas that is cooler than X-ray emitting temperatures ($\sim 10^7$ K) and is not in stars. I present new Spitzer IRAC and MIPS observations of Abell 2597 that reveal significant amounts of warm dust and star formation at the level of 5 solar masses per year.

1 Introduction: Gas Census in the Core

Almost immediately following the discovery of clusters as X-ray sources came the realization that, as thermal X-ray sources, the gas in the cores of many of these systems is radiating energy at a prodigious rate. A simple calculation reveals that the enthalpy ($5nkT/2$) content in the gas in such systems would be expended many times during the lifetime of the system, unless replenished by a source of energy over and above that of gravitational compression. Absent a distributed and significantly large heating source, the gas in this cooling flow picture should cool, lose pressure support, and thus gently settle deeper into the cluster potential cf. [22]).

The rarity of cool gas and of recently formed stars in the cores of clusters of galaxies provided the main counter-evidence for the simple cooling flow model in the cores of clusters throughout the 1990s. The lack of X-ray coronal lines from species such as Fe XVII and O VII was the kiss of death for the simple cooling flow model e.g., [48]. But that discovery begged another question: what supplies the enthalpy radiated by these X-ray sources? The cold gas in cluster cores provides key clues about the processes occurring in the gas there.

Where applicable, I assume that the expansion of the universe can be described with $H_0 = 70h_{70}$ km/s/Mpc. For ease of terminology, I will call these clusters "cool core" clusters, although it is possible for the gas density to be high enough for the cluster to qualify as having a short central cooling time without a temperature gradient. (It is less awkward than referring to them as "clusters formerly known as cooling flow clusters," in any case.)

The core of a cluster of galaxies can usually be defined by the characteristic radius in a beta-model profile of the form density $\rho(R) \propto (1 + (R/R_c)^2)^{-3\beta/2}$. R_c is typically 100-200 h_{70}^{-1} kpc. Inside this region, the hot gas dominates with about 10^{12-13} M$_\odot$, while stars provide $\sim 10^{12}$ M$_\odot$ of the mass. The 10^4K ionized gas, emitting Hα, weighs in at about 10^4 M$_\odot$, and as far as we know this gas seems to be always accompanied by collisionally-excited H$_2$. There can be trace amounts of [OVI]-emitting gas at about 10^6 K. Some systems may hold significant amounts of cold molecular hydrogen ($10^9 - 10^{10}$) M$_\odot$. Recent Spitzer observations, reported here and by [20] show that these systems also host star forming regions – not at the level to explain the old-fashioned mass cooling rates, but enough to produce bright mid-IR Spitzer sources.

I have been invited to discuss the state of knowledge of each of these components in the ICM. I note that this area of observational astronomy is undergoing something of a re-awakening since 2000. The current field is not so much limited by technology, but by limited application of the technology we have to only a few sources. I am dubious as to the existence of a single prototypical cool core cluster. I will attempt to show where my conclusions are based on a detailed observation of a single object, or only on observations of the most extreme examples of the class.

2 Optical Emission Line Nebulae

Many of the cool core clusters host prominent optical emission line nebulae with characteristic radii of about 5-10 kpc. They were first discussed in context with the X-ray clusters by [7] and [23]. This field has been quite active [25, 40, 29, 8, 32, 38, 49, 37, 30, 53, 56, 15, 1, 10, 57, 9, 28, 60]. For a review of the early studies (pre-2000), I have chosen to highlight the results in [30] and [57] below.

The most famous of these nebulae, the nebula in NGC1275, spans almost 100 kpc. They exhibit bright, low-ionization optical emission lines such as Hα, [OII]3727Å, [SII]6716/6730Å, [OI]6300 Å, and [NII]6548/6584Å. The total luminosities are around $10^{41} - 10^{43}$ erg s^{-1}. They are moderately broadened, at 100-200 km s^{-1}. The [SII] lines can be used to derive typical electron densities of 100-300 cm^{-3} e.g., [30], for a variety of sources. [57] studied a single nebula in Abell 2597 with very deep optical spectra. With such specta, they could use a combination of the blue and red [OII] lines and the [SII] lines (the [OIII] triplet has only provided limits to date) [57] indicate that the nebulae are surprisingly hot, $T \sim 10,000 - 12,000$ K. These line ratios indicate a relatively low photoionization parameter $U \sim 10^{-4}$ e.g., [30], where U is the ratio of ionizing photon density to hydrogen particle density.

The line ratio analyses of many of the authors above tend to rule out strong shocks, at least in the high surface brightness nebulae near the center of the source. Most of the line ratios indicate a consistency with an ionization source with a black body temperature of greater than 100,000K, in addition to the expected ionization contributions from hot stars. The lack of He II recombination lines in a deep spectrum of Abell 2597 [57] rules out photons with energies greater than 54 eV.

The radiative contribution of any AGN to the photoionization and heating is local at best. Near an AGN in these sources, one sees enhanced [OIII]/[OII] and prominent peaks in the surface brightness of [NII] and [SII]. Work by [11, 12] has shown that these filaments are not strongly turbulent. An interesting result from [28] regarding the filaments in NGC1275 is that there is an insufficent number of stars to provide the excitation and heating of the nebular gas. (See Crawford, this volume.)

2.1 2A0335+096, a brief case study of a complicated system

We have observed the X-ray cluster 2A0335+096. We show here a new and deep narrow-band Hα image, taken during the early science phase of the SOAR telescope (a new 4-meter telescope located on Paranal, near Gemini South), together with re-reduced VLA data, optimizing the high spatial resolution information. We also present here a long-slit spectrum that was previously analyzed in only a single spectrum mode [15]. But because the slit was aligned along an interesting bar feature in the spectrum, and includes the light from a possibly interacting companion (with its own emission lines) and extended filament gas, it was worth reviewing the kinematic and excitation information [13]. The radio data show a clear lateral source, extending perpendicularly to the bar; the Hα image, while complex, shows filaments arching around the radio source.

We identifed the two peaks in Hα intensity as A and B. These locations are the approximate peaks in the centers of the main galaxy (A) and the companion galaxy (B), separated by about 300 km/s and 4.6 kpc in projection. The peak B corresponds to the broadest lines (~ 500 km/s) in its region, and also to a region with enhanced [NII]/Hα ratios, indicating the presence of a small AGN in B. However, the story is different for A. The velocity peak and an abrupt change in the [NII]/Hα ratio occurs somewhat south of the surface brightness peak. This location (possibly the location of the AGN in galaxy A) is also blueshifted compared to A and B.

There is a gradient of gas velocity from A to B, which may be indicative of stripping. (The suggestion that it might be tidal is unlikely because of the lack of a similar feature in the stars.) From the relative velocities and projected distances, we have estimated that the companion galaxy/B last interacted with the brightest cluster galaxy (BCG, A) about 60 million years ago. The radio source timescale is approximately 50 million years. The estimated star formation rate, from XMM/Optical Monitor UV data and a dust-free Hα estimate is around 5 solar masses per year. We note that the velocity field is not turbulent along the axis of interaction between the two galaxies. The optical line emission is confined to the region of X-ray emitting gas that is less than about 2 keV, while the X-ray peak is well-separated from the BCG.

2A0335+096, while one of the more complex studied, is probably not all that unusual for its evidence of a dynamical interaction. What we find interesting here is that radio source showed evidence for blowing bubbles into the gas in this system and that the timescales for the interaction and the age of the radio source were quite similar. The complexity here may be induced by the rearrangement of injection sources while injection was occurring.

2.2 Integral Field Spectroscopy

2D optical (and IR) spectroscopy is clearly the best way to decompose the velocity fields in these systems. [60] have used the Visible Multiobject Spectrograph on the VLT, with 1600 optical fibers to obtain velocity and line ratio maps for 4 extreme cool-core clusters, Abell 1664, Abell 1835, Abell 2204, and Zw 8193. The relative velocites of the nebular features in these systems were low, around 100-300 km/sec. The biggest disturbances were in galaxies with companions. Remarkably, the [NII]/Hα ratio remains relatively constant across the system. It is possible that the AGN region may not have been well-sampled by the fibers.

3 Infrared Emission Line Nebulae

In the early 1990s, when the near-infrared spectroscopic capabilities were becoming available, [21] began a "cloudy night" project to observe the optical emission line nebulae of cooling flow clusters. They quickly discovered that these objects were bright H_2 sources at 2 microns - nearly as luminous as the X-ray emission coming from the same region. The line ratios of the vibrationally-excited H_2 lines indicated excitation temperatures of around 2000K. HST imaging with NICMOS revealed that the morphology of the vibrationally excited H_2 emission was very similar to that of the optical emission line nebula [14].

Near IR spectroscopy of these objects [34, 24, 58, 41, 36, 19, 59] showed that the spectrum was consistent with heating by a hot $T > 50,000$K stellar continuum, electron densities of over 10^5 cm^{-3}, and line ratios inconsistent with that of shocks. Therefore this gas is over-pressured by a factor of about 100-1000 compared to the pressures derived for the optical and X-ray gas.

[35] detect near IR line emission out to 20 kpc from the centers of these systems, gas dynamics that are highly coupled with that of the optical emission-line gas, and also find evidence for an ionizing source consistent with a black body temperature of over 100,000 K.

I find it extremely interesting that the "Mystery Ionization Sources" for both the optical filaments and the near infrared-emitting molecular gas have the similar property that both require EUV radiation, yet these photons must be less energetic than that required to produce He II recombination emission [57]. More He II searches are needed to see if that deficit in the optical spectrum is common.

4 Cold Molecular Hydrogen and CO

The decade of the 90s experienced many searches for the cold molecular gas associated with the putative cooling flow [27, 43, 45, 2, 46, 26]. However, it wasn't until the technology had progressed such that significant detections and maps were produced [16, 17, 50, 51, 52]. These detections were achieved with the JCMT and the IRAM. They saw between 300 million and 40 billion solar masses of cold ($T \sim 30$K) molecular hydrogen (inferred from the presence of CO) in the $> 10^{42}$ erg/s Hα-luminous systems that they targeted. Their successful detections were most likely in systems with the brightest Hα and 2-micron H$_2$ lines.

5 Hot Gas - [OVI] Line Emission

The existence of the [OVI] emission line doublet at 1032/1035 Å indicates the presence of million K gas. Non-primordial gas at this temperature cools extremely rapidly. It was found most commonly in our own Galaxy, first by the Copernicus satellite and more recently by FUSE. FUSE made long, night-time spectral observations of the central galaxies of several clusters. [47] reported a detection in the cluster Abell 2597, but not Abell 1795. Later, [42] reported only upper limits in Abell 2029 and Abell 3112, while most recently, [3] reported detections in the re-analyzed Abell 1795 data and in NGC1275, but not in the group AWM7.

So far, the detections are only in systems with known optical line emission. None of the reported detections are extremely strong. The [47] Abell 2597 detection is of only one member of the doublet (the fainter one is obscured), while the Abell 1795 re-analysis by [3] shows excess in two locations, at the quoted $3-\sigma$ level, but cut up by H$_2$ and FeII Galactic absorption, and in the presence of significant continuum. It is a very difficult experiment on a small telescope, and appropriate caution should be applied. (The author can say there may be lessons learned from the Einstein Observatory FPCS spectrum of NGC1275, [4], a spectrum obtained with a 1 cm^2 effective collecting area, reporting putative Fe XVII lines in the presence of X-ray continuum light.) That said, these limits and detections of [OVI] are consistent with a rather large cooling rate in the central region of the clusters, of about 30 solar masses per year in Abell 1795. Such a rate, if turned into stars would easily be visible to Spitzer observations.

6 Dust and PAHs: Spitzer's Infrared Vision

While BCGs are not exactly known for their dust emission, even single-orbit HST images show that these systems have dust lanes (e.g. A2052, A2597, PKS0745-191, 2A0335+096, A4059), as noted by [6, 14, 44]. Older ground-based extinction maps of NGC4696 [54] showed evidence for dust in that system that followed the emission-line gas. However, this detailed correspondence between dust and emission-line does not usually exist (e.g., Donahue et al. 2000).

Extinction towards the optical emission line filaments have been estimated from the decrement in a Balmer line series to be $A_V \sim 1$ [57]. IUE-based Lyman-α to ground-based Hα ratios [31] suggest $E(B - V) \sim 0.2$.

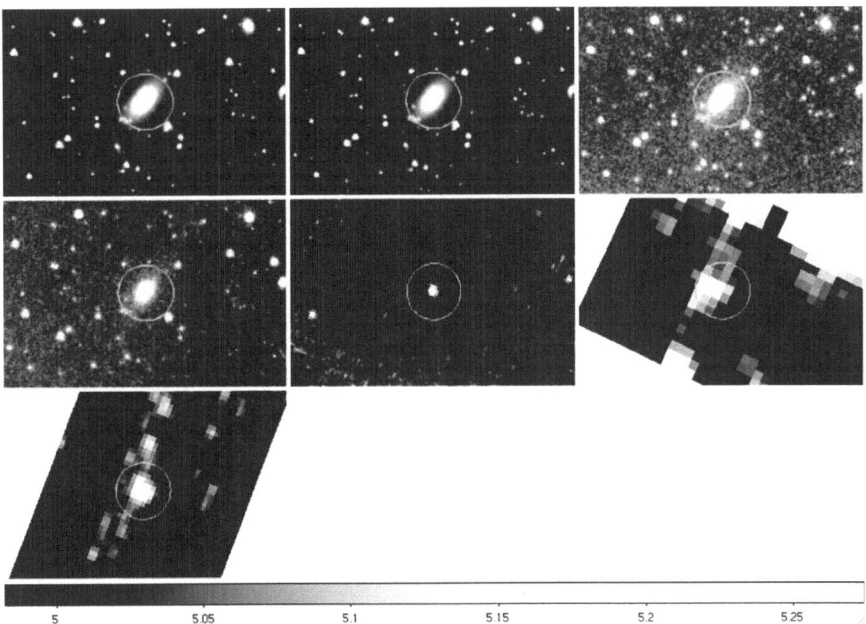

Fig. 1. Images of the BCG in A2597 by Spitzer.

[15] showed that the emission-line gas *itself* is dusty, based on the intensity of calcium emission lines. The lack of calcium lines from the ionized gas shows that calcium has been strongly depleted into grains.

Dust emission was difficult to find, pre-Spitzer. IRAS had poor spatial resolution and low sensitivity, which resulted in only upper limits for dust in clusters. ISO had a scan mode, but here too the low spatial resolution and the low sensitivity limits the constraints. [55] found excesses and deficits of emission at 120 and 180 microns towards 6 nearby clusters. One promising lead may be the development of SCUBA, [18] and, at 14" resolution, [5],

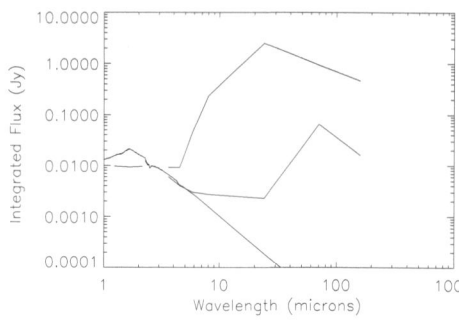

Fig. 2. Spitzer SED for the BCG in Abell 2597. The upper curve is the full flux in the aperture; the middle curve is the background-corrected flux; the lower curve is the expectation based on stellar spectra alone (a RJ tail). The 2MASS photometry points are from the extended source catalog. (Donahue & Sparks 2007, in prep.)

report 2 detections out of 7 clusters observed at 850 microns. NGC1275 has been detected by SCUBA [33]. They report about 60 million solar masses of extended, 20K, dust in NGC1275.

[20] report Spitzer mid-IR (70 micron) detections of 3 clusters, Abell 2390, Zw 3146, and Abell 1835, in the course of their search for high-redshift lensed galaxies behind moderate redshift clusters of galaxies. They picked the 3 BCGs in their sample with the brightest MIPS 24-micron detections to follow up with longer wavelength observations.

We present here new Spitzer data for Abell 2597 (Donahue & Sparks 2007, in prep). The BCG in Abell 2597 is well-detected in both IRAC (3.6, 4.5, 5.8, and 8 microns) and MIPS (24, 70, and 160 micron) observations. The total luminosity in the far infrared is approximately 10^{44} erg s^{-1}. The broad-band spectral energy distribution shows an excess at 70 and 160 microns that well-exceeds the expected contribution from the Rayleigh-Jeans tail of the stellar contribution (see Figure 2). Furthermore, the excesses at 5.8 and 8 microns indicate the presence of strong PAH emission, which should be confirmed by IRS spectroscopy once the extended source analysis is understood.

These observations consisted of 270 minutes in the IRAC band and only 36 minutes in the MIPS. The far infrared luminosity corresponds to about 4 solar masses per year [39], which is inconsistent with the [OVI] cooling luminosity of 20 ± 5 M$_\odot$ yr^{-1} [47]. The inferred star formation rate is consistent with the UV star formation rate inferred from HST/STIS UV observations by O'Dea et al. (2004) and with the Hα star formation rate within about $r = 2$" [14] , about 2 solar masses per year based on the relations described in [39].

7 Conclusions

We have reported a major Spitzer result for BCGs in cool core clusters of galaxies: a 70-160 micron excess well above the Rayleigh-Jeans tail of a stellar contribution. The luminosity and the spectrum are typical of a Luminous

InfraRed Galaxy (LIRG) or a starburst. These spectra are consistent with UV and optical indicators of star formation in Abell 2597.

The various cold constituents of the clusters of galaxies reviewed here suggest a common excitation mechanism for all phases, at least for the brightest parts. The evidence for this includes: common hosts, similar photoionization and heating constraints, a relative lack of line ratio variation across the nebulae. However, these issues are far from settled, and may provide vital clues as to the mechanisms for cooling gas, feedback, and star formation in the local universe, in the largest galaxies known.

The questions remaining include: Are galaxy-galaxy interactions required to stimulate central optical / infrared nebulae? Are optical filaments trails of galaxies punching through a molecular hydrogen reservoir as suggested by [58], or are they lofted by buoyant radio plasma as suggested by Crawford and Fabian? What is the relationship between nebulae and star formation? Are the filaments farther from the stars excited by different mechanism from the filaments close in? What is this mysterious FUV (but not too FUV) ionization source for the optical and infrared filaments?

What next? I would like to see more Spitzer MIPS and MIPS-SED observations done before the cryocoolant is exhausted, at the end of April 2009 and the end of a short Cycle 5. These spectra and images are needed to get critical dust temperatures (see O'Dea cycle 3 program, with many Co-Is in the community). It would be very good to trace the velocity fields of the ICM using IFU and efficient (mapping mode) long-slit observations. Extremely deep optical spectra can provide emission-line diagnostics to get nearly model-independent measures of T, Z, n, U (e.g. [56]). Deep X-ray observations should eventually reveal the faint coronal lines that ought to be present if some cooling gas is fueling the observed star formation. Finally, probing the relationship between the line-emitting gas, the dust emission, and the low-entropy X-ray emitting gas will require multi-wavelength imaging at somewhat similar spatial resolutions in order to test whether the morphologies of the emission regions really are similar.

References

1. Allen, S. W. 1995, MNRAS, 276, 947
2. Braine, J., & Dupraz, C. 1994, A&A, 283, 407
3. Bregman, J. N., Fabian, A. C., Miller, E. D., & Irwin, J. A. 2006, ApJ, 642, 746
4. Canizares, C. R., Markert, T. H., & Donahue, M. E. 1988, in NATO ASIC Proc. 229: Cooling Flows in Clusters and Galaxies, ed. A. C. Fabian, 63–72
5. Chapman, S. C., Scott, D., Borys, C., & Fahlman, G. G. 2002, MNRAS, 330, 92
6. Choi, Y.-Y., Reynolds, C. S., Heinz, S., Rosenberg, J. L., Perlman, E. S., & Yang, J. 2004, ApJ, 606, 185
7. Cowie, L. L., & Binney, J. 1977, ApJ, 215, 723

8. Cowie, L. L., Hu, E. M., Jenkins, E. B., & York, D. G. 1983, ApJ, 272, 29
9. Crawford, C. S., Allen, S. W., Ebeling, H., Edge, A. C., & Fabian, A. C. 1999, MNRAS, 306, 857
10. Crawford, C. S., Edge, A. C., Fabian, A. C., Allen, S. W., Bohringer, H., Ebeling, H., McMahon, R. G., & Voges, W. 1995, MNRAS, 274, 75
11. Crawford, C. S., Hatch, N. A., Fabian, A. C., & Sanders, J. S. 2005a, MNRAS, 363, 216
12. Crawford, C. S., Sanders, J. S., & Fabian, A. C. 2005b, MNRAS, 361, 17
13. Donahue, M., et al., 2007, AJ, submitted
14. Donahue, M., Mack, J., Voit, G. M., Sparks, W., Elston, R., & Maloney, P. R. 2000, ApJ, 545, 670
15. Donahue, M., & Voit, G. M. 1993, ApJL, 414, L17
16. Edge, A. C. 2001, MNRAS, 328, 762
17. Edge, A. C., & Frayer, D. T. 2003, ApJL, 594, L13
18. Edge, A. C., Ivison, R. J., Smail, I., Blain, A. W., & Kneib, J.-P. 1999, MNRAS, 306, 599
19. Edge, A. C., Wilman, R. J., Johnstone, R. M., Crawford, C. S., Fabian, A. C., & Allen, S. W. 2002, MNRAS, 337, 49
20. Egami, E., Misselt, K. A., Rieke, G. H., Wise, M. W., Neugebauer, G., Kneib, J.-P., Le Floc'h, E., Smith, G. P., Blaylock, M., Dole, H., Frayer, D. T., Huang, J.-S., Krause, O., Papovich, C., Pérez-González, P. G., & Rigby, J. R. 2006, ApJ, 647, 922
21. Elston, R., & Maloney, P. 1994, in ASSL Vol. 190: Astronomy with Arrays, The Next Generation, ed. I. S. McLean, 169
22. Fabian, A. C. 1994, ARA&A, 32, 277
23. Fabian, A. C., & Nulsen, P. E. J. 1977, MNRAS, 180, 479
24. Falcke, H., Rieke, M. J., Rieke, G. H., Simpson, C., & Wilson, A. S. 1998, ApJL, 494, L155
25. Ford, H. C., & Butcher, H. 1979, ApJS, 41, 147
26. Fujita, Y., Nagashima, M., & Gouda, N. 2000, PASJ, 52, 743
27. Grabelsky, D. A., & Ulmer, M. P. 1990, ApJ, 355, 401
28. Hatch, N. A., Crawford, C. S., Johnstone, R. M., & Fabian, A. C. 2006, MNRAS, 367, 433
29. Heckman, T. M. 1981, ApJL, 250, L59
30. Heckman, T. M., Baum, S. A., van Breugel, W. J. M., & McCarthy, P. 1989, ApJ, 338, 48
31. Hu, E. M. 1992, ApJ, 391, 608
32. Hu, E. M., Cowie, L. L., & Wang, Z. 1985, ApJS, 59, 447
33. Irwin, J. A., Stil, J. M., & Bridges, T. J. 2001, MNRAS, 328, 359
34. Jaffe, W., & Bremer, M. N. 1997, MNRAS, 284, L1
35. Jaffe, W., Bremer, M. N., & Baker, K. 2005, MNRAS, 360, 748
36. Jaffe, W., Bremer, M. N., & van der Werf, P. P. 2001, MNRAS, 324, 443
37. Johnstone, R. M., & Fabian, A. C. 1988, MNRAS, 233, 581
38. Johnstone, R. M., Fabian, A. C., & Nulsen, P. E. J. 1987, MNRAS, 224, 75
39. Kennicutt, Jr., R. C. 1998, ApJ, 498, 541
40. Kent, S. M., & Sargent, W. L. W. 1979, ApJ, 230, 667
41. Krabbe, A., Sams, III, B. J., Genzel, R., Thatte, N., & Prada, F. 2000, A&A, 354, 439
42. Lecavelier des Etangs, A., Gopal-Krishna, & Durret, F. 2004, A&A, 421, 503

43. McNamara, B. R., & Jaffe, W. 1994, A&A, 281, 673
44. McNamara, B. R., Wise, M., Sarazin, C. L., Jannuzi, B. T., & Elston, R. 1996, ApJL, 466, L9
45. O'Dea, C. P., Baum, S. A., Maloney, P. R., Tacconi, L. J., & Sparks, W. B. 1994, ApJ, 422, 467
46. O'Dea, C. P., Gallimore, J. F., & Baum, S. A. 1995, AJ, 109, 26
47. Oegerle, W. R., Cowie, L., Davidsen, A., Hu, E., Hutchings, J., Murphy, E., Sembach, K., & Woodgate, B. 2001, ApJ, 560, 187
48. Peterson, J. R., Paerels, F. B. S., Kaastra, J. S., Arnaud, M., Reiprich, T. H., Fabian, A. C., Mushotzky, R. F., Jernigan, J. G., & Sakelliou, I. 2001, A&A, 365, L104
49. Romanishin, W., & Hintzen, P. 1988, ApJL, 324, L17
50. Salomé, P., & Combes, F. 2003, A&A, 412, 657
51. —. 2004, A&A, 415, L1
52. Salomé, P., Combes, F., Edge, A. C., Crawford, C., Erlund, M., Fabian, A. C., Hatch, N. A., Johnstone, R. M., Sanders, J. S., & Wilman, R. J. 2006, A&A, 454, 437
53. Sarazin, C. L., O'Connell, R. W., & McNamara, B. R. 1992, ApJL, 397, L31
54. Sparks, W. B., Macchetto, F., & Golombek, D. 1989, ApJ, 345, 153
55. Stickel, M., Klaas, U., Lemke, D., & Mattila, K. 2002, A&A, 383, 367
56. Voit, G. M., & Donahue, M. 1990, ApJL, 360, L15
57. —. 1997, ApJ, 486, 242
58. Wilman, R. J., Edge, A. C., Johnstone, R. M., Crawford, C. S., & Fabian, A. C. 2000, MNRAS, 318, 1232
59. Wilman, R. J., Edge, A. C., Johnstone, R. M., Fabian, A. C., Allen, S. W., & Crawford, C. S. 2002, MNRAS, 337, 63
60. Wilman, R. J., Edge, A. C., & Swinbank, A. M. 2006, MNRAS, 371, 93

Statistics of X-ray Observables of Cooling Core and Non-cooling Core Clusters of Galaxies

Y. Chen[1,2], T.H. Reiprich[3], H. Böhringer[2] and Y. Ikebe[4]

[1] Key Laboratory of Particle Astrophysics, Institute of High Energy Physics, Chinese Academy of Sciences, Beijing 100049, P.R. China
ychen@mail.ihep.ac.cn
[2] Max-Planck-Institut für Extraterrestrische Physik, D-85748 Garching, Germany
[3] Argelander-Institut für Astronomie, Universität Bonn, Auf dem Hügel 71, 53121 Bonn, Germany
[4] National Museum of Emerging Science and Innovation, Tokyo, 135-0064, Japan

Summary. In this poster we study the segregation of Cooling Core (CC) and Non-Cooling Core (NCC) clusters in the L_X–T, L_X–M and M–T relation using the HIFLUGCS sample ([1]). A major goal in this study is to better understand the scatter in these relations, which must be folded into the test of large-scale structure measures and cosmological models.

1 The sample

The extended HIFLUGCS sample of 106 clusters and groups of galaxies is used for the study. We define a class of pronounced cooling core clusters by a lower limit of the ratio of the formal mass deposition rate to the cluster mass, M_{500}. The total sample thus splits up in 36 pronounced CCCs, 16 small to moderate CCCs, and 54 NCCCs.

2 Data analysis

We fit the surface brightness profile with a single β model for 57 clusters and a double β model for other 49 clusters. We find that the fraction of NCCCs clearly increases with M_{500}. In general the main reason may be that the fraction of dynamically young clusters increases with cluster mass, and these clusters in general do not feature cooling cores.

3 Statistical properties

Fig.1 shows the relation of the X-ray temperature of the hot component ([2]), T_h, versus the cluster mass, M_{500}. The self-similar model prediction is a slope with a value of 1.5, consistent with the values we obtain for the total sample. There is no significant influence of cooling cores on this relation. This could

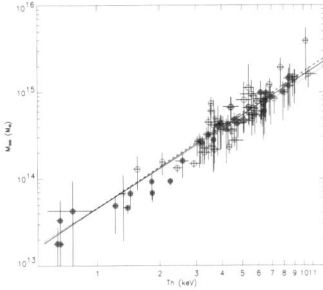

Fig. 1. M_{500}–T_h diagram. In all figures of this poster, the filled circles represent CCCs. and the open circles are NCCCs.

in part be due to the fact that the mass and temperature parameter are correlated. As this is not true for $M_{gas,500}$–T_h (see Table 1), we expect that this is not the main reason, but rather the tight relation shows that mass and temperature are linked in a more fundamental way by simple self-similar gravitational processes than the other relations.

The slopes of L_X–T_h relations (Fig.2), 2.73 ± 0.13, is much higher than that predicted by self-similar models which is 1.5. Remarkable is the clearly higher normalization of the relation for CCCs compared to NCCCs, with offsets of factors of 1.84 with respect to the relation with T_h. The resultant slope of L_X–M relation is again higher than that predicted by a self-similar model, as shown in Fig.3. We see a substantial difference in the normalization of this relation for the CCCs and NCCCs, of about a factor of 2.4.

Table 1. Summary of the fits to the scaling relations.

relation	Number	B	A	comments
$M_{500} - T_h$	88	1.54 ± 0.06	-0.112 ± 0.014	ALL with T_h
$M_{500} - T_h$	47	1.48 ± 0.07	-0.140 ± 0.015	CCCs
$M_{500} - T_h$	41	1.57 ± 0.17	-0.088 ± 0.031	NCCCs
$Mgas_{500} - T_h$	88	2.29 ± 0.09	-0.269 ± 0.015	ALL with T_h
$Mgas_{500} - T_h$	47	2.38 ± 0.10	-0.251 ± 0.017	CCCs
$Mgas_{500} - T_h$	41	2.04 ± 0.14	-0.258 ± 0.025	NCCCs
$L_x - T_h$	88	2.73 ± 0.13	0.363 ± 0.027	ALL with T_h
$L_x - T_h$	47	2.88 ± 0.15	0.492 ± 0.031	CCCs
$L_x - T_h$	41	2.74 ± 0.17	0.227 ± 0.034	NCCCs
$L_x - M_{500}$	106	1.82 ± 0.13	0.521 ± 0.039	ALL
$L_x - M_{500}$	88	1.77 ± 0.12	0.562 ± 0.041	ALL with T_h
$L_x - M_{500}$	47	1.94 ± 0.15	0.763 ± 0.050	CCCs
$L_x - M_{500}$	41	1.75 ± 0.25	0.381 ± 0.062	NCCCs

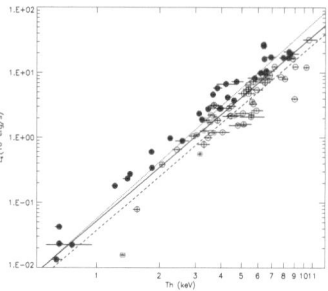

Fig. 2. L_X–T_h diagram.

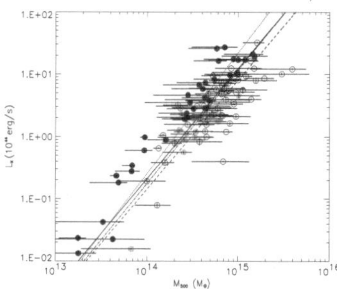

Fig. 3. L_X–M_{500} diagram.

4 Summary

In summary we find from the analysis presented in this poster that: (i) the fraction of NCCCs increases significantly with M_{500} and most of the galaxy groups in HIFLUGCS are cooling core clusters; (ii) among all the observational parameters, the core radius and the X-ray luminosity are most effected by the presence of a cooling core; (iii) the X-ray temperature shows a comparatively small bias for CCCs in comparison to NCCCs.

References

1. Reiprich, T.H. & Böhringer H.. 2002, ApJ, 567, 716
2. Ikebe Y., Reiprich, T.H., Böhringer H. et al.. 2002, A&A, 383, 773

Cold Fronts in Cool Core Clusters

S. Ghizzardi, S. Molendi, M. Rossetti and A. Leccardi

IASF Milano, INAF `simona@iasf-milano.inaf.it`

Summary. Cold fronts have been detected both in merging and in cool core clusters, where little or no sign of a merging event is present. A systematic search for sharp surface brightness discontinuities performed on a sample of 62 galaxy clusters observed with XMM-Newton shows that cold fronts are a common feature in galaxy clusters. Indeed most (if not all) of the nearby clusters ($z < 0.04$) host a cold front. Understanding the origin and the nature of such a frequent phenomenon is clearly important. To gain insight into the nature of cold fronts in cool core clusters we have undertaken a systematic study of all contact discontinuities detected in our sample, measuring surface brightness, temperature and when possible abundance profiles across the fronts. We measure the Mach numbers for the cold fronts, finding values which range from 0.2 to 0.9; we also detect discontinuities in the metal profile of some clusters.

1 Introduction

Chandra and XMM-Newton observations have shown that many clusters host very sharp surface brightness discontinuities. The drop in the X-ray surface brightness is accompanied by a rise of similar magnitude in the gas temperature. These discontinuities, dubbed *cold fronts* [1], have been detected both in merging and in cool core clusters. While the first type have been immediately related to the merging event [2], the origin of the latter is still not fully understood. As we will show, cold fronts are quite common in cool core clusters. Thus, understanding the nature of such a widespread phenomenon is mandatory to characterize the dynamics of galaxy clusters and of their cores.

The most popular scenarios to solve the cooling flow problem concern AGN and the interaction between the radio lobes inflated by the AGN and the ambient gas itself. The mechanical energy transferred from the lobes to the thermal plasma is in many, but not all cases, sufficient to quench the cooling (see [4]).

Another class of heating sources includes mechanisms which provide heat from outside the cool core region. In this context considerable effort has gone into exploring the role of conduction (e.g., [5], [6]). As it turns out there are various difficulties: firstly the conductivity of the ICM is unknown; secondly in some systems the temperature profiles are rather flat (i.e. M87, Ghizzardi et al. 2004 [6]), and thirdly, while heating from conduction scales as $T^{5/2}$, cooling is more efficient at lower temperatures.

Another possible way of heating the flow from outside has to do with the ubiquitous presence of cold fronts in cluster cores. Although the kinetic energy in these cold fronts, a fraction of the thermal energy, is in itself insufficient to offset the cooling, cold fronts could act as an energy reservoir. Hydrodynamical simulations ([7] and [8]) show that cold fronts in cores could be set off by minor, frequent, mergers that would provide a constant supply of energy. Clearly a detailed observational characterization of cold fronts is of primary importance to explore this scenario in greater detail. We study cold fronts in a large sample of galaxy clusters observed with XMM-Newton to provide a precise description of this phenomenon and to better understand the dynamics of cluster cores.

2 Hunting for cold fronts in a large XMM-Newton cluster sample.

We have performed a systematic characterization of cold fronts in a sample of 62 clusters observed with XMM-Newton. The large collecting area of the EPIC telescope onboard the XMM-Newton satellite allows a detailed inspection of the spectral properties of the galaxy clusters, which are important to study the dynamics of the core. The sample includes two different subsamples. The first comprises roughly 20 nearby bright clusters with redshifts in the range $[0.01 - 0.1]$. The second subsample comprises all the clusters available in the XMM-Newton public archive up to March 2005, having redshifts in the range $[0.1 - 0.3]$ (see [9]).

The systematic search for surface brightness and temperature discontinuities in the clusters of our sample resulted in the detection of cold fronts in 21 objects corresponding to a percentage of 34%. It is interesting to study the frequency of cold fronts in different redshift ranges. If we progressively reduce the sample, excluding gradually the more distant clusters, the fraction of clusters having a cold front increases. The occurrence of cold fronts is 41.8% for clusters with redshift $0.01 < z < 0.2$, 50% for clusters with $0.01 < z < 0.1$ and 72.2% for clusters with $0.01 < z < 0.07$. A large fraction (87.5%) of the nearest clusters ($z < 0.04$) host one or more cold fronts. This is in agreement with results derived analyzing a sample of 37 relaxed nearby clusters observed with *Chandra* [10]. Since projection effects and the XMM-Newton resolution can hide a non-negligible fraction of cold fronts, our result implies that probably all the nearby clusters host one (or more) cold fronts.

3 Characterizing cold fronts

The results reported in the previous section show that cold fronts are indeed a common feature in cluster cores. The precise characterization of this phenomenon is a preliminary and necessary step in assessing whether they can play a significant role in the cooling-heating balance within the cluster cores. First, we consider some general features of cold fronts in galaxy clusters (Sec. 3.1); we then characterize the discontinuities by measuring the surface brightness and temperature jump (Sec. 3.2) and finally we derive abundance profiles across cold fronts (Sec. 3.3).

3.1 General aspects of cold fronts in cool core clusters

We have derived surface brightness, temperature and pseudo-pressure (hereafter SB, T and P) maps for some cool core clusters using the *adaptive binning + broad band fitting* method (see [11] for a detailed description of this procedure). Two relevant features have been detected: a surface brightness peak displacement and a spiraling pattern in the temperature maps. As an example we consider 2A0335+096 , which hosts a cold front in the southern direction $\sim 70''$ from the SB peak. As already outlined in [6], the position of the BCG of this cluster matches the P peak, while the SB and the T peaks are shifted in the south direction by 16 arcsec. Similar shifts have been observed in others (e.g., A1795) but not in all (e.g., A496) clusters.

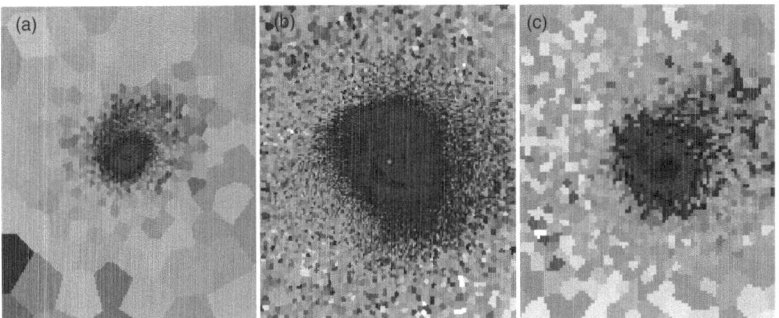

Fig. 1. Temperature maps for (a) 2A0335+096, (b) Perseus and (c) Centaurus. All of them show a spiral pattern.

Another interesting feature is visible in the T maps. In Fig. 1 we show the T maps for 2A0335+096, Perseus and Centaurus: all of them show a spiral structure. Both phenomena have a natural explanation within the scenario proposed by [10] and [8]. According to their model, cold fronts are formed when the central cold gas is subsonically sloshing in the dark matter gravitational potential. Frequent minor mergers can induce gas oscillations in the cluster core, displacing the thermal gas from the bottom of the potential well. Under these circumstances, the P peak and the BCG (which trace the gravitational potential) and the SB and T peaks (which trace the thermal gas) decouple. Simulations by [8] also show that the gas can acquire some angular momentum while oscillating. In this case, some spiraling structure in the T map is induced.

3.2 Measuring discontinuities and velocities of cold fronts

To quantify the dynamics of cold fronts in cool cores we derive the SB and T profiles for each cluster in 15 degrees wide sectors. The sectors have been chosen following the SB contour levels in order to properly characterize the jump. Classic deprojection procedures cannot be applied because of the lack of spherical symmetry. We describe the cold front discontinuity by modeling the electronic density and the gas temperature with power laws inside and outside the cold front edge. We then project these quantities assuming that the cold front has a width of $\Delta\varphi$ and an inclination angle $\Delta\vartheta$ with respect to the line of sight. A detailed description of this procedure will be given in [12]. In Fig. 2 we plot as dots the SB and the projected T for the southern

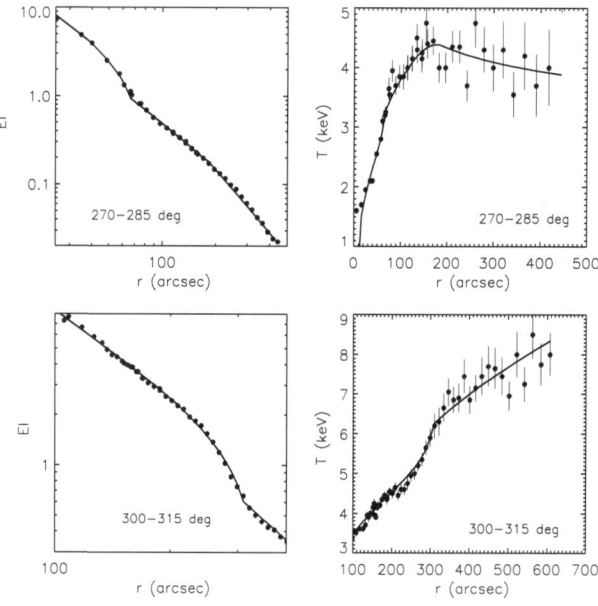

Fig. 2. Surface brightness and temperature profiles for 2A0335+096 (upper panels) and Perseus (lower panels) with the best fits.

sector in 2A0335+096 (top panels) and for the southwestern sector in Perseus (bottom panels); the solid lines are the best fits. For 2A0335+096, we find a density jump $n_{in}/n_{out} = 1.97$ and a temperature jump $T_{in}/T_{out} = 0.88$. This corresponds to a Mach number of $\mathcal{M} = 0.86$. For Perseus we find a density jump $n_{in}/n_{out} = 2.1$ and a temperature jump $T_{in}/T_{out} = 0.79$ corresponding to a Mach number $\mathcal{M} = 0.8$. A more complete and detailed analysis of the discontinuities for all the clusters of our sample will be presented in [12].

3.3 Metal profiles across the cold fronts

We derive the iron abundance profiles across the cold front for some clusters. Fig. 3 shows the profiles for Perseus (northern and southwestern sectors) and for 2A0335+096 (southern sector). The dashed lines mark the position of the cold fronts. The iron abundance has a discontinuity across the cold fronts for the Perseus cluster. Even if the data quality for 2A0335+096 is not as high as for Perseus, an indication for a discontinuity seems to be present also for this cluster. This behavior is expected in a sloshing scenario as the cold metal rich gas is shifted towards more external regions where the iron abundance is lower.

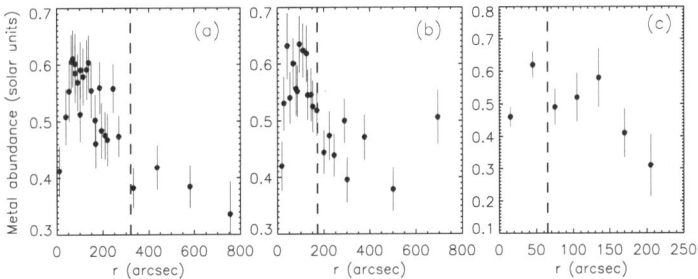

Fig. 3. Metal profiles in some sectors of Perseus (a-b) and 2A0335+096 (c). The dashed lines mark the cold fronts positions. A discontinuity is clearly visible in the Perseus cluster. Some indication of discontinuity can be seen for 2A0335+096.

4 Conclusions

Cold fronts could play an important role in providing heating from the outer regions of the core, contributing to quenching of the cooling flow in cool cores. The main advantage in considering such phenomenon as a possible heat source is twofold: cold fronts are a common feature in the galaxy cluster population; they could have a purely "gravitational origin". Ascasibar and

Markevitch [8] show that frequent minor mergers can induce a disturbance to the gravitational potential well and cause gas sloshing and cold fronts. Each cluster during its formation undergoes several minor merger events. Among the possible heating processes, this mechanism has the advantage of being common to all the clusters and of being unrelated to any particular feature of the cluster (e.g., the presence of an active AGN.)

An analysis of a large sample of clusters observed with XMM-Newton shows that 87.5% of the nearby ($z < 0.04$) clusters host a cold front. The analysis of the surface brightness, temperature and pressure maps for some clusters shows that in some cases the SB and T peaks are decoupled with respect to the P peak. In some clusters, we also observe a spiraling pattern in the temperature map; both these features are expected in a sloshing scenario [8].

We have measured the Mach numbers of cold fronts, finding values that range from 0.2 to 0.9. The analysis of the metal profiles shows that the iron abundance has a sharp discontinuity across the cold front edge in Perseus and some indications of that are present also for 2A0335+096. A detailed characterization of the discontinuities can help to understand the dynamics of the innermost regions of the cluster and to quantify the amount of heating that the cold front can provide to cluster cores.

References

1. Vikhlinin A., Markevitch M. & Murray S.S., 2001, ApJ, 551, 160
2. Markevitch M., et al., 2000, ApJ, 541, 542
3. Markevitch M., Vikhlinin A. & Mazzotta P., 2001, ApJ, 562, L153
4. Rafferty D.A., et al., 2006, ApJ, in press (astro-ph/0605323)
5. Voigt, L.M, and Fabian, A.C, 2004, MNRAS, 347, 1130
6. Ghizzardi, S., Molendi, S., Leccardi, A., Rossetti, M., 2004, *Searching for Sharp Surface Brightness Discontinuities: A Systematic Study of Cold Fronts in Galaxy Clusters*, in proceedings of "The X-ray Universe 2005" (astro-ph/0511445)
7. Tittley E.R. & Henriksen M. 2005, ApJ, 618, 227
8. Ascasibar Y. and Markevitch M., 2006,ApJ, 650, 102
9. Leccardi A., et al., 2006, in preparation
10. Markevitch M., Vikhlinin A. & Forman W.R., 2002, *Matter and energy in clusters of galaxies*, ASP Conference Series, Vol. X, Eds. S. Bowyer & C.-Y. Hwang (astro-ph/0208208)
11. Rossetti M., Ghizzardi S., Molendi S. & Finoguenov A., 2006, A&A in press (astro-ph/0611056)
12. Ghizzardi S., et al., 2006, in preparation

Core structure of Intracluster Gas: Effects of Radiative Cooling on Core Sizes

T. Akahori and K. Masai

Department of Physics, Tokyo Metropolitan University, 1-1 Minami-Ohsawa, Hachioji, Tokyo 192-0397, Japan
akataku@phys.metro-u.ac.jp, masai@phys.metro-u.ac.jp

1 Model and Calculation

From the β-model analyses of intracluster gas in 121 clusters, an interesting result is found that the distribution of core radii exhibits two distinct peaks at ~ 50 kpc and ~ 200 kpc for $H_0 = 70$ km s^{-1} Mpc^{-1} ([6, 7]). The core radii of the large-core clusters ($r_c \sim 200$ kpc) are marginally proportional to the virial radii, and therefore the relation expected from the self-similar collapse is thought to remain valid in large-core clusters ([7, 1]). In the small-core clusters ($r_c \sim 50$ kpc), on the other hand, no clear correlation is found between the core and virial radii. About a half of the small-core clusters possess central cD galaxies, but simulations show that the gas core of cDs is ~ 40 kpc in size at most, and is too small to account for the observed range ~ 40–80 kpc of the small-core clusters ([1]). Another possibility, the effects of radiative cooling on core sizes, is suggested: the central cooling time is significantly shorter than the Hubble time for 48 out of 50 small-core clusters ([1]).

As the core cools, the ambient gas then could inflow to compensate the pressure loss inside. Although the X-ray observations (e.g., [4, 8]) have suggested a much smaller amount of cooled mass than expected for the classical cooling flow model (see [3]), this process increases the gas density toward the cluster center and is likely responsible for the small cores observed. Until a time has elapsed comparable to the cooling timescale, the gas appears to cool rather slowly with temperatures $\sim$ keV at the center (e.g., [9]).

It is suggested that when the gas is cooling with quasi-hydrostatic balancing, the temperature appears to approach a constant value toward the cluster center ([5]). We have investigated the evolution of cooling intracluster gas under the presence of dark matter, for which we apply the King or NFW model as the fixed gravitational potential. We calculate four models of clusters which have the initial core radii, $r_c = 160, 200, 250, 300$ kpc, and satisfy the self-similar relation with their virial radii, r_{vir}, as $r_{\mathrm{vir}}/r_c = 15$, in order to see the evolutions of typical large-core clusters. We apply the β-model distribution for the initial gas with the temperature (isothermal), T, higher than the virial temperature, T_{vir}, as $\beta_{\mathrm{spec}}T \simeq T_{\mathrm{vir}}$ for $\beta_{\mathrm{spec}} = 2/3$ (see [2]). Using an approximate cooling function based on [10], we calculate the evolution until t_{cool} of our interest is reached. Here, $t_{\mathrm{cool}} \equiv 3n_{g0}kT/(n_{g0}^2 \Lambda)$ is the central cooling time, n_{g0} the central density, and Λ is the cooling function.

2 Result and Discussion

To see the evolution of core radii, we apply the double β-model:

$$\rho(r) = \rho_1[1 + (r/r_1)^2]^{-3\beta_1/2} + \rho_2[1 + (r/r_2)^2]^{-3\beta_2/2}, \qquad (1)$$

to simulated clusters. We obtain the best-fit parameters of the "outer" component (labeled by "1"), using the data in the range $1.0 \leq r/r_c \leq 5.0$, and then obtain the "inner" ones ("2"), assuming $\beta_2 = \beta_1$ ([6, 7]), using the data of $0.0 \leq r/r_c \leq 5.0$ including the outer component.

From the time evolution of the core radii of simulated clusters, we calculate the time during which a cluster would have the core radius between r and $r + \Delta r$ and estimate the population or relative number of clusters that would fall into a certain range of core radii. The resultant distribution is represented in Fig. 1. Thermal evolution due to radiative cooling until t_{cool} marginally can reproduce the two peaks in the observed distribution of core radii except for the details such as their widths or the tails.

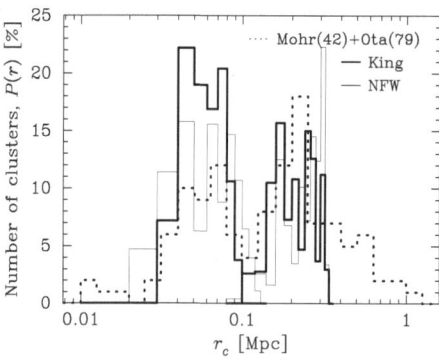

Fig. 1. Distribution of core radii. The thick solid lines represent the relative probabilities within t_{cool} of the four simulated clusters, estimated for the outer and inner cores with bins of $\Delta r = 20$ kpc and 10 kpc, respectively. The thin solid lines represent the probabilities for the case of the NFW potential. The dotted line represents the distribution of core radii of 121 clusters ([1, 2]).

It is interesting that even the small-core peak is produced by the clusters of $t \leq t_{\mathrm{cool}}$, i.e. moderately cooled clusters. Our cooling function underestimates by $\sim 40\,\%$ at $T \sim 1.5$ keV ($\sim t_{\mathrm{cool}}$) compared with the function of [10]. If we would have applied their original function, the small core peak might have a tail toward the small core end. However, this contribution is thought to be small because such a very small core is quite transient. This may be related to the fact that clusters of $t > t_{\mathrm{cool}}$ are out of quasi-hydrostatic balance (see below).

At each step until $\sim t_{\mathrm{cool}}$, we find that $C(r) \equiv (\tilde{M}_r/\tilde{t}_{\mathrm{cool}})/\dot{M}_r$ is nearly constant (Fig. 2) within the initial core radius for all the model clusters, where

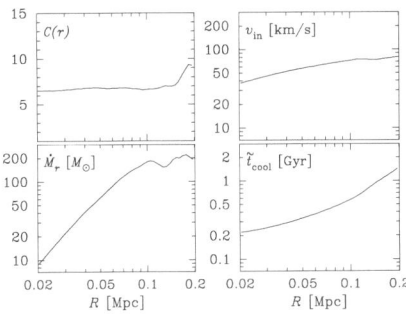

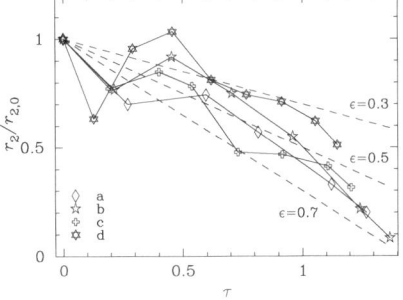

Fig. 2. $C(r)$ ($\equiv (\tilde{M}_r/\tilde{t}_{\rm cool})/\dot{M}_r$), the infall velocity ($v_{\rm in}$), the mass flow rate ($\dot{M}_r$), and the cooling time ($\tilde{t}_{\rm cool}$) at $t = t_{\rm cool}$ of cluster (b).

Fig. 3. Thermal evolutions of inner-core radii. The dashed lines represent $r_2 = r_{2,0}(1 - \epsilon\tau)$ with $\tau \equiv t/t_{\rm cool}$ and $\epsilon = 0.3,\ 0.5,\ 0.7$.

$\dot{M}_r = 4\pi\rho_{\rm g}r^2 dr/dt$ is the continuity equation, $\tilde{M}_r = 4\pi r^3 \rho_{\rm g}/3$ and $\tilde{t}_{\rm cool} = 3n_{\rm g}kT/(n_{\rm g}^2\Lambda)$ are the mass of a uniform gas sphere, and the cooling time for the local values at radius r, respectively. This constant C means that the gas cools while keeping the hydrostatic balance between the gravitational force and pressure gradient [5]. The mass inflow rate $\dot{M}_r(100{\rm kpc}) \sim 200\ M_\odot\ yr^{-1}$ is decreasing to $\dot{M}_r(20{\rm kpc}) \sim 10\ M_\odot\ yr^{-1}$, which is much smaller than expected for the classical cooling flow model and consistent with the quasi-hydrostatic cooling model.

The reproduced inner cores show a self-similar relation normalized by $t_{\rm cool}$ and initial inner-core radius, $r_{2,0}$ (Fig. 3). The relation, however, may be lost in the observed clusters due not only to the different age of clusters, labeled by t, but also the various value of β from 0.6–1.0 in the observations, depending on $\propto \beta^{-1/2}t$ [2]. This may be related to the fact that the observed small-core clusters have cooling times of $\sim$ 1–10 Gyr ([7, 1]).

References

1. T. Akahori, K. Masai, 2005, PASJ, 57, 419
2. T. Akahori, K. Masai, 2006, PASJ, 58, 521
3. A. C. Fabian, 1994, ARA&A, 32, 277
4. K. Makishima, et al., 2001, PASJ, 53, 401
5. K. Masai, T. Kitayama, 2004, A&A, 421, 815
6. N. Ota, K. Mitsuda, 2002, ApJ, 567, L23
7. N. Ota, K. Mitsuda, 2004, A&A, 428, 757
8. J. R. Peterson, et al., 2001, A&A, 365, L104
9. M. Ruszkowski, M. C. Begelman, 2002, ApJ, 581, 223
10. R. S. Sutherland, M. A. Dopita, 1993, ApJS, 88, 253

Investigating the Central Regions
of the *HIFLUGCS* Clusters with Chandra

D. S. Hudson and T. H. Reiprich

Argelander Institut für Astronomie, Auf dem Hügel 71, D-53121 Bonn
dhudson@astro.uni-bonn.de, thomas@reiprich.net

Summary. We present a preliminary analysis of the Chandra follow up to the *HIFLUGCS* sample (Hudson et al, *in prep*). By plotting the central temperature profiles, normalized to the global cluster temperature, against the fraction of the virial radius, we find no evidence for a universal central temperature profile. Similarly by fitting the central temperature profile to a powerlaw, we find a continuous range of values for the slope. Interpreting a positive temperature gradient as an indication of a cooling core, we find that exactly half (32 of 64) of the clusters in the *HIFLUGCS* sample are classified as cooling core clusters. Additionally, we find a flattening of the central entropy, consistent with other observations and predictions from simulations [3]. Perhaps the most interesting feature is the bi-modal distribution in the plot of global temperature versus central entropy. The distributions can be easily divided at 40 keV cm^2. Interpreting the clusters with a low central entropy as cooling core clusters, 27 of the clusters are classified as cooling core clusters with this method.

1 Introduction

Today one of the greatest mysteries of X-ray astronomy is what mechanism prevents the dense gas in the centers of clusters from cooling in significant quantities. To that end, it is important to ask: why do some clusters have cool dense cores and others do not? Is it possible to uniquely divide clusters between cooling core (CC) and non-cooling core (NCC) clusters?

In an attempt to probe these questions, we investigated the properties of the cores of the 64 *HIFLUGCS* clusters. The *HIFLUGCS* sample is a statistically complete sample of the 64 X-ray brightest galaxy clusters outside of the galactic plane [2]. The clusters are therefore bright and close, making them ideal candidates for detailed core studies. Moreover, *Chandra* observations exist for all 64 clusters, allowing a high spatial resolution study of the cores.

2 Central Temperature Profiles

Figure-1 shows the central projected temperature (*left*) and entropy (*right*) profiles for all 64 *HIFLUGCS* clusters. It is clear that there is no universal

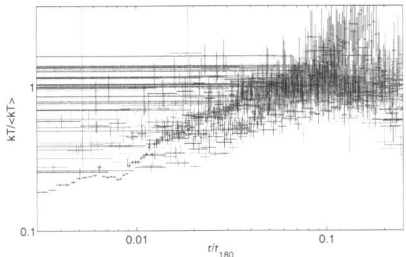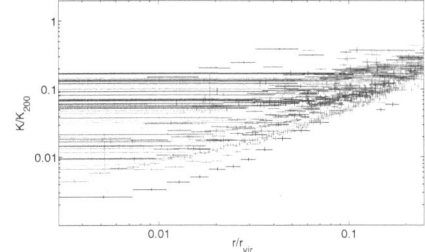

Fig. 1. *Left*: A plot of the radial temperature profile normalized to the global cluster temperature versus the fraction of the virial radius. The plot has been cut at a quarter of the virial radius to emphasize the central temperature profile. The scatter suggests there is no universal central temperature profile. *Right*: This plot shows the central entropy profile scaled to the entropy at r_{200}. The observed flattening and large scatter are consistent with the predictions from simulations. The low central values in this plot can be attributed to projection effects.

temperature or entropy profile in the central regions of clusters. The profiles show a ranges of slopes, from flat to steeply decreasing.

To quantify the steepness of the temperature profile, we took temperatures from all annuli within 0.1 r_{180} or the radius of the third inner-most annulus, whichever was greater, and fit this curtailed profile to a powerlaw. Fig.-2 (*left*) shows the best fit slopes and 1-σ errors. The clusters are ordered from steepest (Abell 2204) to shallowest (Abell 2147). It is clear that there is a continuous spread in the slopes. If there were a universal profile, then one would expect two distributions, one for clusters with cool centers and one for clusters with isothermal centers.

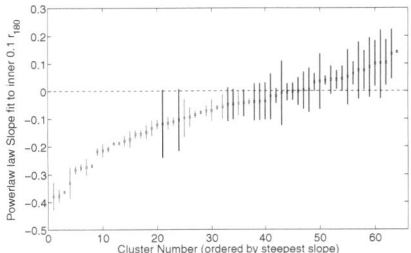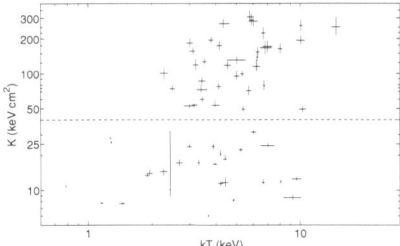

Fig. 2. *Left*:This plot shows the slopes of the power law fit to the cluster temperature profile within 0.1 r_{vir}. The continuous range of values suggests that there is no universal temperature profile. Gray (lighter) points have a slope which is inconsistent (at the 1-σ level) with a flat profile. *Right*: This plot shows the global cluster temperature versus the central entropy. The clear separation of the two distributions suggests this may be a robust method for identifying cooling core clusters.

If we classify cooling core (CC) clusters as those clusters with a slope < 0 (within 1-σ), we find exactly half (32) of our clusters are CC clusters. However, as we pointed out, since there is a continual distribution of slopes, this is an arbitrary cut. In an effort to find a more definitive cut between CC and NCC clusters, we plotted overall cluster temperature versus projected central entropy, K_0, where we define $K_0 \equiv kT\, n_0^{-2/3}$. n_0 is calculated from the normalization of the spectral fit to the central circle assuming density is constant. This rough calculation creates a bias such that smaller regions estimate a higher density than larger regions. However, since CC clusters have denser (and therefore brighter) cores, the central annulus will be smaller than for an NCC cluster[1]. This bias, therefore acts to enhance the separation between CC and NCC clusters. Moreover, this method will classify clusters with dense cores as CC, regardless of their inner temperature profile. Fig-2 (*right*) shows the plot of kT versus K_0. Two populations are apparent and can be separated between clusters with $K_0 > 40$ keV cm^2 and $K_0 < 40$ keV cm^2. Using this classification to separate CC from NCC clusters, we find 27 of 64 *HIFLUGCS* clusters are CC clusters.

Acknowledgement. We would like to thank D. Buote, M. Markevitch, C. Sarazin, and A. Vikhlinin for providing us with proprietary data to complete the Chandra follow-up sample.

References

1. M. Donahue, G. M. Voit, C. P. O'Dea, et al, 2005, ApJ, 630, L1
2. T. H. Reiprich, H. Böhringer, ApJ, 567, 716
3. G. M. Voit, S. T. Kay, G. L. Bryan, 2005, MNRAS, 364, 909

---

[1] We acknowledge there is also a bias on the length of the observation and the proximity of the cluster, but experience has shown that the brightness of the core is a stronger effect. That is, for a short observation of a CC cluster, the central annulus will be smaller than for a long observation of an NCC cluster.

Investigating Heating and Cooling within a Sample of Distant Clusters

R. J. H. Dunn[1] and A. C. Fabian[1]

Institute of Astronomy, Madingley Road, Cambridge, UK, rjhd2@ast.cam.ac.uk

1 Introduction

Since the launch of *Chandra*, many cavities have been found in the intra-cluster medium (ICM). These cavities are created by the jets from the AGN inflating bubbles of relativistic plasma in the thermal gas. They can be used as calorimeters of the mechanical energy output of the AGN [1]. Their effect on the cluster centre is crucial to understanding the way gas accumulates on the central galaxy. See [4, 3, 10] for recent reviews of AGN bubbles in clusters.

2 Nearby Clusters

We select those in the Brightest 55 clusters (B55) with a short central cooling time and a central temperature drop. Such clusters are likely to require some form of heating. The B55 sample is an almost complete flux-limited sample of nearby clusters (only 4 have $z > 0.1$). Of the 20 clusters that satisfied these constraints, at least 14 (70%) have clear bubbles. We compared the bubble energy injection rate to the X-ray cooling rate within the cooling radius, L_{cool} (Fig. 1, r_{cool} defined as the radius where $t_{cool} = 3$ Gyr). On average the bubbles supply enough power to offset the X-ray cooling out to at least $0.86 r_{cool}$ [8].

We also took those clusters from the B55 sample which have a central radio core in the NVSS, but do not appear to require heating. These clusters were combined with those clusters which require heating but do not have clear bubbles. We used the above result of $r_{heat} = 0.86 r_{cool}$ to calculate the expected size of a bubble in these clusters assuming that they expand at the sound speed. In five cases (3C 129.1 (Fig. 2), A2063, A2204, A3112 & A3391) the observed radio emission is of a similar size to that of the expected bubbles [8]. This implies that bubbles are more common than previously thought.

3 Distant Clusters

We select clusters with $z > 0.1$ from the Brightest Cluster Sample (BCS) [2] which also have *Chandra* archive observations (41/51). [6] find cool cores

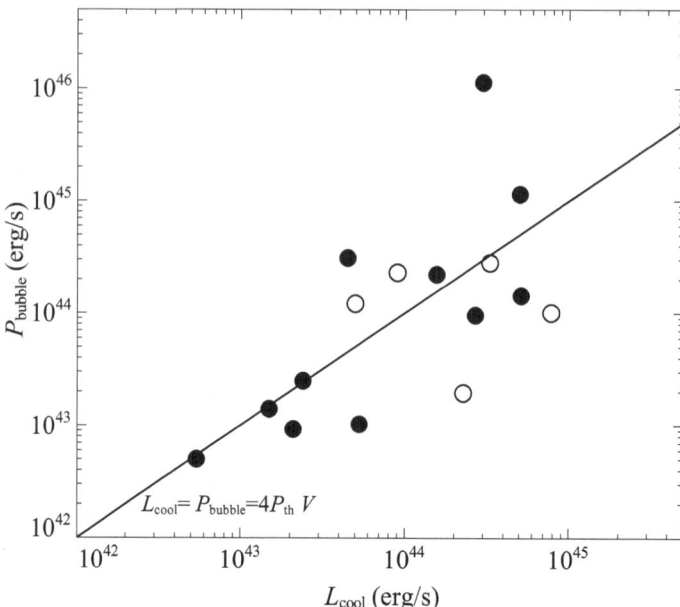

Fig. 1. The X-ray cooling within r_{cool} against the bubble power. The open circles are clusters with ghost bubbles. The line indicates exact balance between the X-ray cooling and the bubble power ($P_{bubble} = 4P_{th}V/t$).

(for $t_{cool} < 2$ Gyr) in 34% of BCS clusters with $z > 0.15$. Selecting those clusters requiring heating (15/41) only five have clear bubbles. Investigating the bubble energy injection rate, we find that in three clusters, the bubbles provide more power than all the X-ray cooling in the analysed regions. The average of the other two give a lower limit of $r_{heat} > 1.53r_{cool}$.

Of the clusters with *Chandra* observations, 22 have radio sources clearly detected in the NVSS, three more have marginal detections; a lower limit of 49%. In the B55 sample at least 29 clusters have radio sources (53%). The low contrast and small angular size of the bubbles, as well as low count rates for some clusters can make detecting bubbles difficult in more distant clusters as well as some from the B55. We intend to search for radio emission from the distant clusters to find more bubbles.

4 Conclusions & Further work

AGN at the centres of clusters can inject significant amounts of energy into the ICM. In some cases this is sufficient to prevent the X-ray emitting gas from cooling. In other clusters the bubbles provide insufficient power, which may be the result of the observations catching the bubbles at the beginning of their evolution and are therefore still small.

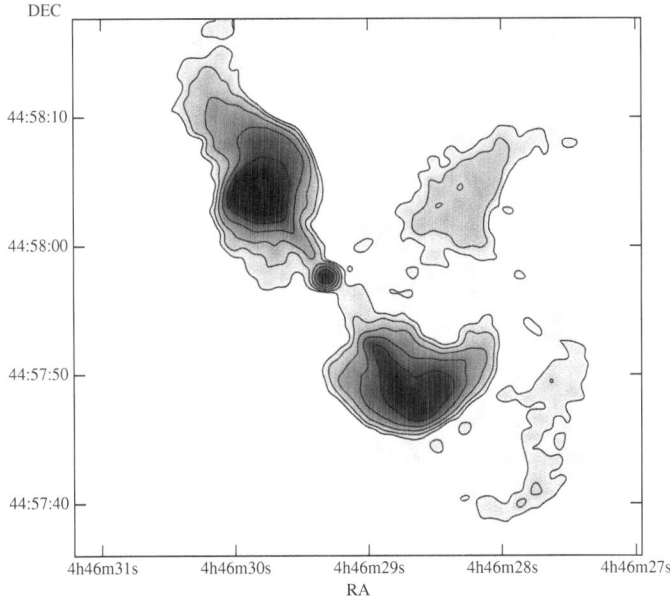

Fig. 2. The 5 GHz radio emission from 3C 129.1 [7]. The morphology of the relativistic plasma indicates clear interaction with the ICM, however X-ray observations of the cluster have poor counts.

Clusters with short observations may harbour undetected bubbles. In more distant clusters, some bubbles may be too small to detect. The existence of more bubbles in the distant cluster sample needs to be investigated to help understand the evolution of the AGN heating process. The exact nature of the energy transfer between the bubbles and ICM also needs to be determined.

References

1. Churazov, E., Forman, W., Jones, C., Böhringer, H., 2000, A&A, 356, 788
2. Ebeling, H., Edge, A. C., Bohringer, H., Allen, S. W. Crawford, C. S., Fabian, A. C., Voges, W., Huchra, J. P., 1998, MNRAS, 301, 881
3. Dunn, R. J. H., Fabian, A. C., Taylor, G. B., 2005, MNRAS, 364, 1343
4. Bîrzan, L., Rafferty, D. A., McNamara, B. R., Wise, M. W., Nulsen, P. E. J., 2004, ApJ, 607, 800
5. Rafferty, D. A., McNamara, B. R., Nulsen, P. E. J., Wise, M. W., 2006, ApJ, in press (astro-ph/0605323)
6. Bauer, F. E., Fabian, A. C., Sanders, J. S., Allen, S. W., Johnstone, R. M., 2005, MNRAS, 359, 1481
7. Taylor, G. B., Govoni, F., Allen, S. W., Fabian, A. C., 2001, MNRAS, 326, 2
8. Dunn, R. J. H., Fabian, A. C., 2006, MNRAS, in press (astro-ph/0609537)

Lack of Cooling Flow Clusters at $z > 0.5$

A. Vikhlinin[1,2], R. Burenin[2], W. R. Forman[1], C. Jones[1], A. Hornstrup[3], S. S. Murray[1] and H. Quintana[4]

[1] Harvard-Smithsonian Center for Astrophysics, Cambridge MA, USA
[2] Space Research Institute, Moscow, Russia
[3] Danish National Space Center
[4] Dep. de Astronomia y Astrofisica, Pontificia Universidad Catolica de Chile

Summary. The goal of this work is to study the incidence rate of "cooling flows" in the high redshift clusters using *Chandra* observations of $z > 0.5$ objects from a new large, X-ray selected catalog [2]. We find that only a very small fraction of high-z objects have cuspy X-ray brightness profiles, which is a characteristic feature of the cooling flow clusters at $z \sim 0$. The observed lack of cooling flows is most likely a consequence of a higher rate of major mergers at $z > 0.5$.

1 Introduction

The central regions in a large fraction of low-redshift clusters are clearly affected by radiative cooling [3]. Some estimates put the fraction of such cooling flow clusters at up to $> 70\%$ (e.g., [7]). A recent study by Bauer et al. suggests that the cooling flow fraction remains high to $z \sim 0.4$ [1]. However, this work is based on the *ROSAT* All-Sky Survey cluster sample, and so it can be strongly affected by Malmquist bias (strongly over-luminous clusters are preferentially selected because of the high flux threshold).

Any evolution in the cooling flow fraction, if detected, must be taken into account in detailed physical models of this phenomenon. We address this important question using a new distant cluster sample, derived from a sensitive survey based on the *ROSAT* pointed observations [2]. All objects were observed with *Chandra*, providing a uniform dataset which should be much less affected by selection effects than the previous samples.

2 Definition of the "cooling flow" cluster

First of all, we need to choose a definition of the "cooling flow" cluster that can be efficiently applied to the X-ray data of various statistical quality. The most common definition is based on the estimated central cooling time: cooling flow clusters have $t_{\text{cool}} \ll t_H$ (e.g., [7]). One could also use the mass deposition rate given by the standard cooling flow model [3]; cooling flow clusters have $\dot{M} \gtrsim (10-100)\, M_\odot\, \text{yr}^{-1}$ [7]. These definitions rely on spatially-resolved spectroscopic measurements which is a serious disadvantage for application at high redshifts. The data of sufficient quality to measure $T(r)$ are available

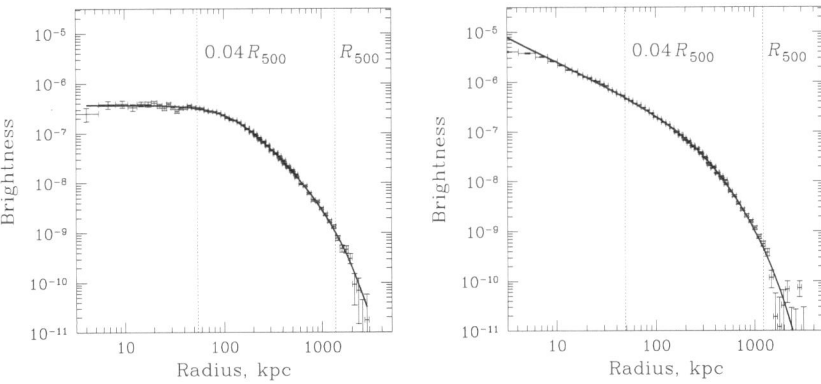

Fig. 1. X-ray surface brightness profiles typical for non-cooling flow (*left*; A401) and cooling flow (*right*; A85) clusters. Solid lines show the model X-ray brightness corresponding to the best-fit gas density model (see [9] for details).

only for a small number of high-z objects. We, therefore, look for a definition based solely on the X-imaging data.

At low redshifts, there is a clear connection between the presence of the cooling flow and the X-ray morphology. Clusters with $t_{\rm cool} \gtrsim t_H$ have X-ray brightness profiles with flat cores while those with $t_{\rm cool} \ll t_H$ have characteristic central cusps in the X-ray brightness distribution (Fig. 1). The central cusp can be characterized by the power-law index of the gas density profile, $\alpha = d \log \rho_g / d \log r$. For uniformity, the radius at which α is measured should be chosen at a fixed fraction of the cluster virial radius. This radius should be sufficiently small so that the effects of cooling are strong. At very small radii, however, the density profiles even in clusters with strong cooling flows can flatten because of the outflows from the central AGN (see many papers in these proceedings). Empirically, a good choice is $r = 0.04 R_{500}$,[5] and so we define the "cuspiness parameter", α, as

$$\alpha \equiv \frac{d \log \rho_g}{d \log r} \qquad \text{at} \quad r = 0.04 R_{500} \tag{1}$$

Cuspiness can be measured by fitting a realistic 3-dimensional gas density model to the observed X-ray surface brightness (our procedure is described in [9]). Examples of the best-fit models are shown by solid lines in Fig. 1. Such modeling is feasible with moderate-exposure *Chandra* observations of high-redshift clusters. R_{500} can be estimated using low-scatter cluster mass proxies such as the average temperature (excluding the central cooling region). We use an even better proxy, the recently proposed Y_X parameter [6], which is remarkably insensitive to the cluster dynamical state and easily measured even in high-redshift clusters.

---

[5] R_{500} is the radius at which the mean enclosed total mass overdensity is 500 relative to the critical density at the object redshift. $R_{500} \approx 0.5 R_{\rm vir}$.

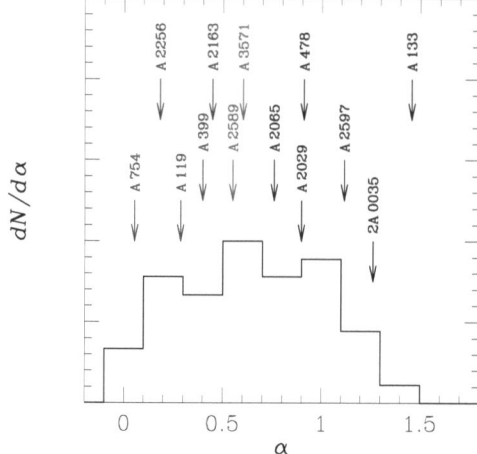

Fig. 2. The distribution of the cuspiness parameter in the low-z cluster sample. Arrows indicate the values for some well-known clusters. The boundary value of $\alpha = 0.5$ approximately separates cooling flow and non-cooling flow clusters.

Our low-redshift cluster sample is a flux-limited subsample of 48 objects from the HIFLUGCS catalog [8], all with the archival *Chandra* observations. The distribution of the cuspiness parameter for these objects is shown in Fig. 2. Clearly, the value of α is closely connected to the more common cooling flow definitions. Clusters with $\alpha > 0.7$ (e.g., A2065, A478, A2029, A2597, 2A 0035, A133) are known to host strong cooling flows. The objects with $\alpha < 0.5$ (e.g., A2163, A399, A119, A2256, A754) are famous non cooling flow clusters. The clusters in the range $0.5 < \alpha < 0.7$ (e.g., A2589, A3571) host weak cooling flows. Therefore, *the cooling flow clusters are those with* $\alpha > 0.5$. Approximately 2/3 of the low-redshift sample (31 of 48 objects) have cuspiness above this value, in line with the previous estimates of the cooling flow incidence rate (e.g. [7]).

3 High-redshift cluster sample

Our high-redshift sample is derived from the recently completed 400 deg^2 *ROSAT* PSPC survey (400d; [2]). This is the largest-area survey based on the *ROSAT* pointed observations. Clusters are detected as extended X-ray sources in the central 17.5′ of the PSPC FOV and required to have fluxes $f_x > 1.4 \times 10^{-13}$ erg s^{-1} cm^{-2}. The X-ray sample is fully identified. It includes 266 optically confirmed clusters (95% of the X-ray candidate list). Spectroscopic redshifts are available for all objects.

A subsample of the high-z 400d clusters has been observed with *Chandra*. The exposure times were chosen to yield at least 2000 source counts, which is

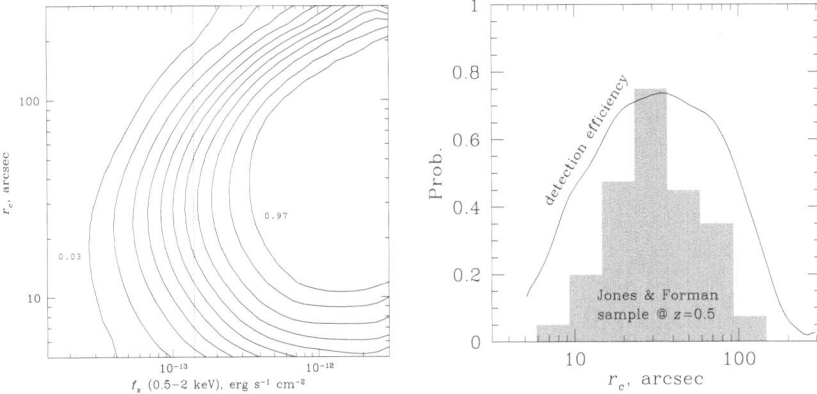

Fig. 3. *(a)* — Detection efficiency of the 400d survey for idealized β-model clusters as a function of total flux and core-radius. Dotted line shows the flux limit of the 400d catalog. Detection efficiency is reduced for $r_c \lesssim 8''$ because the such clusters are hard do distinguish from the point sources. The efficiency is also small for objects with $r_c \gtrsim 150''$ because they are "lost" in the cosmic X-ray background. *(b)* Detection efficiency as a function of core-radius for $f_x = 2 \times 10^{-13}$ erg s^{-1} cm^{-2}. Shaded histogram shows the distribution of core-radii in a low-redshift sample [5] scaled to $z = 0.5$.

sufficient to measure the average cluster temperature with a 15% uncertainty and accurately trace the surface brightness profile to $r \sim R_{500}$. *Chandra*'s angular resolution corresponds to a linear scale of < 8 kpc out to $z = 1$, fully sufficient to measure the cuspiness parameter. In this work, we use 400d clusters at $z > 0.5$, 20 in total. The typical mass of these objects corresponds to today's 4 keV clusters.

The basic characteristics of the X-ray selection in the 400d survey have been extensively calibrated through exhaustive Monte-Carlo simulations (see [2] for details). The aspect most relevant for the present study is the sensitivity of the detection efficiency to the cluster size and structure. A precise two-dimensional map of the detection efficiency as a function of cluster size and core radius was derived for idealized β-model clusters (Fig. 3a). The detection efficiency drops significantly only for $r_c \lesssim 8''$ and $r_c \gtrsim 120''$ (see Fig. 3b which shows the slice through the detection probability map at $f_x = 2 \times 10^{-13}$ erg s^{-1} cm^{-2}, just above the survey flux limit). The angular size range in which the 400d X-ray detection is sensitive encompasses the entire range of core-radii expected for the high-redshift clusters (c.f. shaded histogram in Fig. 3b). Therefore, the 400d selection will not bias the distribution of core-radii for β-model clusters.

The sensitivity of the 400d X-ray detection algorithm to cooling flow clusters with the cuspy X-ray brightness profiles requires a separate study. This issue was addressed by a separate set of Monte-Carlo simulations in which instead of β-models, we used the real X-ray images of a complete

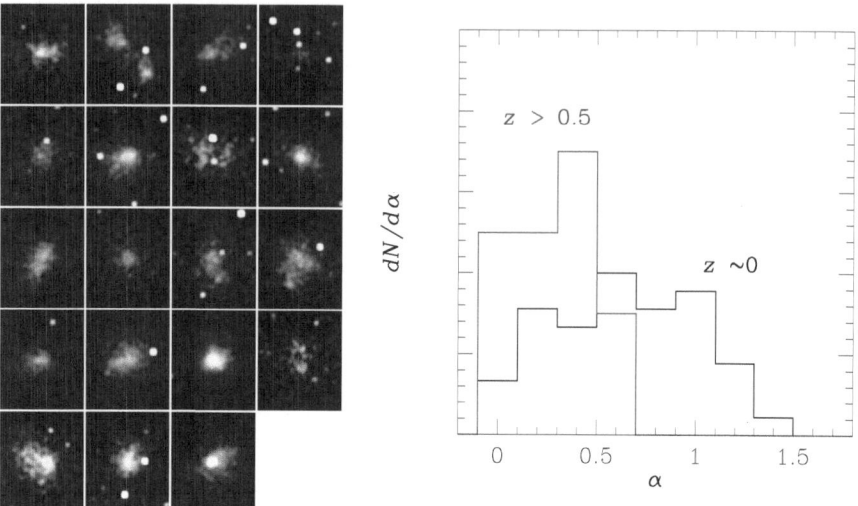

Fig. 4. *(a) Chandra* images of the 400d clusters with $z > 0.5$. Note a high fraction of objects that show clear signs of a major merger. *(b)* Distribution of the cuspiness parameter for the $z > 0.5$ and $z \sim 0$ samples.

sample of low-z clusters, scaled to different redshifts in the range $0.35 < z < 0.8$ (see [2] for details). A short summary of the results from these simulations is that there is no significant difference in the detection efficiency for the β-model and cooling flow clusters (see, e.g., Fig. 16 in [2]). For example, Hydra-A ($\alpha = 1.24$) at $z = 0.45$[6] is detected with the probability 0.67; 2A 0335 ($\alpha = 1.26$) at $z = 0.45$ has $p_{det} = 0.54$; A2029 ($\alpha = 0.90$) at $z = 0.8$ has $p_{det} = 0.69$. These values are near the maximum efficiency for β-model clusters of similar flux (Fig. 3b). Therefore, there should be no discrimination in the 400d survey against the objects similar to today's cooling flow clusters.

4 Observed morphologies and cuspiness parameters of the high-redshift clusters

Chandra images of the $z > 0.5$ objects from the 400d sample show a clear evolution of the cluster X-ray morphologies — at least 15 of 20 objects shows signs of an on-going major merger (Fig. 4a); the corresponding fraction in the low-redshift sample is $\lesssim 30\%$. The same effect is apparent in the distribution of the cuspiness parameter shown in Fig. 4b. Only 3 of 20 high-z clusters have $\alpha > 0.5$ (i.e., above the boundary between cooling flow and non cooling

---

[6] Redshifts here are chosen so that the observed fluxes would correspond to that in Fig. 3b, 2×10^{-13} erg s^{-1} cm^{-2}.

clusters, see § 2), while in the low-z sample this fraction is 31 of 48 (65%). There are no clusters with $\alpha > 0.7$ (strong cooling flows) in the $z > 0.5$ sample, while the fraction of such clusters at $z \sim 0$ is 46% (22 of 48 objects). The statistical significance of the difference in the distribution corresponds to a random fluctuation probability of only $P \simeq 5 \times 10^{-6}$.

Our results provide tantalizing evidence for a strong evolution in the incidence rate of the cluster cooling flows at $z > 0.5$. This is apparently related to the higher cluster merging rate, indeed expected at these redshifts in the Dark Energy dominated, cold dark matter cosmological models (e.g., [4]). Cluster cooling flows thus appear to be a relatively recent phenomenon, which become common only in the past 1/3 of the Hubble time.

References

1. Bauer, F. E., Fabian, A. C., Sanders, J. S., Allen, S. W., Johnstone, R. M., 2005, MNRAS, 359, 1481
2. Burenin, R. A., Vikhlinin, A., Hornstrup, A., Ebeling, H., Quintana, H., Mescheryakov, A., 2006, ApJS, submitted (astro-ph/0610739)
3. Fabian, A. C., 1994, ARA&A, 32, 277
4. Gottlöber, S., Klypin, A., Kravtsov, A. V., 2001, ApJ, 546, 223
5. Jones, C., Forman, W., 1999, ApJ, 511, 65
6. Kravtsov, A. V., Vikhlinin, A., Nagai, D., 2006, ApJ, 650, 128
7. Peres, C. B., Fabian, A. C., Edge, A. C., Allen, S. W., Johnstone, R. M. and White, D. A., 1998, MNRAS, 298, 416
8. Reiprich, T. H., Böhringer, H., 2002, ApJ, 567, 716
9. Vikhlinin, A., Kravtsov, A., Forman, W., Jones, C., Markevitch, M., Murray, S. S., Van Speybroeck, L. , 2006, ApJ, 640, 691

The XDCP Prospects of Finding Cooling Core Clusters at Large Lookback Times

R. Fassbender[1], H. Böhringer[1], J. Santos[1], P. Schuecker[1], G. Lamer[2], A. Schwope[2], J. Kohnert[2], P. Rosati[3], C. Mullis[4] and H. Quintana[5]

[1] Max-Planck Institut für extraterrestrische Physik, Giessenbachstr., 85748 Garching, Germany `rfassben@mpe.mpg.de`
[2] Astrophysikalisches Institut Potsdam, Potsdam, Germany
[3] European Southern Observatory, Garching, Germany
[4] University of Michigan, Ann Arbor, USA
[5] Universidad Catolica de Chile, Santiago, Chile

1 Introduction

Statistical studies of ROSAT selected cluster samples at $z < 1$ show indications that the fraction of strong cooling core clusters (CCC) drops rapidly at redshifts beyond $z = 0.5$ (see Vikhlinin et al., these proceedings). The superb sensitivity and improved resolution of XMM-Newton now enables serendipitous archival X-ray surveys for very distant clusters of galaxies, resolving clusters down to core radii of $50 - 60$ kpc (corresponding to 6-7 arcsecs) at $z > 1$. We are currently engaged in the XMM-Newton Distant Cluster Project (XDCP) [1], a serendipitous X-ray survey with the goal to construct a statistically complete sample of at least 30 galaxy clusters at $z > 1$ for evolutionary and cosmological studies. In these proceedings we show that the prospects of finding very distant cooling core clusters are very promising for ongoing XMM surveys if these systems exist at lookback times of more than 8 Gyrs.

2 Project Status

The X-ray analysis of the first $\sim$40 square degrees of deep ($>$10 ksec), high galactic latitude ($|b| \geq 20$ deg) XMM-Newton archival observations is now complete. All detected extended X-ray sources were carefully screened for low redshift optical counterparts using the DSS and 2MASS all-sky surveys. Only X-ray sources which are basically blank on the DSS enter our distant candidate list, indicating that the cluster is beyond a redshift of $0.5 - 0.6$. We have identified more than 200 distant cluster candidates of which 30-40 are expected to be clusters at redshifts beyond unity. We are currently engaged in optical/NIR follow-up campaigns to obtain two-band imaging for all candidates. The R-z (or alternatively z-H) color of the cluster red-sequence allows an efficient selection of the $z \geq 0.9$ cluster candidates, which will be followed-up spectroscopically to determine accurate redshifts and approximate velocity dispersions.

3 Detectability of Distant Clusters with Compact Cores

Figure 1 shows our distant cluster candidates in the core radius vs. flux plane. Based on cluster studies using ROSAT data, several works, e.g. Chen et al. ([2] and these proceedings), showed that cooling core clusters tend to have significantly smaller core radii and higher luminosities compared to non cool core systems. CCC are thus expected to be located in the well populated encircled region in Fig. 1 which contains more than 50% of our distant cluster candidate sources. The measured core radii (fixed $\beta = 2/3$ model) between 6-15 arcsecs in this parameter region correspond to a physical scale of 50-120 kpc at $z \sim 1$. Compared to the most prominent distant clusters confirmed to date (see triangles in Fig. 1) the detection limits for our distant candidates are about a factor five fainter in terms of flux and roughly 2-3 times smaller for the core radius. Two first examples (indicated by circles in Fig. 1) of recently discovered clusters with $z \sim 1$ and compact core radii of $r_{core} < 100$ kpc are shown Fig. 2.

Even though the currently available data with typically 100-300 X-ray photons per candidate do not allow detailed cooling core diagnostics of high

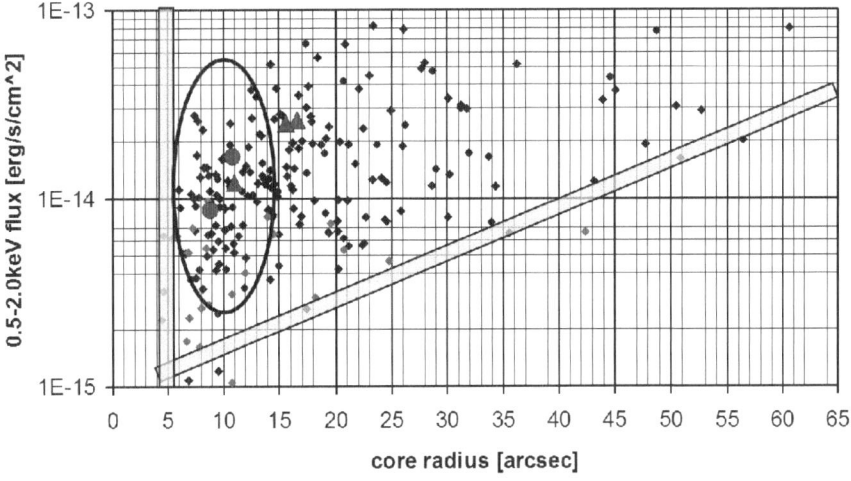

Fig. 1. XDCP distant cluster candidates (*diamonds*) in the core radius vs. flux plane. The parameter space of detected candidates is confined by the XMM resolution limit (*vertical shaded area*) at core radii of about 5 arcsecs (for fixed $\beta = 2/3$) and the background limit (*lower shaded area*), where the cluster surface brightness drops below the detection threshold. The triangles indicate the three most prominent $z > 1$ clusters confirmed to date, from left to right: XCS J2215 at $z = 1.45$ [5], RDCS J1252 at $z = 1.24$ [4], and XMMU J2235 at $z = 1.39$ [3]. Cooling core clusters are expected to be located in the well populated encircled region with small core radii. The red dots indicate the positions of the two distant cluster candidates with compact cores shown in Fig. 2.

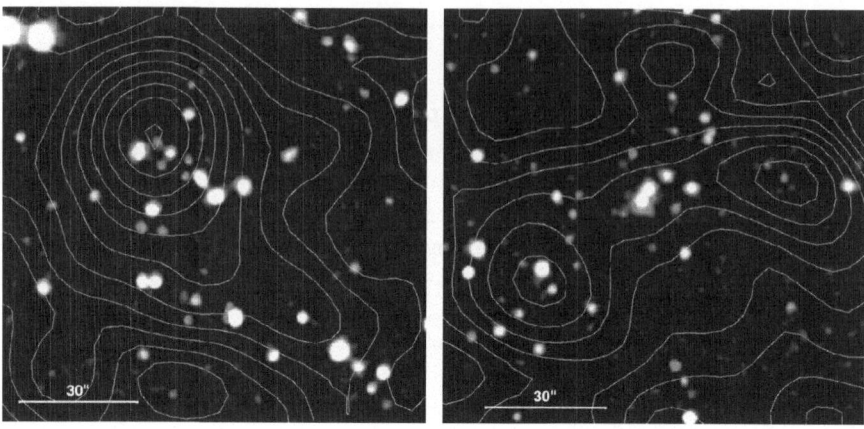

Fig. 2. $z + H$ images with X-ray contours of newly discovered clusters at $z \sim 1$ with compact core radii. The left panel shows a candidate with a measured core radius of 10.4 arcsec (~ 85 kpc) and a $0.5 - 2.0$ keV flux of $1.7 \cdot 10^{-14} ergs/s/cm^2$. The left X-ray center of the candidate cluster in the right panel was detected as an extended source with $r_{core} = 8.8$ arcsec (~ 70 kpc) and a flux of $8.6 \cdot 10^{-15} ergs/s/cm^2$.

redshift cluster candidates, the detection and confirmation of a significant number of $z > 1$ CC clusters seems to be in reach within the next few years.

References

1. Böhringer, H., Mullis, C., Rosati, P., Lamer, G., Fassbender, R., Schwope, A., & Schuecker, P. 2005, The Messenger, 120, 33
2. Chen, Y., Reiprich, T.H., Böhringer, H., & Ikebe, Y., 2006, A&A, submitted
3. Mullis, C. R., Rosati, P., Lamer, G., Böhringer, H., Schwope, A., Schuecker, P., & Fassbender, R. 2005, ApJL, 623, L85
4. Rosati, P., et al. 2004, AJ, 127, 230
5. Stanford, S. A., et al. 2006, ApJL, 646, L13

An XMM-Newton View of Three Clusters of Galaxies

T.F. Laganá[1,2], G.B. Lima Neto[1] and F. Durret[2]

[1] Instituto de Astronomia, Geofísica e Ciências Atmosféricas, USP, Rua do Matão 1226, Cidade Universitária, 05508-900, São Paulo, SP, Brazil.
tflagana@astro.iag.usp.br & gastao@astro.iag.usp.br
[2] Institut d'Astrophysique de Paris, UMR 7095, Université Pierre & Marie Curie, 98bis Bd. Arago, 75014 Paris, France. durret@iap.fr

1 Introduction

Clusters of galaxies occupy a unique position in the hierarchical structure formation scenario. The correlation between physical quantities at different redshifts provides a unique constraint on cosmological scenarios. X-ray studies of galaxy clusters are particularly relevant as they allow the determination of both the baryonic and the total mass components. Based on the application of the hydrostatic equilibrium equation (e.g. [1]), X-ray cluster mass measurements rely on observations of the cluster temperature and brightness profiles. Long-exposure XMM-*Newton* observations are now providing precise spatially resolved measurements of the ICM density and temperature distributions for low and intermediate-redshift galaxy clusters.

In this work, we present the analysis of archival data for three clusters (A496, A1689 and A2667) within the redshift range $0.03 < z < 0.2$ and with temperatures kT between 3 and 8 keV. We evaluated the total and baryonic masses and computed temperature maps for these clusters.

2 Temperature maps and mass determination

The temperature maps are made in a grid, where in each pixel we make a spectral fit (assuming bremsstrahlung + line emission) to determine the temperature. We set a minimum count number (900 net counts) to do a spectral fit with a MEKAL plasma model. If we do not reach this minimum count number in a pixel we gradually enlarge the area up to a 5×5 pixel region (or $128'' \times 128''$ region). The best fit temperature value is attributed to the central pixel. When there are not enough counts in this region, the pixel is ignored and we proceed to the next one. We compute the ARF and the RMF matrices for each pixel in the grid. For more details see [5].

From the temperature maps shown above we see that even in a relaxed cluster such as A496 we see some substructures in the temperature map. This also reveals that the gas is cooler in the central region, as expected for a relaxed cluster which does not show evidence for recent major merging

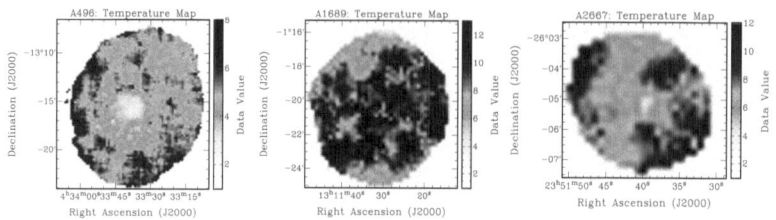

Fig. 1. Temperature maps for the A496, A1689 and A2667, respectively. The temperature is given in keV.

processes. On the other hand, the other two clusters are unrelaxed. A1689 shows temperature variations from 4 to almost 12 keV. This quality of details that are evidenced in the temperature map (like small hotter blobs) is lost in the projected temperature profile where this cluster appears to have an almost isothermal profile. This is due to the determination of an average temperature in each concentric annulus. A2667 reveals possible shock fronts that are correlated with higher X-ray temperatures. In this case the gas was compressed and heated during the shock. This can be seen in the map as brighter pixels at the outskirts. This cluster is probably still in process of building up ([3]). Using the equation of hydrostatic equilibrium together with the temperature and density profiles, we integrated the total mass for these three clusters out to $R = 0.6\ r_{340}$.

The projected temperature profiles of two (A496 and A2667) of our clusters show similar behaviors: the temperature is smaller in the center ($\approx T_{max}/3$), increases up to a broad peak ($\approx 0.3\ R_{max}$) and then decreases towards the outskirts. A1689 has a flat core instead of a cool core. In spite of this difference, we constructed an analytic function for the temperature profile (Eq. 1) in such a way that it can empirically reproduce these general features:

$$T(r) = \frac{T_0 \left[\alpha \sqrt{\frac{r}{r_t}} + \frac{r}{r_t} + 1\right]}{\left(\frac{r}{r_t}\right)^2 + 1}. \tag{1}$$

For the density profiles we used the Sérsic ([4]) and the well known β-model [2]. The mass determination for these models are given in Table 1 where the β-model can be applied for non-relaxed clusters like A1689 and A2667.

3 Conclusions

The β-model may not be a good model for relaxed clusters and it has been used without taking into account the dynamical state of the cluster. For cool-core clusters the temperature profile decreases towards the center and

Table 1. Masses and baryonic fraction for our three clusters.

Sérsic ($R < 0.6\ r_{340}$)	A496	A1689	A2667	β-model ($R < 0.6\ r_{340}$)	A496	A1689	A2667
$M_{\text{tot}}(10^{14} M_\odot)$	1.3	4.5	9.6	$M_{\text{tot}}(10^{14} M_\odot)$	0.97	3.4	3.5
$M_{\text{gas}}(10^{14} M_\odot)$	0.20	0.35	0.51	$M_{\text{gas}}(10^{14} M_\odot)$	0.081	0.34	0.40
f_{b}	**0.15**	0.078	0.053	f_{b}	0.082	**0.10**	**0.11**

if one assumes that in this inner region the derivative of the temperature as a function of the radius is positive and if the surface brightness profile is described by the β-model we would obtain a negative mass in the central part. It is also important to emphasize the deficiency of the standard β-model in modelling satisfactorily the central emission from cooling flow clusters of galaxies. Cooling flows lead to an enhancement of the gas density in the central region and as a consequence the gas density has a cuspy shape in the center instead of a flat core. Due to this feature the well known β-model provides an unacceptable fit.

To avoid these problems, there have been alternative parametrisations to model this excess emission, notably with the use of a double β-model (e.g. [6, 7]). In these models a second term was introduced to take into account this change of slope in the density profile in the core.

In this work we modeled the emission of a relaxed cluster with the Sérsic model. We showed that a single and simple model like the Sérsic-model, which is largely applied to elliptical galaxies, can be adequately applied instead of using a six parameter function.

Acknowledgement. We wish to thank the French-Brazilian cooperation CAPES/ Cofecub (BEX 1468/05-7 and 444/04) and FAPESP (10345-3) for financial support.

References

1. C. L. Sarazin: *X-Ray Emission from Clusters of Galaxies*, Cambridge University Press (1988).
2. Cavaliere & Fusco-Femiano, 1978, A&A, 60, 667
3. G. Covone, C. Adami, F. Durret, J.-P. Kneib, G.B. Lima Neto, E. Slezak, 2006, A&A, in press
4. R. Demarco, F. Magnard, F.Durret, and I. Márquez, 2003, A&A, 407, 437
5. F. Durret, G. B. Lima Neto, W. Forman, 2005, A&A, 432, 809
6. Y. Ikebe, K. Makishima, F. Yasushy, T. Tamura et al., 1999, ApJ, 525, 581
7. A. Vikhinin, A. Kravtsov, W. Forman, C. Jones et al., 2006, ApJ, 640, 991

Exploring Massive Galaxy Clusters: XMM-Newton observations of two morphology unbiased samples at $z \sim 0.2$ and $z \sim 0.3$

Y.-Y. Zhang[1], A. Finoguenov[1], H. Böhringer[1], J.-P. Kneib[2,3], G. P. Smith[4], O. Czoske[5,6], G. Soucail[7], P. Schuecker[1], Y. Ikebe[1,8], K. Matsushita[1,9], L. Guzzo[10] and C. A. Collins[11]

[1] Max-Planck-Institut für extraterrestrische Physik, Garching, Germany
yyzhang@mpe.mpg.de
[2] OAMP, Laboratoire d'Astrophysique de Marseille, traverse du Siphon, 13012 Marseille, France
[3] Caltech-Astronomy, MC105-24, Pasadena, CA 91125, USA
[4] School of Physics and Astronomy, University of Birmingham, Edgbaston, Birmingham, B152TT, UK
[5] AIFA, Universität Bonn, Auf dem Hügel 71, 53121 Bonn, Germany
[6] Kapteyn Astronomical Institute, PO Box 800, 9700AV Groningen, Netherlands
[7] Observatoire Midi-Pyrenees, Laboratorire d'Astrophysique, UMR 5572, 14 Avenue E. Belin, 31400 Toulouse, France
[8] National Museum of Emerging Science and Innovation, Tokyo, Japan
[9] Tokyo University of Science, Tokyo, Japan
[10] INAF - Osservatorio Astronomico di Brera, Merate/Milano, Italy
[11] Liverpool John Moores University, Liverpool, U.K.

Summary. For the spectral determination of the temperature distribution in galaxy clusters we developed a double background subtraction method, and applied it to 2 morphology unbiased samples of X-ray luminous galaxy clusters (12 at $z \sim 0.2$, 13 at $z \sim 0.3$) observed by XMM-Newton. The scaled profiles of the X-ray properties show a self-similar behaviour above $0.2r_{500}$ for both samples. The tight X-ray scaling relations of the 2 samples show no evolution compared to the nearby and more distant samples. The X-ray lensing masses converge for most clusters.

1 Scaled average temperature profiles

We scaled the radial temperature profiles by the volume average global temperatures $T_{0.2-0.5r_{500}}$ and r_{500} for both samples and found a self-similar behaviour at radii above $0.2r_{500}$. The average temperature profiles of both samples agree with the temperature profiles of the cluster samples in [5], [3], [6] and [8] as shown in Fig. 1 (left panel).

2 Scaling relations

To use the mass function of the cluster sample to constrain cosmological parameters, it is important to calibrate the scaling relations between the cluster mass and X-ray observables. The massive clusters selected in a narrow redshift range provide an important means to constrain the normalization and to understand the scatter of the scaling relations such as the L–M relation. For both samples, the cluster masses are uniformly determined from the high quality XMM-Newton data. This guarantees minimum systematics due to the analysis method. Comparing the X-ray scaling relations of such 2 samples to the samples at the other narrow redshift bins (e.g. [4], [1], [9]) shows that the evolution of the scaling relations is accounted for by the redshift evolution. We obtained an overall agreement with the recent studies of the scaling relations within the observational dispersion. This fits into the general opinion that galaxy clusters are self-similar up to $z \sim 1$ (e.g. [1]).

3 X-ray lensing mass comparison

Smith et al. ([7]) present a uniform strong lensing analysis of 9 clusters in this sample. We took these 9 overlapping clusters as a subsample (hereafter the S05 subsample, X-ray selected) and performed the combined studies for strong lensing and X-ray observations. Ten clusters in our $z \sim 0.3$ sample have CFH12k data with weak lensing masses determined in [2]. We took these 10 clusters as a subsample (hereafter the B06 subsample, X-ray selected) and performed a comparison between weak lensing and X-ray results.

For most clusters, the X-ray mass and strong/weak lensing mass converge. The existing mass discrepancy between X-ray and lensing for several individual clusters is due to a combination of the measurement uncertainties and the physics in the individual clusters (details see [11]).

4 Conclusions

Clusters of galaxies are used in a variety of ways in cosmological applications, e.g. by means of their mass function. It is thus important to calibrate the cluster mass determination and mass–observable scaling relations. We performed a systematic analysis to measure the X-ray observables based on the high quality XMM-Newton observations of two samples at $z \sim 0.2$ and $z \sim 0.3$. We found closely universal temperature profiles for the 2 samples. The X-ray scaling relations of the 2 samples agree with the scaling relations of the nearby and more distant samples within the observational dispersion. This fits the opinion (e.g. [1], [9], [10], [11]) that the evolution of massive galaxy clusters up to $z \sim 1$ is well described by a self-similar model. The X-ray – lensing masses converge, though some clusters show mass discrepancy due to their peculiarities. The cluster cores contribute up to 70% of the X-ray luminosity. Using the X-ray luminosities excluding the cluster cores, the L–M relation is insensitive to the exclusion of cool core clusters. We observed a very weak indication that the sample at $Z \sim 0.3$ contains less pronounced cool cores than the sample at $z \sim 0.2$ (Fig. 1, right panel). More details can be found in [10], [11].

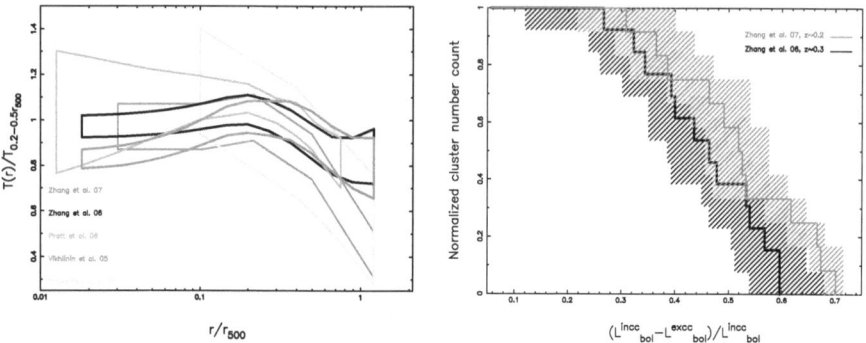

Fig. 1. Scaled average temperature profiles (left) and normalized cumulative cluster number count (right).

References

1. Arnaud, M., Pointecouteau, E., & Pratt, G. W. 2005, A&A, 441, 893
2. Bardeau, S., Soucail, G., Kneib, J.-P., et al. 2006, A&A, in preparation
3. De Grandi, S., & Molendi, S. 2002, ApJ, 567, 163
4. Ettori, S., Tozzi, P., Borgani, S., & Rosati, P. 2004, A&A, 417, 13
5. Markevitch, M., Forman, W. R., Sarazin, C. L., & Vikhlinin, A. 1998, ApJ, 503, 77
6. Pratt, G. W., Böhringer, H., Croston, J. H., et al. 2007, A&A, 461, 71
7. Smith, G. P., Kneib, J.-P., Smail, I., et al. 2005, MNRAS, 359, 417
8. Vikhlinin, A., Markevitch, M., Murray, S. S., et al. 2005, ApJ, 628, 655
9. Vikhlinin, A., Kravtsov, A., Forman, W., et al. 2006, ApJ, 640, 691
10. Zhang, Y.-Y., Böhringer, H., Finoguenov, A., et al. 2006, A&A, 456, 55
11. Zhang, Y.-Y., Finoguenov, A., Böhringer, H., et al. 2007, A&A, accepted

Part II

AGN-ICM Interaction in Cool Cores

Heating and Cooling in the Perseus Cluster Core

A.C. Fabian and J.S. Sanders

Institute of Astronomy, University of Cambridge, Madingley Road, Cambridge, CB3 0HA, UK. acf@ast.cam.ac.uk, jss@ast.cam.ac.uk

1 Introduction

It is well known that the radiative cooling time of the hot X-ray emitting gas in the cores of most clusters of galaxies is less than 10^{10} yr (Fig. 1). In many clusters the gas temperature also drops towards the centre. If we draw a causal connection between these two properties then we infer the presence of a cooling flow onto the central galaxy (e.g.[8]). High spectral resolution XMM-Newton data [26][34] and high spatial resolution Chandra data, e.g. [1], show however a lack of X-ray emitting gas below about one third of the cluster virial temperature. The explanation is that some form of heating balances cooling. The smoothness and similarity of the cooling time profiles and the flatness of the required heating profiles all indicate that we must seek a relatively gentle, quasi-continuous (on timescales $< 10^8$ yr), distributed heat source. The likely such source is the central black hole and its powerful jets which create bubble-like cavities in the inner hot gas ([5], see [25] for a review).

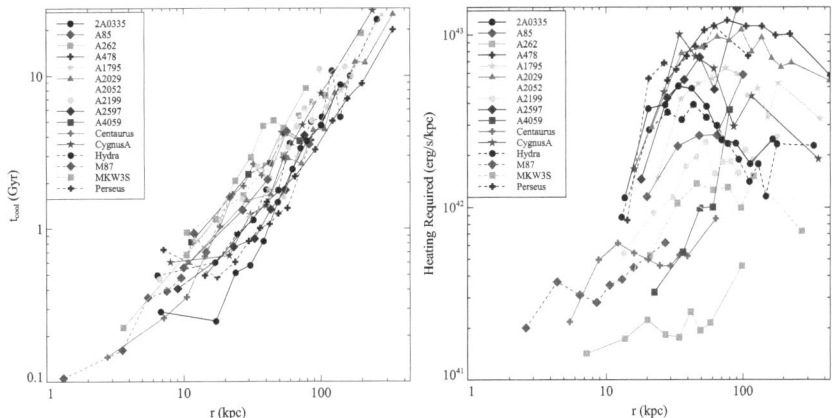

Fig. 1. Heating and cooling properties of the intracluster gas in the cores of 16 X-ray bright clusters [7]. (Left) radiative cooling times and (Right) heating rate required per kpc to balance cooling.

We briefly review the *general* heating and cooling statistics in an X-ray bright sample of cluster before we discuss the *detailed* situation in the Perseus cluster, the X-ray brightest cluster in the Sky. The Chandra count rate from the Perseus cluster is twice that of any other cool core cluster (the Virgo and Centaurus clusters come next) and the total exposure, of almost 1 Ms, is almost twice as long (compare with 500 ks for Virgo and 200 ks for Centaurus). In many ways therefore the Perseus cluster data are the best, and possibly unique, for determining the details of the energy flow in a cluster core.

2 General conclusions from the Brightest Clusters

Here we summarize the general situation in low redshift clusters using the Chandra analysis of the Brightest 55 sample from [7]. The results, where relevant, are in agreement with the earlier ROSAT analysis of [24]. Fig. 2 lists 30 sources of which only one (AWM7) does not have a radio source associated with the central galaxy. 14 of them have clear bubbles and another 5 have possible bubbles or interaction between the radio source and the hot gas. Only 10 out of the 30 do not need any heating (i.e. either the central cooling time exceeds 3 Gyr, see Fig. 2, and/or any central temperature drop is less than a factor of two).

Clear Bubbles	NVSS Radio		No Radio
Heating	Heating	No Heating	Heating
A85	2A 0335+096[†]	3C 129.1	AWM7
A262	A496	A1644	
A478	A2204	A1651	
A1795	PKS 0745-191	A1689	
A2029	MKW3s[†]	A2063	
A2052		A3112	
A2199		A3391	
A2597		A3558	
A4059		Klem44	
Centaurus		Ophiuchus	
Cygnus A			
Hydra A			
M87			
Perseus			
14	5	10	1

Fig. 2. Status of the cores in clusters from the Brightest 55 sample with Chandra data [7]. Objects in the column on the left show clear bubbles, those in the 2 centre columns have a central radio source and the one on the right has no reported radio source at the centre. Heating is defined as required if the central cooling time is less than 3 Gyr and the central temperature drop exceeds a factor of two.

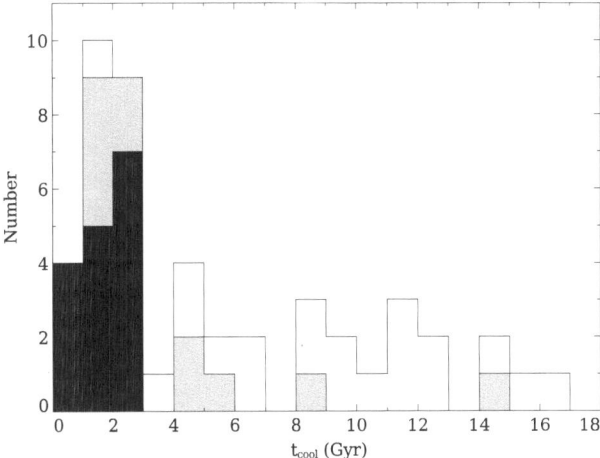

Fig. 3. Frequency histogram of the clusters tabulated in Fig. 2. Black shading indicates that clear bubbles are seen; grey means that there is a plausible radio source at the centre [7].

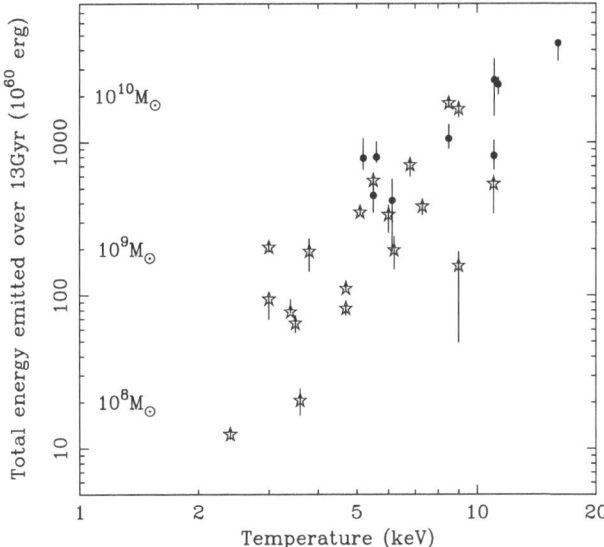

Fig. 4. An approximate estimate of the total energy emitted within the cooling radius over 13 Gyr plotted versus cluster temperature [14].The accreted black hole mass, assuming a radiative efficiency of 0.1, is also shown. The results reduce roughly as $(time)^2$, so drop by 7 if a cluster age of only 5 Gyr is assumed.The accreted masses are then still considerable (up to $3 \times 10^9\, M_\odot$).

The common occurrence of bubbles in the clusters which require heating means that the duty cycle of bubbling is between 75 and 90 per cent – the jets are 'on' most of the time. Such a mechanism satisfies the quasi-continuous requirement.

The steep abundance gradient observed in many clusters requires that the heating mechanism be relatively gentle and not produce much turbulence [28][19]. The relatively smooth and similar cooling time profiles support this.

The energy requirements are considerable (Fig. 4 from [14]). The PdV work done by the bubbles on the surrounding gas is nevertheless of the right order of magnitude [2][27][7]. It appears inescapable that the central radio source is pumping energy into the intracluster gas at roughly the required level. What is not clear from a general study of clusters, albeit the brighter ones, is how that energy is distributed on the right spatial scale to balance radiative cooling. Gas is not piling up at any particular radius or temperature. An even more challenging issue is how the feedback required for a tight heating / cooling balance operates.

Highly energetic events, such as inferred for Hydra A [23], whilst spectacular, are not typical of the norm. They also dump most of their energy well beyond the cooling region.

3 The Perseus Cluster

The Perseus cluster, A 426, is the X-ray brightest cluster in the Sky. The X-ray emission is sharply peaked on the cluster core, centred on the cD galaxy NGC 1275. Jets from the nucleus of that galaxy have inflated bubbles to the immediate N and S, displacing the ICM [3][12]. Ghost bubbles devoid of radio-emitting electrons, presumably from past activity, are seen to the NW and S. The radiative cooling time of the gas in the inner few tens of kpc is 2–3 hundred Myr, leading to a cooling flow of a few $100 M_\odot \, yr^{-1}$ if there is no balancing heat input. Energy from the bubbles or the bubble inflation process is a likely source of heat but the energy transport and dissipation mechanisms have been uncertain.

With 200 ks Chandra exposure we discovered both cool gas and shocks surrounding the inner bubbles as well as quasi-circular ripples in the surrounding gas which we interpreted as sound waves generated by the cyclical bubbling of the central radio source. Related features have been seen in the Virgo cluster [16][17]. The NW ghost bubble has a horseshoe-shaped Hα filament trailing it which shows the streamlines in the hot gas. (The velocities along the filament are reported by [21] and are consistent with the streamline hypothesis.) On this basis we concluded that the ICM is not highly turbulent and thus that viscosity (and possibly conduction) is high enough to dissipate the energy carried by the sound waves [10][11]. Such an energy transport and dissipation mechanism is roughly isotropic and can thereby provide the required gently distributed heat source [30][31][32][29][13]. The energy flux in

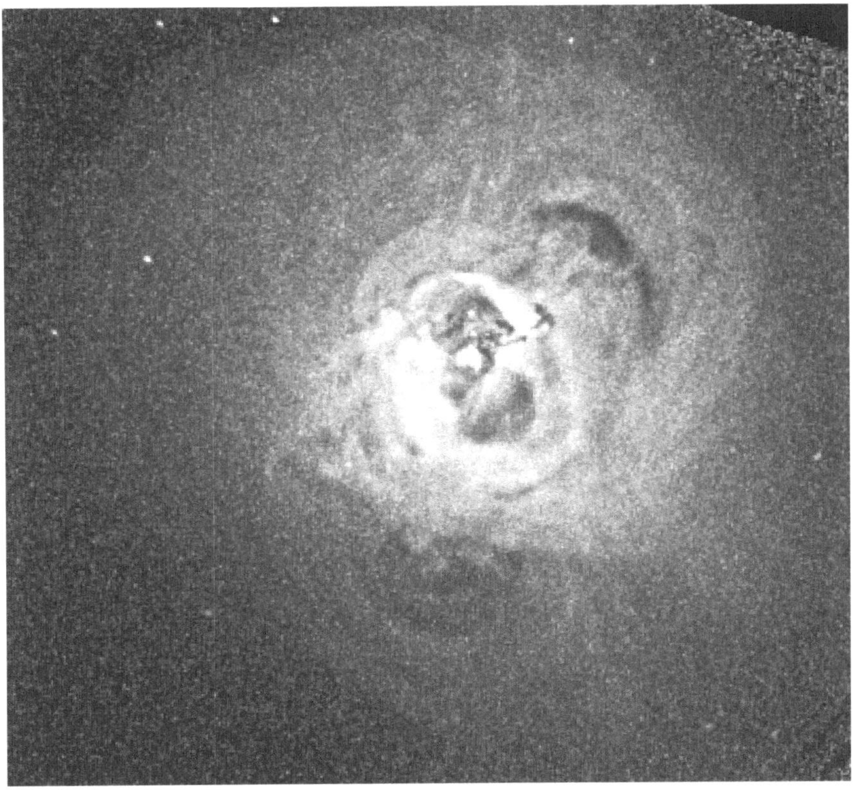

![Full 900 ks Chandra image of the centre of the Perseus cluster]

Fig. 5. Full 900 ks Chandra image of the centre of the Perseus cluster with red, green and blue representing soft, medium and hard X-rays within the Chandra band [15].

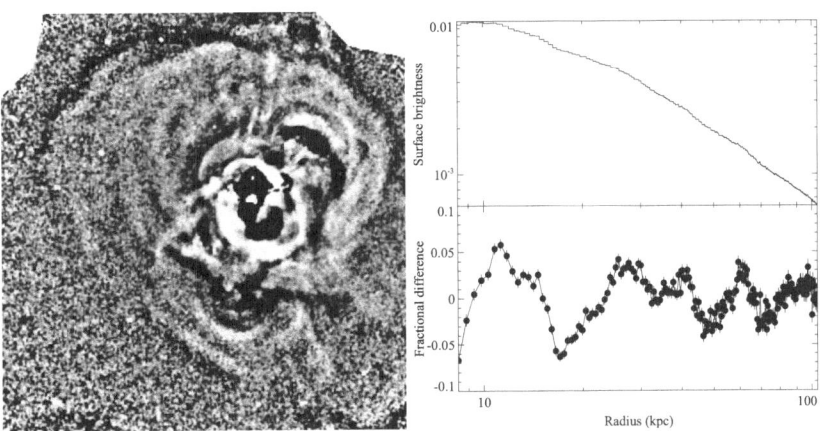

Fig. 6. Left: unsharp-mask image showing the ripples, Right: profile of the raw data to the E with deviations from a simple beta-model fit to that profile.

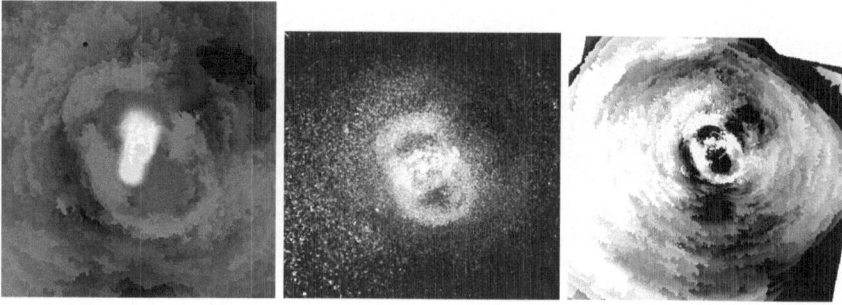

Fig. 7. Perseus pressure maps. Left: with radio overlay, Centre: showing that the high pressure regions are thick and circular around the radio bubbles, Right: pressure difference map showing a series of outer bubbles and channels to S and N.

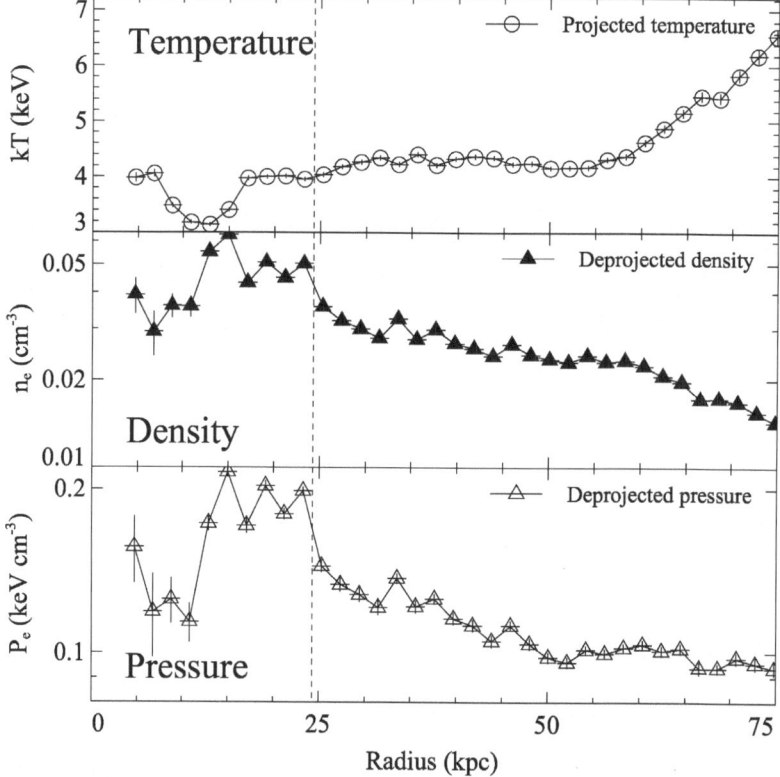

Fig. 8. Temperature, density and pressure profiles in a 30 deg wide sector to the NE of the Northern bubble. The position of the edge of the higher pressure ring (the shock front) is marked by a dashed line.

the sound waves seen equals the energy required to balance cooling within a radius of about 100 kpc.

A ripple is presumably made each time a new bubble grows. Close to the bubble the amplitude must be high so a shock is likely to form, which could dissipate so much of the energy that there is little left for significant sound waves and would overheat the innermost regions [18][22]. We now have 900 ks of good Chandra data and can see the details of what is happening here. The ripples (Fig. 6) have an amplitude of ±5 per cent or less (a level probably not detectable in any other cluster). Pressure maps show a thick high pressure rim to the bubbles (Fig. 6) with a sharp outer edge (best seen to the NE in Fig. 5). We measure the density jump at this edge as 40%. If the gas has an adiabatic index of 5/3 this should lead to a temperature rise from 4 to 5 keV. No such temperature jump is seen. Indeed the gas temperature, if anything, drops slightly across the edge [15]. (The sharp edge appears all round the bubbles so it is not a cold front.)

Observationally it seems that the violent pumping action of bubble growth does *not* lead to excessive heating of the innermost regions and allows sound waves to propagate outward, where the energy can be dissipated by viscosity and conduction in a more distributed manner, as required. The lack of a temperature jump is puzzling, but can be understood if energy is spread by thermal conduction [15].

4 Multiphase effects and cooling

The central galaxy in the Perseus cluster, NGC 1275, is surrounded by a spectacular filamentary optical nebulosity (see [6]). This is detected in the emission from molecular hydrogen [20] and CO [33] implying temperatures ranging from several 100 to several 1000 K. Soft X-ray emission is seen coincident with the filaments requiring a significant mass of gas at 0.7 keV (Fig. 9). There is also OVI emission detected by FUSE indicating gas at $\sim 5 \times 10^5$ K [4]. Assuming that this last emission is typical and attributing it all to cooling, then the cooling rate in the central 6 kpc is $\sim 30 \, M_\odot \, yr^{-1}$ [4]. The cooling rate implied at 1 keV by Fig. 9 is also only about $10 \, M_\odot \, yr^{-1}$. Some mixing, conduction etc could affect these estimates but it would be difficult to argue for any simple cooling flow over the central region exceeding about 10% of that deduced by the hotter X-ray emitting gas, in the absence of heating. The heating / cooling balance in the Perseus cluster core, as is the case in most such objects, must be tightly controlled.

5 Discussion

Heating by the repetitive formation and growth of bubbles has the quasi-continuous form required to balance radiative cooling. The deep imaging of

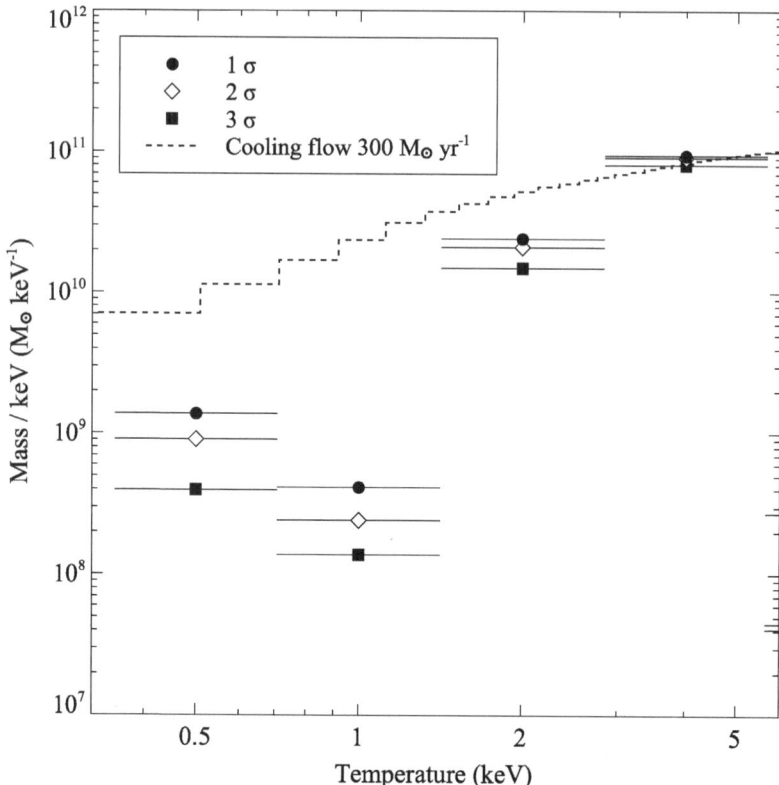

Fig. 9. Gas mass as a function of temperature within the central 1.5 arcmin radius of the Perseus cluster [15]. The expectation for a constant pressure cooling flow is shown for comparison.

the Perseus cluster core reveals the sound waves produced and shows that the inner shock is isothermal. More work is required to understand this, but electron conduction is a plausible explanation.

A major remaining problem is how the necessary tight feedback is achieved. Just how does the central accretion rate adjust to make heating balance cooling over a $\sim$ 100 kpc radius region?

Acknowledgement. We thank Robert Dunn and others for collaboration. ACF thanks the Royal Society for support.

References

1. Allen SW, et al, MNRAS, 324, 842 (2001)
2. Birzan, L., Rafferty, D. A., McNamara, B. R., Wise, M. W., Nulsen P. E. J. ApJ 607, 800 (2004)

3. Böhringer H., Voges, W., Fabian, A. C., Edge, A. C. and Neumann, D. M., MNRAS, 264, 25 (1993)
4. Bregman JD, Fabian AC, MIller E, Irwin JA, ApJ, 642, 746 (2006)
5. Churazov E., Forman, W., Jones, C., and Böhringer, H., A&A, 356, 788 (2000)
6. Conselice CJ, Gallagher JS, Wyse RFG, AJ, 122, 2281 (2001)
7. Dunn RJH, Fabian, A. C., MNRAS, 373, 959 (2006)
8. Fabian A.C., ARAA, 32, 277 (1994)
9. Fabian A.C., Celotti A., Blundell K.M., Kassim N.E., Perley R.A., MNRAS, 331, 369 (2002)
10. Fabian AC., Sanders, J.S., Allen, S.W., Crawford, C.S., Iwasawa, K., Johnstone, R.M., Schmidt, R.W., and Taylor, G.B., MNRAS, 344L, 43F (2003a)
11. Fabian AC., Sanders, J.S., Crawford, C.S., Conselice, C.J., Gallagher, J.S., and Wyse, R.F.G., MNRAS, 344L, 48 (2003b)
12. Fabian, A. C., Sanders, J. S., Ettori, S., Taylor, G. B., Allen, S. W., Crawford, C. S., Iwasawa, K., Johnstone, R. M. and Ogle, P. M., MNRAS 318, L65 (2000)
13. Fabian AC., Reynolds CS., Taylor GB., Dunn RJH., 2005, MNRAS, 363, 891
14. Fabian AC, Allen SW, Crawford CS, Johnstone RM, Morris GM, Sanders JS, Schmidt RG., 2002, MNRAS, 332, 0L50
15. Fabian, A. C., Sanders, JS, Taylor GB, Allen SW, Crawford CS, Johnstone RM, Iwasawa K, MNRAS, 366, 417 (2006)
16. Forman W. et al, ApJ, 635,894 (2005)
17. Forman W. et al, astro-ph/0604583 (2006)
18. Fujita Y., Suzuki TK., ApJ, 630, L1 (2005)
19. Graham J, Fabian AC, Sanders JS, Morris RG, MNRAS 368, 1369, (2006)
20. Hatch NA, Crwaford CS, Fabian AC, Johnstone RM, MNRAS, 358, 765 (2005)
21. Hatch NA., Crawford CS, Johnstone RM., Fabian AC., 367, 433 (2006)
22. Mathews, W.G., Faltenbacher, A., Brighenti, F., ApJ, 638, 659 (2006)
23. Nulsen PEJ, McNamara BJ, Wise M, David LP, 2005, ApJ, 628, 629 (2005)
24. Peres C., Fabian AC, Edge, AC, Allen SW, Johsntone RM, White DA, MNRAS 298, 416 (1998)
25. Peterson J., Fabian A.C., Phys. Rep., 427, 1 (2006)
26. Peterson J. Parels FBS, Kaastra JS, Arnaud M, Reiprich TH, Fabian AC, Mushotzky RF, Jernigan JG., Sakelliou I, A&A, 365, L104 (2001)
27. Rafferty DA, McNamara BR, Nulsen PEJ, Wise MW, ApJ, 652, 216 (2006)
28. Rebusco P, Churazov E, Böhringer H, Forman W, MNRAS, 359, 1041, (2005)
29. Reynolds., McKernan, B., Fabian, A.C., Stone J.M., and Vernaleo,J.C., MNRAS,
30. Ruszkowski M., Brüggen M., Begelman M.C., ApJ, 611,158 (2004a)
31. Ruszkowski M., Brüggen M., Begelman M.C., ApJ, 615,675 (2004b)
32. Ruszkowski M., Brüggen M., Hallman E., astro-ph/0501175 (2005)
33. Salomé P et al, A& A, 454, 437 (2005)
34. Tamura T et al, A&A, 365, L87 (2001)

Hard X-ray emission from the core of the Perseus cluster and the thermal content of the radio bubbles

J.S. Sanders and A.C. Fabian

Institute of Astronomy, University of Cambridge, Madingley Road, Cambridge, CB3 0HA. UK. jss@ast.cam.ac.uk, acf@ast.cam.ac.uk

Summary. We use a very deep 900 ks *Chandra* X-ray observation of the core of the Perseus cluster to measure and confirm the hard X-ray emission detected from a previous analysis. By fitting a model made up of multiple temperature components plus a powerlaw or hot thermal component, we map the spatial distribution of the hard flux. We confirm there is a strong hard excess within the central regions. The total luminosity in the 2-10 keV band inside 3 arcmin radius is $\sim 5 \times 10^{43}$ erg s^{-1}. As a second project we place limits on the thermal gas content of the X-ray cavities in the cluster core. This is done by fitting a model made up of multiple components to spectra from inside and outside of the bubbles, and looking at the the difference in strength of a component at a particular temperature. This approach avoids assumptions about the geometry of the core of the cluster. Only up to 50 per cent of the volume of the cavities can be filled with thermal gas with a temperature of 50 keV.

1 Introduction

We found evidence for an additional hard emission from the core of the Perseus cluster in a 200 ks *Chandra* observation [1, 2], which we interpreted as nonthermal X-ray emission due to inverse Compton scattering of cosmic microwave background or infrared seed photons. We now verify the detection of the hard flux and characterise it better with a longer 900 ks *Chandra* observation [3].

The first *Chandra* observations of the cluster also showed that the X-ray cavities likely generated by the lobes the central AGN 3C 84 do not appear to contain thermal volume filling gas below 11 keV [4]. Analysis of the 200 ks data suggested that further limits were difficult to obtain due to the nonspherical nature of the cluster core preventing deprojection analyses [1]. To limit the thermal content further we here examine the deep 900 ks data without geometric assumptions.

2 Hard X-ray emission

In our previous analysis of a 200 ks observation of the cluster we used a simple thermal plus powerlaw model to search for any hard emission in the centre of the cluster [2]. The thermal component accounts for thermal gas, and the powerlaw accounts for hot thermal or nonthermal components. The disadvantage of this approach is that if there are multiphase components or if projected hot gas is significant, this will lead to a false powerlaw signal. In the previous analysis, we simulated the effect of projected emission, to account for its effects. There is, however, an extremely extended Hα nebulosity around the central galaxy NGC 1275 [5, 6], associated with $\sim 10^9$ M$_\odot$ of cool X-ray emitting gas [3]. It may therefore be important to include contributions from cool X-ray gas in the spectral modelling, otherwise the X-ray emission at low temperatures may be detected by the powerlaw component.

Radio observations [7] indicate that the radio photon index of the central regions is steep ($\Gamma \geq 2$). If inverse Compton is the emission mechanism and the electron distribution continues without a break to $\gamma \sim 1000$, we would predict X-ray powerlaw emission with a similar steep photon index. Hot thermal gas would give flatter photon indices if fitted with a powerlaw model.

2.1 Analysis

We here examine a 900 ks long *Chandra* dataset [3] of the core of the Perseus cluster. We used the Contour Binning algorithm [8] to select regions containing a signal to noise ratio greater than 500 in the 0.5 to 7 keV band on the ACIS-S3 CCD. The method takes a smoothed map (here we smoothed the input X-ray image with a top hat kernel with a radius to give a signal to noise ratio greater than 60), and creates bins following the surface brightness. A geometric constraint factor of 2 was used to stop the bins becoming too elongated. The regions contain approximately 250,000 foreground counts.

We extracted spectra from the foreground event files for each observation from each of the spatial regions. We also extracted background spectra from blank sky background files, and additional "background" spectra to correct for out-of-time events (these were generated from the foreground event files where the CHIPY coordinate of each event on the CCD had been randomised). The foreground spectra were added together, their responses averaged, and their backgrounds combined using our previous prescription [3]. We constructed a model made up of photoelectrically absorbed APEC [9] thermal components at 0.5, 1, 2, 3, 4 and 8 keV, accounting for the range of thermal gas observed in the cluster, plus a powerlaw component. The thermal components were all fixed in temperature, with free normalisations. The metallicities of the components were tied together and free in the fit. The absorption was allowed to vary from bin to bin. The powerlaw photon index was frozen at a value of 2, with a free normalisation. The model was fitted to the spectra from each region between 0.6 and 8 keV.

Fig. 1 shows images of the normalisations per unit area in the central region for each of the components. As in our previous analysis [2], the normal-

Fig. 1. Normalisations of the different temperature components per unit area from the spectral fits, plus the normalisation per unit area of the powerlaw component with photon index 2. Also shown is an X-ray image in the 0.3 to 7 keV band. The images measure roughly 4.5 arcmin vertically (100 kpc using if $H_0 = 70 \, \mathrm{km \, s^{-1} \, kpc^{-1}}$).

isation of the powerlaw component strongly increases in the central regions, indicating it is not due to a background subtraction problem.

The $\Gamma = 2$ fits require that the photoelectric absorption in the very centre of the cluster, where the powerlaw is strong, is significantly larger than in the outer regions (an increase of N_{H} of $\sim 2 - 3 \times 10^{20} \, \mathrm{cm^{-2}}$). This is because the powerlaw component becomes strong at low X-ray energies releative to the thermal components.

If the powerlaw photon index is allowed to be free, steep powerlaws (if the absorption is free) are preferred over flatter powerlaws. However, we also attempted to fit models with a flatter powerlaw photon index ($\Gamma = 1.5$) and using a hot 16 keV plasma to replace the powerlaw component. These fits do not appear to require any excess absorption if it is free in the fit. We show profiles of the flux in the 2-10 keV band for the three different models in Fig. 2. The plot shows that the models give very similar fluxes for the central regions of the cluster, indicating the hard flux measurement is fairly model independent. We also plot a radial profile measured from the earlier 200 ks observation, using the simple thermal plus powerlaw model with variable photon index, but subtracting the modelled contribution from projected gas. The new results match the old results well.

In the 2-10 keV band, we calculate deabsorbed total fluxes in the inner 3 arcmin of 5.9×10^{-11} erg cm^{-2} s^{-1} for the $\Gamma = 2$ model, and 6.9×10^{-11} erg cm^{-2} s^{-1} for the 16 keV thermal model. These fluxes correspond to a 2-10 keV luminosity of $\sim 5 \times 10^{43}$ erg s^{-1} if $H_0 = 70$ km s^{-1} kpc^{-1}. This is larger than the luminosity of the central nucleus in this band [10].

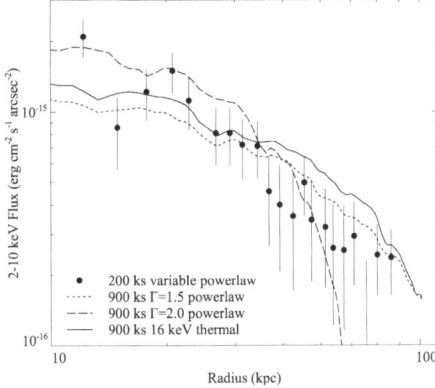

Fig. 2. Profile showing the 2-10 keV flux of powerlaw (fixed to $\Gamma = 2$ or 1.5) or hot thermal (16 keV) component from the fits to the new data and the flux from the variable powerlaw fits to the old data. Distances assume 0.372 kpc per arcsec.

2.2 Conclusions

We confirm a hard thermal component centred on the core of the Perseus cluster using a multicomponent spectral model. Its 2-10 keV luminosity is of the order of $\sim 5 \times 10^{43}$ erg s^{-1}. If the hard flux is nonthermal in origin with a steep photon index similar to the radio emission, then additional photoelectric absorption is required in the centre of the cluster. Flatter nonthermal or hot thermal models require no additional absorption. In future work we will explore the different emission mechanisms and their consequences, and examine other data to limit the possible models.

3 Limiting hot thermal bubble contents

To limit the presence of hot thermal gas in the X-ray cavities in the cluster, we fitted a multicomponent model to spectra extracted from regions in the holes and other regions at the same radius. We compared the normalisation of a component of gas a particular temperature between the two regions. This differential method avoids geometric assumptions, only assuming that the cluster is similar at similar radii. The model consisted of fixed APEC components at 0.5, 1, 2, 3, 4 and 5 keV, a powerlaw fixed to $\Gamma = 2$ and free absorption. These components were designed to model the emission from the cluster plus the hard component above. We also included an additional component to represent any hot thermal emission in the bubbles. This component representing the hot gas was stepped in temperature between 6 and 60 keV. We compared the normalisation per unit area (for a particular bubble temperature) with another regions at the same radii, and computed a limit on the difference in normalisation between the two for each temperature.

The normalisation difference can be converted to a limit on the volume filling fraction simply. We assume a geometry for the regions extracted from the bubbles to calculate a volume. The regions on the sky are 0.12 arcmin in radius, and we assume a cylinder depth 0.42 and 0.6 arcmin for the inner SW and ghost NW bubble, respectively. Using the volume and the difference in normalisation we can estimate an upper limit for the density if the thermal gas is volume filling. This is multiplied by the temperature currently examined to calculate a volume filling electron pressure upper limit. We took the ratio of this pressure to the thermal electron pressure of the gas surrounding the X-ray holes from existing deprojected electron pressure maps [1] to calculate the upper limit on the volume filling fraction.

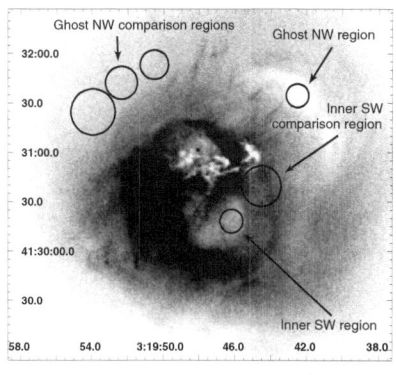
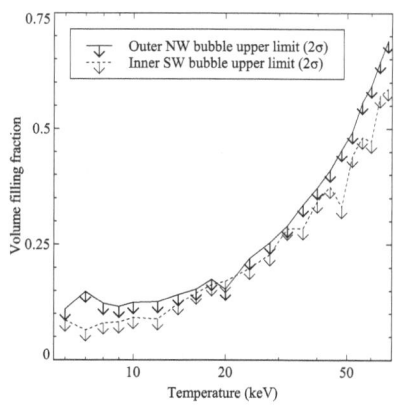

Fig. 3. (Left) Bubble and comparison regions in spectral analysis. (Right) 2σ upper limits for the volume filling fraction of hot thermal gas within the inner SW radio bubble and outer NW ghost cavity.

Fig. 3 shows the resulting upper limits as a function of temperature for the inner SW radio bubble and outer NW ghost cavity. Note that the limits are not independent. At most half of the volume of the bubbles is filled with thermal gas with temperatures less than 50 keV.

3.1 Conclusions

We place limits for thermal content of two of the X-ray holes in the core of the Perseus cluster. By fitting multitemperature models to regions within the holes and nearby, and comparing the results, we deduce that at most 50 per cent of the volume of the bubbles is occupied by thermal gas below 50 keV.

References

1. J.S. Sanders, A.C. Fabian, S.W. Allen, R.W. Schmidt: MNRAS **349**, 952 (2004)
2. J.S. Sanders, A.C. Fabian, R.J.H. Dunn: MNRAS **360**, 133 (2005)
3. A.C. Fabian, J.S. Sanders, G.B. Taylor, S.W. Allen, C.S. Crawford, R.M. Johnstone, K. Iwasawa: MNRAS **366**, 417 (2006)
4. R.W. Schmidt, A.C. Fabian, J.S. Sanders: MNRAS **337**, 71 (2002)
5. R. Minkowski: Optical investigations of radio sources. In: *Proc. IAU Symp. 4*, ed by H.C. Van de Hulst (Cambridge Univ. Press, Cambridge 1957), p. 107
6. R. Lynds: ApJ **159**, L151 (1970)
7. D. Sijbring: A Radio Continuum and HI Line Study of the Perseus Cluster. PhD Thesis, University of Groningen (1993)
8. J.S. Sanders: MNRAS **371** 829 (2006)
9. R.K. Smith, N.S. Brickhouse, D.A. Liedahl, J.C. Raymond: ApJ **556**, L91 (2001)
10. E. Churazov, W. Forman, C. Jones, H. Böhringer: ApJ **590**, 225 (2003)

Outbursts from Supermassive Black Holes: Shocks, Bubbles, and Filaments Around M87

W. Forman[1] E. Churazov[2,3], C. Jones[1], P. Nulsen[1], R. Kraft[1], H. Böhringer[4], A. Vikhlinin[1,2], M. Markevitch[1], J. Eilek[5,6], F. Owen[6], M. Begelman[7] and S. Heinz[8]

[1] Smithsonian Observatory, 60 Garden St., Cambridge, MA, USA
[2] Space Research Institute, Moscow, Russia
[3] Max Planck Institute for Astrophysics, Garching, Germany
[4] Max Planck Institute for Extraterrestrial Physics, Garching, Germany
[5] New Mexico Tech, Socorro, NM, USA
[6] National Radio Astronomy Observatory, Socorro, NM, USA
[7] University of Colorado, Boulder, CO, USA
[8] University of Wisconsin, Madison, WI, USA

1 Shocks, Bubbles, and Filaments

M87 is the central, dominant galaxy in the core of the Virgo cluster (D=16 Mpc, $1' = 4.65$kpc) and hosts a 3×10^9 $M_\odot$ supermassive black hole (SMBH) [18] with a $20''$ long jet seen at radio, optical, and X-ray wavelengths (e.g.,[19, 15] and references therein). It is typical of many collapsed dark matter halos with luminous, early type central galaxies that also contain a hot, virialized X-ray emitting plasma for which simple "cooling flow" models (e.g., [9, 5]) predict significant mass deposition rates from cooling gas (e.g., [8]). For M87, such calculations, that assume no energy is (re)supplied to the plasma, predict a mass deposition rate of ~ 20 $M_\odot$ yr^{-1}. However, XMM-Newton and Chandra observations have shown that mass deposition rates are at least five times smaller than expected (e.g., [23, 7]). The dramatic reduction in the mass deposition rate requires significant energy input to compensate for radiative losses. In this contribution, we focus on the role of the SMBH in M87 in heating its surrounding hot atmosphere.

As we describe below, the M87 observations show the first unambiguous signature of a classical shock, powered by an AGN, in the atmosphere around a central cluster galaxy. The density and temperature discontinuities *independently* yield consistent values for the shock Mach number, $M \sim 1.2$ (for monatomic gas with $\gamma = 5/3$). The shock has an outer radius of $2.8'$ (13 kpc) in the hot atmosphere around M87. With the new Chandra observation, we derive the outburst energy, $E \sim 5 \times 10^{57}$ ergs, the age of the event, $\tau \sim 14$ Myrs, and the duration of the outburst, $\sim 2 - 5$ Myrs. At soft energies (0.5-1.0 keV), we detect a complex filamentary web. This filamentary structure is particularly striking in the eastern arm where we suggest the filaments are the outer edges of a series of buoyant bubbles [1, 24]. The energy output from the main outburst, combined with the additional energy from buoyant bubbles, is comparable to the energy radiated from the gas surrounding M87 which suggests that the central black hole is able to maintain the energy balance in the hot gaseous atmosphere surrounding M87 [3, 4, 1, 24, 12, 13, 14].

2 A Classical Shock Powered by the Supermassive Black Hole in M87

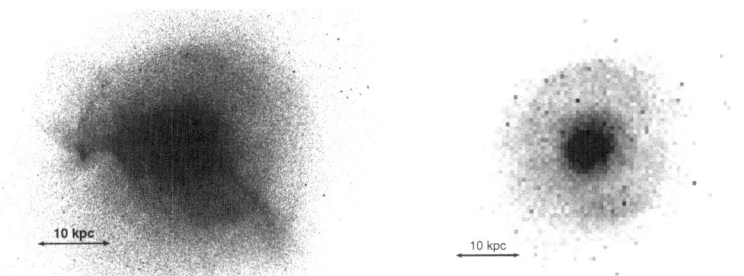

Fig. 1. a) The 1.2-2.5 keV band image of M87. Over this energy band, the Chandra count rate is relatively insensitive to gas temperatures above 1.0 keV. Therefore, this image is a projected map of $\int n^2 dl$. b) The 3.5-7.5 keV band image. Over this energy range, the observed Chandra counts depend on $\int p^2 dl$, where p is the pressure (see text). Hence, we are directly imaging the nearly circular, overpressured shock with outer radius 13 kpc.

The two Chandra images in different energy bands, Fig. 1a and b, show remarkably different views of M87 (for details of this discussion see [13]). The Chandra count rate in the energy band 1.2-2.5 keV depends only weakly on gas temperature for temperatures above about 0.75 keV. Therefore, Fig. 1a is a "gas density" map (proportional to n_e^2 projected along the line of sight) and shows the bright (overexposed) core region harboring the bright nucleus and the jet as well as arms of thermal plasma (see also [3, 2, 21, 12]) extending to the east and southwest as well as a hint of a circular structure at a radius of about 13 kpc (see [25, 12, 13] for more details). In the 3.5-7.5 keV energy band, the count rate depends on the pressure, p. [9] Hence, Fig. 1b provides a pressure map in which we clearly see a nearly circular pressure enhancement (at a radius of $\sim$ 13 kpc), a weak shock, centered on the nuclear region of M87. Note that we assume no strong variations in elemental abundances.

---

[9] The photon flux per unit volume F, can be expressed as $F \propto n^2 \epsilon(T) \propto (p/T)^2 \epsilon(T) \propto (p^2 \epsilon(T)/T^2)$. In the hard 3.5-7.5 keV energy band, $\epsilon(T)/T^2$ depends only weakly on T (for temperatures from 1-4 keV), and hence, the 3.5-7.5 keV band image is approximately an image of the square of the pressure (projected along the line of sight), i.e., total photon flux $\propto \int p^2 dl$.

To derive quantitative outburst parameters, we extracted radial surface brightness profiles in the two energy bands 1.2-2.5 and 3.5-7.5 keV (see Fig. 2). These were in turn used to derive deprojected (3-D) gas density (from the 1.2-2.5 keV band) and gas temperature (from the combined 1.2-2.5 keV and 3.5-7.5 keV data) distributions. These radial profiles of gas temperature and gas density yield values of the temperature and density jumps associated with the 2.8' (13 kpc) shock. The two discontinuities yield *independent* measures of the shock strength using the Rankine-Hugoniot shock jump conditions. The values are in good agreement and are consistent with a weak shock of Mach number $M \sim 1.2$.

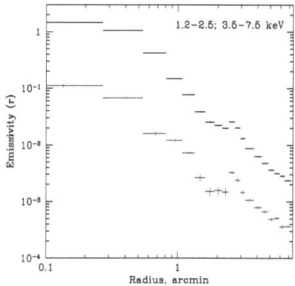

Fig. 2. The 1.2-2.5 keV (upper curve) and 3.5-7.5 keV (lower curve) band deprojected emissivity profiles. The shock at 2.8' (13 kpc) is visible in both energy bands, but most pronounced in the harder energy band.

We have modelled this shock using a one-dimensional numerical model and are able to derive robust estimates for the shock parameters [14]. We assume that the shock was powered by an AGN outburst from M87's SMBH which inflated the central cavity/cocoon (Fig. 3). Using the observed size of the radio cavity (the piston driving the shock), the present radius of the shock, and the amplitude of the density and temperature discontinuities at the shock, we compute an outburst energy of $\sim 5 \times 10^{57}$ ergs and an age of 14×10^6 years.

In addition, we note that M87 appears to be currently undergoing another episode of activity with the current jet power at about 10^{44} ergs s^{-1} [22]. Thus, if the current age of the shock represents the typical time between outbursts, then the average power produced by the SMBH is $\sim 10^{43}$ ergs s^{-1}, comparable to the energy radiated from the cooling core and sufficient to balance radiative cooling.

One additional property of the outburst is noteworthy. The observed properties of the gas surrounding the driving piston, the central radio bubble (see Fig. 3), suggest that the outburst was of moderately long duration. An impulsive outburst would drive a strong shock and produce a hot, low density plasma surrounding the radio synchrotron cocoon. Alternatively, a longer duration outburst, with a more gradual inflation of the central piston, would produce cool rims surrounding the central cavity. This latter scenario is supported by the observations since we see cool gas rims rather than a hot, low gas density (faint surface brightness) region surrounding the central cocoon (see Fig. 2 and Fig. 3). Detailed calculations yield an outburst duration in the range 2-5 Myr [14].

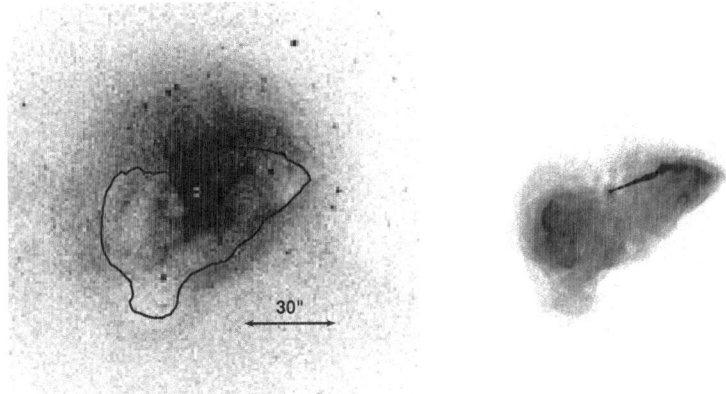

Fig. 3. a) The Chandra image of the core of M87. b) Radio emission (6 cm) in the core of M87 [16] shows the relativistic plasma that drove the 13 kpc shock. The contour on the X-ray image is derived from the outer boundary of the radio image and shows that beyond the radio emitting relativistic plasma, the X-ray gas is bright, indicating a moderate duration outburst.

We note that the above scenario relies on 1-D models while the actual situation is undoubtedly more complex. Measurements of the velocity of jet knots argue for a jet that is inclined by about 30° from the line of sight. Hence, the central core, although appearing azimuthally symmetric, is most likely significantly elongated, warranting more complex models.

3 Buoyant Bubbles - Large and Small

In addition to the shock and the central cocoon/bubble of relativistic plasma, M87 hosts a series of buoyant bubbles (Fig. 4, see [22, 3, 12, 13] for details). The oldest bubble system, seen as two large low surface brightness, nearly circular radio features, have ages of roughly 10^8 years. The next oldest system is that associated with the eastern arm seen most clearly in the radio and X-ray. The X-ray images show a cool column of thermal plasma that has been produced as gas is uplifted by the buoyant radio emitting plasma bubble that appears as a torus with a trailing column of plasma. [3, 22, 2, 21, 12, 13].

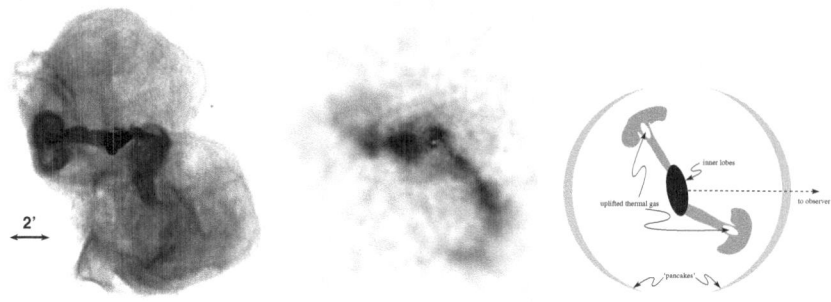

Fig. 4. a) VLA 90 cm radio map[22] showing the remarkable radio arms and mushroom-shaped structure to the east. b) XMM-Newton temperature map (darkest regions are 1.3 keV and lightest grey regions are 2 keV; the whole system is embedded in 2.5 keV cluster plasma [2, 21, 12]. c) Schematic model of the Virgo core[3] showing the two large flattened bubbles, seen in projection as circular features in the radio image, the mushroom-shaped buoyant bubbles whose cool stems are visible in the X-ray images and temperature maps, and the central radio cocoon that drives the shock.

The new 500 ksec Chandra observation shows that the eastern arm, co-incident with the radio column appears as a series of narrow filaments (see Fig. 5). In addition, to the south of these long filaments, we see what may be a series of small ($\sim$ 1 kpc) bubbles surrounded by cool filaments. The youngest of these is the "bud" (see Fig. 5 and [12, 13] for details) which emanates from the radio cocoon to the southeast. This entire structure of small buoyant bubbles, rising in a stream, is very similar to the "effervescent" heating proposed to explain the distributed heating in the absence of strong shocks [3, 4, 1, 24].

The southwestern arm is apparent both in radio and X-rays (Fig. 4, 5 and [12, 13]) and differs from its eastern counterpart. The thermal X-ray plasma is anti-correlated with the relativistic radio emitting plasma which appears to wrap around the very fine, long X-ray arm. Clearly magnetic fields must play a significant role in the structures especially along the southwestern arm.

4 Conclusion

Many of the features seen in M87 were first noted in detailed studies of the Perseus cluster and its central galaxy NGC1275. In the hot atmosphere surrounding NGC1275, a series of weak shocks have been detected[11]. These weak shocks, if there is sufficient viscosity, can maintain the thermal energy of the gas in the cooling core of the Perseus cluster (see detailed studies including [11, 10]).

The buoyant bubbles and shocks seen in Perseus and M87 which arise from SMBH outbursts, are seen in a wide range of systems whose gaseous

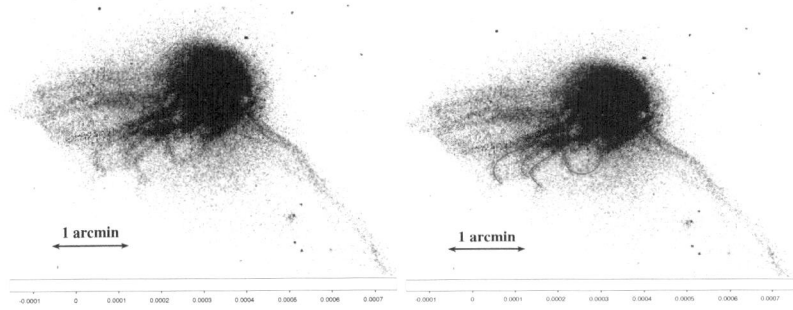

Fig. 5. a) The 0.5-1.0 keV band image showing the web of filaments. The southwestern arm appears as a single long filament while the eastern arm, coincident with the radio "mushroom" stem and cap (see [22, 3]) appears as a complex of filaments. b) Same as a) but outlining the innermost "bud", a buoyant bubble with an age of about 10^7 years that coincides with radio emission seen at 6 cm [16]. A sequence of possible bubbles, all older than the bud, are also marked. The long filaments composing the eastern arm could be a still older series of smaller bubbles uplifted as the large buoyant bubble seen as the radio mushroom cap (Fig. 4) rose in M87's atmosphere.

atmospheres provide a record of past activity. On the smallest scales of individual galaxies, NGC4636 shows a shock of radius 5 kpc, age 3 Myr, and total energy of 6×10^{56} ergs [17]. On the largest scales, Hydra A , a hot, luminous cluster, shows evidence for a shock with radius 160 kpc, an age of 60 Myr, and a total outburst energy of $\times 10^{61}$ ergs [20].

SMBHs are providing significant energy input to their surroundings and appear capable of replacing radiative losses from hot gas in the dark matter halos of individual early type galaxies, groups and clusters [6]. The X-ray emitting gaseous atmospheres provide a record of SMBH activity. In M87, this activity has produced essentially all the observed surface brightness asymmetries including shocks, bubbles, and a rich web of filaments.

References

1. M. Begelman: ASP **240**, 363 (2001)
2. E. Belsole et al: A&A **365**, L188 (2001)
3. E. Churazov, M. Bruggen, C. Kaiser, H. Böhringer & W. Forman: ApJ **554**, 261 (2001)
4. E. Churazov et al: MNRAS **332**, 729 (2002)
5. L. Cowie, J. Binney: ApJ **215**, 723, (1977)
6. R. Dunn, A. Fabian: MNRAS in press, astro-ph/0609537
7. L. David et al: ApJ **557**, 546 (2001)
8. A. Edge et al: MNRAS **270**, L1 (1994)
9. A. Fabian, P. Nulsen: MNRAS **180**, 479 (1977)

10. A. Fabian, J. Sanders, G. Taylor, S. Allen, C. Crawford, R. Johnstone, K. Iwasawa: MNRAS **366**, 471 (2006)
11. A. Fabian et al: MNRAS **334**, L43 (2003)
12. W. Forman et al: ApJ, **635**, 894 (2005)
13. W. Forman et al: submitted to ApJ, (2006) astro-ph/0604583
14. W. Forman et al: in preparation (2007)
15. D. Harris: NewAR **47**, 617 (2003)
16. D. Hines, F. Owen, J. Eilek: ApJ **347**, 713 (1989)
17. C. Jones et al: ApJ, **567**, 115 (2002)
18. F. Macchetto, A. Marconi, D. Axon, A. Capetti, W. Sparks, P. Crane: ApJ **489**, 579 (1997)
19. H. Marshall, B. Miller, D. Davis, E. Perlman, M. Wise, C. Canizares, D. Harris: ApJ **564**, 683 (2002)
20. P. Nulsen, B. McNamara, M. Wise, L. David: ApJ **628**, 629
21. S. Molendi: ApJ **580**, 815 (2002)
22. F. Owen, J. Eilek, N. Kassim: ApJ **543**, 611 (2000)
23. J. Peterson et al: Proceedings of "The Riddle of Cooling Flows in Clusters of Galaxies" eds. Reiprich, Kempner, Soker, astro-ph/0310008 (2003)
24. M. Ruszkowski, M. Begelman: ApJ **581**, 223 (2001)
25. A. Young, A. Wilson, C. Mundell: ApJ **579**, 560 (2002)

The Gaseous Atmosphere of M87 seen with XMM-Newton

A. Simionescu[1], H. Böhringer[1], M. Brüggen[2] and A. Finoguenov[1]

[1] Max Planck Institute for Extraterrestial Physics, Giessenbachstr, 85748 Garching, Germany, `aurora@mpe.mpg.de`
[2] International University Bremen, Campus Ring 1, 28759 Bremen, Germany

Summary. Known as the closest cooling-core hosting an active nucleus, M87 is among the best candidates for detailed studies of the mechanisms behind the heating versus cooling balance. The deepest XMM-Newton observation of M87 to date enables generating detailed temperature, entropy and pressure maps of its X-ray halo. We discuss some physical explanations for the features seen in these maps. The pressure distribution shows an overall ellipticity, the presence of a weak shock with a radius of 3' and possibly of relativistic plasma in the regions corresponding to radio lobes. A NW/SE asymmetry in the entropy map as well as in the temperature and metallicity maps strongly suggests the motion of M87 relative to the Virgo intra-cluster medium.

1 Observations and data reduction

We focus on the results of a 109 ksec observation of M87 performed with XMM-Newton on January 10th, 2005. Using the method implemented by [1], we adaptively binned the combined counts image from the three XMM-Newton EPIC detectors to a signal-to-noise ratio of 100 in the 0.5-7.5 keV energy band. We extracted the spectra and computed the response files for each bin and each detector using the standard XMM Science Analysis System (SAS). The spectra were fit with a single-temperature MEKAL model. From the temperature and spectrum normalizations we determined the pseudo-deprojected gas density, entropy and pressure, which we subsequently divided by a radially symmetric, non-parametric smooth model to reveal small deviations. A more detailed account of the data reduction methods is given in [3].

2 The pressure distribution

Overall ellipticity

A map of the pressure deviations from a radially symmetric model (Figure) shows a relative pressure increase towards the NW and SE, which can be accounted for by applying an elliptically symmetric model with the semimajor axis oriented along the NW/SE direction. The overall ellipticity of the pressure distribution suggests, under the assumption of hydrostatic equilibrium, a corresponding ellipticity of the underlying dark matter distribution.

Lowered pressure in the X-ray arms

In the direction of the E and SW X-ray arms we find a pressure decrease, which even upon subtracting an elliptical model remains 5-10% lower than the ambient value. Assuming an overall pressure balance, this suggests the possible additional contribution to the pressure from relativistic electrons injected by the AGN. Moreover, the radio lobes rise in directions orthogonal to the semimajor axis of the elliptical pressure, thereby presumably following the steepest dark-matter potential gradient as one would expect.

The 3' shock

The pressure map further shows a ring of enhanced pressure with a radius of roughly 3', coinciding with the position of the weak shock proposed by [2]. With the XMM-Newton data we are able to place a lower limit of 1.04 on the Mach number of this shock, conditioned by projection effects and the XMM-Newton PSF.

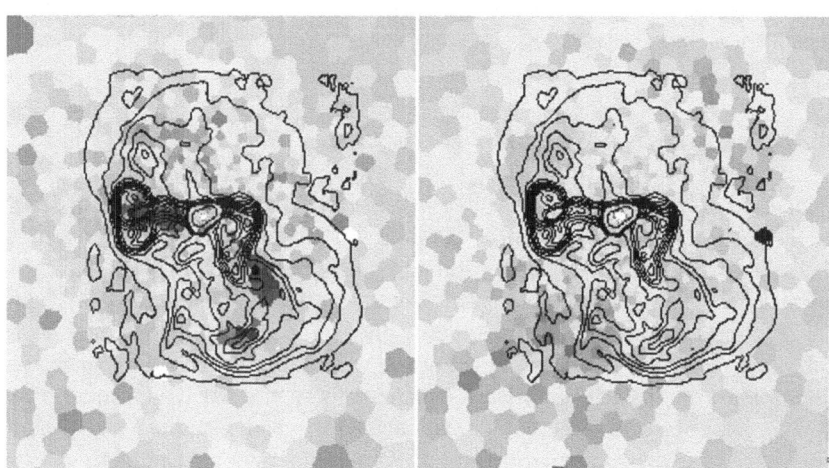

Fig. 1. Entropy (left) and pressure (right) deviations from a smooth radially symmetric model overlaid with radio contours (90 cm). Radio map kindly provided by F. Owen. The E and SW radio lobes coincide well with the regions of low entropy in the X-ray arms. Also, the edge of the large radiolobe to the north roughly coincides with a NW edge in the entropy map.

3 The entropy distribution

The X-ray arms

By far the most striking features in both the temperature and entropy maps are the E and SW arms characterized by lower temperature and lower entropy with respect to the surroundings. Both arms are seen to curve clockwise, thereby connecting to the larger-scale radio halos north and south of the nucleus.

Motion of M87 to the NW

We find, for the first time, a clear NW/SE asymmetry in the entropy map. Within a radius of 6' the entropy values to the NW are clearly more elevated than in the SE. A further analysis also shows more elevated metallicities and lower temperatures to the SE compared to the NW part. As a physical interpretation of this, our data favor a scenario in which, due to a relative velocity between the galaxy nucleus and the gas halo, bubbles emitted initially to the north-west are advected downstream, a fact confirmed by the bubbles observed with Chandra [2]. Much more mixing is therefore induced towards the SE, favoring the transport of metals and cool gas, which explains the relative iron and silicon abundance increase and the lower temperature in this direction.

Possible core oscillations

The model subtracted entropy map reveals an edge at about 6' SE of the core which is suggestive of a cold front (Figure), with the pressure across the front staying roughly constant and the temperature and entropy being lower on the more X-ray luminous side. This feature corresponds in the metallicity map to a visible sharp edge in the abundance distribution, emphasizing the presence of a contact discontinuity here. Another relatively sharp edge in the metallicity can also be seen at about 3' NW of the core. Moreover, we also find a region of lower entropy beyond 6' to the NW. The opposite and staggered placement of these low-entropy regions and abundance edges suggests that the current relative motion of the gas and the galaxy may be part of a longer process of oscillatory motions along the NW/SE direction.

Correlations with radio emission

The edge to the NW where the entropy decreases corresponds spatially very well to the edge of the NW large-scale radio bubble (Figure). This spatial coincidence suggests a link between the injection of radio-loud plasma into the ICM and the processes leading to entropy generation and gas heating. Such a link is at the basis of the feedback models in which mechanical input, usually by a radio source, offsets the cooling flow, heating the gas at the cluster center and flattening the central entropy profiles.

References

1. Diehl, S., & Statler, T. S., 2006, MNRAS, 368, 497
2. Forman, W., et al., 2006, ApJ, submitted (astro-ph/0604583)
3. Simionescu, A., Böhringer, H., Brüggen, M., & Finoguenov, A., 2006, A&A, in press (astro-ph/0610874)

The *XMM-Newton* View of MS0735+7421: the Most Energetic AGN Outburst in a Galaxy Cluster

M. Gitti[1], B. R. McNamara[1,2], P. E. J. Nulsen[3] and M. W. Wise[4]

[1] Dept. of Physics and Astronomy, Ohio University, Clippinger Labs, Athens, OH 45701 (USA) gitti@phy.ohiou.edu
[2] University of Waterloo, 200 University Avenue West, Waterloo, Ontario N2L 3G1 (Canada) mcnamara@uwaterloo.ca
[3] Harvard-Smithsonian Center for Astrophysics, 60 Garden Street, Cambridge, MA 02138 (USA) pnulsen@head.cfa.harvard.edu
[4] Astronomical Institute, University of Amsterdam, Kruislaan 403, 1098 SJ Amsterdam (The Netherlands) wise@science.uva.nl

Summary. We discuss the possible cosmological effects of powerful AGN outbursts in galaxy clusters by starting from the results of an *XMM-Newton* observation of the supercavity cluster MS0735+7421.

1 Introduction

The majority of cooling flow clusters contain powerful radio sources associated with the central cD galaxies [1]. As indicated by high resolution X-ray images, these radio sources have a profound impact on the intra-cluster medium (ICM) – the radio lobes displace the X-ray emitting gas, creating X-ray deficient cavities ([10], [2], [6]). The recent discovery of giant cavities and associated large-scale shocks in three galaxy clusters (MS0735+7241 [11], Hercules A [12], Hydra A [13]) has shown that AGN outbursts can not only affect the central regions, but may also have an impact on cluster-wide scales.

This new development may have significant consequences for several fundamental problems in astrophysics. The non-gravitational heating supplied by AGNs could represent an important contribution to the extra energy necessary to "pre-heat" galaxy clusters, thus explaining the steepening of the observed luminosity vs. temperature relation with respect to theoretical predictions that include gravity alone ([8], [15]). Powerful central AGN outbursts may also affect the general properties of the ICM (e.g., temperature and metallicity profiles, X-ray luminosity, gas mass fraction). It is essential to understand well the ICM physics and evaluate the potential impact of AGN outbursts on the mass vs. temperature (M-T) and luminosity vs. temperature (L-T) relations, which are the foundation to construct the cluster mass function and use galaxy clusters as cosmological probes.

We address these problems by studying the X-ray properties of the most energetic outburst known in a galaxy cluster. MS0735+7421 (hereafter

MS0735) is at a redshift of 0.216. With $H_0 = 70$ km s^{-1} Mpc^{-1}, $\Omega_M = 1 - \Omega_\Lambda = 0.3$, the angular scale is 3.5 kpc per arcsec.

2 *XMM-Newton* view of MS0735+7421

MS0735 was observed by *XMM–Newton* in April 2005 for a total clean exposure time of about 50 ksec. The X-ray image of the central region of the cluster (see Fig. 1, top) shows twin giant cavities having ~ 200 kpc diameter each, in agreement with results from the previous *Chandra* observation [11].

The presence of the radio source at the position coincident with the holes in the X-ray emission [11] implies that the cavities are filled with a population of relativistic electrons radiating at low radio frequencies. The cavities may also be filled with a shock-heated thermal gas that can contribute to the internal pressure necessary to sustain them. In order to investigate this hypothesis we performed a detailed spectral analysis by modeling the spectra extracted in the cavity regions (see Fig. 1, top) as the sum of ambient cluster emission and a hot thermal plasma, each with a characteristic temperature. Hints of a second thermal component with $kT > 10$ keV were found in the northern cavity, in agreement with similar estimates in other clusters ([2] , [5], [9]).

We also attempted a study of the shock properties. *Chandra* observations reveal a feature in the X-ray surface brightness that has been interpreted as a weak cocoon shock driven by the expansion of the radio lobes that inflate the cavities [11]. Our results from a spectral analysis of the pre-shocked and post-shocked gas are consistent those from *Chandra*, although *XMM–Newton*'s spatial resolution is insufficient to unambiguously measure the temperature across the shock.

3 Do cavities affect the average cluster properties?

From the *XMM–Newton* observation of MS0735 we can get new insights to evaluate the impact of energetic AGN explosions on general cluster properties and scaling relations, which are fundamental to use galaxy clusters as cosmological probes. The main results can be summarized as follows (see [7] for a more detailed discussion):

– The total energy in cavities and shock is $\sim 6 \times 10^{61}$ erg, making MS0735 the most energetic AGN outburst known so far. This energy is enough to quench the nominal $\sim 260\,\mathrm{M}_\odot\mathrm{yr}^{-1}$ cooling flow and, since most of the energy is deposited outside the cooling region (~ 100 kpc), to heat the gas within 1 Mpc by $\sim 1/4$ keV per particle. It thus contributes a substantial fraction of the 1 to 3 keV per particle of excess energy required to preheat the cluster [15].

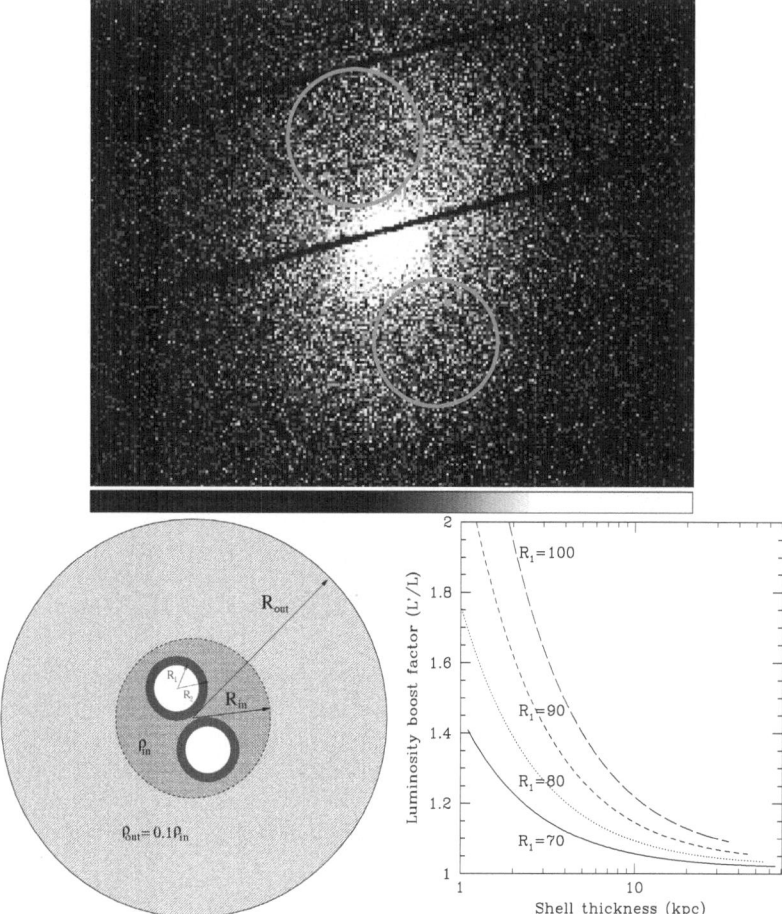

Fig. 1. *(Top):* MOS1 image of MS0735 in the [0.3-10] keV energy band. The image is corrected for vignetting and exposure. The circular regions considered for the spectral analysis of the cavities are indicated. The N and S cavities have a radius of ∼ 100 kpc and are located at a distance of ∼ 170 and ∼ 180 kpc from the cluster center, respectively. *(Bottom, Left):* Geometry of the simplified phenomenological model considered in estimating the luminosity boost factor. During the cavity expansion, all the gas filling the cavities is assumed to be compressed into the bright shells. *(Bottom, Right):* Estimated luminosity boost factor due to cavity expansion and gas compression in the shells, as a function of the thickness of the shell (see left panel for the geometry considered). Different curves refer to different radius R_1 of the cavities (units in kpc). For an adiabatic index $\gamma = 5/3$, the maximum compression in a normal shock is a factor $f = 4$, so that the ratio of the shell thickness to the cavity radius is $R_2/R_1 - 1 > (1 + 1/f)^{1/3} - 1 \simeq 0.077$. For the cavities observed in MS0735 ($R_1 \sim 100$ kpc), the shell thickness is ∼ 8 kpc, leading to a luminosity boost factor ∼ 25%.

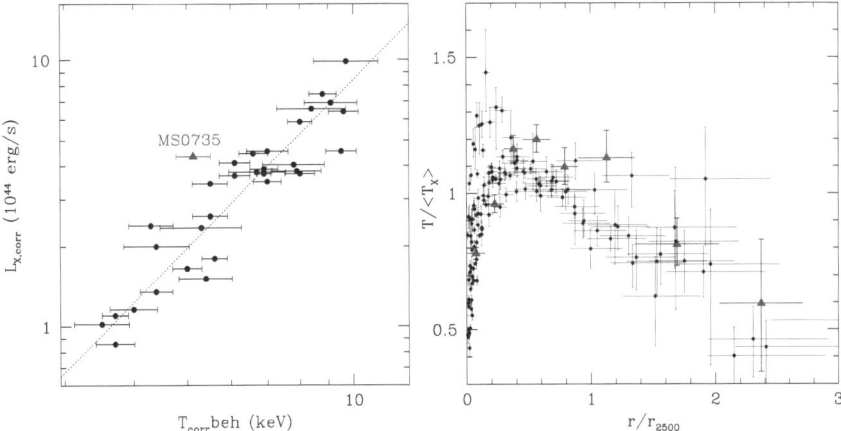

Fig. 2. *(Left):* Observed X-ray luminosities (corrected for the effect of cooling flows in the central 70 kpc) vs. emission-weighted temperatures (derived excluding cooling flow components) for a sample of nearby galaxy clusters [8]. The red triangle represents MS0735 data from the present observations. Note that, since the cooling flow region in MS0735 is bigger than the one adopted by [8] for the cluster sample, the effect of the cooling flow is not completely corrected for. *(Right):* Temperature profile measured for MS0735 (red triangles) overlaid onto the temperature profile observed for a sample of 12 relaxed clusters [14]. The profiles for all clusters are projected and scaled in radial units of r_{2500}. The temperatures are scaled to the cluster emission-weighted temperature excluding the central 70 kpc regions.

– The energetic outburst and the consequently rising cavities uplift the central cool, low-entropy gas up to large radii, and at the same time the compression in the shells increases the ICM density. This results in an increase of emissivity and thus luminosity. An attempt to estimate the expected boost in luminosity is presented in Fig. 1 (bottom left), which shows a sketch of the simple phenomenological model of the structure and emissivity of the gas considered to evaluate the effect of the cavity expansion and compression of the ICM in the bright shells. We estimate that the luminosity is boosted by a factor which depends upon the cavity radius and shell thickness. For the particular configuration of the cavities observed in MS0735, we expect an increase in luminosity by a factor of the order of about 25% (see Fig. 1, bottom right), consistent with our measurements. The unabsorbed X-ray luminosity ([0.1-2.4] keV) estimated by the spectra extracted after excising the cooling flow region is $\sim 3.8 \times 10^{44}$ erg/s. To evaluate the effect of the cavities on the luminosity, this has to be compared with the value estimated by the spectra extracted after excising the cavity regions (the cavities lie outside the cooling region), which is $\sim 3.0 \times 10^{44}$ erg/s. In both estimates the missing luminosity expected from a β-model profile inside the masked regions is added back in. The increased emissivity due to the cavities could also lead to an overestimate of the gas mass fraction, in qualitative agreement with the high value ($f_{gas,2500} = 0.165 \pm 0.040$) that we measure for MS0735.

– MS0735 is a factor ~ 2 more luminous than expected from its average temperature on the basis of the observed L-T relation for galaxy clusters ([8], see Fig. 2, left). This effect may be partially explained by the boost in luminosity due to the cavities. Besides this, no obvious immediate impact on properties such as the scaled temperature profile (see Fig. 2, right) and scaled metallicity profile show up from our analysis. Also, the quantities we measure for MS0735 are consistent with the M-T relation predicted by the cluster scaling laws. We conclude that violent outbursts such as the one in MS0735 do not cause gross instantaneous departures from cluster scaling relations (other than the L-T relation). However, if they are relatively common they may play a role in shaping these relations.

Acknowledgement. We thank A. Vikhlinin for sending the data used to make Fig. 2, right. This research is supported by NASA grant NNG05GK876 and by NASA Long Term Space Astrophysics Grant NA64-11025.

References

1. Burns, J. O. 1990, AJ, 99, 14
2. Blanton, E. L., Sarazin, C. L., & McNamara, B. R. 2003, ApJ, 585, 227
3. De Grandi, S., & Molendi, S. 2001, ApJ, 551, 153
4. Evrard, A. E., Metzler, C. A., & Navarro, J. F. 1996, ApJ, 469, 494
5. Fabian, A. C., Celotti, A., Blundell, K. M., Kassim, N. E., & Perley, R. A. 2002, MNRAS, 331, 369
6. Gitti, M., Feretti, L., & Schindler, S. 2006, A&A, 448, 853
7. Gitti, M., McNamara, B. R., Nulsen, P. E. J., & Wise M. 2006, submitted to ApJ
8. Markevitch, M. 1998, ApJ, 504, 27
9. Mazzotta, P., Brunetti, G., Giacintucci, S., Venturi, T., & Bardelli, S. 2004, Journal of Korean Astronomical Society, 37, 381
10. McNamara, B. R., et al. 2000, ApJL, 534, L135
11. McNamara, B. R., Nulsen, P. E. J., Wise, M. W., Rafferty, D. A., Carilli, C., Sarazin, C. L., & Blanton, E. L. 2005, Nature, 433, 45
12. Nulsen, P. E. J., Hambrick, D. C., McNamara, B. R., Rafferty, D., Birzan, L., Wise, M. W., & David, L. P. 2005a, ApJL, 625, L9
13. Nulsen, P. E. J., McNamara, B. R., Wise, M. W., & David, L. P. 2005b, ApJ, 628, 629
14. Vikhlinin, A., Markevitch, M., Murray, S. S., Jones, C., Forman, W., & Van Speybroeck, L. 2005, ApJ, 628, 655
15. Wu, K. K. S., Fabian, A. C., & Nulsen, P. E. J. 2000, MNRAS, 318, 889

Shock Heating by Nearby AGN

J.H. Croston[1], R.P. Kraft[2] and M.J. Hardcastle[1]

[1] School of Physics, Astronomy and Mathematics, University of Hertfordshire, College Lane, Hatfield AL10 9AB, UK j.h.croston@herts.ac.uk
[2] Harvard-Smithsonian Center for Astrophysics, 60 Garden Street, Cambridge, MA 02138, USA

Summary. Shock heating by radio jets is potentially an important process in a range of environments as it will increase the entropy of the heated gas. Although this process is expected to occur in the most powerful radio-loud AGN, strong shocks have so far only been detected in nearby low-power radio galaxies. Here we discuss X-ray detections of strong shocks in nearby galaxies, including a new detection of shocked gas around both lobes of the nearby radio galaxy NGC 3801 with inferred Mach numbers of $3 - 6$ and a total injected energy comparable to the thermal energy of the ISM within 11 kpc. We discuss possible links between shock heating, AGN fuelling and galaxy mergers and the role of this type of system in feedback models.

1 When and where is shock heating important?

All radio-loud AGN, whatever their radio luminosity and eventual morphology, are expected to go through an initial phase of supersonic expansion before coming into pressure balance (see, for example, [1]). The length of this phase, the amount of energy injected into the external medium during this phase, and the location of the energy injection depend on the jet power and density of the environment. In a poor environment, the radio source will remain overpressured for longer, so that the shock heating phase will be longer lived. It therefore seems likely that the two places where shock heating will be easiest to detect are in the poorest environments, and in the environments of the most powerful AGN. Indeed, the first (and until recently only) direct detection of radio-galaxy shock heating was in the galaxy halo of the nearest radio galaxy Centaurus A [2], which has a low radio luminosity and FRI morphology, but whose inner lobes are still in the supersonic expansion phase (see Section 2).

More recently weak shocks have been detected in the cluster environments of several more powerful radio galaxies, e.g. M87 [4], and the FRII sources Cygnus A [3, 5] and Hydra A [6]; however, there remains no convincing case of a strong shock associated with an FRII radio galaxy. In addition, FRII radio galaxies for which measurements exist of both the internal pressure (via lobe inverse Compton emission) and the external pressure appear to be close to pressure balance rather than strongly overpressured [7, 8, 9], so that lobe expansion is not likely to be highly supersonic. While it is not possible to rule out an important role for strong shocks produced by powerful radio galaxies, the observational evidence suggests that it is in the early stages of radio-source evolution, for both FRI and FRII sources, that shock heating is most important. Although the main emphasis of most work on radio-source impact has been on the group and cluster scale effects of radio galaxies, the impact of shock heating on the ISM of AGN host galaxies is likely to be dramatic, as we demonstrate below.

In the following sections we review the first detection of radio-source shock heating in Centaurus A before presenting a new example of strongly shocked gas shells in the ISM of NGC 3801 that share some characteristics with Cen A but also show some important differences. Finally, based on the nuclear properties and host galaxy characteristics of systems with detected strong shock heating, we discuss the links between shock heating and AGN fuelling and possible implications for the role of shock heating in feedback models.

2 Cen A and NGC 1052

Kraft et al. (2003) [2] detected a bright shell of hot gas surrounding the south-west inner lobe of the nearest radio galaxy Centaurus A. Fig. 1 shows more recent *Chandra* data [10] illustrating the sharp X-ray shell. The shell has a temperature ten times higher than that of the surrounding interstellar medium, and the total thermal energy of the shell is a significant fraction of the energy of the ISM. Centaurus A has an FRI morphology, so would traditionally have been expected to have subsonically expanding lobes; the detection of strongly shocked gas in this system has highlighted the fact that energy input via shocks is likely to be important in the early stages of expansion for all types of radio galaxies.

NGC 1052 is another nearby galaxy where it has been suggested that the small radio source could be shocking and heating its hot ISM [11, 12]. A recent, deep *Chandra* observation reveals in detail the radio-related X-ray structure hinted at by the earlier snapshot observation, but does not show clear evidence for shocked shells, suggesting that the radio-source/environment interaction in this system may be considerably more complex than the shock heating seen in Cen A.

3 A new detection of shock heating in NGC 3801

Our recent *Chandra* observations of NGC 3801 [13] revealed a second definite example of strong shocks produced by a small FRI source on galaxy scales. Fig shows the *Chandra*-detected emission from NGC 3801, which traces well the outer edges of the radio lobes.

We can rule out a non-thermal model for the X-ray emission based on its spectrum, and find best-fitting *mekal* temperatures of 1.0 keV and 0.7

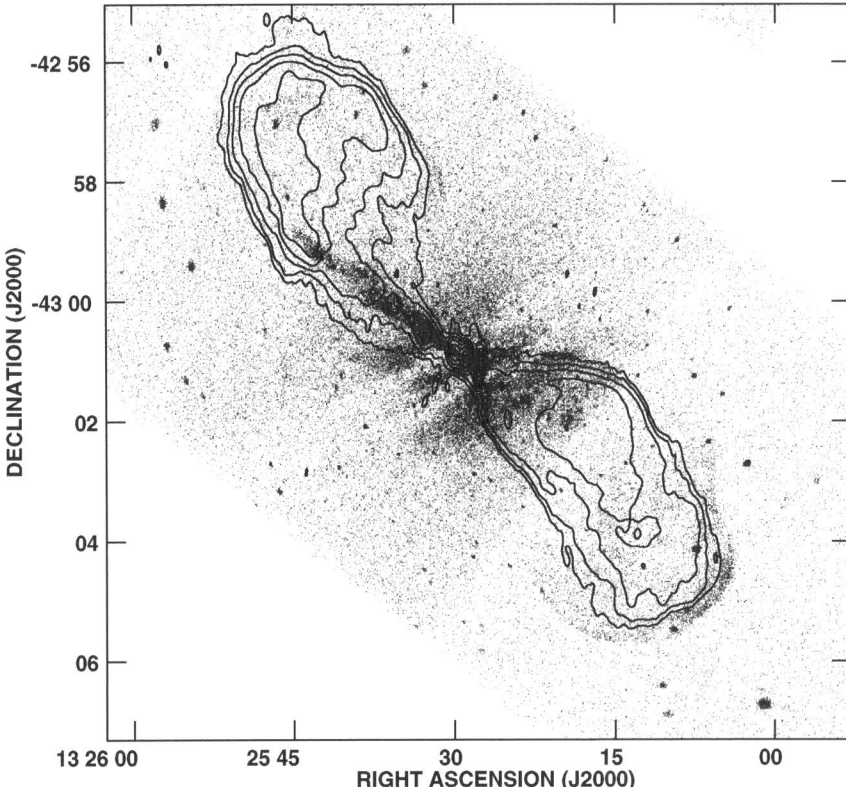

Fig. 1. The shell of shocked gas around the south-western lobe of Centaurus A as observed with *Chandra* [2, 10]. 20cm radio contours [17] are overlaid.

keV for the West and East lobes, respectively. The undisturbed interstellar medium has a temperature of 0.23 keV. We find that the observed density contrast is consistent with the value of 4 expected for a strong shock, using the mean properties of the shell and the ISM density halfway along the lobe. The shells are overpressured by a factor of 13 - 20 and the shell pressure is ~ 7 times the synchrotron minimum internal lobe pressure (consistent with the general finding that FRI minimum pressures are typically an order of magnitude lower than external pressure [14, 15, 16]).

We estimated the shock Mach number using two methods, as descibed in more detail in [13]: applying the Rankine-Hugoniot jump conditions using the observed temperature jump gives $\mathcal{M} \sim 3 - 4$; alternatively, ram pressure balance gives $\mathcal{M} \sim 5 - 6$. The discrepancy between the two methods is probably due to the expected temperature and density structure of the shell and the interstellar medium (in both cases the data are too poor to constrain

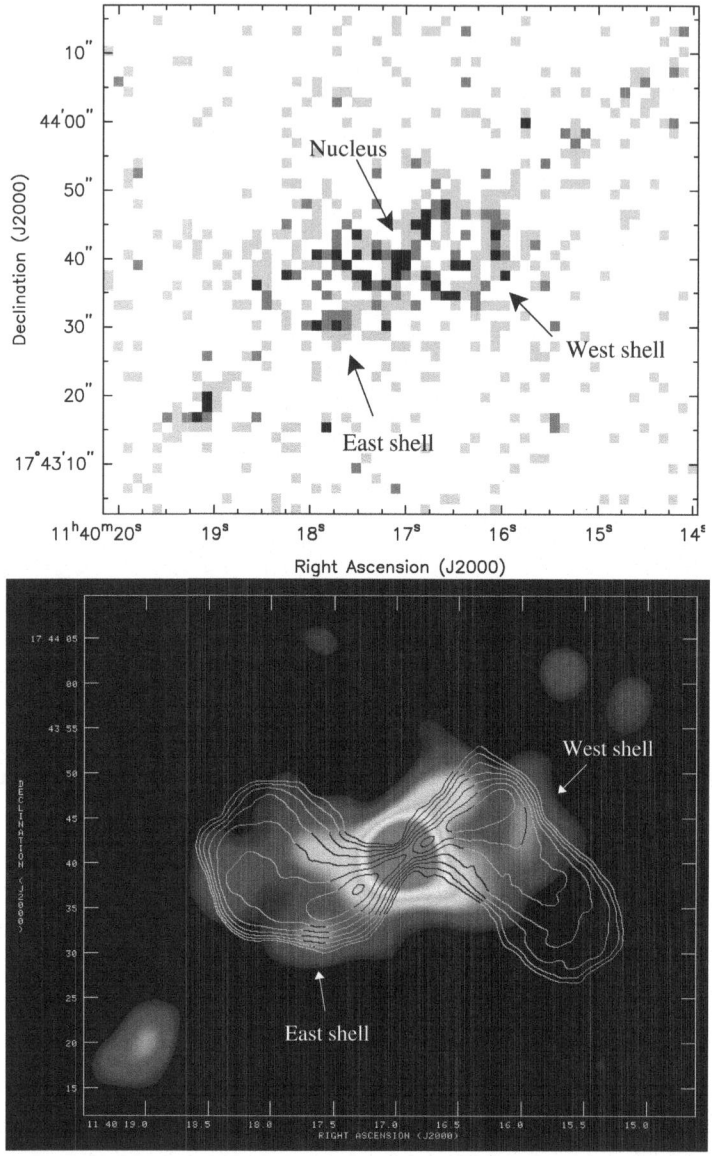

Fig. 2. The shocked gas shells of NGC 3801 as observed by *Chandra*. Top: binned 0.5 - 2 keV counts; bottom: Gaussian smoothed 0.5 - 2 keV image with 20 cm radio contours overlaid from new VLA data.

these). Nevertheless this is a clear detection of strongly shocked gas with $\mathcal{M} \sim 3 - 6$, which implies a lobe expansion speed of $\sim 600 - 1200$ km s^{-1}.

The total thermal energy stored in the hot gas shells is $\sim 8 \times 10^{55}$ ergs, and for $\mathcal{M} \sim 4$, the total kinetic energy of the shells is $\sim 9 \times 10^{55}$ ergs. The total energy of the shells, 1.7×10^{56} ergs, is comparable to $P_{int} V$, the approximate total energy available from the radio source as work; however, it is ~ 25 times the minimum work required to inflate the lobe cavities ($\sim 7 \times 10^{54}$ ergs), so that a simple calculation of the radio-source energy input from the cavity size would be a significant underestimate. The total energy is also equivalent to the thermal energy of the ISM within 11 kpc (or 25 percent of the thermal energy within 30 kpc). Shock heating is therefore the dominant energy transfer mechanism during this phase of radio-source activity, and will have dramatic long term effects: part or all of the ISM may be expelled from the galaxy, and the entropy of the gas will be permanently increased. The internal energy of the radio source ($\sim 4 \times 10^{56}$ ergs) must also eventually be transferred to the environment.

The age of the radio source in NGC 3801 is estimated to be $\sim 2 \times 10^{6}$ y from radio spectral ageing and dynamical arguments, which implies an energy injection rate of $\sim 3 \times 10^{42}$ ergs s^{-1}. This should correspond to a considerable fraction of the jet power, which is consistent with a rough estimate of its jet power based on scaling that of 3C 31 [18] by the ratio of radio luminosities of NGC 3801 and 3C 31. The rate of mechanical energy extracted is roughly an order of magnitude higher than the accretion-related X-ray luminosity, so that the AGN is more efficiently converting energy into jet production than radiation. We also find that the Bondi accretion rate from hot gas would be sufficient to power this radio outburst, for $\eta \sim 0.05$.

4 A link between shock heating and AGN fuelling?

Both Cen A and NGC 3801 are disturbed ellipticals with evidence for fairly recent mergers. Another property that the two sources have in common is that their nuclear X-ray spectra show a component of emission with heavy intrinsic absorption ($N_H > 5 \times 10^{22}$ cm^{-2} in both cases) as seen in high-excitation FRII radio-galaxy X-ray spectra [20]. This is in contrast to the vast majority of FRI radio galaxies, which possess no direct evidence for accretion-related X-ray emission or a torus [19, 20]. It is therefore interesting to speculate that these systems represent a particular class of FRI radio outburst fuelled by cold gas that may be driven into the centre during gas-rich mergers, a mechanism that is unlikely to operate in rich group or cluster-centre FRI radio sources. If this is true, then the shock heating process is not self-regulating, as most of the AGN energy goes into the hot phase of the ISM, so that the accretion rate of cold material is not directly affected. Cen A and NGC 3801 may represent a class of systems at the massive end of the galaxy luminosity function that experience extreme heating effects.

5 Conclusions

We have recently found a second example of strong shocks associated with the radio lobes of a nearby galaxy [13], with a total energy in the shock-heated shells ~ 25 times the minimum that would have been required to inflate the cavities subsonically: shock heating is therefore the dominant energy transfer mechanism for this source. Young radio galaxies should all go through an early stage of supersonic expansion, and the examples of Cen A and NGC 3801 show that this stage can have dramatic effects on the host galaxy ISM. As this stage is comparatively short-lived, and outbursts of the luminosity of NGC 3801 and Cen A are currently only detectable to $z \sim 0.04$ in the radio, further examples of this process may be difficult to find with current generation instruments; however, they are expected to be orders of magnitude more common than Cygnus A type radio outbursts. The nuclear and host galaxy properties of NGC 3801 and Cen A suggest that the shock heating in these galaxies may be directly related to their merger history; we suggest that merger-triggered radio outbursts could be an important galaxy feedback mechanism.

References

1. S. Heinz, C.S. Reynolds, M.C. Begelman, 1998, ApJ, 501, 126
2. R.P. Kraft, S.E. Vázquez, W.R. Forman et al., 2003, ApJ, 592, 129
3. A.S. Wilson, D.A. Smith, A.J. Young, 2006, ApJL, 644, 9
4. W.R. Forman, P.E.J. Nulsen, S. Heinz et al., 2005, ApJ, 635, 894
5. E. Belsole et al., these proceedings
6. P.E.J. Nulsen, B.R. McNamara, M.W. Wise et al., 2005, ApJ, 628, 629
7. M.J. Hardcastle, M. Birkinshaw, R.A. Cameron et al., 2002, ApJ, 581, 948
8. J.H. Croston, M. Birkinshaw, M.J. Hardcastle et al., 2004, MNRAS, 353, 879
9. E. Belsole, D.M. Worrall, M.J. Hardcastle et al., 2004, MNRAS, 352, 924
10. R.P. Kraft et al., in prep.
11. M. Kadler, J. Kerp, E. Ros et al., 2004, A&A, 420, 467
12. J.H. Croston, M.J. Hardcastle, M. Birkinshaw, 2005, MNRAS, 357, 279
13. J.H. Croston, R.P. Kraft, M.J. Hardcastle, 2006, ApJ, submitted
14. R. Morganti, R. Fanti, I.M. Gioia et al., 1998, A&A, 189, 11
15. D.M. Worrall, M. Birkinshaw, 2000, ApJ, 530, 719
16. J.H. Croston, M.J. Hardcastle, M. Birkinshaw et al., 2003, MNRAS, 346, 1041
17. M.J. Hardcastle, R.P. Kraft, D.M. Worrall, 2006, MNRAS, 368, L15
18. R.A. Laing, A.H. Bridle, 2002, MNRAS, 336, 1141
19. D.A. Evans, D.M. Worrall, M.J. Hardcastle et al., 2006, ApJ, 642, 96
20. M.J. Hardcastle, D.A. Evans, J.H. Croston, MNRAS, 2006, 370, 1893

Radio Source Heating in the ICM:
The Example of Cygnus A

E. Belsole and A.C. Fabian

Institute of Astronomy, University of Cambridge, Madingley Road, Cambridge
CB3 0HA, UK, elena@ast.cam.ac.uk, acf@ast.cam.ac.uk

Summary. One of the most promising solutions for the cooling flow problem in-
volves energy injection from the central AGN. However it is still not clear how
collimated jets can heat the ICM at large scale, and very little is known concerning
the effect of radio lobe expansion as they enter into pressure equilibrium with the
surrounding cluster gas. Cygnus A is one of the best examples of a nearby powerful radio galaxy for which the synchrotron emitting plasma and thermal emitting
intracluster medium can be mapped in fine detail, and previous observations have
inferred possible shock structure at the location of the cocoon. We use new *XMM-
Newton* observations of Cygnus A, in combination with deep *Chandra* observations,
to measure the temperature of the intracluster medium around the expanding radio
cavities. We investigate how inflation of the cavities may relate to shock heating of
the intracluster gas, and whether such a mechanism is sufficient to provide enough
energy to offset cooling to the extent observed.

1 Introduction

Cygnus A (3C 405) is the most radio-luminous Active Galactic Nucleus
(AGN) to a redshift of ~ 1, and is the third-brightest source in the radio sky.
Its power ($L_{178\mathrm{MHz}} = 6 \times 10^{27}$ W Hz sr^{-1}) and its proximity ($z = 0.0562$)
have made it one of the most studied extra-galactic sources.

In X-rays Cygnus A was first observed with the *Uhuru* satellite. Later
Einstein Observatory [1] and *ROSAT* observations [7, 13] have shown that
the X-ray emission is dominated by the thermal radiation from intra-cluster
medium (ICM) and found that the gas in the inner 50 kpc is significantly
cooler than the ICM at larger scale, with an inferred cooling flow rate of
~ 250 M$_\odot$ yr^{-1}.

The large scale temperature distribution of the ICM was mapped with
ASCA [10], and suggests a merging event between two similar clusters, the
one hosting Cyg A and a secondary cluster detected to the north-west (NW),
at a projected distance of 11.5 arcmin ($\sim 740 h^{-1}$ kpc) . The merger scenario
is in agreement with the galaxy distribution ([11, 9]).

More recently, Cygnus A has been studied with *Chandra*. The jet and
hotspots are clearly detected and coincide with the radio. The magnetic field
properties of the hotspots of the radio galaxy were studied by [15], while [17]
focused on the X-ray emission from the nucleus. The large scale structure

and the interaction between the radio source and the ICM was discussed by
[14] and [16]. The central 2.5 arcmin ($\sim 160h^{-1}$ kpc) are characterised by a
filamentary structure of spiral-like shape with the nucleus at its centre (named
"belts" by [14]). *Chandra* also detects, with a finer resolution than *ROSAT*
HRI, the edge-brightened emission around the cavity coincident with the
radio lobes ([5]). This elliptically shaped structure encompassing the whole
radio source has a major axis of ~ 1.1 arcmin ($\sim 70h^{-1}$ kpc) coincident
with the direction of the radio jet, and has been interpreted as the cocoon of
shocked gas due to the expansion of the radio galaxy ([16]).

In this paper we present new *XMM-Newton* data of Cygnus A. The larger
field of view (FoV) and sensitivity of *XMM-Newton* allow us to reveal new
features in the X-ray emission from the cluster environment of this object.
We discuss the effect of the radio galaxy on heating and cooling of the ICM.

2 X-ray observations

Cygnus A was observed with *XMM-Newton* in October 2005 in two separate
exposures of 22 ks (ObsIDs 0302800101) and 19 ks (ObsIDs 0302800201)
respectively. In this preliminary work we use local background for the analysis
of the central 5 arcminutes, while the X-ray background is modelled for the
large scale structure analysis using blank-sky background.

In this paper we also present a new analysis of the *Chandra* data. *Chandra*
observed Cyg A in different exposures ranging from 0.8 to 59 ks, for a total
combined exposure of 213 ks. Only one of the observations was obtained with
the ACIS-S as primary instrument (Obs ID 360). Here we only use 120 ks, by
excluding the shorter exposures and the ACIS-S observation for simplicity.

3 Heating

[14] and [16] used 20 and 50 ks *Chandra* observations to measure variations
of temperature across the edge-brightened structure suggesting the detection
of the shock of Mach number 1.3 associated with the cocoon as expected in
models of radio bubble expansion (e.g. [2]). This implies that the simple model
in which the expansion of the radio lobes is able to heat the surrounding gas,
with the effect of a shock at the edge of the cocoon, is reasonably verified for
Cygnus A. The analysis of [16] measures a temperature of 6 keV within the
cocoon and a temperature of $4.6(\pm0.5)$ keV outside it (see Figure 3 of [16]).

3.1 Results

We used the higher sensitivity of *XMM-Newton* to try to verify the *Chandra*
results. In both *Chandra* and *XMM-Newton* observations there is a rather

sharp surface brightness edge at the location of the possible cocoon, which is more clear in the *Chandra* image thanks to the 1 arcsec Point Spread Function (PSF).

We generated a temperature map of Cygus A using the wavelength algorithm described in [4]. (The new version of this method includes a model of the X-ray background in addition to the particle background). Figure shows the temperature map of the Cygnus A cluster in the whole field of view of *XMM-Newton*. The large scale structure is clearly complex and shows regions of high temperature gas, mostly associated with the merging of the main cluster with the object detected to the NW. We will discuss the merging event in detail in Belsole et al. (in prep.). In Figure , right, we show the temperature distribution of the central 5 arcmin. The contours are the radio emission at 1.4 GHz. The core and hotspots were masked while generating the temperature map and thus appear as holes in the figure.

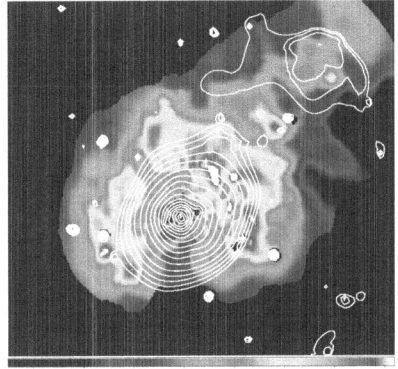
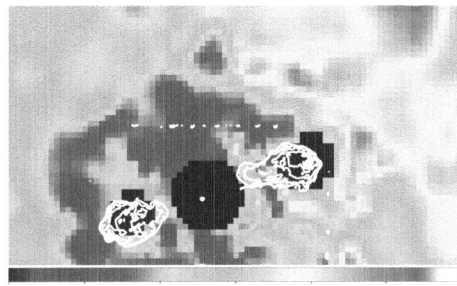

Fig. 1. Left: Temperature map of the Cygnus cluster generated with the wavelet algorithm described in [4]. Contours are the X-ray surface brightness in the energy range 0.5-7.0 keV. Right: zoom on the temperature map in the central 5 arcmin. Contours are from the 1.4 GHz radio map.

Figure 2 left, shows a combined *Chandra* image in the energy band 0.5-7.0 keV. The superimposed elliptical regions were used to evaluate the temperature inside and outside the cocoon with *XMM-Newton*. We measure a temperature inside the cocoon (inner ellipse) of 4.68 ± 0.12 keV and a temperature in the elliptical annulus just outside the cocoon of 4.51 ± 0.2 keV. The temperature of the ICM further out is found to be around 7 keV.

We observe that there is an arc-shaped region at the termination of the NW lobe (Fig. , right). This region is found to be at $kT > 12$ keV, which is significantly higher than the temperature of the surrounding areas both closer to the centre (at the location of the hotspot and within the cocoon) and further out.

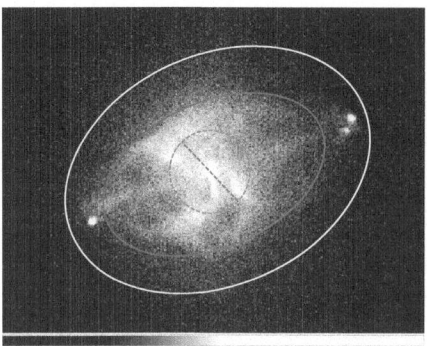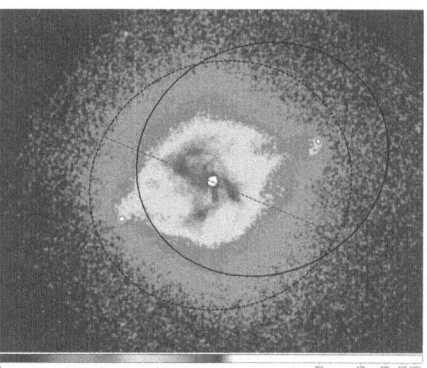

Fig. 2. Left: Chandra image in the 0.5-7.0 keV energy band. The elliptical regions indicate the areas used to extract the spectrum inside and outside the cocoon. Right: zoom of the region used to evaluate the temperature of the bow shock detected at the edge of the NW hotspot. This should be compared with Figure 1, right.

3.2 Discussion

The higher sensitivity of *XMM-Newton* allows a more precise measurement of the temperature, but some features may be washed out because of the PSF. The surface brightness edge observed neatly with *Chandra* is not as sharp with *XMM-Newton* and the radial profile does not show any obvious discontinuity. At a distance of ~ 2.5 arcmin from the centre, we detect a hot, $kT=[9\text{-}11]$ keV, horse-shoe-shape area that we associate with the merging event. This may partially explain why we measure a higher temperature far from the cocoon. However, the temperature of the *belts* (Fig. 2) is significantly cooler than the immediate surrounding, in agreement with [16]. With the *XMM-Newton* data we show that the temperature inside and outside the cocoon are in fact in agreement. This is not in disagreement with the results of [16], as once statistical errors in their analysis are taken into account, the two temperatures they measure (inside and outside the cocoon) are in fact consistent.

The arc-shaped hot region may be associated with the bow shock due to the expansion of the radio lobes in the NW direction. Assuming Rankine-Hugoniot jump conditions, the shock (12.8 ± 2.3 keV) and pre-shock (8.6 ± 0.6 keV) temperature yield a Mach number of 1.62, or gas moving at a velocity of 2450 km s^{-1}. However, we are not able to explain why we observe a shock in the NW direction while the SE lobe shows a cooled region in front of it.

4 Cooling

In this section we discuss a possible detection of cool gas in the centre of the Cygnus A cluster. The predicted mass of cool gas in "cooling core" clusters is inconsistent with the observed mass contained in stars and molecular gas (see e.g., McNamara and Crawford, these proceedings). The hypothesis which is mostly supported today is that the gas is prevented from cooling by some heating mechanisms that appear to be related to the AGN activity (e.g., [4], [12] for a review). However, there is the possibility that cold material is present, but is not easy to detect or may be detected under special conditions only. Clouds of cold material can be observed by their fluorescent line at 6.4 keV, but their detection may depend on the optical depths, the shape of the emitting region and the covering fraction (e.g., [18, 6]).

4.1 Results

We extracted spectra of the central region of Cygnus A cluster at increasing distance from the centre and in different directions: taking as the axis of symmetry the jet line, spectra were extracted in 3 elliptical annular sectors (crescent-shaped regions) in the direction orthogonal to the jet and at a mean distance of 25, 50, and 65 arcsec, to the NE and the SW. Two more spectra were extracted in the cavities occupied by the radio lobes. Figure 3 shows the most distant annular sector to the NE. The spectrum of this region is fitted with a thermal model of temperature $kT = (6.52 \pm 0.16)$ keV. The fit is not excellent and the reduced χ^2 of 1.4 is due to a significant residual at the position of the neutral Fe line at 6.4 keV (Figure 3, top left). The addition of a power law and a neutral 6.4 keV line (of equivalent width 210 eV) to the model significantly improves the fit and accounts for the residual at the Fe line position (Figure 3, bottom right).

The spectrum obtained in the SE lobe cavity (Figure 3), which is at a slightly closer distance of 40 arcsec to the nucleus, does not show the emission line at 6.4 and is adequately fitted with a thermal model at $kT \sim 5$ keV (we also account for a possible inverse-Compton component associated with the radio lobes).

4.2 Discussion

The results described above suggest the presence of cold gas observed through its fluorescent line at 6.4 keV [6]. The *XMM-Newton* PSF is a major concern here as the nucleus spectrum also has a strong neutral Fe line emission. We will discuss in detail the core spectrum in a separate paper. However, the *XMM-Newton* PSF is rather circular (with the MOS 2 camera showing a more triangular shape), and it would be difficult to explain why we detect Fe 6.4 keV line emission from a region at 65 arcsec to the NE of the nucleus but not from a closer region in the jet direction, corresponding to the cavity of the SE radio lobe. Inspection of *Chandra* data is a natural choice to confirm our detection. Using the ACIS-I observations, we generated a merged image in the energy bands 5335-5665 eV (A), 5820-6150 (B, or the Fr 6.4 keV line),

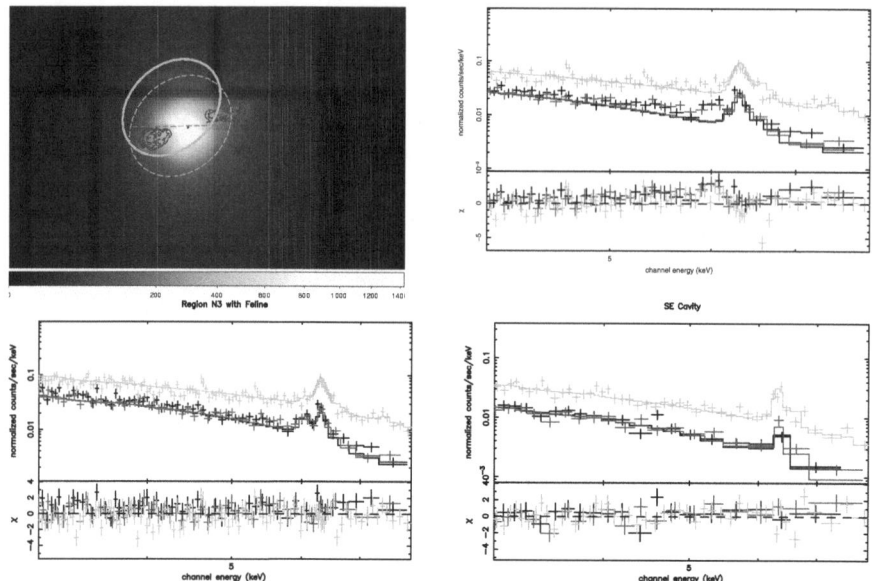

Fig. 3. Top left: The region (N3) superimposed on the epic-pn image is one of the crescent-shape area used for the spectral analysis described above. It is at an average distance of 65 arcsec from the centre; Top right: spectrum of region N3 fitted with a mekal model. The residuals at the position of the Fe neutral line is evident. Bottom left: spectrum of region N3 fitted with a mekal+power law+neutral 6.4 keV line. Bottom right: spectrum of a region at a mean distance of 40 arcsec corresponding to the SE lobe cavity. The spectrum is well fitted with a simple mekal model of $kT \sim 5$ keV.

6180-6410 eV (C, or the Fe 6.7 keV line), and 6835-7165 eV (D). We then interpolated the continuum under the lines using images A and D, and by subtracting the continuum to the line image, we generated the images of the neutral (fluorescent) Fe line and the ionised (ICM) Fe K line in order to localise the regions with significant Fe line emission. We find that there are several, patchy regions where the neutral line emission is detected at ~ 4 sigma above the background and some of these regions correspond to a lack of emission at 6.7 keV. However, a precise spectral analysis is not possible with the low photon statistics (we will discuss these results in a forthcoming paper). The areas which shows the patchy 6.4 keV emission are at a distance from the nucleus which is still affected by the *XMM-Newton* PSF, so no conclusive answer can yet be drawn from this analysis. However, we find that the luminosity of the neutral emission line in region N3 is consistent with the expected ratio between the fluorecent line and the K-edge [8] for reflecting clouds illuminated by the AGN (assuming a covering fraction of $\sim 3\%$ at their distance from the core). We notice that the global *Chandra* spectrum described in [14], which excludes the nuclear area, also shows excess emission at the position of the 6.4 emission line (see Figure 7 of [14]). However, the authors do not discuss their residuals. A deeper investigation of this detection is needed using both *XMM-Newton* and *Chandra*.

5 Summary and preliminary conclusions

We have presented new *XMM-Newton* data of Cygnus A and we have re-analysed 120 ks of *Chandra* observations. We find that the ICM temperature distribution of the cluster hosting the radio galaxy Cygnus A is complex and at least some of this complexity is due to the merging event with a secondary cluster detected to the NW. With both *XMM-Newton* and *Chandra* we detect an elliptically-shaped sharp edge in the gas surface brightness, with a major axis similar to the distance between the two hotspots. This is interpreted as the predicted cocoon of shocked gas due to the expansion of the radio lobes. However no significantly different temperature is detected inside and outside the cocoon.

We may have detected a shock associated with the expansion of the lobes along the major axis at the location of the NW lobe, with a bow shock of Mach number 1.6, which may be a lower limit due to the projected hot gas at large scale. However we are unable to explain the apparent cold gas associated with the edge of the SE lobe.

Using *XMM-Newton* we may have spectrally detected excess emission at 6.4 keV in the preferred axis orthogonal to the jet direction. This may be due to fluorescent line from geometrically small, optically thin cold clouds reflecting photons from the core and the hot gas in the cocoon. Although contamination of the Cygnus A core due to the *XMM-Newton* PSF is likely, this cannot explain the axis-asymmetry we observe, since in the direction of the jet no 6.4 keV emission line is detected.

Acknowledgement. We are very grateful to the local committee for this very well organised conference with plenty of interesting results. We also thank Giovanni Miniutti and Gabriel W. Pratt for useful discussion, Hervé Bourdin for providing a user-friendly version of the temperature map algorithm and Roderick Johnstone for technical support.

References

1. Arnaud, K.A., Fabian, A.C., Eales, S.A., Jones, C., Forman, W., 1984, MN-RAS, 211, 981
2. Begelman, Mitchell C., Cioffi, Denis F., 1989, ApJ, 345, 21
3. Birẑan L., Rafferty, D.A., McNamara, B.R., Wise, M.W., Nulsen, P.E.J., 2004, ApJ, 607, 800
4. Bourdin, H., Sauvageot, J.-L., Slezak, E., Bijaoui, A.; Teyssier, R., 2004, A&A, 414, 429

5. Carilli, C. L., Perley, R. A., Harris, D. E., 1994, MNRAS, 270. 173
6. Churazov, E., Sunyaev, R., Gilfanov, M., Forman, W., Jones, C., 1998, MN-RAS, 297, 127
7. Harris, D.E., Carilli, C.L., Perely, R.A., 1994, Nat., 367, 713
8. Krolik, J.H., Kallman, T. R., 1987, 320, L5
9. Ledlow, M.J., Owen, N.O., Millar, N.A., 2005, AJ, 130, 47
10. Markevitch, M., Sarazin, C.L., Vikhlinin, A., 1999, ApJ, 521, 526
11. Owen, F.N., Ledlow, M.J., Morrison, G.E., Hill, J.M., 1997, ApJ, 488, L15
12. Peterson, J.R., Fabian, A.C., 2005,
13. Reynolds, C.S, Fabian, A.C., 1996, MNRAS, 278, 479
14. Smith, D.A., Wilson, A.S., Arnaud, K.A., Terashima, Y., Young, A.J., 2002, ApJ, 565, 195
15. Wilson, A.S., Young, A.J., Shopbell, P.L., 2000, ApJ, 544, L27
16. Wilson, A.S.,Smith, D.A., Young, A.J., 2006, ApJ, 644, L9
17. Young, A.J., Wilson, A.S., Terashima, Y., Arnaud, K.A. Smith, D.A., 2002, ApJ, 564, 176
18. White, D. A., Fabian, A. C., Johnstone, R. M., Mushotzky, R. F., Arnaud, K. A., 1991, MNRAS, 252, 72

A Deep Chandra Observation of A2052

E. L. Blanton[1], E. M. Douglass[1], C. L. Sarazin[2], T. E. Clarke[3] and
B. R. McNamara[4]

[1] Boston University eblanton@bu.edu
[2] University of Virginia
[3] Naval Research Laboratory and Interferometrics, Inc.
[4] University of Waterloo

Summary. We present initial results from a long (125 ksec) Chandra observation of Abell 2052. A2052 is a bright, nearby, cooling core cluster at a redshift of z=0.0348. It was previously observed for 36 ksec with Chandra [3, 4]. The longer observation reveals ripples in the surface brightness, similar to what has been seen in e.g., the Perseus cluster [5] and M87/Virgo [6]. The southern cavity now appears to be split into two cavities with the southernmost cavity likely representing a ghost bubble from earlier radio activity. There also appears to be a ghost bubble present to the NW of the cluster center. Bright emission in the X-ray corresponds very well with optical line emission, and the correlated X-ray emission is seen to continue from the N bubble edge closer to the AGN in this longer exposure, tracking the H-α emission. The energy deposited by the radio source, as determined by measuring the pressure in the bright, X-ray shells, averaged over the repetition rate of the radio source (determined from either the ripple separation or the ghost cavity distances) can easily offset the cooling in the core of the cluster.

1 Chandra Image

An adaptively-smoothed image of the central region of A2052, with a minimum signal-to-noise ratio of 3 per smoothing beam is shown in Fig. 1 (left). Clear bubbles are seen to the N and S of the AGN. A filament extending into the N bubble corresponds with optical line emission. The S bubble seems to be split into an inner and outer cavity. The optical DSS image of the central region of A2052 with contours from the adaptively-smoothed Chandra image superposed is displayed in Fig. 1 (right). Most of the structures revealed in the X-ray emission are contained within the central cD galaxy. At $z = 0.0348$, $1'' = 0.69$ kpc, assuming $\Omega_M = 0.3$, $\Omega_\Lambda = 0.7$, and $H_\circ = 70$ km s^{-1} Mpc^{-1}.

1.1 Unsharp-masked Image

An unsharp-masked image of A2052, created by subtracting a 9.$''$8 Gaussian-smoothed image from a 0.$''$98 Gaussian-smoothed image is shown in Fig. 2 (left). This process reveals potential ghost cavities, to the SE and NW of the cluster, as well as ripples that may indicate the propagation of sound waves into the ICM. The ripple features, although less prominent, are highly

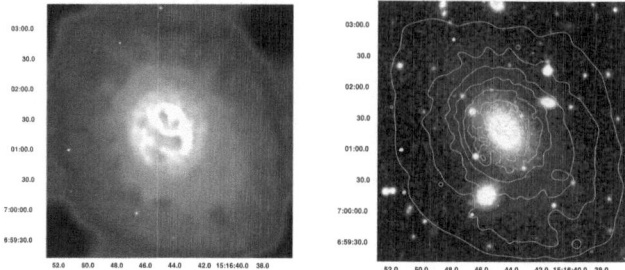

Fig. 1. Adaptively-smoothed Chandra image of A2052 (left). Optical DSS image superposed with contours of X-ray emission (right).

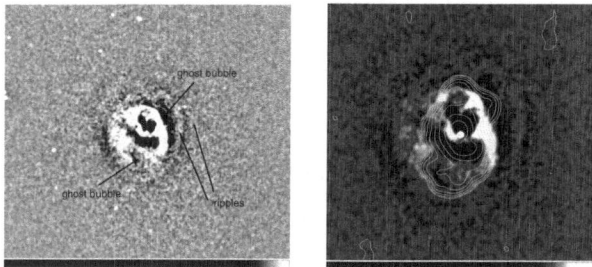

Fig. 2. Unsharp-masked image (left), with radio contours overlaid (right).

reminiscent of the features seen in the unsharp-masked image of the Perseus cluster [5]. The projected location of the SE ghost cavity translates to a repetition rate for the radio source of $\sim 2.5 \times 10^7$ yr, assuming it rose buoyantly at the sound speed in this region of ~ 500 km s^{-1}. The ripple separation to the W corresponds to a radio repetition rate of $\sim 1 \times 10^7$ yr, similar to that found for the Perseus cluster. Such a repetition rate, combined with the energy deposited in the bubbles (measured using the pressure within the bright shells and the bubble volumes) easily allows the radio source to offset the cooling rate in the core of ~ 40 $M_\odot$ yr^{-1}, as determined in [4]. The unsharp-masked image of A2052 with 20 cm radio contours from the VLA FIRST survey [2] superposed is shown in Fig. 2 (right). The radio emission fills the N and S cavities and extensions go into the NW and SE ghost cavities.

1.2 Correlation with Optical Line Emission

The Gaussian-smoothed (0.″98) image of the central region of A2052 is shown in Fig. 3 (left panel). Contours of H-α emission [1], corresponding to emission from $\sim 10^4$ K gas, are superposed on the X-ray emission. The H-α emission matches very well with the brightest regions of the X-ray shells, as well as with a filament that runs from the N shell toward the AGN. In the shorter

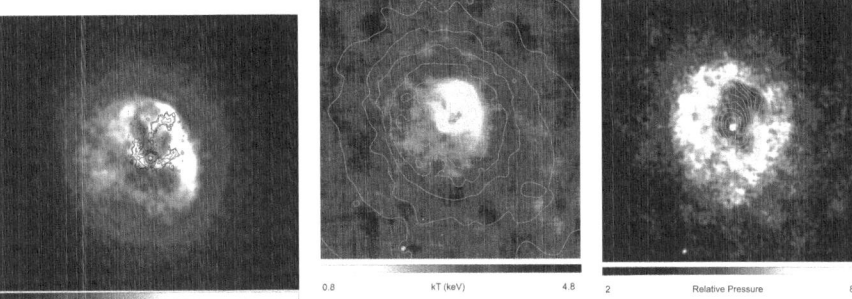

Fig. 3. Gaussian-smoothed Chandra image of A2052 with optical emission line contours superposed (left). Maps of temperature (with X-ray contours, center) and pressure (with radio contours, right).

Chandra exposure of A2052 [3, 4], there appeared to be a gap in this filament in the X-ray, however with the longer exposure, the X-ray emission matches that in H-α all the way from the shell to the AGN.

2 Temperature and Pressure Maps

Temperature and relative pressure maps of the central region of A2052 are shown in the center and right panels of Fig. 3, respectively. The temperature map shows that the bright, X-ray shells are the coolest regions of the cluster. The SW ghost cavity appears to coincide with a region of higher temperature. The pressure map was constructed using the temperature map and the 0.''98 Gaussian-smoothed image, approximating that the surface brightness is proportional to gas density squared. The pressure map shows that the radio source is confined by the higher pressure surrounding ICM. There may be a contribution from hot, diffuse gas within the holes that contribute to their pressure support. Also, magnetic fields may supply pressure to the holes.

Acknowledgement. E. L. B. was supported by NASA through Chandra award GO5-6137X.

References

1. S.A. Baum et al., 1988, ApJS, 68, 643
2. R.H. Becker, R.L. White, D.J. Helfand, 1995, ApJ, 450, 559
3. E.L. Blanton, C.L. Sarazin, B.R. McNamara, M. W. Wise, 2001, ApJ, 558, L15
4. E.L. Blanton, C. L. Sarazin, B.R. McNamara, 2003, ApJ, 585, 227
5. A.C. Fabian et al., 2003, MNRAS, 344, 43
6. W.R. Forman et al., 2007, ApJ, in press (astro-ph/0604583)

Chandra and XMM-Newton Observations of the Group of Galaxies HCG 62

T. Ohashi[1], U. Morita[1], Y. Ishisaki[1], N. Y. Yamasaki[2], N. Ota[3], N. Kawano[4] and Y. Fukazawa[4]

[1] Department of Physics, Tokyo Metropolitan University,
1-1 Minami-Osawa, Hachioji, Tokyo 192-0397
ohashi@phys.metro-u.ac.jp
[2] Institute of Space and Astronautical Science (ISAS), Japan Aerospace Exploration Agency,
3-1-1 Yoshinodai, Sagamihara, Kanagawa 229-8510
[3] Cosmic Radiation Laboratory, The Institute of Physical and Chemical Research (RIKEN), 2-1 Hirosawa, Wako, Saitama 351-0198
[4] Department of Physical Science, School of Science, Hiroshima University,
1-3-1 Kagamiyama, Higashi-Hiroshima, Hiroshima 739-8526

Summary. We present results from Chandra and XMM-Newton observations of the bright group of galaxies HCG 62. The two cavities in the central region show no significant change of temperature compared with that in the surrounding region. We studied radial distributions of temperature and metal abundance. Two temperatures are required in the inner $r < 2'$ (35 kpc) region, and a sharp drop of temperature is seen at $r \sim 5'$ where the gas may not be in hydrostatic equilibrium. The metal distribution suggests that iron and silicon are produced by type Ia supernova in the central galaxy, while galactic winds by type II supernova have caused a wider distribution of oxygen. The pressure due to electrons and magnetic fields is too low to displace the group hot gas, and other pressure contributions from high energy protons or by galaxy-scale dynamical motions are nearly 700 times higher. Detailed accounts are given in [5].

1 Introduction

HCG 62 is a bright group of galaxies at $z = 0.0145$ ($1' = 18$ kpc) discovered in X-rays with ROSAT [6], and there was a report of hard X-ray emission with ASCA [2]. The total X-ray luminosity is $L_X = 2 \times 10^{42}$ erg s^{-1}, and the radio luminosity is $L_{\mathrm{radio}} = 1.8 \times 10^{38}$ erg s^{-1} (10 MHz–5 GHz), respectively. The 1.4 GHz radio image is taken with only $45''$ resolution. In this work, archive data from Chandra (48 ksec observation) and XMM-Newton (~ 10 ksec) are analyzed, following the previous studies [1, 7]. Prominent ghost cavities are noted in the central region of HCG62, and the present study (also see [5]) shows the detailed X-ray properties and their implications.

2 Results

2.1 Mass distribution

Spectral fits indicate that 2 temperatures are necessary in the region $r < 4'$. We assumed pressure balance between cool and hot components, and radial distributions of density and the filling factor of the two gas components are examined. The cool component is dominant within 10 kpc, and the stellar mass density exceeds the hot gas density within 30 kpc. Also, a sharp drop of gas temperature is noted in the outer region around $r = 70$ kpc, which leads to a feature that the gas density is even larger than the gravitational mass density. It is likely that the gas is not under a hydrostatic equilibrium in the outer region where a freshly accreting gas may have dynamical energy.

2.2 Metallicity distribution

Deprojected metal abundances are examined for the combined Chandra and XMM data. Three elements, Mg, Si, and Fe, show similar abundance gradients with more than one solar in the center, however O abundance is about half a solar in the centre with a slow drop with radius. This is similar to the feature observed in M87 (by XMM: [3]) and the Fornax cluster (by Suzaku: [4]).

The excess Fe and Si are thought to be mainly produced by the central galaxy HCG62a, while Type II supernovae would have caused wide O distribution through galactic winds. Detailed abundance features in the outer region will be studied with Suzaku. The mass density profiles of both Fe and

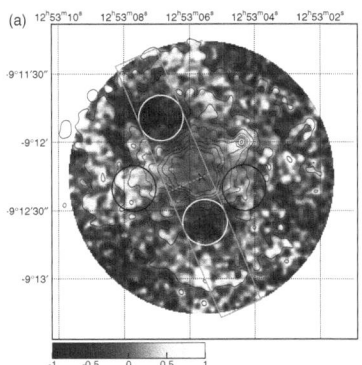
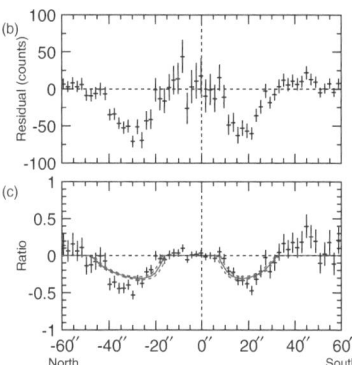

Fig. 1. (a) Deviation from a smooth 2-β model for the entire group in 0.5–4 keV. X-ray intensity is shown with contours, and 2 white circles indicate the cavity positions. (b) Residual intensity along the rectangular region in (a). (c) Ratio for the residual counts to the 2-β model. Smooth curves show expected curve for hollow cavities.

O are flatter than the stellar profile from the group center to the outer region, suggesting that metals may have blown out from the central galaxies rather than being due to enrichment by ram-pressure stripping.

2.3 Cavities

As shown in Figure 1(a), the cavities are seen in the northeast and southwest region of HCG 62a, with diameters 7–9 kpc ($25'' - 30''$). The cavity bottoms show about 50% depression in the count rate, as shown in figures 1(b). The cavity regions show no significant temperature change: 2 cavities and 2 ambient regions all show $kT = 0.74 - 0.76$ keV (± 0.05 keV). Absorption by cold medium is unlikely to be the origin, since the required N_H($1.2 - 1.8 \times 10^{21}$ cm^{-2}) for the flux reduction exceeds the total observed hydrogen ($< 10^9 M_\odot$) in the galaxy group. The count rate profile implies that cavities are almost hollow with $r = 4 - 5$ kpc, as shown by the smooth line in figure 1(c).

3 On the origin of the cavities

The constraint on the central point source (AGN) is $L_X < 10^{39}$ erg s^{-1} in 0.5-4 keV, much lower than the required mechanical luminosity $L_{mech} = 1.8 \times 10^{42}$ erg s^{-1}. At least there is no energy input from AGN now. The radio data indicates $L_{radio} < 9 \times 10^{37}$ erg s^{-1} per cavity, and the equipartition condition gives $P_{non-th} = 6$ eV cm^{-3}, nearly 1/3 of $P_{gas} \sim 17$ eV cm^{-3}. With K as proton to electron pressure ratio and f as filling factor, $K/f > 690$ is required to offset the hot gas. The synchrotron life of electrons is $t_{synch} = 8$ Myr $(K/f)^{-4/7}$: comparable to the expansion time, $t_{exp} \sim 9$ Myr, and the lift-up time, $t_{lift} \sim 15$ Myr. These are much shorter than the life of the galaxy group, so the cavities are produced recently and have to be maintained by some non-thermal mechanism. For possible supporting mechanisms, a high-temperature gas with $n_e \sim 10^{-3}$ cm^{-3}, $kT > 2$ keV gives undetectable low emissivity but the thermal conduction time is 0.5 Myr under the Spitzer conductivity. If the HCG62a galaxy is orbiting around the group core and we look through its trail, the path may be seen as a cavity. The Kepler period is 130 Myr, so the tunnel should exist for a long enough time to keep the 2 cavities visible.

References

1. Birzan, L. et al. 2004, ApJ 607, 800
2. Fukazawa, Y. et al. 2001, ApJ 546, L87
3. Matsushita, K., Finoguenov, A. & Böhringer, H. 2003, A&A 401, 443
4. Matsushita, K. et al. 2006, PASJ 59, in press, astro-ph/0609065
5. Morita, U. et al. 2006, PASJ 58, 719
6. Ponman, T. J. & Bertram, D. 1993, Nature 363, 51
7. Vrtilek, J. M. et al. 2002, APS Meeting B17.107

Radio Properties of Cavities in the ICM: Imprints of AGN Activity

L. Bîrzan[1], B. R. McNamara[2], C. L. Carilli[3], P. E. J. Nulsen[4] and M. W. Wise[5]

[1] Ohio University birzan@helios.phy.ohiou.edu
[2] University of Waterloo mcnamara@sciborg.uwaterloo.ca
[3] NRAO
[4] Harvard-Smithsonian Center for Astrophysics
[5] University of Amsterdam

1 Introduction

The cooling time of the intracluster gas in the cores of many galaxy clusters is shorter than 1 Gyr. In the absence of heating a "cooling flow" is expected to form, but *Chandra*, ASCA and *XMM-Newton* spectra do not show the expected signature of cooling below 2 keV [1]. At the same time, *Chandra* images have shown that radio jets interact with the intracluster medium (ICM) and are energetically able to offset the radiative losses in many systems [2, 3]. In recent years it has also been recognized that active galactic nuclei (AGN) may prevent the formation of extremely bright galaxies [4, 5].

Motivated by the impact that AGN have on large scale structure formation and on the ICM, we investigate the properties of AGN in cluster cores. We have constructed a radio data set at 327 MHz, 1400 MHz, 4500 MHz and 8500 MHz using the Very Large Array (VLA). We show that 327 MHz is a better tool than higher frequencies for studyng the history of AGN activity in the cores of clusters over the past several hundred million years.

2 Sample

We present a sample of 16 objects taken from the samples of [2, 3] of systems with well-defined surface brightness depressions associated with their radio sources. Our sample consist of 15 galaxy clusters and one elliptical galaxy, M84. There is a broad range in redshift from 0.0035 (M84) to 0.35 (RBS 797).

3 Data Reduction and Analysis

3.1 Radio Observations, Data Reduction and Radio Analysis

In order to have data at 327 MHZ, 1.4 GHz, 4.5 GHz and 8.5 GHz we use the VLA archive and new measurements made when the archive data was insufficient. The new observations date from December 2004 to December 2006. Using only the flux in the lobes and by fitting the data with a power law electron injection model, we computed the break frequency (ν_C) above which the spectrum steepens. We computed the bolometric radio luminosity by integrating the spectrum (calibrated at 327 MHz) between 10 MHZ and 10 GHZ, using the spectral index between 327 MHz and 1400 MHz [2]. We also computed the synchrotron ages, which depend on the break frequency and the magnetic field: $t_{\mathrm{syn}} \propto B^{-3/2}\nu_C-1/2$ [6].

3.2 X-ray Analysis

The total energy required to create the X-ray cavities is estimated as the work done by the jets against the ICM plus the internal energy [2]: $E_{\mathrm{cav}} = pV + (\gamma - 1)^{-1}pV$, where p is the gas pressure, V is the cavity volume and γ, the ratio of the gas specific heats, is assumed to be 4/3 (relativistic gas). The cavities' sizes and positions were measured from the X-ray images and the pressures are those at the cavities' centers. The age for each cavity (t_{cav}) was calculated in three different ways: the sound speed age, buoyancy age and refill age [2]. The cavity power is defined as: $P_{\mathrm{cav}} = E_{\mathrm{cav}}/t_{\mathrm{cav}}$. The X-ray properties that we used in this work are from [3].

4 Results and Discussion

4.1 Radiative Efficiencies

An important problem is the degree of coupling between jet power (traced by the cavities) and synchrotron power (the bolometric radio luminosity). This coupling was discussed in detail in [2], where we plotted jet power (P_{cav}) versus synchrotron power. In that paper, we used data from the literature to calculate the bolometric radio luminosity. Figure 1 is an upgraded version of Figure 1 from [2]. Here we analyzed all of the radio sources in a consistent way in order to calculate the bolometric radio luminosity. For consistency, we separate radio filled cavities from radio ghost cavities as we did in our previous work [2].

Figure 1 shows the same trend between bolometric radio luminosity and cavity power as in [2]. The most luminous radio objects generally have the largest cavity powers. This trend is shared again by both radio filled cavities and radio ghost cavities. However, there is a great deal of scatter in this relation, similar to our previous finding. We conclude that the scatter is intrinsic to the radio data, for reasons that include radio aging and adiabatic expansion.

Figure 1 differs from our previous finding only in the details, but we see again that the ratio of cavity power to radio power ranges from a few to a few thousand. Theoretical arguments (e.g., [7]) predict that the ratio of jet power

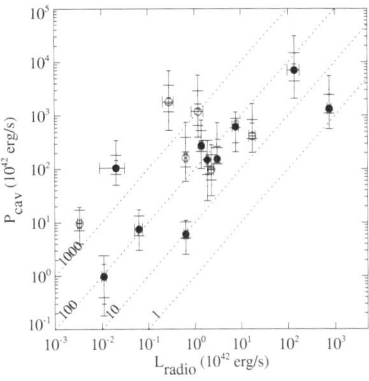 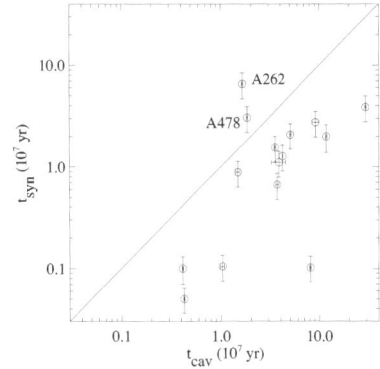

Fig. 1. *Left*: Cavity power versus total bolometric radio power. The errors in cavity power reflect the range in cavity ages [2]. Filled symbols denote radio-filled cavities and open symbols denote ghost cavities. The lines denotes different ratios of cavity power to bolometric radio luminosity. *Right*: Synchrotron age versus buoyancy age for the entire sample.

to radio power should range between $10 - 100$. Most objects in our sample fall within this range, with the exception of A1835, A262, A478 and RBS 797, which are very inefficient radiators (radiative efficiencies of ~ 0.001). We don't see a clear separation between the radio filled cavities and the ghost cavities, but in general the ghost cavities have a higher ratio of cavity power to radio power (i.e. they are less efficient).

We conclude that the scatter that we see in Figure 1 is intrinsic to the radio data. It is important to note that we didn't include the contribution of shocks in the cavity power calculation for any of the objects (e.g., Hydra A [8] and M87 [9], both presented during this conference). In cases where shocks are present, the cavity powers are lower limits to the jet powers, and as a consequence the radiative efficiencies are overestimated.

4.2 Age Estimates

Using both X-ray and radio data, we can compare the synchrotron ages of the radio lobes to the buoyancy and sound speed ages of the cavities. The synchrotron age depends on the break frequency, which was derived using multifrequency radio data, and the magnetic field strength. In order to calculate the magnetic field we assumed equipartition, equal energy in protons and electrons ($k = 1$), and that relativistic particles and magnetic fields occupy the same volume (filling factor $\phi = 1$). Most of the objects range between $t_{syn} = t_{cav}$ and $6t_{syn} = t_{cav}$. Uncertainties in the magnetic field strengths and break frequencies may be the cause of this wide range. In a strong field, particles lose energy rapidly, but in a weak field the decay of the particles can be significantly slower. As a consequence, the particles can be much older than they look; in such cases the synchrotron age is a lower limit on the true age. Additionally, some of the systems may be in a driving state (the cavities are continuously pumped by the radio source) in which the buoyancy assumption doesn't apply.

On the other hand, A262 and A478 are completely different from the rest of the objects. In these cases the synchrotron age is higher than the X-ray age by a factor of 5. It is possible that the X-ray age was underestimated due to projection effects, but that would account for only a factor of $\sqrt{2}$ on average. A more plausible explanation for these cases is that the break frequency is in error. A262 and A478 are two out of three objects for which we have lobe emission at only 2 frequencies (327 MHz and 1.4 GHz). In these cases the break frequency was estimated using the method of [10]. Therefore, break frequency and magnetic field misestimates may be the reasons for the discrepancy between the synchrotron ages and the buoyancy ages.

4.3 Particle Content

We calculated the ratio of the proton energy to electron energy (k), assuming $\phi = 1$ for three different scenarios: that equipartition applies, that the synchrotron age equals the buoyancy age, and that the synchrotron age equals the sound speed age. In each scenario, we first used the volume from the X-ray maps (cavity volume) and then repeated the calculations using the volume from the 327 MHz radio maps (lobe volume). From theoretical estimates, k is predicted to range between 1 and 2000, depending on the mechanism generating the electrons [11]. The typical values used in the literature are $k = 1$ for an electron-positron plasma and $k = 100$ from cosmic rays assumptions. We found that for most of the cavities, k is on the order of hundreds, but there is a wide range from a few to a few thousand or more. Our range is similar to that of [12, 13, 14], although our analyses differ in details. Such high values of k imply that in these systems other particles beside electrons are providing additional pressure support, for example protons [15].

4.4 Radio Observations as a Tracer of Cavity Sizes

Because synchrotron losses are smaller at 327 MHz than at higher frequencies, 327 MHz radio maps have proved to be crucial for detecting the outer, fainter cavities in Hydra A [8, 16]. In order to evaluate whether 327 MHz emission is generally a better predictor of cavity size than 1.4 GHz emission, we compare pV_{cav}, where V_{cav} is the volume of the cavities from the X-ray maps against $pV_{1.4}$, where $V_{1.4}$ is the volume of the lobes from 1.4 GHz radio maps, in Figure 2, left. We also plot pV_{cav} against pV_{327} (Figure 2, right). The pressure on each axes is measured at the position of the radio lobe's center or cavity's center, and can therefore differ from one axis to another.

Figure 2 suggests that 1.4 GHz emission is a good tracer of cavity size in most cases, but there is a significant scatter, with many points falling below

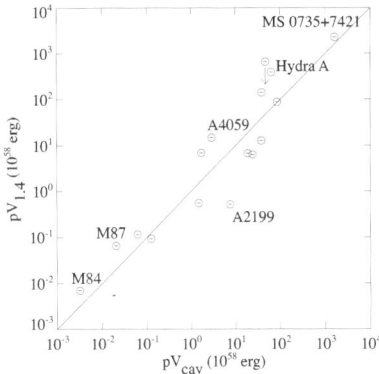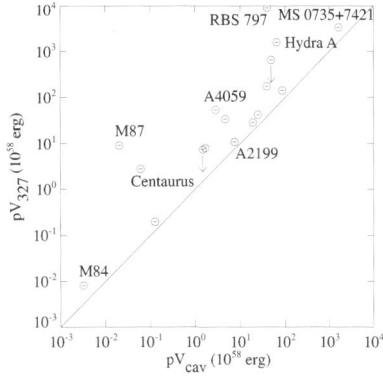

Fig. 2. *Left*: $pV_{1.4}$ versus pV_{cav}. For Hydra A, $V_{1.4}$ is calculated from the radio map of [16]. *Right*: pV_{327} versus pV_{cav}. For M87 and Hydra A, V_{327} is calculated from the maps of [17] and [16], respectively. The upper limits in both plots denote unresolved radio sources.

the equality line. Assuming that P_{cav} traces the minimum jet power, we note that 1.4 GHz underestimates the cavity power in some systems. On the other hand, at 327 MHz all of the points are above the line. It appears that 327 MHz recovers the energy better and is more sensitive to extended emission than 1.4 GHz. 327 MHz can be used as a predictor for the existence of X-ray cavities, such as in those objects that are well above the equality line in Figure 2, right (e.g., Centaurus and RBS 797). In both plots, some of the objects are on the equality line (e.g., M84 and MS 0735+74). In these cases, both 327 MHz and 1.4 GHz emission fills the cavities, and both frequencies are good tracers of cavity size. However, some of the objects that at 1.4 GHz lie below the equality line lie on the line at 327 MHz (e.g., A2199). In these cases, 327 MHz emission fills the cavities more fully. Therefore, for these objects 327 MHz emission is a better tracer of cavity size than 1.4 GHz emission.

Other objects are above the equality line in both plots (e.g., Abell 4059, Hydra A and M87). In these cases, the radio lobes are larger than the cavities. It is possible that the X-ray measurements underestimated the size of these cavities or that the radio plasma has diffused beyond the cavities.

In Hydra A, the inner cavities that we measured from X-ray images are filled with 4.5 GHz radio emission. However, from [8] we know that Hydra A has outer cavities that are filled with 327 MHz radio emission [16]. With this information in mind we note that in Hydra A, pV_{cav} will increase when we add the contribution from these outer cavities. A similar situation is true in M87, where [9] discovered large outer cavities filled with 327 MHz radio emission [17]. There are other objects similar to Hydra A where the pV inferred from the 327 MHz radio emission is well above the pV we measure from X-ray images (e.g., Centaurus and RBS 797). In these objects we expect to see outer cavities in deeper Chandra images which will increase their total pV_{cav} estimates.

5 Conclusions

We have presented an analysis of new, high-resolution radio data for a sample of 16 systems with X-ray cavities located in cluster cores. By combining X-ray data and VLA radio data taken at multiple frequencies, we find that the radiative efficiency in these systems is $\sim1\%$, but can be much lower. We find that the synchrotron ages of the lobes are generally less than the X-ray ages of the cavities, suggesting that the cavities are being pumped. We find that most of the lobes are not in equipartition. We conclude from the high ratios of proton to electron energy that other particles besides electrons are needed to provide additional pressure support. By comparing the size of the cavities from the X-ray data with the size of the lobes from the radio maps, we conclude that 327 MHz maps are a better tool than higher frequency maps for studying the history of AGN activity in the cores of clusters over the past several hundred million years. Finally, we note that 327 MHz radio maps can be a good proxy for X-ray cavities in systems where it is difficult to image the cavities directly in X-rays.

References

1. J. R. Peterson, F.B. S. Paerels, J. S. Kaastra et al, 2001, A&A , 365, L104
2. L. Bîrzan, D. A. Rafferty,B. R. McNamara et al, 2004, ApJ , 607, 800
3. D. A. Rafferty, B. R. McNamara, P. E. J. Nulsen et al., 2006, ApJ, 651
4. A. J. Benson, R. G. Bower, C. S. Frenk et al., 2003, ApJ , 599, 38
5. R. G. Bower, A. J. Benson, R. Malbon et al., 2006, MNRAS , 370, 645
6. P. Alexander, & J. P. Leahy, 1987, MNRAS , 225, 1
7. G. V. Bicknell, M. A. Dopita, & C. P. O'Dea, 1997, ApJ, 485, 112
8. M. W. Wise, this volume
9. W. Forman, this volume
10. S. T. Myers, & S. R. Spangler, 1985, ApJ, 291, 52
11. A. G. Pacholczyk, 1970, Radio Astrophysics, (Freeman, San Francisco)
12. A. C. Fabian, A. Celotti, K. M. Blundell et al., 2002, MNRAS , 331, 369
13. R. J. H. Dunn, & A. C. Fabian, 2002, MNRAS , 355, 862
14. R. J. H. Dunn, A. C. Fabian, & G. B. Taylor, 2005, MNRAS , 364, 1343
15. D. S. De Young, 2006, ApJ, in press (astro-ph/0605734)
16. W. M. Lane, T. E. Clarke, G. B. Taylor et al., 2004, ApJ, 127, 48
17. F. N. Owen, J. A. Eilek, & N. E. Kassim, 2000, ApJ, 543, 611

The Growth of Black Holes and Bulges at the Cores of Cooling Flows

D. A. Rafferty[1], B. R. McNamara[2], P. E. J. Nulsen[3], and M. W. Wise[4]

[1] Ohio University, rafferty@ohio.edu
[2] University of Waterloo, mcnamara@uwaterloo.edu
[3] Harvard-Smithsonian Center for Astrophysics
[4] University of Amsterdam

1 Introduction

The intracluster medium (ICM) at the center of a majority of galaxy clusters has a cooling time less than 10^{10} yr [1]. In the absence of a source of heat, this gas should cool, resulting in a slow inward flow of material know as a "cooling flow". However, recent high-resolution X-ray spectra from XMM-*Newton* do not show the features expected if large amounts of gas are cooling below $kT \sim 2$ keV [2]. The emerging picture of cooling flows is one in which most of the cooling is roughly balanced by heating from the active galactic nucleus (AGN) due to accretion onto the central supermassive black hole (BH). In this regulated-cooling scenario, net cooling from the ICM would lead to condensation of gas onto the central galaxy, driving the star formation observed in many systems (e.g., [3]). We test this scenario using star formation rates (SFRs), ICM cooling rates, and AGN heating and BH growth rates for a sample of cooling flows with AGN-created X-ray cavities.

2 BH Growth, Cooling, and Star Formation

Cavities seen in the X-ray emission of clusters allow a direct measurement of the non-radiative energy output via jets from the AGN. The total energy required to create a cavity is equal to its enthalpy ($E_{\mathrm{cav}} = \gamma/[\gamma-1]pV$), which depends on the pressure (p) and volume (V) of the surrounding gas and the ratio of specific heats (γ) of the gas filling the cavity. The jets are powered by accretion, through the conversion (with efficiency ϵ) of the gravitational binding energy of the accreting material into outburst energy. Since some of the accreting material's mass goes to power the jets, the BH's mass grows by $\Delta M_{\mathrm{BH}} = (1 - \epsilon)E_{\mathrm{cav}}/\epsilon c^2$ during the outburst.

Figure 1 (*left*) shows the BH growth rate (found by dividing ΔM_{BH} by the cavity's buoyant rise time) versus the bulge growth rate (traced by star formation rates taken from the literature, see [4] for details) for the systems in our sample with reliable star formation rate estimates. The trend in Figure 1 may indicate that, in a time-averaged sense, the growth of the bulges and BHs in our sample proceeds roughly along the Magorrian relation. The

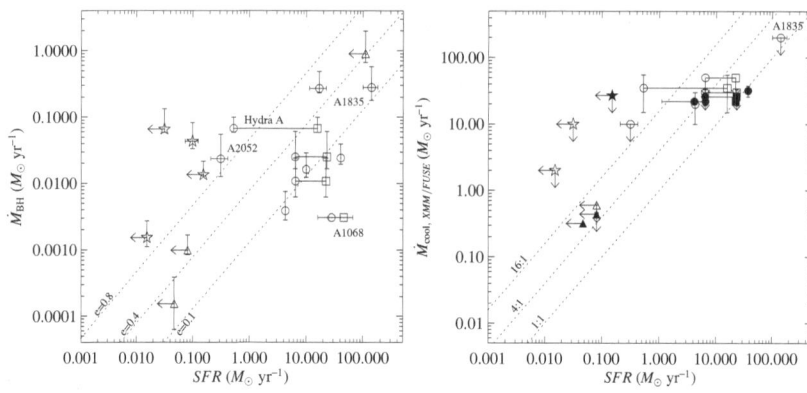

Fig. 1. *Left:* Black hole versus bulge growth rate (SFR). Lines show the time derivative of the Magorrian relation (from [5]) for various efficiencies (ϵ). *Right:* Cooling rate versus SFR. Lines denote ratios of cooling to SFR, filled symbols denote *FUSE* rates, and open symbols denote XMM rates. In both plots, circles and stars denote continuous SFRs derived from optical images or spectra, respectively; squares denote burst SFRs. Triangles denote far-infrared rates.

scatter indicates that present-day growth is occurring in spurts, with periods star formation (as in A1068) in which the bulge grows quickly with little commensurate BH growth, while during periods of heating (as in Hydra A or A2052) the BH grows more quickly than the bulge. A1835 is a system in which the rates follow the Magorrian relation.

In the classical cooling flow problem, the X–ray-derived cooling rates were factors of $10-100$ in excess of the star formation rates in most systems. Figure 1 (*right*) shows that the star formation and cooling rates have converged greatly and are in rough agreement in some systems (e.g., A1835). However, until line emission that is uniquely due to cooling below ~ 2 keV is identified, cooling through this temperature at any level cannot be confirmed.

To investigate whether the AGN outbursts are powerful enough to balance cooling, we plot in Figure 2 (*left*) the cavity power of the central AGN against the luminosity of the ICM within the cooling radius. Remarkably, half of the systems in our sample have cavity powers sufficient to balance the entire radiative losses of the ICM within the cooling radius (as found ealier by [6]). However, we note that the time-dependent nature of AGN feedback does not require that cooling is always balanced by heating [7].

Finally, we plot in Figure 2 (*right*) the ratio of the accretion rate to the Bondi rate versus the semi-major axis of the central region from which the Bondi rates were calculated. Objects near or below the overplotted lines could reasonably have ratios of order unity or less and thus be consistent with Bondi accretion. While Bondi accretion can feed the less powerful outbursts easily (see also [8]), the most powerful outbursts (such as in MS0735.6+7421) are generally inconsistent with Bondi accretion.

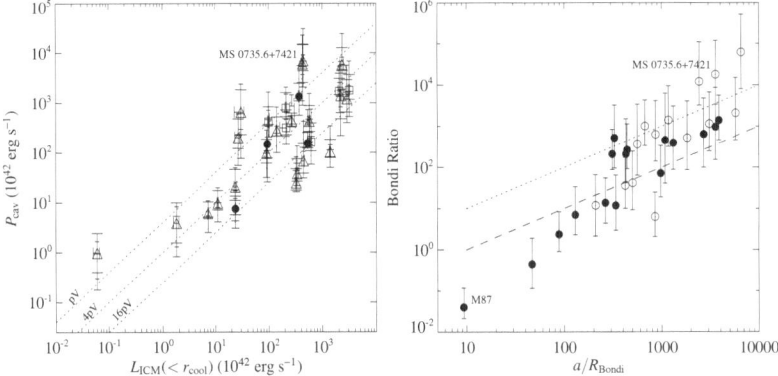

Fig. 2. *Left:* Cavity power versus cooling luminosity. Lines denote equality for different values of γ. Circles, triangles, and squares denote well-defined, intermediate, and poorly-defined cavities, respectively. *Right:* Bondi ratio versus size of the extracted region normalized to the Bondi radius. Lines denote the scaling of the measured Bondi ratio with radius, assuming a true Bondi ratio of unity at the Bondi radius and a density profile that rises as $\rho \propto r^{-1}$ to the Bondi radius (upper line) or flattens inside $a/R_{\mathrm{Bondi}} = 10$ (lower line), as observed in M87 [9].

3 Conclusions

We have presented an analysis of the star formation and AGN properties in the central galaxies in the cores of cooling flows (discussed in detail in [4]). We find that AGN outbursts in most systems with cavities have enough energy to offset much of the radiative losses of the ICM, and to severely reduce cooling to levels that approach the star formation rates in the central galaxy. Using the cavities to infer BH growth and star formation to infer bulge growth, we find that bulge and BH growth rates scale with each other on average in accordance with the slope of the Magorrian relation, but with large scatter that may indicate that growth occurs in spurts.

References

1. C. B. Peres, A. C. Fabian, A. C. Edge et al, 1998, MNRAS, 298, 416
2. J. R. Peterson, S. M. Kahn, F. B. Paerels et al, 2003, ApJ, 590, 207
3. B. R. McNamara, & R.W. O'Connell, 1989, AJ, 98, 2018
4. D. A. Rafferty, B. R. McNamara, P. E. J. Nulsen et al, 2006, ApJ,, 652, 216
5. N. Häring, & H. Rix, 2004, ApJ, 604, L89
6. L. Bîrzan, D. A. Rafferty, B. R. McNamara, et al, 2004, ApJ,, 607, 800
7. H. Omma, & J. Binney, 2004, MNRAS, 350, L15
8. S. W. Allen, R. J. H. Dunn, A. C. Fabian et al, 2006, MNRAS, 327, 21
9. T. di Matteo, S. W. Allen, A. C. Fabian et al, 2003, ApJ, 582, 133

Tracing Ghost Cavities with Low Frequency Radio Observations

T. Clarke[1,2], E. Blanton[3], C.L. Sarazin[4], N. Kassim[1], L. Anderson[3],
H. Schmitt[1,2], Gopal-Krishna[5] and D.M. Neumann[6]

[1] Naval Research Laboratory, 4555 Overlook Ave SW, Washington, DC USA
[2] Interferometrics, Inc., 13454 Sunrise Valley Drive, Herndon, VA USA
   tracy.clarke@nrl.navy.mil
[3] Boston University, 725 Commonwealth Ave., Boston, MA USA
[4] University of Virginia, 530 McCormick Rd., Charlottesville, VA USA
[5] NCRA-TIFR, Pune University Campus, Pune India
[6] CEA/Saclay, L'Orme des Merisiers, Gif-sur-Yvette, France

Summary. We present X-ray and multi-frequency radio observations of the central radio sources in several X-ray cavity systems. We show that targeted radio observations are key to determining if the lobes are being actively fed by the central AGN. Low frequency observations provide a unique way to study both the lifecycle of the central radio source as well as its energy input into the ICM over several outburst episodes.

1 Introduction

The radiative cooling time in the central regions of many dense clusters is less than the age of the cluster. Without any outside disturbance this gas should continue to cool to very low temperatures creating a "cooling flow" in the cluster center [7]. The search for cool gas through X-ray spectroscopic observations has revealed the surprising fact that the temperature drop in cluster cores seems to halt at temperatures around one-third of the maximum cluster temperature [16]. An obvious candidate for energy input to offset significant cooling is the cluster-center AGN, which is typically radio-loud in dense cooling core systems.

X-ray images of several cooling core clusters reveal the presence of significant interaction between the central radio source and the intracluster medium (ICM), thus supporting the idea that the AGN may be a source of energy input to the surrounding gas. The ICM shows evidence of cavities surrounded by (at least partial) rims in several systems. Associated radio observations at 1.4 GHz reveal that many of the cavities are filled with radio plasma from the lobes of the central AGN. In some systems, however, the X-ray cavities are not associated with emission at this frequency. These cavities are generally referred to as "ghost cavities" and are thought to be the result of buoyantly rising lobes from past radio outbursts. One such system, Perseus [8], shows low frequency radio spurs toward the ghost cavities which suggests that at

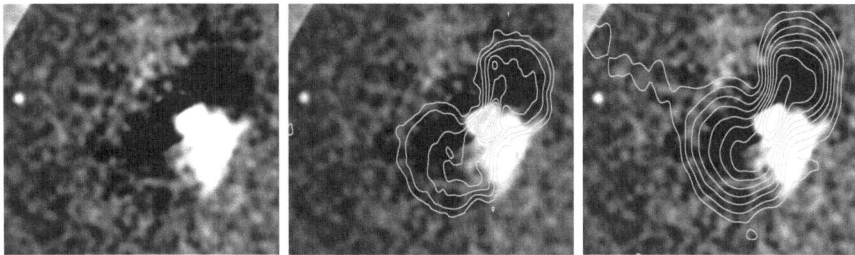

Fig. 1. Left Unsharp masked *Chandra* image of the central $\sim 120 \times 150$ kpc region of Abell 4059 showing the X-ray bar across the cluster center and the two cavities located to the N and SE of the cluster core. **Middle** VLA 1.4 GHz contours overlaid on the *Chandra* image. The radio lobes bend to completely fills both cavities. **Right** VLA 330 MHz radio contours show somewhat larger lobes and a possible trail of emission running to the NE of the eastern lobe.

least some ghost cavities are filled with old radio plasma. In order to under-stand the details of the AGN energy input into the ICM is it necessary to combine the X-ray observations with targeted radio observations over a range of different frequencies. Here we present a few examples of new results from multi-frequency radio observations of cavity systems.

2 Abell 4059

X-ray cavities in Abell 4059 were first detected with the High Resolution Imager on the ROSAT satellite [10]. Subsequent observations with *Chandra* revealed the presence of a central X-ray bar and showed that the two cavities (Figure 1 left) were not symmetric about the cluster center [9, 3]. Radio observations at 4.8 and 1.4 GHz showed that the emission associated with the central AGN was not aligned with the two cavities and did not extend as far as either cavity [3]. In this context the cavities in Abell 4059 were thought to be buoyant ghost cavities from a previous outburst.

New VLA observations of Abell 4059 at frequencies of 1400 and 330 MHz show that the X-ray cavities are filled with radio plasma (Figure 1 middle & right). The radio emission to the north extends beyond that seen by [3] and bends to the west to fill the cavity. Similarly the southern radio lobe bends to the east to fill the other X-ray cavity. In fact, radio emission within the cavities is seen at frequencies as high as 4.8 GHz, suggesting that the cavities are still being actively fed by the central AGN.

3 Abell 2597

Abell 2597 is well known to host a compact C-shaped radio source in the cluster core [17]. X-ray observations reveal the presence of ghost cavities [12]

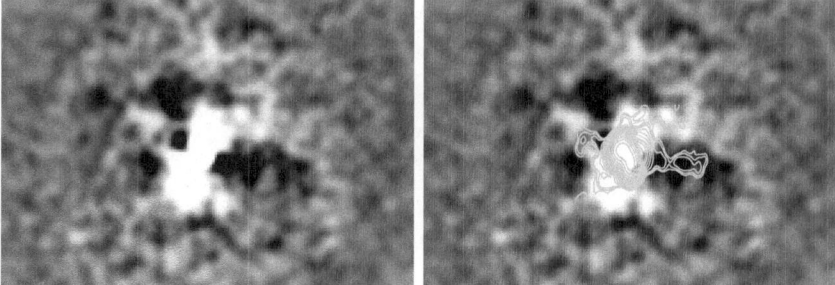

Fig. 2. Left: Unsharp masked *Chandra* image of the central $\sim 85 \times 125$ kpc region of Abell 2597. This deep (112 ks) image confirms the X-ray tunnel and cavities to the NE and shows evidence of a bright rim surrounding the tunnel. **Right** VLA 1.4 GHz radio contours overlaid on the unsharp masked *Chandra* image.

as well as a tunnel in the ICM gas connecting the core to the western ghost cavity [5]. Evidence for cool gas is seen in HST FUV observations which show diffuse emission as well as filaments and knots [13], while further evidence from cool gas in the cluster core comes from FUSE OVI observations [14].

Radio observations of the cluster core at frequencies of 4.8 GHz and below reveal emission extended beyond the compact C-shaped source. Low frequency (330 MHz) observations show synchrotron emission filling the tunnel [5] while at 1.4 GHz (Figure 2 right) the emission appears to be clumpy and only fills a portion of the tunnel. The 1.4 GHz radio contours also trace emission extended to the NW, as well as an arc of emission to the NE which is associated with the inner NE cavity as well as a bright Lyα filament seen in the HST images.

4 Abell 262

Radio observations of the core of Abell 262 show that it is host to a weak double lobed source [15] which displays an S-shaped morphology. This radio source appears to be interacting with the surrounding thermal gas as seen from *Chandra* observations of the system which revealed an X-ray cavity associated with the eastern radio lobe [2]. An analysis of the energy required to create the X-ray cavity compared to the cooling luminosity in the system suggested that the current outburst in Abell 262 is too weak to provide sufficient energy to offset cooling [2].

Our unsharp masked analysis of the *Chandra* data of Abell 262 [1] reveals that the eastern cavity is surrounded by a complete X-ray rim. In addition, the cluster is also host to an X-ray tunnel running westward from the AGN (Figure 3 left). Multi-frequency radio observations at frequencies below 1.4 GHz reveal that the central radio source is more than three times larger than previous observations showed [4]. The 610 MHz emission (Figure 3 right)

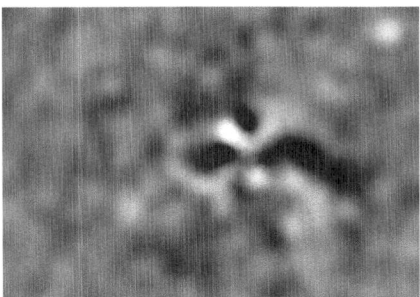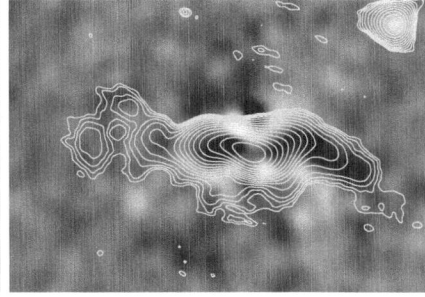

Fig. 3. Left: Unsharp masked *Chandra* image of the central $\sim 50 \times 80$ kpc region of Abell 262 showing the eastern cavity as well as the tunnel to the west of the cluster core. **Right:** GMRT 610 MHz contours of the radio emission from the central AGN in Abell 262 overlaid on the *Chandra* residual image.

fills the western tunnel, eastern X-ray cavity, and reveals three distinct radio features further eastward of the previously detected X-ray cavity. In fact the radio feature closest to the eastern X-ray cavity falls on top of a low significance X-ray deficit seen in the original *Chandra* images [2]. The other more eastern radio features are not detected as X-ray deficits but this may simply be a result of having insufficiently deep X-ray images to detect the depression. If the observed radio features all correspond to separate AGN outbursts (repetition timescale $\sim 3 \times 10^7$ yr) that have created X-ray cavities, then we find that the total energy input from those outbursts is within a factor of two of the X-ray cooling luminosity.

5 NGC 507

The galaxy group NGC 507 displays significant evidence of disturbance to the central X-ray emission [11]. The core shows two central X-ray clumps surrounded by an extended diffuse emission which displays a sharp surface brightness edge running from the northeast to the southeast (Figure 4 left). The sharp edge may be related to a metallicity gradient in the gas where the brighter emission is associated with cooler, higher abundance material [11]. A suggested origin for this material is gas which has been displaced from the central regions by the expanding radio lobe [11]. In the right panel of Figure 4 we show new high resolution VLA 1.4 GHz contours overlaid on the X-ray surface brightness image. The eastern radio lobe is seen to take a sharp bend to the south to trace the inner edge of the X-ray discontinuity, while the western radio lobe seems to be strongly interacting with the central ICM, creating a possible depression between the two central X-ray peaks.

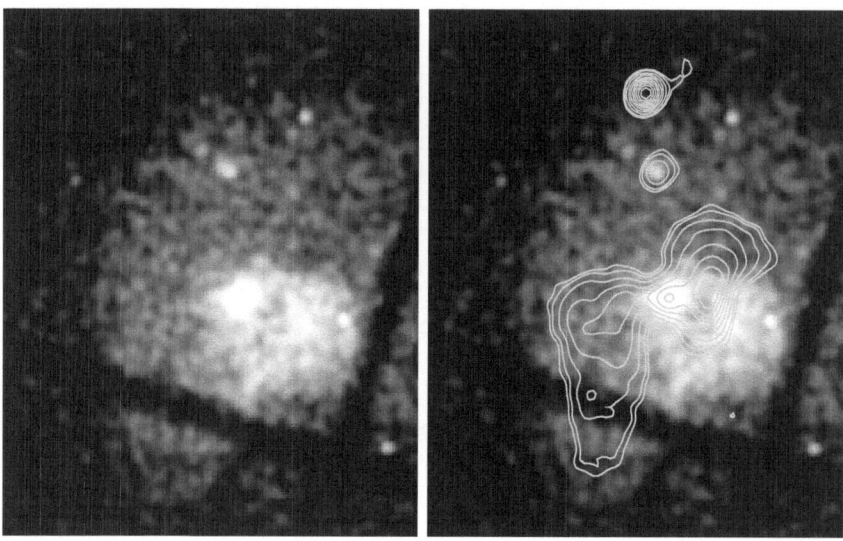

Fig. 4. Left: Gaussian smoothed *Chandra* X-ray image of the central ∼ 115 × 90 kpc region of NGC 507. The thermal gas is separated into two clumps in the core and a sharp-edged extension running from the NE to the SE. **Right:** VLA 1.4 GHz radio contours show that radio emission traces morphology of the X-ray surface brightness edge to the SE.

6 Discussion

There is no question that the observations obtained from *Chandra* and XMM have significantly advanced our understanding of the interactions between the central AGN and the thermal gas in galaxy groups and clusters. It is clear that over a wide range of system masses the central AGN has the ability to disrupt the thermal gas in the dense cores of these systems. A detailed understanding of these central interactions requires the addition of targeted multi-frequency radio data that is selected to probe both the relevant spatial scales for the system in question as well as a wide frequency range (including low frequencies) to track the radio outburst history of the central AGN.

In this paper we presented a few cases of sources where new radio observations have changed the view of the system under study. In the case of Abell 4059 it appears that the X-ray cavities are likely still undergoing active injection from the central AGN and thus are not buoyantly rising detached lobes. The radio observations of the central source in Abell 2597 reveal extended structures along several different position angles which are suggestive of multiple outbursts along possibly different initial directions.

In the case of Abell 262 the new radio observations reveal multiple structures along roughly the same position angle. These distinct structures may be signatures of different outburst episodes. Adding up the total energy input into the ICM from all radio outbursts seen in Abell 262 suggests that the AGN is more powerful than previously thought and is within a factor of two of being powerful enough to offset cooling over several outburst episodes.

The presence of the extended radio emission toward the eastern region of Abell 262 suggests that the low frequency radio flux is originating from detached buoyant lobes from past AGN outbursts. Using the observed projected offset of the radio structures from the cluster cores in Abell 2597 and Abell 262 together with the buoyant velocity in the systems we estimate a repetition timescale for the central AGN of a few $\times 10^7$ yr. This is significantly shorter than the typical assumption of 10^8 yr. Although there is some evidence of an X-ray deficit associated with one of the eastern emission regions in Abell 262 we note that it is relatively difficult to detect X-ray cavities once they are located beyond the densest parts of the cluster cores [6]. In these cases, low frequency radio observations provide an important tool for tracing the total energy input into the system to determine if it is sufficient to offset the radiative cooling.

Finally, in the case of NGC 507 we note that sensitive, high resolution radio images allow us to trace the interactions of the western radio lobe with the central X-ray structure. These observations also show the clear correspondence between the eastern radio lobe and the southern portion of the sharp surface-brightness discontinuity in the system.

Acknowledgement. T. E. C. acknowledges support from NASA through *Chandra* award GO6-7115B. Basic research in radio astronomy at the Naval Research Laboratory is supported by 6.1 base funding.

References

1. L. Anderson, E. L. Blanton, T. E. Clarke et al., in preparation
2. E. L. Blanton, C. L. Sarazin, B. R. McNamara et al., 2004, ApJ, 612, 817
3. Y.-Y. Choi, C. S. Reynolds, S. Heinz et al., 2004, ApJ, 606, 185
4. T. E. Clarke, E. L. Blanton, C. L. Sarazin et al., in preparation
5. T. E. Clarke, C. L. Sarazin, E. L. Blanton et al., 2005, ApJ, 625, 748
6. T. A. Ensslin, S. Heinz, 2002, A&A, 384, 27
7. A.C. Fabian et al., 1994, ApJ, 436, 63
8. A. C. Fabian, A. Celotti, K. M. Blundell et al., 2002, MNRAS, 331, 369
9. S. Heinz, Y.-Y. Choi, C. S. Reynolds et al., 2002, ApJ, 569, 79
10. Z. Huang, C. L. Sarazin, 198, ApJ, 496, 728
11. R. P. Kraft, W. R. Forman, E. Churazov et al., 2004, ApJ, 601, 221
12. B. R. McNamara, M. W. Wise, P. E. J. Nulsen et al., 2001, ApJ, 562, 149
13. C. P. O'Dea, S. A. Baum, J. Mack, et al., 2004, ApJ, 612, 131
14. W. R. Oegerle, L. Cowie, A. Davidsen, et al., 2001, ApJ, 560, 187
15. P. Parma, H. R. de Ruiter, C. Fanti et al., 1986, A&AS, 64, 135
16. J. R. Peterson, A. C. Fabian, 2006, Physics Reports, 427, 1
17. C. L. Sarazin, J. O. Burns, K. Roettiger et al., 1995, ApJ, 447, 559

High Sensitivity Low Frequency Radio Observations of cD Galaxies

S. Giacintucci[1,2], T. Venturi[1], S. Bardelli[2], D. Dallacasa[1], P. Mazzotta[3,4] and D.J. Saikia[5]

[1] INAF - IRA, via Gobetti 101, I-40129, Bologna, Italy sgiaci_s@ira.inaf.it
[2] INAF - OAB, via Ranzani 1, I-40126, Bologna, Italy
[3] Univ. Roma Tor Vergata, via della Ricerca Scientifica 1, I-00133 Roma, Italy
[4] Harvard-Smithsonian CfA, 60 Garden Street, Cambridge, MA 02138, USA
[5] NCRA, Pune University campus, Ganeshkhind, Pune, 411 007, India

1 Introduction

cD galaxies are the most luminous and massive galaxies known. They are found at the centre of both rich and poor galaxy clusters, usually at the peak of the X–ray emission. Their radio emission shows a large variety of morphologies: compact, FRI and FRI/FRII morphology, wide–angle–tail (WAT), core–halo and more peculiar structures. The advent of the high resolution X–ray imaging with *Chandra* and XMM–Newton posed new questions related to the interaction between the central radio source and the intracluster medium (ICM). It is now clear that the radio lobes may displace the external medium and create depressions (*cavities*) in the ICM. These can be filled with radio emission (*radio–filled cavities*); alternatively a misplacement between radio lobes and cavities can be observed (*ghost cavities*). Both features may be present in the same cluster (e.g. Perseus; Fabian et al. [1]). In particular ghost cavities tend to be lacated more in the external regions than the radio filled ones. This led to the suggestion that repeated outbursts of the central radio galaxy may play a role in the cavity formation (McNamara et al. [4]): ghost cavities would be the result of a previous AGN burst, and thus filled with aged radio plasma emitting at very low radio frequencies. All recent investigations of the radio galaxy–ICM interaction in clusters have mostly concentrated on single objects, based on the X–ray information. An unbiased comprehensive study of a statistical sample of clusters is still missing.

1.1 cD sample and GMRT observations

In order to overcome possible biases induced by radio and/or X–ray selections, we defined a statistical sample of cDs starting from the optical information. We selected a sample of 132 cDs (109 in Abell clusters and 23 in poor clusters; for further details see Giacintucci et al. [2], hereinafter G06). A fraction of $\sim 40\%$ is radio quiet at the sensitivity limit of the 1.4 GHz NVSS survey. We inspected all cDs with radio emission and selected 13 sources still lacking high sensitivity and high resolution radio imaging in the literature.

For these clusters we performed observations with the GMRT at 1.28 GHz to complete the radio information for the whole sample at this frequency. Furthermore we observed those sources promising for the presence of relic emission related to a previous AGN burst at 235 and 610 MHz. These low frequencies are suitable for the detection of old (i.e. steep spectrum) radio emitting plasma; moreover, combined with the 1.4 GHz data, they allow us to study the source spectral index (total and point–to–point) over a wide frequency range, providing crucial information on the nature of the observed emission.

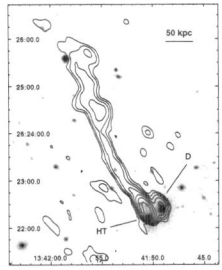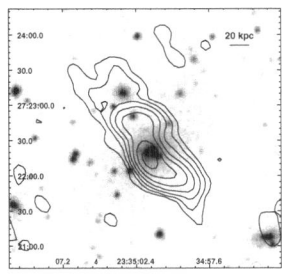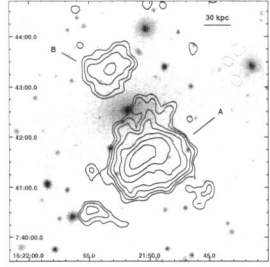

Fig. 1. GMRT 235 MHz contours (logarithmic, starting from $\pm 3\sigma$) on the POSS–2 optical image. *Left panel*: A 1775, 1σ=1.5 mJy/b, HPWB=12.7″ × 9.1″. *Central panel*: A 2622, 1σ=0.8 mJy/b, HPWB=17.1″ × 11.4″; *Right panel*: MKW 03s, 1σ= 2 mJy/b, HPWB=12.1″ × 9.1″.

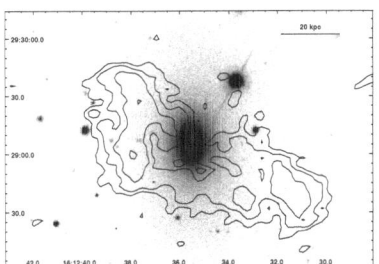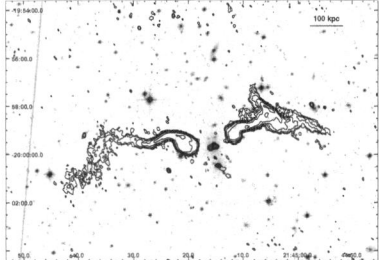

Fig. 2. GMRT 610 MHz contours (logarithmic, starting from $\pm 3\sigma$) on the POSS–2 optical image. *Left panel*: A 2162, 1σ=0.15 mJy/b, HPWB=7.1″×4.9″. *Right panel*: A 2372, 1σ= 0.065 mJy/b, HPWB=7.5″ × 5.7″.

2 Preliminary results

We found 8 active radio galaxies and 5 sources with candidate relic emission (steep spectrum and lack of central nuclear activity). Here we present the GMRT 235 MHz images for three clusters and 610 MHz images for two sources of the sample.

A 1775 – A 1775 hosts a Dumbell galaxy at its centre. Both galaxies are radio sources: a double source (labelled as D in Fig.1) is associated with the North–Western component, while an head–tail (HT) is associated with the companion galaxy. We measured a total flux density of $S_{(HT+D)}$=2.04 Jy at 235 MHz. We observed the source also at 610 MHz (G06), and found a morphology in very good agreement with the 235 MHz image. Using our data, and literature data, we determined a total spectral index of $\alpha = 1.15$ ($S \propto \nu^{-\alpha}$) between 80 MHz and 4.85 MHz.

A 2622 – A double radio source is associated with the cD at 235 MHz (Fig.1). Our $5''$ resolution observation at 610 MHz (G06) reveals a complex structure for this source. Its spectral index is very steep ($\alpha = 1.42$) between 74 MHz and 4.5 GHz. The 235–610 MHz spectral index distribution over the source shows that $\alpha \sim 0.6 - 0.7$ in the central region, and steepens along the lobes up to $\alpha \sim 1.8 - 2$ (G06). This confirms the presence of aged radio emission in this galaxy.

MKW 03s – A combined GMRT 1.28 GHz/610 MHz and *Chandra* study of this source was presented in Mazzotta et al. [3]. In the new 235 MHz image (Fig.2) the source shows two lobes, the southern lobe being brighter than the northern one. We do not detect the compact component between the lobes observed at higher frequency, and associated with the cD nucleus. The source 235 MHz flux density is 8.44 Jy. The spectral index between 235 MHz and 1.28 GHz is very steep ($\alpha = 2.49$), confirming the presence of relic emission in this source.

A 2162 – The 610 MHz radio emission is weak, and shows two radio lobes and no hint of a central compact component (Fig.2). The source flux density at 610 MHz is 240 mJy. Using this value and literature data, we found $\alpha = 0.81$ in the 74 MHz–4.7 GHz frequency range.

A 2372 – This cD hosts a very large WAT radio source, with a total flux density of 1.02 Jy at 610 MHz (Fig.2). We observed A 2372 also at 235 MHz (G06). Using our data and literature values, we found α=1.13 between 235 and 1400 MHz. The image of the 235–610 MHz spectral index distribution shows that the $\alpha \sim -0.5$ in the core, and steepens from $\sim$0.7 to $\sim$1.0 along the jets. In the outer regions of the tails $\alpha \sim 2$ (G06).

References

1. Fabian A.C., Sanders J.S., Ettori S. et al., 2000, MNRAS 318, L65
2. S.Giacintucci, T.Venturi, R.Athreya, S.Bardelli, D.Dallacasa, P.Mazzotta, D.J.Saikia, A&A, to be submitted – (G06)
3. Mazzotta P., Brunetti G., Giacintucci S., Venturi T., Bardelli S., 2004, JKAS 37 381
4. McNamara B., 1996, *Solid-State Physics*, 2nd edition, (Springer, Berlin), p45

Radio Galaxies in Cooling Cores: Insights from a Complete Sample

J. A. Eilek[1] and F. N. Owen[2]

[1] New Mexico Tech, Socorro NM USA jeilek@aoc.nrao.edu
[2] NRAO, Socorro NM USA fowen@aoc.nrao.edu

Summary. We have observed a new, complete, cooling-core sample with the VLA, in order to understand how the massive black hole in the central galaxy interacts with the local cluster plasma. We find that every cooling core is currently being energized by an active radio jet, which has probably been destabilized by its inter-action with the cooling core. We argue that current models of cooling-core radio galaxies need to be improved before they can be used to determine the rate at which the jet is heating the cooling core. We also argue that the extended radio haloes we see in many cooling-core clusters need extended, *in situ* re-energization, which cannot be supplied solely by the central galaxy.

1 Introduction

What heats cooling cores? The spotlight has turned on active galactic nuclei (AGN), driven by massive black holes in the heart of the galaxy at the center of the cooling core (CC). The jets in a few bright, well-studied radio galaxies (RGs) (*e.g.*, M87 [17]; A2052 [2]; Perseus A [9]) seem to be pouring out more than enough energy to offset radiative cooling in the CC in these clusters. But this is not the full answer; questions remain. Does every CC have a central RG? Is a typical cooling-core radio galaxy (CCRG) strong enough to offset local cooling? Does the energy carried by the jet couple effectively to the intracluster medium (ICM)? How can we use radio and X-ray data to estimate the jet power and energy input to the CC? To answer these questions, we must study more than the brightest few CCRGs, and must also look critically at dynamical models of the RG. We therefore carried out deep radio observations of a complete sample of CCRGs. In this paper we summarize our results and speculate on how to extend current models; more details will be given in [6].

2 The Data: What We Did

We formed a complete, X-ray selected sample of CCs in nearby Abell clusters. We started with ROSAT All-Sky Survey images of nearby ($z < 0.09$) Abell clusters [14]. We identified clusters which are X-ray bright ($L_x > 3 \times 10^{43}$erg/s within a 500 kpc aperture), centered on a massive galaxy, and with a centrally concentrated X-ray atmosphere (ratio of flux within 500 and 62.5 kpc apertures no larger than ~ 14). These criteria correlate well with strong CCs found in other, more detailed deprojection analyses (e.g., [19].) From these we selected clusters favorably placed in the sky for nightime VLA observations in 2002. This procedure gave us 22 clusters: A85, A133, A193, A426, A496, A780, A1644, A1650, A1651, A1668, A1795, A1927, A2029, A2052, A2063, A2142, A2199, A2428, A2495, A2597, A2626 and A2670. Good, deep radio data already exist for 3 of these (A85; A426, the Perseus A cluster; A780, the Hydra A cluster). We observed the rest with the VLA. Because high-resolution radio data exist for many of these objects, we designed our observations to detect faint, extended radio emission [15]. We note that M87 is not in our formal sample, because it is not in an Abell cluster, and its CC is on the weak side. We include it in much of our analysis, however, because it is so well studied [17, 10], and it is an important example of the interaction between CCs and their embedded RGs. In addition, nine of our clusters were also included in our VLA search [15] for cluster-scale radio haloes, giving us additional information on extended emission from these objects.

3 The Data: what we have learned

Our data show that the story is more complicated than has been thought. We find evidence that every cooling core is being disrupted, and probably energized, by an AGN. Our data also suggest that the radio-loud plasma does mix with the ICM, at least on large scales, and that the AGN may well *not* be the only driver for the ICM in a cooling-core cluster.

3.1 Every Cooling Core Contains a Radio Source

Every cooling core in our sample contains a currently-active radio core (some too faint to have been detected in previous work). This means that the central AGN are active 100% of the time; they do not have any "off" periods. However, if currrent dynamical estimates of source ages (~ 100 Myr) are close to correct, the central AGN is probably variable, cycling through high-power and low-power states. The jet and inner halo of M87 [11] appear to be an example of a recently "reinvigorated" AGN.

It follows that every RG in a cooling-core is currently being driven by an active jet. In particular, our deep radio images sometimes reveal faint jets connecting the central AGN to what were previously thought to be offset "relics" (e.g., A133, A2199). We therefore argue that very few CCRGs are simply passive, buoyant bubbles. The situation is more complex; we hope the data can guide us toward improved models.

3.2 The Radio Galaxies are Unusually Disturbed

Cooling-core RGs are characterized by unusual morphologies. Most of them are neither Fanaroff-Riley Type I (tailed), nor Type II (classical doubles). Sixteen CCRGs in our sample are well enough imaged to reveal their structure (the remaining 7 are too faint and too small). Three of these 16 are standard tailed sources (including Hydra A). The rest are diffuse and amorphous. Such a large fraction (80% of the set) is far too many to be "normal" RGs seen end-on. Furthermore, amorphous sources such as these are rare in the general radio-galaxy population; only 5 of the ~ 200 well-imaged cluster radio galaxies in the Owen-Ledlow set [18] are amorphous, and *all* of those sit in the centers of strong cooling cores.

These data show that an unusually strong interaction occurs between the radio jet and the dense cooling-core into which it tries to propagate. The interaction seems to destabilize the jet, on a scale of only a few kpc (the short jets in M87, Perseus A and A2052 are good examples here). It follows that the evolution of a cluster-core RG is not governed by directed momentum flux, in a large-scale jet, as is the case with most RGs. Instead, isotropized energy flow from a disrupted jet creates the amorphous haloes that we see. The strong RG-CC interaction is also suggestive of an effective energy transfer between the jet and the local cooling core; however the details of the process remain unclear.

3.3 The Radio Haloes Extend to Large Scales

Many of our CCRG are much larger than was previously known. Most of the amorphous sources have two scales of radio emission: a smaller, brighter source (often previously studied in higher resolution observations) is embedded within a larger, faint, extended mini-halo. Typical sizes of these mini-haloes range from ~ 70 to 200 kpc. For instance, Per A [3] and A2029 [15] can be traced to ~ 200 kpc, nearly as large as the long radio tails of Hyd A [13]. If the AGN is always "on", but cycles between strong and weak states, the mini-haloes may be relics of previous activity cycles. In addition, we have detected Mpc-scale radio haloes in two clusters, A2328 and A2495, which refutes the current idea that Mpc-scale haloes avoid cooling cores.

To put these sizes in context, recall that the size of the cooling core is typically $\sim 50-100$ kpc. The size of the cluster's potential well, as measured by the Navarro-Frenk-White scale radius, is only a few hundred kpc in CC clusters. Furthermore, the haloes do not obviously have clear edges. The sizes we measure are limited by the sensitivity of the observations, and may not be the true extent of the radio emission. With such large scales, the radio haloes may better be regarded as part of the entire ICM, not just a byproduct of the central AGN.

3.4 The Radio and X-ray Plasmas Must Mix

Our sample contains a variety of mixing states. Some of our clusters have small, clear X-ray cavities in the inner CC, approximately coincident with the RG. These cavities are very likely filled by radio-loud plasma, with little or no X-ray plasma. In other clusters (*e.g.*, M87 [10], A1795 [7]), the interaction between the two plasmas is more complex. The X-ray plasma is clearly interacting with, but not evacuated by, the radio plasma. In still others, the X-ray plasma appears smooth and undisturbed on the scale of the RG, at the best current X-ray resolution and sensitivity. We suspect the two plasmas have at least partially mixed in these CCs. In addition, because the ICM is not dramatically disturbed on hundred-kpc scales, the larger radio haloes must be effectively mixed with the ICM.

It seems, therefore, that the radio and X-ray plasmas mix effectively during the lifetime of the CCRG. Just how this occurs is not clear, given the stabilizing effects of even a small magnetic field in the ICM [12]. Deep, high resolution radio images (*e.g.*, M87, [17]; A2199, in preparation) may provide hints. We see radio-loud filaments in these sources that appear to be escaping from the RG and penetrating the CC plasma. These filaments may be similar to the magnetic flux ropes which are known to penetrate the terrestrial magnetopause. Such flux ropes, once formed, may decouple from the main body of the radio source and rise bouyantly through the ICM, giving rise to an extended radio halo coincident with a relatively smooth X-ray atmosphere.

4 The Models: Where to Go Next

The important question in the context of this meeting is, how much energy does a "typical" central AGN deposit in the cooling core plasma? To answer this, we must determine the mean jet power, P_j, averaged over the lifetime of a typical CCRG, and how effectively that power is deposited in the local plasma. We emphasize that we have *no* direct measure of P_j. The radio power of the CCRG is a poor tracer of the jet power [5]. The best we can do directly from observations is to use minimum-pressure arguments, which are possible if the jet is resolved (*e.g.* M87 [17]). This gives us a lower bound on P_j. To go further, we must choose a dynamical model for the RG and its interaction with the local ICM. This sounds simple, but the devil is in the details.

4.1 Calorimetry

The simplest cases are cooling cores which have clear X-ray cavities that coincide with an extended RG. In these the mean jet power can be found from the energy within the cavity and the age of the source. This is a simple, attractive appproach, which has been applied by various authors (*e.g.*, [1, 4]). However, it has complications. One is that measuring the energy content of the cavity is not straightforward, because it is hard to know the extent to which the radio and X-ray plasmas have mixed in most of these clusters. A second concern is how to estimate the age of the source. Most authors currently assume the radio source is passive, having been previously inflated by an AGN which has since turned off. If this holds, the source age is its size divided by the bouyant speed, v_b. But what is v_b? The sound speed is no more than an optimistic upper limit, because v_b is quite subsonic for small structures. In addition, magnetic tension from even a very small intracluster field can exceed hydrodynamic drag, and reduce v_b even more [12].

A more serious concern, however, is the evidence from our work that every CCRG is currently being driven by jets from an active AGN, and that large-scale radio haloes have mixed with the ICM. It follows that very few CCRGs are well described as isolated, passive, buoyant bubbles (although buoyancy surely plays some role in the evolution of the RG). New models are needed.

4.2 Possible Dynamical Models

As a first step towards such new models, we suggest that CCRG evolution can be broken into two stages. We envision an early stage in which young, driven sources interact with and expand into the ICM, and a later stage in which the RGs have grown into extended mini-haloes mixed with the ICM. If AGN activity is cyclic, an older mini-halo could coexist with a younger, restarted inner core. We note that our ideas here are no more than toy models; they need to be developed and tested against real, well-observed CCRGs.

Because the observations show that the AGN in every CCRG is "alive", and because CCRGs are often amorphous, we suggest that small, young sources are being driven by a quasi-isotropic energy flux (as from an unstable jet). Such evolution can be approximated by a self-similar analysis [8]. However, because the edges of the X-ray cavities are not strong shocks (*e.g.*, [9]), we know the expansion is slow; this suggests the expansion proceeds at approximate pressure balance [17]. Such a model predicts the source size $R(t) \propto (P_j t)^x$, where x depends on the ambient pressure gradient. We emphasize that P_j and t cannot be determined separately in this model; the best we can do is the limit $\dot{R} < c_s$, which gives an upper limit to P_j.

Because the data also show the radio and X-ray plasmas are well mixed for larger RGs, we further suggest that CCRGs eventually fragment and mix with the ICM. The fragmentation may occur *via* MHD surface effects (such as the tearing-mode instability) which create magnetic filaments or flux ropes. Alternatively, the small-scale flux ropes which we know exist in MHD turbulence may retain coherence and diffuse into the extended ICM in late stages of CCRG evolution. (The ubiquity of filaments in well-imaged RGs suggests such structures are common in general; why should CCRGs be different?) We expect the flux ropes to rise slowly under buoyancy, and to retain their identity for awhile, after which they probably dissipate and merge with the local ICM. In principle, P_j could be estimated for such a source from the energy content of the radio plasma and its buoyant rise time, but uncertainties in filling factors and flux rope sizes limit the quantitative usefulness of this approach.

4.3 The Large Radio Haloes

Some of our radio haloes are large enough to raise the question, where does "CCRG" end and "cluster halo" begin? That is, on what scale is the physics of the full cluster more important than the influence of the AGN? The synchrotron size is one criterion: how large can the radio halo be without needing *in situ* energization? We are skeptical of simple synchrotron-aging estimates, because magnetic fields in the radio source and the ICM are almost certainly inhomogeneous. One can, however, derive a useful limit. The lowest loss rate for the radio-loud electrons is that of inverse Compton losses on the cosmic microwave background. If the electrons spend most of their time in sub-μG magnetic fields, and occasionally migrate into high-field regions (probably a few μG) where they become radio-loud, we can find an upper limit to their synchrotron life.

This cartoon predicts the radio plasma in a buoyant flux rope can reach ~ 100 kpc before it fades away. Radio sources larger than this must be undergoing extended, *in situ* re-energization. It follows that some driver other than the AGN must exist on large scales. Ongoing minor mergers are thought to support radio haloes in large, non-CC clusters. They may be important in CC clusters as well (*e.g.*, [16]), and may be driving the larger haloes. But then, if we admit the need for non-AGN heating of CC clusters on large scales, can we be sure that the cooling core itself is heated only by the AGN?

References

1. L. Birzan, D. Rafferty, B. McNamara et al, 2004, ApJ, 607, 800
2. E. Blanton, C. Sarazin, B. McNamara etal, 2001, ApJ, 558, 15
3. J. Burns, M. Sulkanen, G. Gisler et al, 1992, ApJ, 388, 49
4. D. De Young, 2006, ApJ, in press (astro-ph/0605734)
5. J. Eilek, 2004, in *The Riddle of Cooling Flows in Galaxies and Clusters of Galaxies*, ed. by T. Reiprich et al, published electronically at *http://www.astro.virginia.edu/coolflow*
6. J. Eilek, F. Owen, T. Marković, ApJ, in preparation
7. S. Ettori, A. Fabian, S. Allen et al, 2002, MNRAS, 331, 635
8. S. Falle, 1991, MNRAS, 250, 581
9. A. Fabian, A. Celotti, K. Blundell et al, 2002, MNRAS, 331, 369
10. W. Forman, P. Nulsen, S. Heinz et al, 2005, ApJ, 635, 894
11. D. Hines, F. Owen, J. Eilek, 1989, ApJ, 347, 713
12. T. Jones, D. De Young, 2005, ApJ, 624, 586
13. W. Lane, T. Clarke, G. Taylor et al, 2004, AJ, 127, 48
14. M. Ledlow, W. Voges, F. Owen et al, 2003, AJ, 126, 2740

15. T. Marković, 2004, Dynamics of the Thermal and Nonthermal Components of the Intra-Cluster Medium, PhD thesis, New Mexico Tech
16. P. Motl, J. Burns, C. Loken et al, 2004, ApJ, 606, 635
17. F. Owen, J. Eilek, N. Kassim, 2000, ApJ, 543, 611
18. F. Owen, M. Ledlow, 1997, ApJS, 493, 73
19. C. Peres, A. Fabian, A. Edge et al, 1998, MNRAS, 298, 416

Supernova Feedback in Groups and Low-Mass Clusters of Galaxies

J. Rasmussen and T. J. Ponman

University of Birmingham, School of Physics and Astronomy, Edgbaston, B15 2TT, United Kingdom, jesper@star.sr.bham.ac.uk (JR)

1 Introduction

The chemical enrichment of the intracluster medium (ICM) in groups and clusters of galaxies is believed to originate mainly in material ejected from their member galaxies by supernovae. The spatial distribution of ICM metals therefore provides insight into some of the processes, other than gravity, that have shaped the thermodynamic properties of the ICM. While abundance profiles of massive clusters have been extensively studied, the situation in the much more numerous lower-mass systems is less clear. Here we present radial profiles of the abundance of Fe and Si from high-quality X-ray data of 15 groups, selected from the *Chandra* archive on the criteria that the data should feature more than 6,000 net counts and the groups appear reasonably relaxed. The sample size allows us to explore statistical trends within the sample, and assess the role of supernova metal and energy injection, even in the X-ray faint group outskirts where most of the intragroup gas resides.

2 Fe and Si Radial Profiles

Profiles of Fe and Si abundance for all groups combined are shown in Fig. 1, with all radii normalized to r_{500} (inside which the mean density is 500 times the critical density), and with the data binned into radial bins of 20 measurements. Fe is seen to decline outside the group core, converging towards a value of $\sim 0.1\,Z_\odot$ at r_{500}. We note that this value is lower than that seen in cluster outskirts by a factor of ~ 2 (e.g. [1]), even though the Fe abundance in group cores is comparable to cluster levels. Si is more spatially uniform than Fe, declining less steeply outside the core. Intrinsic scatter is most prominent in the core, where non-gravitational processes are expected to be important.

Adopting the SN model yields of [2, 3], the Si/Fe ratios suggest that the enrichment in group cores can be attributed to a mixture of SN Ia and SN II which is similar to that in the Solar neighbourhood. At large radii the Si/Fe ratio suggests an increasing predominance of SN II, with the abundance pattern being consistent with pure SN II enrichment at r_{500}. As the intrinsic scatter outside the group cores is small, we made simple parametrizations of

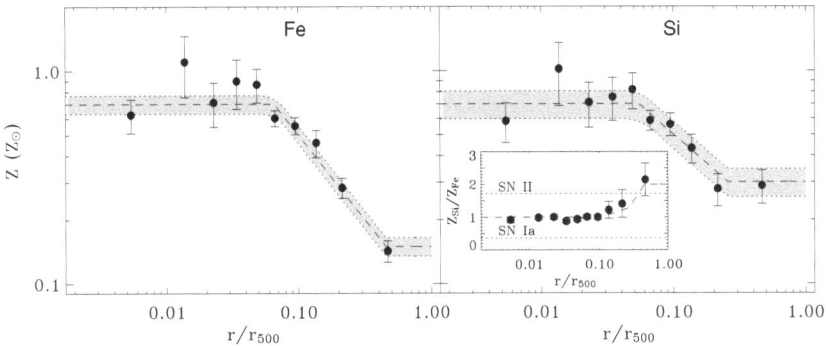

Fig. 1. Combined radial abundance profiles for all 15 groups, binned into radial bins of 20 data points. The inset shows the resulting Si/Fe ratio, with dotted lines marking the expectations from pure SN Ia and SN II enrichment

the binned profiles, in order to obtain prescriptions for $Z_{Fe}(r)$ and $Z_{Si}(r)$ that can be seen as representative for the entire group sample. Their 1σ error envelopes are shown as shaded areas in Fig. 1, with the resulting (parametrized) ratio $Z_{Si}(r)/Z_{Fe}(r)$ conforming to the range allowed by the adopted SN yields.

3 Implications: Supernova metal and energy injection

Fig. 2a shows the parametrized profiles decomposed into contributions from SN Ia and SN II. This indicates that about half of the iron in group cores is provided by SN Ia, but that SN II products are much more widely distributed than those of SN Ia, similarly to results for clusters. The total iron mass released by SN within r_{500} spans the range 3×10^7–3×10^9 $M_\odot$ across the sample. Assuming each SN releases 10^{51} erg into the ICM, the equivalent injected energy of 2×10^{59}–2×10^{61} erg is comparable to the total thermal energy in the ICM, with the clear majority of it ($\sim 95\%$) provided by SN II.

Total iron mass-to-light ratios within r_{500} are shown in Fig. 2b (results for Si are very similar). The low Fe abundance in group outskirts compared to clusters is reflected in lower Fe M/L ratios. Even across the fairly narrow range in total group mass studied here (estimated assuming a mass-temperature relation $M \propto \langle T \rangle^{1.5}$), there is a clear tendency for lower-mass groups to contain relatively lower amounts of enriched material for their optical luminosity.

Figures 1 and 2 demonstrate that the metals are not completely mixed throughout the ICM. This applies particularly to the metals attributed to SN Ia, of which a considerable fraction is likely to be associated with prolonged enrichment from the stellar population of the central galaxy present in all the groups. As also discussed by [1] on the basis of a small sample

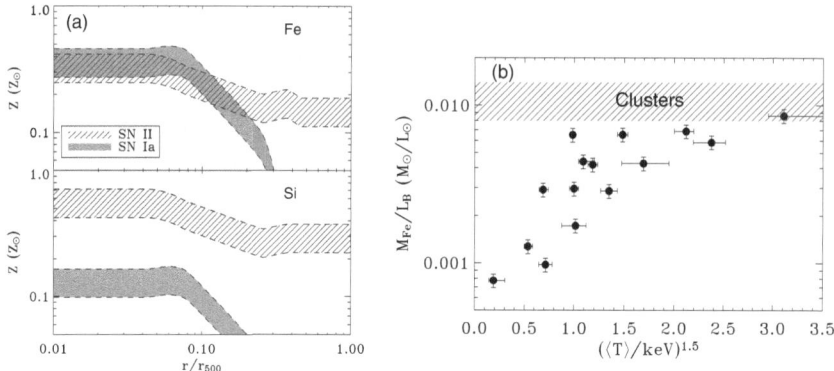

Fig. 2. (a) Contribution to the total abundance of Fe and Si from SN Ia and SN II. (b) Fe mass-to-light ratio within r_{500} as a function of group 'mass' $M \propto \langle T \rangle^{1.5}$. The dashed region outlines the typical range covered by clusters [1]

of three groups, the much wider distribution of SN II products suggests a scenario of SN II–dominated enrichment at an early stage in the formation of the groups, leaving time for the enriched material to mix well throughout the ICM.

In the group outskirts, the inferred energy per particle imparted to the IGM by SN II is of order 0.5 keV. Coupled with the above, this indicates that galactic winds associated with an early phase of strong starburst activity could have contributed substantially to pre-heating of the ICM in groups. It is likely that much of this preheating and the associated chemical enrichment took place before the gas collapsed into groups, while still located in filaments of low overdensities [4]. Some fraction of the SN ejecta should have been able to escape from low-mass filaments, giving rise to the trend of increasing iron mass-to-light ratio with present-day group mass. The inferred central rise in SN II products can then be explained if some of these metals were released after the group collapsed, possibly facilitated by galaxy–galaxy and galaxy–ICM interactions, and potentially with an additional contribution associated with stellar wind loss from the central elliptical galaxy.

Acknowledgement. JR acknowledges the support of the European Community through a Marie Curie Intra-European Fellowship.

References

1. A. Finoguenov, L.P. David, T.J. Ponman, 2000, ApJ , 544, 188
2. K. Nomoto, et al., Nucl. Phys. A, 1997, 616, 79
3. K. Nomoto, et al., Nucl. Phys. A, 1997, 621, 467
4. T.J. Ponman, D.B. Cannon, J.F. Navarro, 1999, Nature , 397, 135

Part III

Cooling Cores in Elliptical Galaxies

Cooling and Feedback in Galaxies and Groups

C. Jones[1], W. Forman[1], E. Churazov[2,3], P. Nulsen[1], R. Kraft[1] and S. Murray[1]

[1] Smithsonian Observatory, 60 Garden St., Cambridge, MA, USA
[2] Space Research Institute, Moscow, Russia
[3] Max Planck Institute for Astrophysics, Garching, Germany

1 Hot Gas and Cooling Flows in Early Type Galaxies

Prior to the first Einstein X-ray observations of early type galaxies (e.g. [8, 9, 3]), early type galaxies were generally considered to be gas free, with the gas from stellar mass loss thought to be removed by supernova driven winds (e.g. [6]). However the Einstein observations showed extended hot gas halos surrounding optically luminous elliptical and lenticular galaxies. The extensive galaxy surveys carried out with ROSAT [1, 23] detected X-ray emission from hundred of galaxies and showed that lower temperature systems generally had lower gas fractions.

The high central gas densities and short central cooling times of the gas in these early-type galaxies led to the suggestion that galaxies should have cooling flows with infall rates of $0.02 - 3 M_\odot$ yr^{-1} [22, 29]. Since the cooling flow rates in the standard model are modest, AGN outbursts with energies on the order of 10^{57} ergs would need to occur only every fifty million years or so to prevent the accumulation of large amounts of cold gas in the galaxy cores.

As described in this review, with Chandra and XMM-Newton observations, we can directly observe features produced through AGN outbursts, including X-ray cavities, nuclear and jet emission, and shock heated gas in X-ray coronae. Such features are common, with ~80% of early-type galaxies with detected nuclear emission and ~30% of the brighter systems having X-ray cavities. While the energy outbursts in galaxies are small compared to those seen in the central galaxies in clusters, the impact of the outbursts on the surrounding gas can be dramatic, even in some cases driving much of the gas out of the galaxy. Although extensive cool coronae are not found around galaxies in the cores of hot massive clusters, even in these environments, galaxies can maintain a small core of cool (1 keV) gas. In these compact coronae, the cooling gas is likely being reheated through conduction from the surrounding hot cluster medium. AGN outbursts in these compact coronae show how energy is transported from the nucleus.

2 Feedback and the Presence of X-ray Cavities

As shown in Fig. 1, one of the first galaxies observed by Chandra to have a very disturbed X-ray corona was M84 (NGC4374) in the core of the Virgo cluster [7]. The radio lobes of 3C272.1 define the structure of the X-ray gas. The brightest X-ray emission is found along the sides of the radio lobes, and is cooler than the surrounding gas, implying that the expansion of the radio lobes is not supersonic.

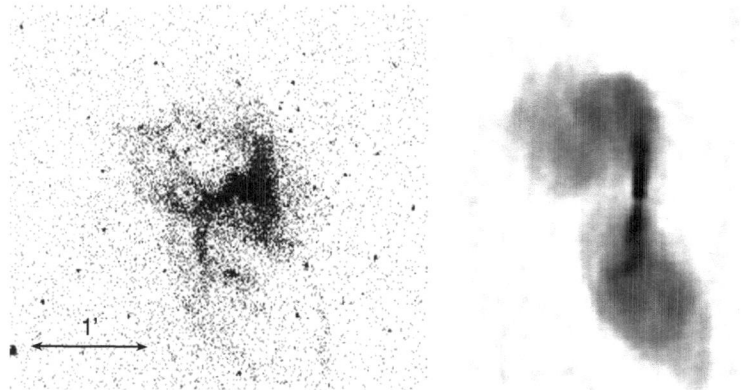

Fig. 1. (left) The Chandra 0.5-2 keV image of M84 (NGC4374) shows two pronounced X-ray cavities, one north and one south of the nucleus. Each cavity is filled with radio emission from 3C272.1. The X-ray rims of the cavities are cooler than the surrounding gas and, unlike the shell in Cen A, the X-ray emission is brighter on the edges of the rims than at the "ends". (right) The VLA radio image of M84=3C272.1 shows the nucleus, jets and lobes.

At a distance of only 3.4 Mpc, Cen A is our nearest active galactic nucleus, hosting a low power FRI radio source, as well as the brightest "steady" extragalactic Gamma-ray source. Cen A appears as an elliptical galaxy crossed by a dust lane, believed to be the result of a merger with a small spiral about 10^9 years ago. Both radio and X-ray observations show nuclear emission and, on kiloparsec scales, a predominantly one-sided jet. The Chandra image of Cen A (Fig. 2) also shows a warm (0.3 keV) ISM, hundreds of point sources and a shell surrounding the southwest inner radio lobe [14, 15]. Unlike M84 where the brightest X-ray emission lies along the sides of the radio lobes, in Cen A, the brightest emission is at the end of the southwest lobe. With 3×10^6 $M_\odot$ of swept-up ISM in the shell, Cen A is the best known example of a shell of gas compressed and shock heated by the supersonic expansion of a radio lobe. As Kraft et al. [15] determined, the X-ray gas in the shell is ten times hotter and twelve times denser than the surrounding interstellar gas. Thus the thermal pressure in the shell is ~100 times the pressure in the surrounding medium. One can maintain this highly over-pressured shell by balancing its pressure by the ram pressure of the expanding lobe. The required ram pressure corresponds to the shell expanding at 2400 km sec^{-1}, or Mach 8.5. Eventually most of the kinetic energy in the shell, which is about six times its thermal energy, should be converted into heating the ISM through conduction and gas mixing. Thus the expansion of high Mach number jets/lobes provides another mechanism for AGN to reheat the cooling gas in the cores of galaxies.

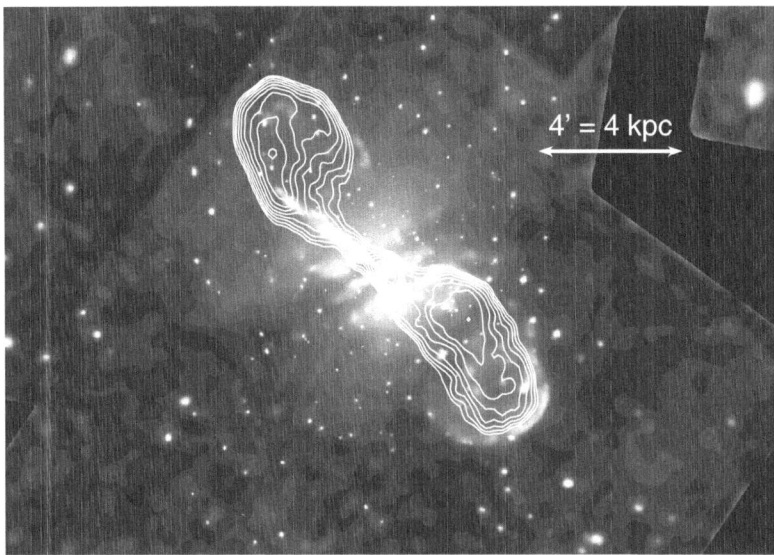

Fig. 2. The Chandra image of Cen A (NGC5128) with contours of the radio emission superposed shows the jet northeast of the nucleus, and the two inner radio lobes. A bright shell of X-ray emission can be seen along the far edge of the southwest radio lobe.

While the presense of radio lobes in M84 and Cen A should have suggested that the X-ray gas in these galaxies would be morphologically disturbed, the general expectation for other "normal" early-type galaxies was that their X-ray emission would be composed of "relaxed," symmetric, hot gaseous coronae, punctuated with bright LMXBs. Instead Chandra observations of early type galaxies have often shown X-ray cavities in their interstellar gas; e.g. NGC 4636 ([17]), NGC4472 ([2]), NGC507 ([16]) and NGC4552 ([20]). The galaxy NGC5813 shows a double set of X-ray cavities, as shown in Fig. 3, produced in two separate episodes of outburst activity. From VLA-FIRST observations, NGC5813, like many of the other galaxies with cavities, shows only weak radio emission from the nucleus.

While the X-ray rims around the cavities are cooler than the surrounding gas in most galaxies, in a few galaxies (e.g. Cen A, NGC4636, and

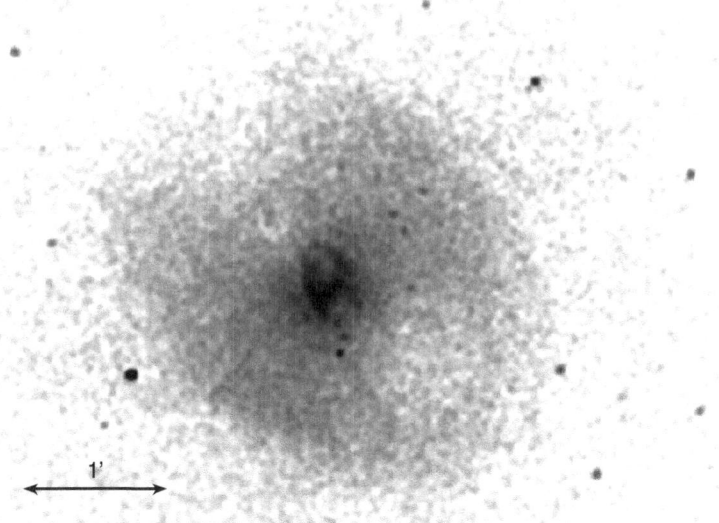

Fig. 3. The 0.3-2 keV smoothed Chandra image of NGC5813 shows two sets of cavities to the northeast and southwest of the galaxy nucleus. The inner cavities are outlined by X-ray bright rims of emission. A sharp change in the X-ray surface brightness, in all directions at the radius of the outer cavities, likely corresponds to a shock.

NGC4552), the X-ray emission around the cavities is hotter than the surrounding medium, likely the result of shock heating by the supersonic expansion of the lobes. Thus in these galaxies, the cooling gas can be reheated through conduction and mixing with the hotter gas in the rims of the cavities. Shocks, as found in clusters (see, for example, papers in this volume by Fabian, Forman, and McNamara) are also found in the ISM of galaxies. In NGC5813 (Fig. 3) the sharp transition in the surface brightness at the radius of the outer cavities is likely produced by a shock.

We analyzed Chandra observations for a sample of 160 early-type galaxies, determining the presence of X-ray cavities in the gas, the age of the cavities and the energy required to inflate them, as well as the current X-ray emission from the nucleus [18]. We identified 27 galaxies with X-ray cavities. All but three of these systems were in the more X-ray luminous galaxies ($L_x > 10^{40}$ ergs s^{-1}), where the brighter surface brightness of the gas allows cavities to be more easily detected. Fig. 4 shows a histogram of the ages of the cavities and the energy required to produce them. As this figure shows the cavity ages range from 10^6 years to nearly 10^8 years, with the largest number at $\sim 10^7$ years. Energies to produce the cavities range from 10^{52} ergs to 10^{58} ergs, with a peak at $\sim 10^{56}$ ergs. These energies can be compared to the much larger energies associated with outbursts in clusters. For example M87's outburst energy is nearly 10^{59} ergs [10], while the energy required to produce the giant cavities in MSO0735.6+7421 is 6×10^{61} ergs [21]. However, while the galaxy outbursts are modest by comparison to those in clusters, they can have dramatic effects on the gas and on the star formation.

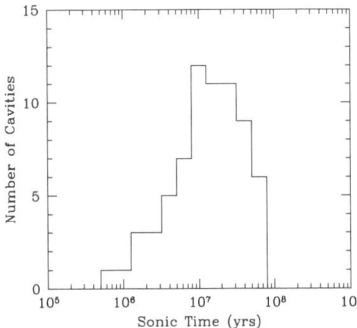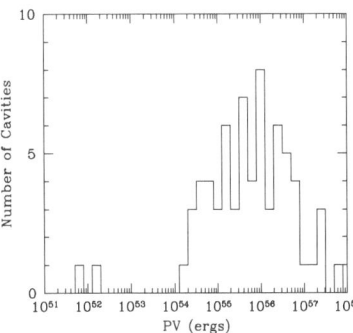

Fig. 4. (left) For all 27 galaxies with detected X-ray cavities, a histogram of the time since the outburst, as determined from the distance of a cavity from the galaxy nucleus. (right) A histogram of the energy in each outburst that is required to inflate each cavity, again for all 27 galaxies with detected cavities.

3 X-ray Emission from the Supermassive Black Holes in Normal Galaxies

With few exceptions (e.g. Cen A and M87), most nearby galactic nuclei exhibit only low levels of activity. In particular in their spectroscopic survey of a magnitude limited sample of 486 galaxies, [12] found emission lines in the spectra of nearly every spiral galaxy and in 60% of the 145 elliptical and S0 galaxies. Although the bolometric luminosities of the weakest Seyfert galaxies are more than eight orders of magnitude lower than the most powerful quasars, it is widely believed that these low luminosity AGN also share the same production mechanism, in particular, accretion onto a supermassive black hole.

To reliably detect and quantify weak nuclear emission, at any wavelength, requires observations with both good sensitivity and high angular resolution to distinguish the nucleus from the bright bulge. Thus the availability of Chandra observations with a ~0.3" FWHM PSF [30] for a large sample of nearby early type galaxies permits a detailed study of their nuclear emission.

Since X-ray emission is a common and sensitive signature of nuclear activity in galaxies, a number of studies, primarily using ROSAT or Chandra, have been undertaken. In particular in a sample of 39 galaxies observed with ROSAT, [4] found compact X-ray sources within 2' of the nucleus in 21 galaxies, only four of which were ellipticals. Also [11] and [25] detected X-ray emission from half or more of the low luminosity AGN in Seyfert or LINER galaxies that they studied, while [13] used Chandra observations to detect X-ray emission from the nuclei in 15 of 24 low luminosity AGN selected from the Palomar spectroscopic survey [12]. [24] collected published luminosities and analyzed the nuclear X-ray emission for a sample of 50 galaxies ranging in type from E0 to Sbc. [26] found soft X-ray sources associated with the nuclei in the massive elliptical galaxies NGC4472 and NGC4649, while [19] placed strong limits on the hard X-ray emission from the three giant ellipticals NGC1399, NGC4472 and NGC4636.

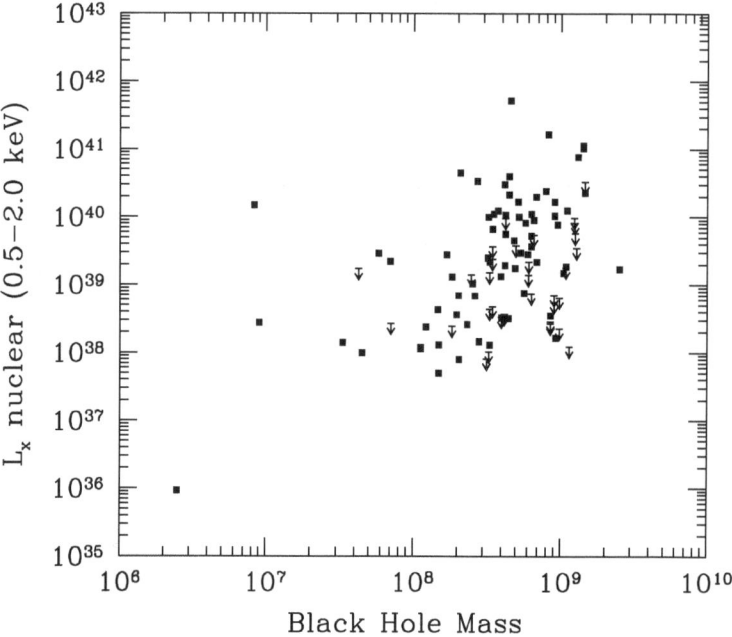

Fig. 5. The X-ray luminosity of the emission from the central low luminosity AGN is each galaxy is plotted against the mass of the central black hole, estimated from the galaxy velocity dispersion.

In preparation for our analysis of the Chandra observations of 160 early type galaxies, we determined the galaxy centers primarily from the 2MASS images, supplemented with radio positions, when the nucleus is "radio-loud". Since many of these galaxies contain bright X-ray gas, we measured the X-ray emission from the nucleus by modelling the radial surface brightness profiles with two components, one due to the extended X-ray emission and a second due to the nuclear point source. In particular we fit the surface brightness profile in the soft energy band (below 2 keV) after detected point sources were omitted with a beta model to describe the extended emission, along with a point source at the nucleus. Since the extended emission is primarily due to hot gas in the corona of the galaxy, in the hard energy band (2-6 keV), the diffuse emission is generally faint. We fit the hard surface brightness profile with the same beta model determined in the soft energy band, but allowed its normalization to be free and included a point source component at the nucleus.

As Fig. 5 shows, $\sim 80\%$ of the galaxies in our sample have detected X-ray emission from their nuclei. In this sample of "normal" early-type galaxies, the nuclear emission spans a wide range in luminosity, with most falling between 10^{38} ergs s^{-1} and 10^{41} ergs s^{-1}. Weak X-ray emission is often detected from the nuclei, even when X-ray cavities are not found.

4 Driving Gas from Galaxies through AGN Outbursts

The correlation of galaxy absolute magnitude with X-ray luminosity has always shown significant dispersion (e.g. [9, 23]), which has been attributed to various sources, most often those related to galaxy environment, particularly the ram-pressure stripping or evaporation of gas by a surrounding cluster medium. Galaxies with low X-ray luminosities, compared to their optical luminosity, were dubbed "X-ray faint." Their total gas mass is less than the total that would be expected from stellar mass loss over cosmological times. The realization that supermassive black holes reside in the centers of all massive bulges and that accretion, generally through cooling gas, but also through the occasional galaxy merger, can fuel AGN outbursts provides another mechanism for altering the gas mass content of galaxies. In particular, AGN outbursts can produce large X-ray cavities, and also may heat the gas so that it is no longer bound by the gravitational potential of the galaxy. The X-ray image of the X-ray faint elliptical NGC1316=Fornax A in Fig. 6 (left) shows that the complex emission from the central region extends only $\sim 1'$. Most of the gas from stellar mass loss has been removed from this galaxy, likely due to the energy input from the outburst that produced the radio source Fornax A.

Another galaxy which shows a striking deficit of X-ray emission from the central region is UGC408=NGC193 . The X-ray emission from UGC408, shown in Fig. 6 (right) shows a bright ring of emission at a radius of 20–100". If the ring and lack of bright central emission are produced by expanding lobes, these lobes and the resulting X-ray cavities likely lie nearly along our line of sight.

The Millenium Run simulations discussed by [5] show that in galaxies where AGN outbursts are included, star formation can be truncated, thus preventing the growth of simulated galaxies that are more massive than those

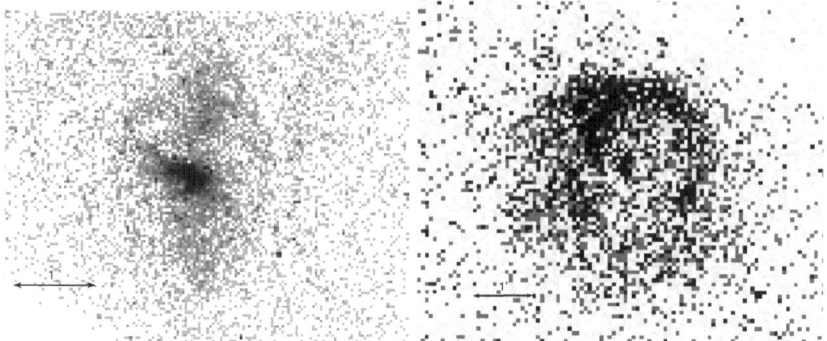

Fig. 6. (left) The 0.3-2 keV Chandra image of the core of NGC1316=Fornax A. The hot ISM is structured with little X-ray gas emission beyond a radius of 1', likely due to the removal of gas by the Fornax A outburst. (right) The 0.3-2 keV Chandra image of UGC408 shows emission from a ring of diffuse emission along with emission from the nucleus.

actually observed. As seen in NGC1316, large AGN outbursts which are more common at earlier epochs, may not only heat the gas in the galaxy cores, in which case the timescale for new stars to form is approximately the cooling time of the gas, but may drive much of the gas from the galaxy, in which case the timescale for new stars to form is the time for significant stellar mass to accumulate in the galaxy core and cool.

5 Channeling Outburst Energy through Jets to Maintain Small, Cool Coronae

Finally, examining the X-ray coronae in early-type galaxies with radio jets provides insight into how the energy from the nucleus is transported into the galaxy and beyond. While most luminous early type galaxies in poor environments have extended hot gas coronae, galaxies in rich clusters generally do not. However, work by [31, 27, 28, 32] shows that, even in rich clusters, very small gas coronae still survive in the cores of massive early type galaxies. The small size of these coronae likely results from the evaporation of their outer atmospheres by the surrounding hot cluster gas, which also is responsible for reheating the remaining core gas. The high gas densities and 1 keV gas in these small coronae imply short cooling times, although heat conduction from the hot ICM can significantly reheat the coronae, although the conduction rates must be significantly reduced from the Spitzer rate so as not to completely evaporate the gas.

Of particular interest are NGC3842 in A1367 and NGC4874 in the Coma cluster, which host double lobed radio sources as shown in Fig. 7. If the total energy from the radio outbursts were deposited in the coronae, the coronae would be destroyed. These small, fragile coronae apparently survive the AGN outbursts because the energy, as it exits the nucleus, is first channeled through a jet and then deposited primarily at larger radii. If this process is the same in all galaxies, then a small amount of gas in the core would remain even following an outburst, allowing accretion to continue at a low level and explaining the existence of low-luminosity AGN in the majority of early-type galaxies.

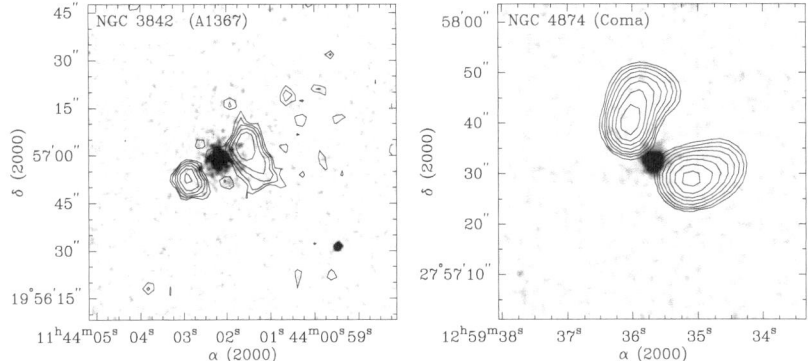

Fig. 7. (left) The Chandra image of NGC3842 in the A1367 cluster with contours from the VLA FIRST 20 cm image superposed. (right) The Chandra image of NGC4874, one of the large cD galaxies in the Coma cluster. These images show that the radio lobes lie outside the region filled by the X-ray coronae and thus that the active AGN release most of their energy at large radii, beyond the galaxy core. Image from [27].

Acknowledgement. We thank Hans Bohringer and others for organizing an outstanding and stimulating conference. We also acknowledge support from the Smithsonian Institution, the Chandra X-ray Center and the Max Planck Institute for Astrophysics.

References

1. J. Beuing, S. Dobereiner, H. Böhringer, R. Bender: MNRAS **302**, 209 (1999)
2. B. Biller et al: ApJ 613, 238 (2004)
3. C. Canizares, G. Fabbiano, G. Trinchieri: ApJ **312**, 503 (1987)
4. E. Colbert, R. Mushotzky: ApJ **519** , 89 (1999)
5. D. Croton et al: MNRAS **365**, 11 (2006)
6. S. Faber, J. Gallagher: ARA&A **17**, 135 (1979)
7. A. Finoguenov, C. Jones: ApJ (Letters) **547**, 107 (2001)

8. W. Forman et al: ApJ(Letters) **234**, 27 (1979)
9. W. Forman, C. Jones, W. Tucker: ApJ **61**, 33 (1985)
10. W. Forman et al: ApJ **635**, 894 (2005)
11. E. Halderson et al.: AJ **122**, 637 (2001)
12. L. Ho, A. Filippenko, W. Sargent: ApJ **487**, 568 (1997)
13. L. Ho et al: ApJ(Letters) **549**, 51 (2001)
14. R. Kraft et al: ApJ **569**, 54 (2002)
15. R. Kraft et al: ApJ **592**, 129 (2003)
16. R. Kraft et al: ApJ **601**, 221 (2004)
17. C. Jones et al: ApJ **567**, 115 (2002)
18. C. Jones et al: in preparation
19. M. Lowenstein et al: ApJ **555**, 21 (2001)
20. M. Machacek et al: ApJ **648**, 947 (2006)
21. B. McNamara et al: Nature **433**, 45 (2005)
22. P. Nulsen, G. Stewart, A. Fabian: MNRAS **208**, 185 (1984)
23. E. O'Sullivan, D. Forbes, T. Ponman: MNRAS **328**, 461 (2001)
24. S. Pellegrini: ApJ **624**, 155 (2005)
25. S. Satyapal: ApJ **633**, 86 (2005)
26. D. Soldatenkov, A. Vikhlinin, M. Pavlinsky: AstL **29**, 298 (2003)
27. M. Sun et al: ApJ **619**, 169 (2005)
28. M. Sun et al: ApJ in press (2007)
29. P. Thomas et al: MNRAS **222**, 655 (1986)
30. L. Van Speybroeck et al: SPIE **3113**, 89 (1997)
31. A. Vikhlinin et al: ApJ (Letters) **555**, 87 (2001)
32. N. Yamasaki, T. Ohashi, T. Furusho: ApJ **578**, 833 (2002)

Morphological Signatures of AGN Feedback in the Hot Interstellar Medium of Normal Elliptical Galaxies

S. Diehl[1] and T. S. Statler[2]

[1] Los Alamos National Laboratory, diehl@lanl.gov
[2] Ohio University, statler@ohio.edu

Summary. We present the results of a systematic morphological analysis of the X-ray emission of the hot interstellar medium (ISM) in 54 normal elliptical galaxies observed with the *Chandra* satellite. We find evidence for the gas not to be resting completely in hydrostatic equilibrium. In particular, we find that optical and X-ray ellipticities are uncorrelated, contrary to expectations from hydrostatic equilibrium. We also demonstrate that the degree of asymmetry in the gas correlates positively with both the 20 cm radio luminosity and the X-ray luminosity of the central point source, emphasizing the general importance of AGN feedback even in normal elliptical galaxies.

1 Introduction

With the launch of the new generation X-ray satellites *Chandra* and *XMM-Newton*, our understanding of normal elliptical galaxies has leaped forward. Detailed studies of individual X-ray bright elliptical galaxies have revealed an intricate interplay between the central radio source and the hot ISM. E.g. in NGC 4374 [8] and NGC 4472 [3], radio lobes fill cavities in the hot gas, similar to the situation in clusters of galaxies [4]. O'Sullivan et al. find evidence for an AGN triggered plume in NGC 4636 [11], entraining material from the core. In addition, some recent work by Allen et al. [1] on X-ray bright ellipticals suggests a tight correlation between Bondi accretion rate and jet power, suggesting a comparatively simple accretion mode.

Thus, for the case of X-ray bright elliptical galaxies, it is becoming evident now that they are not simple hydrostatic objects, but far more complex, and that their central AGNs can have a profound impact on the energetics and morphology the hot ISM. However, Best et al. [2] have shown that intermittent AGN activity has at least the potential to offset cooling in elliptical galaxies in general. Yet no *systematic* effort has been undertaken to use an ensemble of these new high-resolution observations to characterize morphologies of elliptical galaxies as a whole.

We present a morphological survey of a sample of 54 elliptical galaxies available in the *Chandra* archive and try to address some fundamental open questions that have been left unanswered so far: If we restrict our in-depth analysis to X-ray bright galaxies, are we biasing our view toward "interesting object"? Is there evidence for AGN feedback in galaxies that are not strongly disturbed in the X-ray? If there is significant interaction with the AGN, what are the implications for hydrostatic equilibrium and mass determinations? And finally, is the assumption of hydrostatic equilibrium testable?

2 Methods

2.1 Isolating the Hot Gas Emission

The X-ray emission from normal elliptical galaxies is a composite of two distinct components: point sources, mainly consisting of low-mass X-ray binaries (LMXBs) and diffuse emission from a hot interstellar medium. As the integrated luminosities of both components are comparable in many objects, it is essential to cleanly spatially separate these two components and to reveal the gas morphology alone.

We first remove resolved point sources and fill the holes by interpolating the local diffuse emission. However a large fraction of the point source component is generally below *Chandra's* detection limit and has to be removed in a different manner. To achieve this goal, we have developed a new technique, based on the fact that the observed X-ray image is a linear combination of the two components. By splitting the observation into a soft band S (0.3−1.2keV, dominated by the gas) and a hard band H (1.2 − 5keV, dominated by LMXB emission), we are then able to completely disentangle the hot gas G from the remaining unresolved point sources P:

$$S = \gamma P + \delta G, \tag{1}$$
$$H = (1 - \gamma)P + (1 - \delta)G; \tag{2}$$
$$\text{thus,} \quad G = \frac{1 - \gamma}{\delta - \gamma} \left[S - \left(\frac{\gamma}{1 - \gamma} \right) H \right]. \tag{3}$$

Here, S and H denote the background and exposure map corrected soft and hard band images, while P and G are the unresolved LMXB and gas components. The linear coefficients δ and γ are the softness parameters (soft/hard flux) for gas and point sources, respectively, and determined by spectral fitting. For more details on this technique, please refer to [7].

To be able to extract any morphological information out of these sparse images containing both positive and negative data, we apply an adaptive binning technique based on weighted Voronoi tesselations [6, 5], which is also publicly available at http://www.phy.ohiou.edu/~diehl/WVT/.

2.2 Quantifying Morphology

We first characterize the overall shape of the emission by fitting the hot gas image with elliptical isophotes. This procedures enables us to investigate ellipticity and position angle as a function of radius and to compare them to available optical photometry.

While ellipticities are often a good first approximations, many objects exhibit signs of morphological disturbances. Figure 1 shows only a small subset of our sample, but already indicates the variety and wide range of morphologies that we are finding. To measure the deviation from a smooth simple elliptical model, we define the asymmetry index η the following way:

$$\eta = \frac{1}{N} \sum_{i=1}^{N} \left[\left(\frac{G_i - M_i}{M_i} \right)^2 - \left(\frac{\sigma_{G,i}}{M_i} \right)^2 \right]. \tag{4}$$

Thus, this asymmetry index is the deviation of the gas image G_i from the model M_i, corrected for the statistical noise $\sigma_{G,i}$ expected due to the Poissonian nature of the data, and averaged over all pixels i. Tests show that η is a good measure of asymmetry and independent of non-morphological parameters such as luminosity, exposure time, radial extent, etc.

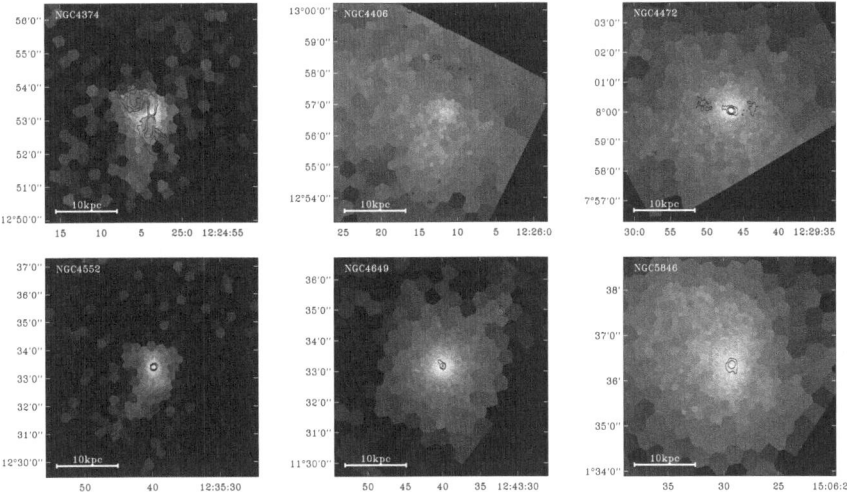

Fig. 1. A subsample of 6 galaxies that have radio data from the FIRST survey, showing a wide variety of X-ray morphologies. The adaptively binned X-ray gas images are all scaled to the same physical scale (35 kpc across), with radio contours overlaid. Only in few cases, the depth of the radio observations are sufficient to unambiguously identify interaction of the radio source and the interstellar medium.

3 Results

3.1 Gas is not completely hydrostatic

Although Figure 1 only shows a small subset of our 54 galaxy sample, it already gives a taste of the various morphologies that can be seen. The full sample is shown in [7]. A comparison of optical and X-ray morphologies shows that they are essentially decoupled. It is *impossible* to predict the distribution of the X-ray emitting gas from the optical appearance of the galaxy and vice versa.

To substantiate this claim, we extract ellipticities between 0.8 and 1.2 2MASS J-band effective radii (R_J). Within this radius, the gravitational potential is generally believed to be dominated by the stellar component [10, 9]. One can show that, for a variety of mass models, one expects $\epsilon_X \approx \epsilon_{\mathrm{opt}}$ if stars dominate the mass within R_J. The data are inconsistent with the expected strength of correlation (Figure 2) unless there is a substantial dark matter component whose shape is utterly unrelated to that of the stars. This is unlikely, since stars and dark matter obey identical collisionless dynamics, and therefore should have similar distributions following a major event, such as a merger. A more likely alternative is that hydrostatic equilibrium does not hold precisely enough to allow the shape of the potential to be inferred from the shape of the gas isophotes.

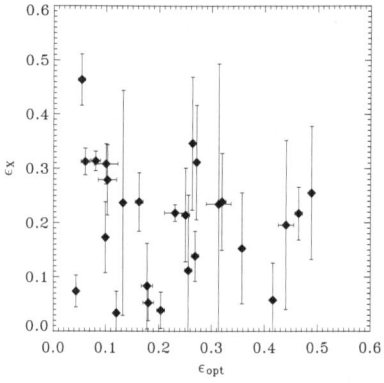

Fig. 2. X-ray gas ellipticitiy as a function of optical ellipticity, both evaluated between $0.8 - -1.2$ J-band effective radii from 2MASS. For simple hydrostatic models in stellar dominated potentials, a $\epsilon_X \propto \epsilon_{\mathrm{opt}}^{1/3}$ correlation is expected

3.2 AGN disturbs the Gas

To understand the lack of correlation between optical and X-ray ellipticities, we look at the asymmetry index η. We find that the asymmetry index is correlated with two independent measures of the central AGN activity: the *Chandra* X-ray luminosity of the central point source and the 20 cm radio luminosity from NVSS. This correlation suggests that the central AGN is responsible for pushing the hot gas around. However, as Figure 1 shows, there are only few cases where the X-ray and radio morphologies are tied together in an obvious way. We already show the majority of galaxies with (shallow) high-resolution VLA radio data available; deeper radio observations are still needed to confirm the direct interaction of the radio sources with the interstellar medium. The relatively large scatter in both AGN-asymmetry correlations could be interpreted as a sign of the intermittent nature of this interaction process.

We also find that the ellipticity of the X-ray emitting gas actually correlates with its degree of disturbance, indicating a common cause for both morphological parameters. This also emphasizes that simple elliptical models are not appropriate to characterize the shape of the gas emission.

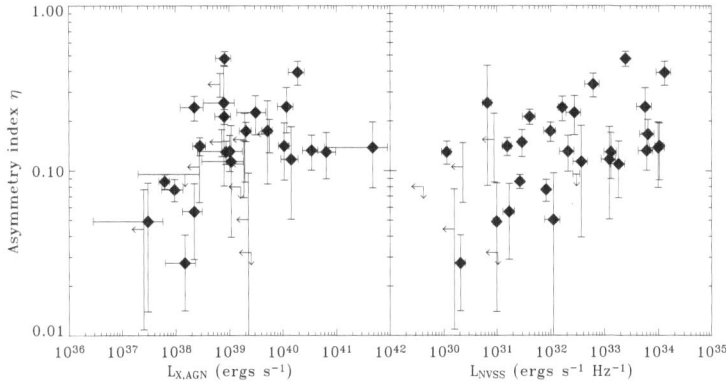

Fig. 3. Asymmetry index η as a function of AGN luminosity (left: central *Chandra* X-ray point source, right: 20 cm radio luminosity from NVSS). In both cases, one can see a clear positive correlation, indicating that galaxies with stronger AGN are more disturbed. A combined spearman rank analysis on both plots rejects the null-hypothesis on the $> 99.99\%$ level.

4 Conclusions

By investigating the morphology of the hot gas in a large sample of elliptical galaxies, we find that the hot gas is in general not completely in hydrostatic equilibrium. Instead we find signs that the hot gas is continually stirred and redistributed by intermittent outbursts of the central AGN. This may affect individual total mass measurements based on the assumption of hydrostatic equilibrium.

Taken together with the result by Statler & Diehl ([12], this volume), linking the central thermodynamic structure of the hot gas to AGN properties as well, this AGN–morphology connection may imply that the AGN may indeed be generally responsible not only for shaping, but also for reheating the gas in normal elliptical galaxies.

References

1. Allen, S. W., Dunn, R. J. H., Fabian, A. C., Taylor, G. B., & Reynolds, C. S., 2006, MNRAS, 372, 21
2. Best, P. N., Kaiser, C. R., Heckman, T. M., & Kauffmann, G., 2006, MNRAS, 368, L67
3. Biller, B. A., Jones, C., Forman, W. R., Kraft, R., & Ensslin, T., 2004, ApJ, 613, 238
4. Bîrzan, L., Rafferty, D. A., McNamara, B. R., Wise, M. W., & Nulsen, P. E. J., 2004, ApJ, 607, 800
5. Cappellari, M. & Copin, Y., 2003, MNRAS, 342, 345
6. Diehl, S. & Statler, T. S., 2006a, MNRAS, 368, 497
7. Diehl, S. & Statler, T. S., 2006b, ApJ, submitted
8. Finoguenov, A. & Jones, C., 2001, ApJL, 547, L107
9. Humphrey, P. J., Buote, D. A., Gastaldello, F., Zappacosta, L., Bullock, J. S., Brighenti, F., & Mathews, W. G., 2006, ApJ, 646, 899
10. Mamon, G. A. & Łokas, E. L., 2005, MNRAS, 362, 95
11. O'Sullivan, E., Vrtilek, J. M., & Kempner, J. C., 2005, ApJL, 624, L77
12. Statler, T. S., & Diehl, S., 2006, ESO Astrophysics Symposia: Heating vs. Cooling in Galaxies and Clusters of Galaxies (Springer Verlag)

Thermal Structure of the Hot ISM in Normal Ellipticals: Evidence for Local and Distributed AGN Heating

T. S. Statler[1] and S. Diehl[1,2]

[1] Astrophysical Institute, Ohio University; statler@ohio.edu
[2] Los Alamos National Laboratory; diehl@lanl.gov

Summary. A comprehensive survey of the morphology of the hot interstellar medium (ISM) in normal elliptical galaxies shows that these systems are, at best, only approximately in hydrostatic equilibrium, and that the degree of morphological disturbance is tightly correlated with the radiative luminosity of the central active galactic nucleus (AGN). In this contribution we show that the temperature profiles in these systems can be outwardly rising, outwardly falling, both, or neither; and that the sign of the temperature gradient in the main bodies of the galaxies is strongly correlated with AGN power. This strengthens the view that AGN are responsible for stirring up the hot ISM in nearly all ellipticals, and further suggests that the role and/or mode of AGN feedback qualitatively differs in systems with weak and strong AGN.

1 Introduction

The primary bottleneck in understanding the formation and evolution of galaxies is the physics of feedback. Feedback, from stellar winds, supernovae, or accreting supermassive black holes (BHs), acts to oppose radiative cooling, and thus inhibits flow of baryonic matter into dark matter (DM) potential wells, limiting star formation and BH growth. Feedback impinges directly on the interstellar medium (ISM) in galaxies and on the intracluster medium in clusters, determining, in large part, the entropy distribution of the gas and the observable X-ray properties.

In normal elliptical galaxies—those not at the centers of rich clusters—the effects of feedback are not always conspicuous. These systems have often been thought of as quiescent, with the hot ISM in hydrostatic equilibrium [12, 4, 13, and references therein]. This view has become increasingly untenable, due in part to *Chandra* observations of morphological disturbances associated with radio jets and lobes in X-ray-bright galaxies. In addition, a comprehensive study of gas morphology that includes lower luminosity and lower radio power objects shows that, in general, hydrostatic equilibrium holds at best only approximately, and that morphology is correlated with AGN power [8, 5, 6]. In this contribution we discuss the thermal profiles of the hot ISM in these galaxies. We show that systematic changes in these profiles as a function of AGN power signal a qualitative change in the role or mode of AGN feedback.

2 A *Chandra* Archive Survey of Normal Ellipticals

We have studied the X-ray gas morphology in 54 normal ellipticals in the *Chandra* archive [5], spanning a range of environments, from isolated field objects to group members, group-dominant galaxies and galaxies in poor clusters. We isolate the diffuse emission from the hot ISM by a novel technique that removes the contribution from both resolved and unresolved X-ray binaries. We find that highly disturbed systems with relatively luminous AGN are just the high-power end of a continuum. The morphological asymmetry is strongly correlated with 20 cm radio power and the X-ray luminosity of the central point source, both of which are AGN indicators. This correlation persists all the way down to very weak AGN, suggesting that, even in these systems, intermittent AGN outbursts continually stir the gas [5, 8].

3 Temperature Gradients and AGN Power

Clusters, groups, and X-ray luminous galaxies have long been known to have cool centers and outwardly rising temperatures. But not all normal ellipticals have such positive radial temperature gradients [7]. The left panel of Figure 1 shows examples of positive, negative, hybrid (sign-changing), and nearly isothermal profiles. These differences have been noted previously: Fukazawa et al. [9] suggest that they are linked to a difference in radial structure (extended vs. compact), with rising temperatures produced by intragroup gas; Humphrey et al. [10] argue that the transition from positive to negative gradients happens at a particular mass scale that marks a division between normal isolated galaxies and fossil groups.

We do not see support for either of these claims in the data. In particular, we find that environment plays a role only in the temperature gradient at large radii. The upper right panel of Figure 1 shows a clear correlation between the gradient measured between 2 and 4 J-band effective radii and the number density of neighbor galaxies. By contrast, the gradient measured in the main body of the galaxies ($< 2R_J$) is uncorrelated with local galaxy density (lower right of Figure 1).

The inner temperature gradients are most tightly correlated with AGN power. Figure 2 shows the inner gradient plotted against 20 cm power from the NVSS, which itself is strongly correlated with gas morphology. Thus, systems with higher AGN luminosities and morphologically disturbed gas have positive (outwardly rising) temperature gradients; systems with low AGN luminosities and less disturbed gas have negative (outwardly falling) gradients. And yet, both positive- and negative-gradient galaxies have radiative cooling times shorter than a Hubble time, and shorter than 10^8 yr inside a few kpc. Thus the gas must be continually reheated. The AGN-gradient relation implies that the AGN plays an important, and yet systematically varying, role along the sequence of normal ellipticals.

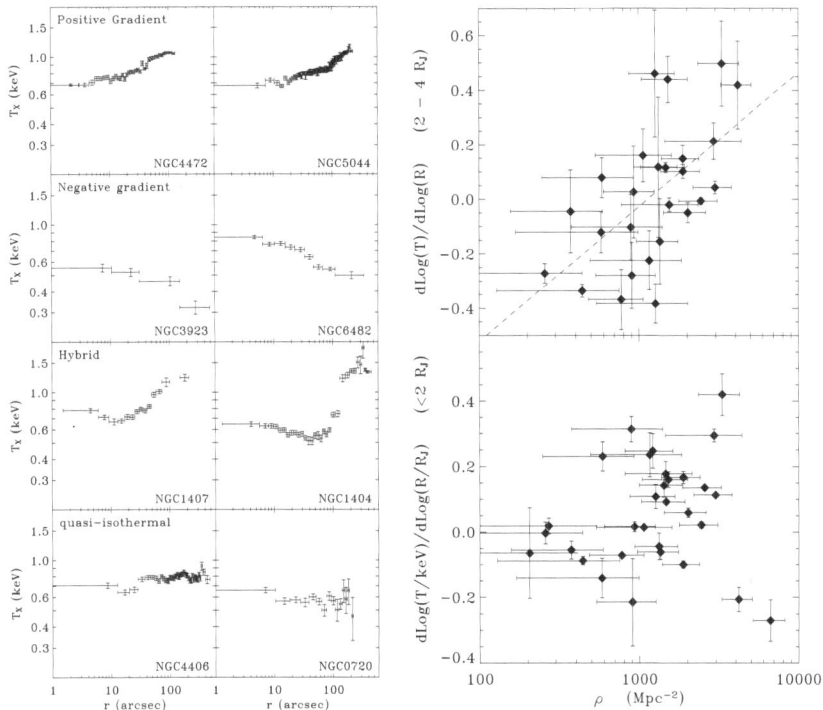

Fig. 1. *Left*: Examples of projected radial temperature profiles, divided into 4 major groups. *Right*: Gas temperature gradient plotted against projected number density of neighbor galaxies. *Upper Right*: Outer temperature gradient, measured between 2 and 4 J-band effective radii, is a strong function of environment, suggesting intragroup or intracluster gas dominating around $4R_J$. *Lower Right*: Inner gradient, measured inside $2R_J$, is not dependent on environment. (Figures from [7].)

4 Discussion

Chandra has revealed that the hot ISM in normal elliptical galaxies is not hydrostatic, and almost always disturbed to some degree. In these systems the dynamical times, sound crossing times, cooling times, and AGN cycle times are all $\sim 10^8$ yr. This is different from the situation in clusters, where the characteristic timescales of the observable features can be much longer, and consequently the AGN can look like a steady heat source to the bulk of the intracluster medium. In normal ellipticals, the observed morphological disturbances are dynamically evolving, together with the cooling of the gas and the refueling of the AGN. AGN may not need to drive massive outflows in order to keep the gas mechanically and thermally agitated.

What, then, is the difference between the weak-AGN systems with negative temperature gradients and the stronger-AGN systems with positive gradients? We can suggest four possibilities [7]:

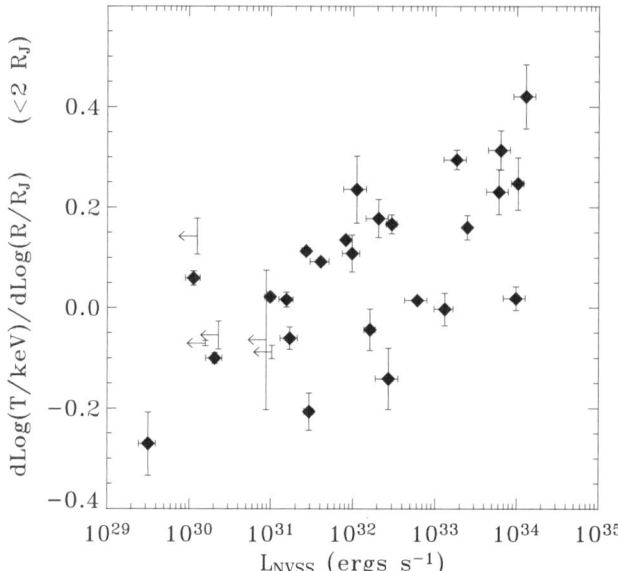

Fig. 2. Temperature gradient measured interior to $2R_J$ plotted against 20 cm radio power from the NVSS. The correlation is significant at the 99.9% confidence level, strongly suggesting that the temperature profile in the main bodies of normal ellipticals is influenced by AGN. (From [7].)

1. Compressive heating in a slowly cooling inflow may be able to offset radiative losses for cool gas in steep gravitational potentials [13]. This counter-intuitively produces a cooling flow that heats its own center. Perhaps the least disturbed systems settle into such a slow evolution, with AGN feedback perpetually insignificant. Khoshroshahi et al. [11] show that the temperature profile of NGC 6482 (the roundest and most symmetric object in our sample) can be reproduced with a steady-state cooling flow model.

2. Energetically, type Ia supernovae could play a significant role in heating the gas. A back-of-the-envelope estimate shows that the SNIa mechanical luminosity is comparable to the X-ray luminosity of the gas for systems near the transition from positive to negative gradients. Early models involving SN feedback [3] were actually dismissed on the grounds that they produced negative gradients. However, these models are sensitive to the assumed SN rate, and require fine-tuning [13].

3. AGN may heat the ISM cyclically, and the negative-gradient systems may represent the "AGN-off" phase of such a heating/cooling cycle. Evidence for this scenario comes from estimates of the power required to inflate cavities in the X-ray gas and its correlation with black hole mass [1, 2]. In this picture we would expect individual galaxies to follow cyclic tracks in the parameter space of Figure 2.

4. The spatial deposition of AGN energy in the ISM may be directly related to AGN luminosity, with weak AGN heating locally and higher-luminosity sources feeding jets that inflate large cavities and distribute the heat globally. This is consistent with the observation that smaller galaxies have weaker AGN and less extended radio emission. In this scenario, weak AGN would still be disturbing the gas, but on a scale and at a surface brightness level that is simply less detectable.

5 Conclusions

The hot ISM in normal elliptical galaxies is only approximately in hydrostatic equilibrium. Its morphology is being continually modified by the action of AGN. The gas in galaxies witih less luminous AGN is on average cooler, rounder, and more symmetric than that in systems with more powerful AGN. Moreover, negative temperature gradients are found in the main bodies of galaxies with weak AGN and small morphological disturbance. These gradients are not determined by environment, and do not indicate a special mass scale. Instead, the transition from positive to negative gradient most likely signals a qualitative change in the role or mode of AGN feedback.

Acknowledgement. This work was supported by the National Aeronautics and Space Administration through Chandra Awards G01-2094X and AR3-4011X, issued by the Chandra X-Ray Observatory Center, which is operated by the Smithsonian Astrophysical Observatory for and on behalf of NASA under contract NAS8-39073, and by National Science Foundation grant AST-0407152.

References

1. Allen, S. W., Dunn, R. J. H., Fabian, A. C., Taylor, G. B., and Reynolds, C. S, 2006, MNARS, 372, 21
2. Best, P. N., Kaiser, C. R., Heckman, T. M., and Kauffmann, G, 2006, MNRAS, 368, L67.
3. Binney, J., & Tabor, G., 1995, MNRAS, 276, 663.
4. Buote, D. A., & Canizares, C. R., 1994, ApJ, 427, 86.
5. Diehl, S. and Statler, T. S., 2006, ApJ, submitted.
6. Diehl, S. and Statler, T. S., 2007b ApJ, in preparation.
7. Diehl, S. and Statler, T. S., 2007c, ApJ, in preparation.
8. Diehl, S. and Statler, T. S., 2007a, ESO Astrophysics Symposia: Heating vs. Cooling in Galaxies and Clusters of Galaxies, (Springer Verlag).
9. Fukazawa, Y., Botoya-Nonesa, J. G., Pu, J., Ohto, A., & Kawano, N., 2006, ApJ, 636, 698.
10. Humphrey, P. J., Buote, D. A., Gastaldello, F., Zappacosta, L., Bullock, J. S., Brighenti, F., and Mathews, W. G., 2006, ApJ, 646, 899.
11. Khosroshahi, H. G., Jones, L. R., & Ponman, T. J., 2004, MNRAS, 349, 1240.
12. Mathews, W. G., 1978, ApJ, 219, 413.
13. Mathews, W. G., & Brighenti, F., 2003 ARAA, 41, 191.

Radio-Loud AGN Heating in Elliptical Galaxies and Clusters

P. N. Best

Institute for Astronomy, University of Edinburgh, Royal Observatory, Blackford Hill, Edinburgh EH9 3HJ, UK pnb@roe.ac.uk

Summary. The role of radio-loud active galactic nuclei (AGN) in continually re-heating the cooling gas in the haloes surrounding elliptical galaxies, and in clusters of galaxies, is investigated. The prevalence of radio-loud AGN activity amongst elliptical galaxies is found to scale strongly with the mass of the central black hole (as $M_{\rm BH}^{1.6}$). Combining this result with estimates of the mechanical energy output of these radio sources, the time-averaged energy output associated with recurrent radio source activity is derived: for elliptical galaxies of all masses, this is shown to balance the radiative energy losses from the hot gas surrounding the galaxy. Recurrent radio-loud AGN activity may therefore provide a self-regulating feedback mechanism capable of controlling the rate of growth of galaxies. The prevalence of radio-loud AGN activity amongst brightest cluster galaxies, and other cluster members is also determined, to investigate AGN feedback in galaxy clusters. Unless the efficiency of converting AGN mechanical energy into heating increases by 2–3 orders of magnitude between groups and rich clusters, radio–mode heating will not balance radiative cooling in systems of all masses.

1 Introduction

In recent years it has become apparent that active galactic nuclei (AGN) may play an important role in the process of galaxy formation and evolution. Essentially all nearby galaxies contain a massive black hole at their centre, the mass of which is tightly correlated with that of the stellar bulge of the galaxy (e.g. [12]). Furthermore, the cosmic evolution of the star formation rate, and that of the black hole accretion rate, match each other remarkably well back to at least $z \sim 2$. These results imply that the build-up of a galaxy and that of its central black hole are fundamentally linked. Theoretical interpretations of this favour models in which 'feedback' from the growing black holes is responsible for controlling star formation in the host galaxy (e.g. [16, 10]).

A related issue is that models of galaxy formation have had a long–standing problem in explaining the properties of the most massive galaxies. Unless the cooling of gas is somehow switched off, then semi–analytic models of galaxy formation predict that massive galaxies will continue to accrete gas and form stars in the nearby Universe, leading to much larger masses and bluer colours than observations dictate. The incorporation of AGN feedback into these models has now been shown to provide a good solution to these problems (e.g. [7, 9]), and evidence is growing which suggests that it is the radio–loud AGN which play the dominant feedback role. A number of critical questions remain, however. If radio-loud AGN feedback is responsible for controlling galaxy growth, what are the physical processes involved, and by what mechanism is the activity triggered? Do the energetics work out correctly: does the energy provided by AGN match that required to balance the radiative cooling losses? In what galaxies is this feedback process important, and how do the properties change with galaxy mass? Is radio-loud AGN feedback also energetically feasible on the scales of galaxy groups and clusters? Addressing these questions is the goal of the current contribution.

2 The host galaxies of radio-loud AGN

It has long been known that radio–loud AGN are preferentially hosted by massive elliptical galaxies. The availability of the SDSS (e.g. [19, 17]), together with the large–area radio surveys NVSS [8] and FIRST [1], has for the first time allowed much more detailed statistical study of these host galaxies: not only can large samples of radio–loud AGN be constructed, but it is also possible to compare the host galaxies of the radio–loud AGN with those of the rest of the galaxy population. In this way, both the origin of the radio–loud AGN activity, and the role that radio–loud AGN can have on the evolution of their host galaxies, can be investigated.

Best et al. [2] cross-compared the main galaxy sample of the second data release of the SDSS, with a combination of the NVSS and FIRST radio surveys designed to optimise the advantages of each radio survey. They defined a sample of 2215 radio–loud AGN[1], which forms the basis sample for the analysis here. The left panel of Figure 1 shows the fraction of galaxies with redshifts $0.03 < z < 0.1$ that are classified as radio–loud AGN (with $1.4\,\mathrm{GHz}$ radio luminosity above 10^{23}, 10^{24} and $10^{25}\,\mathrm{W\,Hz^{-1}}$), as a function of the stellar mass of the galaxy. The fraction rises from 0.01% of galaxies with stellar mass $3 \times 10^{10} M_\odot$ up to over 30% of galaxies more massive than $5 \times 10^{11} M_\odot$. The right panel shows the equivalent relations as a function of black hole mass. Once again, a strong trend with mass is seen; the slope of the relation is $f_{\mathrm{radio-loud}} \propto M_{\mathrm{BH}}^{1.6}$.

Figure 2 shows the fraction of the SDSS galaxies which are radio–loud AGN brighter than a given radio luminosity, as a function of radio luminosity, for six bins in black hole mass. A remarkable feature is that the shape of the functions are very similar for all mass ranges: there is no evidence for any dependence of the break luminosity on mass. Thus, whilst the *probability* of a galaxy becoming a radio source is a very strong function of its mass, the *luminosity* of the radio source that results is independent of mass.

---

[1] These mostly have low radio luminosities, $L_{1.4\mathrm{GHz}} < 10^{25}\,\mathrm{W\,Hz^{-1}}$, typical of Fanaroff & Riley Class 1 sources. The results presented in this paper are only relevant for these sources, and may not be applicable to the more powerful FR 2s.

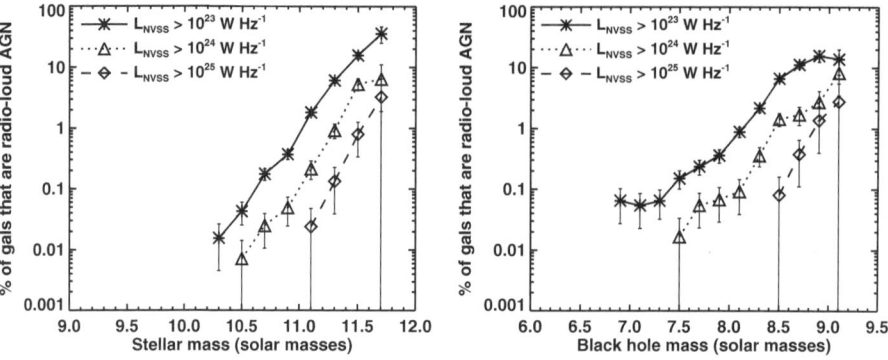

Fig. 1. The fraction of galaxies which are radio–loud AGN, as a function of stellar mass (left) and black hole mass (right), for different cuts in radio luminosity.

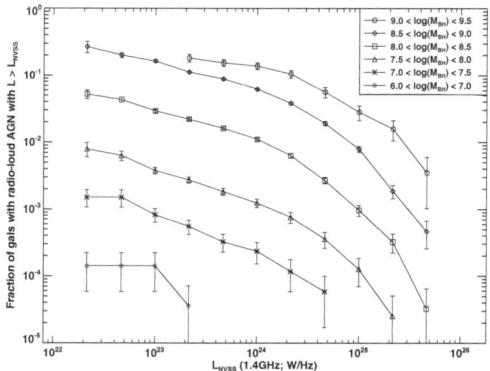

Fig. 2. The fraction of radio–loud AGN brighter than a given radio luminosity, as a function of black hole mass.

3 Re–triggered activity, and AGN heating rates

The strong mass dependence of the radio–loud AGN fraction contrasts starkly with the lack of mass dependence in the fraction of galaxies which harbour AGN selected from optical emission lines [11, 3]. In addition, Best et al. [3] show that the probability of a galaxy hosting a radio–loud AGN is independent of whether or not it is classified as an emission–line AGN. It can thus be concluded that emission–line AGN activity and low–luminosity radio activity are independent physical processes. Low luminosity radio activity appears to be associated with the re-fuelling of already well–formed massive black holes. Best et al. [3] argue that this activity is associated with the accretion, through a radiatively inefficient accretion flow, of hot gas cooling out of the hot X–ray haloes that surround the massive galaxies (or groups / clusters of galaxies).

They show that this provides a natural explanation for the steep power law dependence of the radio–loud AGN fraction with black holes mass.

Radio–loud AGN are believed to live for only 10^7 to 10^8 years, and yet at the highest masses the fraction of galaxies that are radio–loud rises to $> 25\%$. This implies that radio source activity must be constantly re–triggered. The data of Figure 2 can therefore be interpreted *probabilistically* as the *fraction of its time* that a galaxy of given black hole mass spends as a radio source brighter than a given radio luminosity.

To determine the heating effects of radio–loud AGN, what is then needed is an estimate of the mechanical energy output of radio sources. Although monochromatic radio luminosity accounts for typically less than 1% of the radio source energy, it is possible to use this to make a good first–order estimate of the mechanical energy. Bîrzan et al. [6] studied a large sample of cavities and bubbles produced in clusters and groups of galaxies due to interactions between radio sources and the surrounding hot gas. They derived the (pV) energy associated with each cavity, and also estimated the cavity ages. Best et al. [4] showed that although there is no single scaling factor between the mechanical luminosities so derived and the radio luminosities of the radio sources producing the bubbles, it is nonetheless possible to obtain a reasonable fit to the data using a simple power-law relation: $L_{\mathrm{mech}}/10^{36}\mathrm{W} = (3.0 \pm 0.2)(L_{1.4\mathrm{GHz}}/10^{25}\mathrm{WHz}^{-1})^{0.40\pm0.13}$

Using this conversion, the probability function for the time that a galaxy of given black hole mass spends as a radio source with a given radio luminosity can be converted into a probability function for mechanical luminosity. Integrating across this gives the time–averaged heating rate for black holes due to radio–loud AGN activity (see [4] for more details): $\bar{H} = 10^{21.4}(M_{\mathrm{BH}}/M_\odot)^{1.6}$ W. Note that the black hole mass dependence of this heating rate arises entirely from the variation of the radio–loud fraction with black hole mass (which is well determined from the observations) whilst the normalisation term encompasses the shape of the radio luminosity and the radio to mechanical luminosity conversion. Uncertainties in these can therefore be accounted for by introducing an uncertainty factor f (which should have a value close to unity) into the equation: $\bar{H} = 10^{21.4}f(M_{\mathrm{BH}}/M_\odot)^{1.6}\mathrm{W}$.

4 The heating–cooling balance in elliptical galaxies

This heating rate due to radio–loud AGN activity can be compared with the radiative cooling rates of the haloes of hot gas surrounding elliptical galaxies, as determined using X–ray observations. Figure 3 shows the bolometric X–ray luminosities (L_X) of the 401 early–type galaxies from the sample of O'Sullivan et al. [14], compared with their optical luminosities (L_B). Also shown are the predicted radio–loud AGN heating rates for three different assumed values of f. It is remarkable how well the predicted radio–loud AGN heating rates for $f = 1$ match the radiative cooling rates at all optical luminosities. Provided that all or most of the energetic output of the radio–loud AGN can be coupled effectively to their host galaxy then recurrent radio–loud AGN activity provides a sufficient feedback effect to suppress the cooling of gas and thus control the rate of growth of massive elliptical galaxies.

5 The heating–cooling balance in groups and clusters

Radio–loud AGN activity in brightest cluster galaxies (BCGs) has been proposed as a potential solution to the cooling–flow problem: the amount of gas cooling in cluster cores is only about 10% of that predicted for a classical cooling flow, implying that some heating source must balance the radiative cooling losses, preventing further gas from cooling. Is radio–loud AGN feedback an energetically viable candidate for this heating source?

von der Linden et al. [18] derived a robust sample of 625 groups and clusters from the SDSS, based upon the 'C4' cluster catalogue of Miller et al. [13]. Best et al. [5] showed that the radio-loud fraction of these BCGs is higher than that of 'all galaxies' at all stellar masses, by an order of magnitude for stellar masses below $10^{11} M_\odot$, but less than a factor of two higher at stellar masses above $5 \times 10^{11} M_\odot$. They also showed that the distribution of radio luminosities for the BCGs is the same as that for 'all galaxies', and also doesn't vary significantly with the velocity dispersion of the host group or cluster [5].

The same approach as discussed above can be adopted to determine the time–averaged heating rate of the BCGs: $\bar{H}_{\mathrm{BCG}} = 1.7 \times 10^{35} f (M_*/10^{11} M_\odot)$W. The combination of the boosting of the radio-loud fraction for BCGs and their typically higher masses compared to other cluster galaxies, means that the radio–loud AGN heating rate of BCGs exceeds that of all other galaxies within the cluster cooling radius combined (see [5] for a more detailed analysis). The AGN heating rate due to the BCG and all other cluster galaxies has only a weak dependence on cluster velocity dispersion, and so scales much more slowly with cluster mass than the radiative cooling losses (since X–ray luminosity scales roughly as σ^4; e.g. [15]). Thus, as shown in Figure 4, if the efficiency factor f does not change then AGN heating cannot precisely offset cooling flows in groups and clusters of all masses.

For a value of $f \approx 1$, the radio–loud AGN heating falls short of balancing cooling in the most massive systems, whilst overheating smaller clusters and groups [5]. In this case, thermal conduction could provide an additional source of heat in the most massive clusters, which would account for the shortfall of AGN heating in those systems; the overheating effect of AGN in smaller systems may be the origin of the break seen between the luminosity–temperature relation of intracluster gas and that of intragroup gas. Alternatively, the efficiency factor f would have to be dramatically lower in smaller systems, with much of the radio source energy deposited beyond the cooling radius.

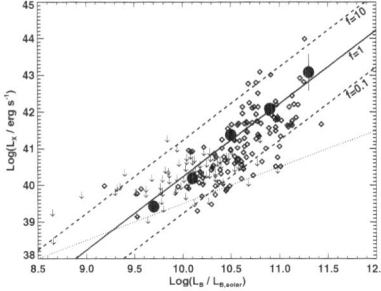 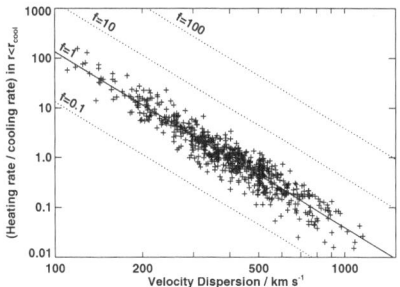

Fig. 3. A plot of L_X versus L_B for the elliptical galaxies in the sample of O'Sullivan et al. [14]. The large filled circles show the mean values of L_X for galaxies in 5 bins of L_B. The solid and dashed lines shows the predicted X–ray emission from the hot haloes of the ellipticals, assuming that heating from radio–loud AGN balances the cooling, for different values of f.

Fig. 4. The data points show the ratio of the time–averaged radio–AGN heating rate of the BCG, assuming $f = 1$, to the radiative cooling losses within the cooling radius, as a function of cluster velocity dispersion, for the 625 SDSS clusters. The solid line represents the fit to these points, and the dotted lines indicate where the data would be located for different values of f.

Acknowledgement. The author thanks Guinevere Kauffmann, Tim Heckman, Christian Kaiser and Anja von der Linden for their collaborative work on the various papers around which this paper is based.

References

1. Becker R. H., White R. L., Helfand D. J., 1995, ApJ, 450, 559
2. Best P. N., Kauffmann G., Heckman T. M., Ivezić Ž. 2005a, MNRAS, 362, 9
3. Best P.N. et al., 2005b, MNRAS, 362, 25
4. Best P.N., Kaiser C., Heckman T., Kauffmann G., 2006, MNRAS, 368, L67
5. Best P.N. et al., MNRAS, submitted; astro-ph/0611197
6. Bîrzan L. et al., 2004, ApJ, 607, 800
7. Bower R. G. et al., 2006, MNRAS, 370, 645
8. Condon J. J. et al., 1998, AJ, 115, 1693
9. Croton D. et al., 2006, MNRAS, 365, 11
10. Fabian A. C., 1999, MNRAS, 308, L39
11. Kauffmann G. et al., 2003, MNRAS, 346, 1055
12. Magorrian J. et al. 1998, AJ, 115, 2285
13. Miller C. J. et al., 2005, AJ, 130, 968
14. O'Sullivan E., Forbes D.A., Ponman T.J., 2001, MNRAS, 328, 461
15. Popesso P. et al., 2005, A&A, 433, 431
16. Silk J., Rees M., 1998, A&A, 331, L1
17. Stoughton C. et al., 2002, AJ, 123, 485
18. von der Linden A., et al., 2006, MNRAS, submitted; astro-ph/0611196
19. York D. G. et al., 2000, AJ, 120, 1579

"Radio–active" Brightest Cluster Galaxies

A. von der Linden[1], P. N. Best[2] and G. Kauffmann[1]

[1] Max Planck Institut für Astrophysik, Karl-Schwarzschild-Str. 1, Postfach 1317, 85741 Garching, Germany anja@mpa-garching.mpg.de

[2] SUPA, Institute for Astronomy, Royal Observatory Edinburgh, Blackford Hill, Edinburgh EH9 3HJ

Summary. We present a sample of clusters drawn from the C4 cluster catalog of the Sloan Digital Sky Survey, where particular emphasis has been placed on the correct identification of the Brightest Cluster Galaxy (BCG) and the determination of the velocity dispersion within the virial radius. In this sample, we investigate the occurence of radio-AGN activity in the BCGs. We find that the radio-loud fraction is enhanced by a factor of up to 10, compared to galaxies of the same mass that are not BCGs.

1 Selection of BCGs and Clusters

The basis of our cluster sample is the C4 cluster catalog [1], which is based on the spectroscopic data of SDSS and is currently available for DR3. It identifies clusters in a parameter space of position, redshift, and color. Due to fiber collisions in the SDSS (no two fibers can be placed within $55''$ of each other), the true BCG is not included in the SDSS spectroscopic data for about 30% of the dense clusters and is thus missed by the C4 algorithm. Although an earlier version of the C4 catalog for DR2 selected a BCG based on the photometric catalog, visual checks revealed that this did not provide a reliable BCG, e.g. the sample was contaminated by stars misclassified as galaxies, as well as spiral galaxies located at the edge of the cluster). In the following, we describe our method to improve the BCG identification.

To identify the BCGs for the C4 clusters at $z \leq 0.1$ (833 clusters), we draw a list of candidates, made up of the BCG(s) identified by C4, as well as bright bulge–dominated galaxies in the vicinity of the cluster center. By eye-balling thumbnail images of these candidates, we choose the BCG, based on its brightness and location in an area of strong clustering. With an iterative algorithm, we then determine the redshift, velocity dispersion, and R_{200} of the cluster, as measured from galaxies in the spectroscopic SDSS catalog within $\pm 3\sigma_v$ and R_{200} of the BCG position. After accounting for clusters associated with identical BCGs, clusters for which the iteration failed to keep more than 4 galaxies in the cluster, and after visual inspection of all the clusters, our cluster sample consists of 625 clusters.

2 Radio–Active BCGs

In [2], the radio–emitting galaxies within the main spectroscopic sample of the SDSS were identified by cross–comparing these galaxies with a combination of the NVSS and FIRST radio surveys. It is found that the fraction of galaxies that are radio–loud AGN is a strong function of the stellar mass of the galaxy (stellar masses taken from [4]). We have extended this study to our sample of BCGs and the non-BCG cluster galaxies, and find that at all masses, BCGs

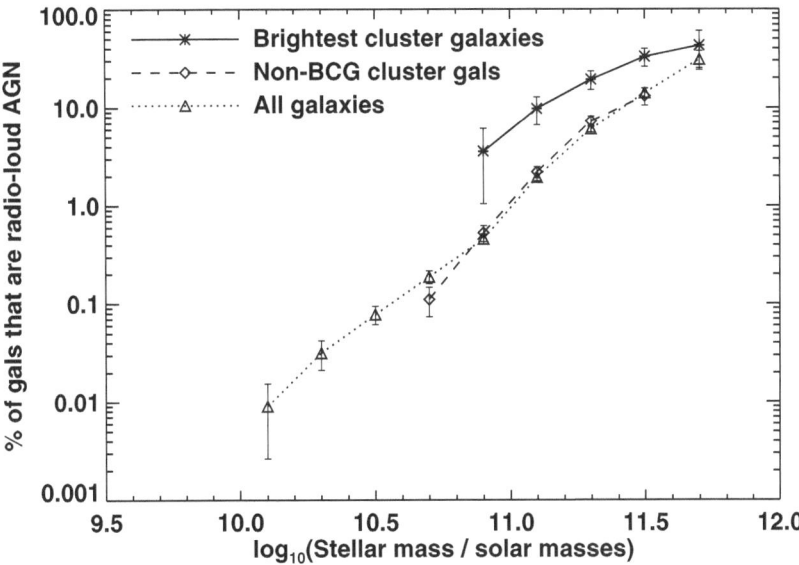

Fig. 1. The fraction of galaxies that are radio loud AGN, as a function of stellar mass. Shown are all galaxies at $z < 0.1$ (open triangles), and the BCGs (solid circles). Galaxies are considered radio–loud if their 1.4 Ghz radio luminosity is greater than 10^{23} W/Hz, and they are not classified as star-forming.

are more likely to be radio–loud than other galaxies of the same stellar mass (Fig. 1). This enhancement ranges from a factor of 10 at masses of $5 \times 10^{10} M_{\odot}$ to less than a factor of two in the highest mass bin of $5 \times 10^{11} M_{\odot}$.

3 Discussion

The properties of the central galaxies of clusters are tightly connected to their parent clusters, an aspect underlined by our findings: galaxy clusters efficiently provide gas for their central galaxy to fuel its central black hole.

Other cluster galaxies, not at the bottom of the potential well, are not affected. It is tempting to identify this additional gas supply for BCGs with the cooling flows observed in massive clusters and thus speculate that the heating due to the AGN activity balances the cooling rate, providing a solution to the cooling problem. However, we find no correlation of the radio–loud fraction with cluster velocity dispersion, whereas the cluster cooling rate (as indicated by their X-ray luminosity) scales steeply with it. While the AGN activity provides enough heating to balance cooling in isolated ellipticals (as demonstrated by [3]), the factor of 10 increase in radio-AGN activity in BCGs might be enough to balance cooling in galaxy groups and poor clusters, but not in the most massive clusters (which would require factors of 10–100). Thus, the increased radio–AGN activity of BCGs cannot account for (all) the heating needed to balance cooling flows in clusters.

References

1. C. J. Miller, R. C. Nichol, D. Reichart et al.: AJ **130**, 968 (2005)
2. P. N. Best, G. Kauffmann, T. M. Heckman, Ž. Ivezić: MNRAS, **362**, 9 (2005)
3. P. N. Best, C. R. Kaiser, T. M. Heckman, G. Kauffmann: MNRAS, **368**, L67 (2006)
4. G. Kauffmann, T. M. Heckman, S. D. M. White et al.: MNRAS, **341**, 33 (2003)

Part IV

Modelling the AGN-ICM Interaction

Review of Heating Mechanisms in Clusters of Galaxies

M. Ruszkowski

Max Planck Institute for Astrophysics, Karl-Schwarzschild-Str. 1, Postfach 1317, 85741 Garching, Germany mr@mpa-garching.mpg.de

Summary. There is growing consensus that active galactic nuclei (AGN) play a leading role in solving the riddle of cool core clusters of galaxies. I review the subject of heating of clusters paying particular attention to AGN, but also discuss alternative (or supplementary) heating mechanisms. I am arguing that while the bulk properties of cool flow clusters can be explained by invoking AGN, the precise nature of heating still remains to be fully understood. To address these problems from the theoretical point of view will require proper modelling of microphysics (plasma physics, MHD, transport processes) as well as connecting "small scale" central engine physics to cluster scales via large dynamical range numerical simulations.

1 Overall balance of heat

1.1 Cooling vs. heating. Are black holes big enough ?

In a significant fraction of clusters of galaxies the central cooling time of the intracluster gas (ICM) is shorter than the Hubble time. If unimpeded, cooling will lead to large mass accretion rates onto the cluster centers. However, *Chandra* and XMM *Newton* observations demonstrated that this is not the case. A likely explanation for this discrepancy (known as the "cooling flow problem") is that some source of heating is offsetting cooling and, thus, preventing uncontrolled accumulation of the gas and star formation. The most promising candidate for the heating source are Active Galactic Nuclei (AGN). One gram of matter accreted on a supermasive black hole (SMBH) in the centers of these objects can accelerate $10^4 (e_{kin}/0.01)\sigma^{-2}$ grams of surrounding material to escape speed [3]. This, combined with the Magorrian relation, $(M_{bulge} \sim 10^3 M_{bh})$ shows that growing black holes have the potential to significantly affect their environment and limit galaxy luminosity function ([14]). It is also quite likely that the same outflows can provide enough energy to solve both the cooling flow and and the entropy excess problem (see below). Nevertheless, it has been suggested that if black holes in the brightest cluster galaxies (BCGs) follow the $M - \sigma$ relation then their masses may be too small to offset cooling unless the efficiency for converting accreted mass to energy is extremely high [26]. However, there is theoretical (e.g., [6]) and observational evidence ([69], [39], [4], [68]) to suggest that BCGs have higher stellar luminosities than predicted from the standard Faber-Jackson relation. This suggests that black holes in BCGs could have masses larger than previously thought, although this problem is still a matter of debate (see, e.g., [2]). Direct measurements of black hole masses will settle this issue (see, e.g., [15])

1.2 Linking small and large scales

For the black hole heating mechanism to be sucessful there must be a physical link between the mechanical power of the AGN outflow and the supply of gas toward the cluster center. [1] demonstrate that there is a linear (in log space) relation between jet power and Bondi accretion rate. It remains to be verified if such a relation exists for BCGs and, if so, whether its properties can be reconciled with the AGN heating paradigm in clusters and groups. Nevertheless, it is remarkable that accretion rate prescription based on such simple physics holds for the galaxies that they considered. Irrespective of the connection between the power of SMBH jets and the amount of accreted mass, there is a clear trend on larger scales for the mechanical power in buoyant bubbles inflated by the jets to scale with cooling luminosity of the cluster gas. The most recent compilation of these measurements [53] shows that the mechanical power contained in these bubbles can successfuly compete with the cooling rate, although this conclusion depends somewhat on the equation of state inside the bubbles. On even larger scales, the mechanical energy injected by SMBH can distort scaling relations in groups. In an important paper, [13] decompose their group sample into objects that contain radio loud and radio quiet AGN and show that these two populations are clearly separated, with radio loud ones having larger temperatures (or lower luminosities) just as expected in case of AGN heating. This clearly demonstrates that AGN are the primary mechanism responsible for departures from the standard cluster/group scaling relations that result from virialization.

1.3 Semianalytic models

Semianalytic models have been devised to explain bulk properties of cool core clusters. An example of such a model is a "thermostat model" of [12] (see also [5]) where self-regulation of the cooling flow comes about as a result of coupling of injected mechanical energy via Bondi accretion rate. In a related approach, the effervescent heating model, [60]) show that models broadly consistend with the main observational properties of clusters can be constructed when heating comes from a combination of work associated with bubble expansion and buoyant motion and thermal conduction. This model predicts low mass deposition rates and temperature profiles consistent with observations. Recent attempts to compare the predictions of this simple model with observations ([29], [50]) show that successful fits can be obtained in 75% of objects. Other semi-analytical models have also been proposed and give reasonable agreement with observations (e.g., [36], [7]). The effervescent heating model has also been extended to address the entropy excess problem in clusters. [59] show that entropy measurements can be fitted by this model while satisfying the shapes of entropy profiles (i.e., the negative central entropy profiles are only transient features). An interesting prediction of this model is that masses of BCGs are expected to scale as $M_{bh} \sim M_{dm}^{\alpha}$, where $\alpha \sim 5/3$ and M_{dm} is the mass of dark matter halo. The model predicts SMBH masses somewhat higher than predicted from the $M - \sigma$ relation.

2 The devil is in the details

Despite the overall success of the AGN paradigm in explaining the bulk properties of cooling core clusters there remain a number of issues that need further examination. I will now discuss these issues in some detail.

2.1 Jets - momentum-driven stage

The initial stage of energy injection by AGN is the momentum-driven phase, where collimated jets pump mechanical energy into the cluster. Existing simulations still lack the dynamic range necessary to simultaneously model both the jet physics and outflows on cluster scales. However, some progress has been made in this direction. In an original contribution [55] studied jetted outflows in cluster atmospheres and the transition from the jet to the buoyant stage. One of the potential problems with the heating of clusters by unidirectional jets is that heat is not distributed uniformly and, thus, the cooling catastrophe may not be prevented. [66] considered jets in cluster atmospheres including cooling and found that such outflows are unable to stop accumulation of large amounts of gas in the cluster center unless some additional physics is included. Only the initial outburst has a significant effect on the cluster atmosphere but subsequent outflows propagate along the channel carved by the original outburst without heating the cluster. This work has been recently extended by [33] to cluster environments extracted from cosmological simulation. They showed that realistic initial conditions prevent the jet from following the "chimney" as it gets easily destroyed by random gas motions in the cluster. Thus, most outbursts are likely to encounter "pristine" cluster environment and efficiently heat the gas as long as the frequency of outbursts is not too high.

2.2 Waves - energy-driven stage

The momentum driven stage is followed by the energy-driven one. It is in this stage that significant amount of energy can be chanelled to weak shocks or sound waves. The exact fraction of the energy injected by SMBH that goes into this mode of heating depends on the details of accretion models but essentially boils down to how overpressured the initial injection region can get. Sound waves in clusters were first detected by [19] who proposed that their dissipation may provide significant source of heating. Other notable detections of sound waves or weak shocks are in Virgo cluster [24], MS0735.6+7421 [42] and Hercules A [46].

Transport processes and dissipation

In order for the energy of the wave to be efficiently converted to heat, a dissipating mechanism (such as viscosity) must be present. Some evidence for viscosity in the ICM comes from: (1) the suggestion that the flow around buoyant bubbles is not fully turbulent [20]: H-α emitting filaments that act as "particle" tracers appear to move along ordered (laminal) streamlines resambling those seen in laboratory experiments of low Reynolds number gas and (2) some bubbles appear to be coherent even far from the location of the AGN suggesting that some mechanism (viscosity?) supresses instabilities on their surfaces. Both arguments are not without their flaws. For example, the filaments are about 10^3 times denser than the the surrounding ICM and may stay laminar even if the ICM is turbulent, and a mechanism other than viscosity may suppress instabilities. Nevertheless, given current evidence, the "pro" arguments cannot be ruled out and viscosity, if present, has very appealing consequences. Interestingly, if viscosity is of the order of the Braginskii value then the characteristic dissipation length of sound waves is $L \sim 26\lambda_{10}^2 n_{0.02} T_4^{-2}$ kpc, where λ_{10} is the wavelength in tens of kpc, $n_{0.02}$ is the electron number density in 0.02cm^{-3} and T_4 is the temperature units of in 4 keV (Landau & Lifshitz). This means that most of the dissipation should take place right where it is needed, i.e., within the cooling cores of clusters. First simulations of this mechanism were performed by [61], [62] who showed numerically that this mode of heating provides spatially-distributed dissipation at the level comparable to local radiative cooling just as suggested by [19] and [21]. Other properties of the simulated weak shocks also are consistent with observations. For example, typical Mach numbers are slightly above unity. There is also tentative evidence that the characteristic interference pattern of waves originating from the bipolar AGN outflow have also been observed. (see temperature maps in [42]).

A natural consequence of viscosity should be non-negligible thermal conduction. In case of unmagnetized plasma, the ratio of conductive to viscous dissipation of waves scales as $(m_p/m_e)^{0.5}$ ([8], [21]) and conductive energy dissipation will dominate. Recent observations of shocks in the Perseus [22] and Abell 2199 [64] suggest that shocks may be isothermal. Isothermality could be achieved if significant thermal conduction operates in the ICM. However, counter examples are also known. [24] (see also this volume) show that temperature gradients exist across the shocks in the Virgo cluster (but obviously this observation does not rule out conduction). It has been suggested that wave dissipation model that incoporates Braginskii viscosity as the sole dissipation mechanism leads to unrealistic temperature profiles when integrated over long times [41]. However, [27] show that this problem can be alleviated when viscosity acts in conjunction with thermal conduction. Another possiblility is that the actual functional form of the transport coefficient is very different from the classical Braginskii formula.

2.3 Bubbles - buoyancy-driven stage

Where does the energy go to ?

Following the pressure-driven stage, bubbles begin to move due to buoyant force. There have been a number of numerical studies of this phase (see, e.g., [28] for a review). It is instructive to consider energy transfer between the bubble and the ICM during this stage. The total energy transferred from the bubble to the ICM is $\Delta W = -V(p)dp$, where V is the bubble volume and p denotes pressure (approximate pressure equilibrium between the bubble and the ICM defines the transition from the energy-driven stage to the buoyant one). This work includes both the pdV work and the drag force contribution. The fact that these two contributions are independent can be realized by considering two extreme example cases: (1) a soccer ball rising from the bottom of a swimming pool in which case no pdV work is done, and the energy release comes from lowering of the potential energy of water. In this case dissipation takes place almost entirely in the turbulent wake behind the ball. (2) An anchored baloon being inflated in which case only pdV work is done. This pdV work is first stored as the potential energy of the surrounding gas/water and then released. Whereas in the former case the presence of substantial viscosity is not crucial (the amount of energy dissipated in the turbulent wake does not crucially depend of the precise level of viscosity as long as it is low), in the latter case disssipation of pdV work can be converted to heat more efficiently when stronger microscopic dissipation mechanisms are present (note also that $\Delta W_{\mathrm{drag}} = (\gamma - 1)pdV$ and $W_{\mathrm{exp}} = pdV$ so, for $\gamma = 5/3$, both channels of energy dissipation are comparable).

Bubble stability - interplay between viscosity and magnetic fields

As mentioned above some fraction of the available energy can be dissipated even if viscosity is low. This can happen due to Richtmyer-Meshkov instability [32]. This instability occurs when a shock wave intercepts a pre-existing hot bubble. When this happens the wave produces a pressure gradient inside the bubble and propagates faster than its bulk velocity. This creates a vortex flow that traps the wave energy. This prevents wave energy from escaping the cluster cool core even if transport processes are weak. The efficiency of this mechanism depends on the volume filling factor of the bubbles and requires sufficiently frequent outbursts to guarantee that turbulence is fully developed. The latter requirement is necessary if one wants turbulence to dissipate energy efficiently; if this requirement is not met then subsequent wave passages through the bubbles just increase kinetic energy in the vortex and some viscosity is still needed to convert it to heat. It is also possible that buoyant bubbles are prevented from disruption by viscous forces. [56] performed a series of numerical experiments, compared inviscid and viscous cases and quantified how much viscosity is needed to prevent bubble disruption. For viscosity at the level of 25% of the Braginskii value they obtained results consistent with observations. The same problem has been considered analytically by [37] who computed instability growth rate as a function of scale and viscosity coefficient. An alternative possibility is that bubbles are made more stable due to significant deceleration during their initial evolution [51].

Whether or not buoyant bubbles are made stable by dynamical or viscous effects in a realistic situation is not entirely clear. However, the ICM and the bubbles themselves are known to be magnetized. Although magnetic fields in clusters are known to have plasma $\beta > 1$, they may in principle have a strong effect on suppressing Kelvin-Helmholz and Rayleigh-Taylor instabilities. [58] and [34] considered bubble evolution in a magnetized ICM and found that bubbles could be prevented from getting shredded even when β is as low as ~ 120. However, their simulations were performed for very idealized magnetic field configurations such as donout-shaped fields with symmetry axis parallel to the directon of motion. In such a case a vortex-like velocity field of rising bubbles is less likely to change the overall topology of the field and the bubbles may not be shredded. However, [63] found that only when the power spectrum cutoff of magnetic field fluctuations is larger than the bubble size can bubble shredding be suppressed. It is quite possible that such a draping case is not representative of typical cluster fields ([67]), in which case another mechanism, such as viscosity, would be required to keep the bubbles stable.

Transport processes in the presence of magnetic fields

Magnetic fields suppress transport processes. If the bubble instabilities are prevented by magnetic fields then they also reduce the amount of dissipation in weak shocks and mechanical energy transferred from the bubbles to the ICM via pdV work. This means that we place somewhat contradictory "demands" on our heating model. Is there any way to keep viscosity and conductivity at a high level even in the presence of magnetic fields ? In an influential paper, [44] estimated the level of thermal conductivity within the framework of Goldreich-Shidhar [30] theory and found out that suppression below the Braginskii value may be as little as factor of 5. This approach has been extended by [10] and [40] to include advective transport of heat. They show that turbulent motions can transport heat more efficiently than thermal electron motions. This may happen even when magnetic fields are present and turbulence is subsonic. For example, for stochastic fields of order a μGauss, gas velocity at the turbulence injection scale ~ 500km/s and turbulence injection length scale ~ 50kpc, we get $\kappa_e/\kappa_{Brag} \sim 0.1$ and $\kappa_{turb}/\kappa_{Brag} \sim 7$, where κ_{Brag} is the classical conduction coefficient for unmagnetized plasma. The turbulent conduction idea has recently been tested with *Chandra* by [54], who discuss diffusion of metallicity gradients near BCGs. They note that metallicity peaks are more extended than the distribution of light coming from BCGs and argue that this is due to turbulence driven by the central AGN. A similar approach has been adopted in the case of Centaurus cluster by [31], who find that length scales have to be ten times larger than those expected if inflating and rising bubbles are driving turbulence. They also show that turbulent conductivity alone is unable to quench cooling. However, there remains a possibility that substructure mergers could additionally contribute to driving turbulence, thus changing characteristic length scales.

Turbulence and turbulent heat transport could also be driven by magneto-thermal instability (MTI; [49], [9]). Even the presence of weak magnetic fields can dramatically alter convective stability criterion of stratified atmospheres. This instability can be thought of as an analog of well studied magneto-rotational instability (MRI) in the context of accretion disks. In both cases strong, magnetic fields stabilize short wavelength perturbations through magnetic tension. For large wavelength perturbations, the istability criterion is that pressure and temperature gradients are in the same direction. This means that the standard Schwarzschild instability criterion may not be adequate! Whereas in the centers of cooling flow clusters the MTI instability is not expected to operate (because the temperature gradient is positive there), it may be important in the outer layers. However, since AGN are copious sources of cosmic rays and cosmic ray presssure, $p_{\rm cr}$ further modifies the MTI criterion to $nk_B \nabla T + \nabla p_{cr} < 0$. This means that AGN may further contribute to turbulent transport in cool cores if the ICM is polluted by cosmic rays diffusing out of buoyant bubbles. Whether this happens would need to be checked in future numerical experiments. Diffusing cosmic ray protons will interact with thermal nucleons of the ICM to produce pions. The decay of pions will lead to γ-rays that could be detectable with *GLAST*. If cosmic rays are not efficiently diffused from bubbles then shock waves passing through them will reenergize the plasma inside them, producing "fossil" radio plasma [17]. Cosmic rays have also been considered as a possible (simultaneous) explanation for anomalous lithium abundance and entropy excess in clusters [45].

2.4 Mixing

In the final stages of bubble evolution, heating of ICM will take place via mixing. This has been considered by, e.g., [16] who were able to find equilibrium states (albeit at the expense of fine tuning of the heating parameters). In a related approach, [48] showed how weak jets form fast-expanding cavities that push colder material from the center of the cluster to larger distances, where it mixes with the hotter ICM. Hot jet material also mixes with the ICM, thus offseting cooling. In addition to these two processes, g-modes are excited but damp on long timescales. Physical viscosity could damp such modes on a shorter timescale. Interestingly, g-modes of a frequency $\omega > \omega_{BV}(r)$, where $\omega_{BV}(r)$ is the Brunt-Vaisala frequency at a given radius r, are trapped within this radius. This means that their eventual dissipation will likely be confined to the cool core. The problem of fine-tuning of AGN feedback was addressed by [47]. They considered jet activity triggered by accretion of gas onto the center and estimated the energy release from the Bondi rate and found out that this led to structural stability of cluster atmospheres. This work has been extended by Sijacki & Springel (these proceedings) to include self-regulating AGN feedback in galaxy clusters extracted from cosmological simulations.

2.5 Alternative/supplementary processes

In realistic galaxy clusters, bubble mixing will be helped by random gas motions due to substucture mergers. There is strong theoretical evidence that such motions should be present ([43], [57]). Mergers themselves can also excite large scale sound and shock waves. It has been suggested early on that dissipation of such waves could heat cluster cores [52]. An appealing feature of this model is that waves generated on the outskirts of clusters will tend to focus on the center as the sound speed in cool cores decreases toward the center. The cooler the central regions are the stonger this effect is and, thus, this mechanism has the potential to be self-regulating. The interaction of such waves in cooling cluster atmospheres has been investigated numerically by [25]. They find that focusing of waves eventually leads to turbulence in the core that stops catastrophic cooling. Such effects should in principle be present in cosmological simulations of cluster formation. However, the spatial resolution achieved by [25] (22 pc) exceeds by far even that in the highest dynamical range gasdynamical simulations of cluster formation.

Substructure motions also constitute a significant source of energy that could be tapped. Galaxies moving through the ICM typically have supersonic velocities because (1) part of the pressure support against gravity comes from turbulent motions and (2) in equlibrium $P_{\text{gas}} = \sigma_{\text{gas}}^2/3$, where $\sigma_{\text{gas}} \sim v_{\text{gal}}$ is the velocity disperssion of the microscopic thermal motions of the gas. For slightly supersonic galactic velocities, the gas behind the galaxies forms a wake that exerts dynamical friction and converts kinetic energy to heat. This problem has been studied in detail by [23], [18], [38]. The efficiency of conversion of galactic motions into heat was found to be low. Three factors may have contributed to this conclusion: (1) it is possible that turbulence may have destroyed wakes, (2) possibly much higher numerical resolution is needed to model this effect and (3) transport processes have been neglected. Interestingly, recent simulations that include viscosity in cosmological clusters ([65]), show that it tends to "evaporate" gas from shallow substructure potential wells forming wakes. The effects of thermal conduction on cluster heating in cosmological simulations has also been studied by [35].

3 Open questions

There is growing consensus on the basic principles behind solving the cooling flow riddle but a number of challenging theoretical problems remain. Fortunately, most of these issues seem to be converging or be related to one main mechanism – AGN heating. Some of the main problems are:

(1) Why are black holes in BCGs more massive then predicted from the $M - \sigma$ relation?

(2) Can the relation between accretion rate and jet power found by [1] be explained in ab initio cluster simulations of large dynamical range?

(3) What initial conditions for energy injection are appropriate and how much injected energy should go into $p-$ vs. $g-$ modes ?

(4) What is the true nature of viscosity and conductivity ?

(5) What is the role of magneto-thermal instability in clusters ?

(6) What is the equation of state of the ICM, how "leaky" are the bubbles to cosmic rays and how efficient is mixing ?

(9) How important are non-AGN heating mechanisms ?

Acknowledgement. I wish to thank Marcus Bruggen and Mitch Begelman for fruitful and continuing collaboration on the projects mentioned in this review.

References

1. Allen, S. W., et al. 2006, MNRAS, 372, 21
2. Batcheldor, D. et al. 2006, astro-ph/0610264
3. Begelman, M. C. 2004, Coevolution of Black Holes and Galaxies, 374
4. Bernardi, M. et al., 2006, astro-ph/0607117
5. Böhringer, H. et al. 2002, A&A, 382, 804
6. Bower, R. G. et al. 2006, MNRAS, 370, 645
7. Brighenti, F., & Mathews, W. G. 2003, ApJ, 587, 580
8. Brüggen, M., Ruszkowski, M., & Hallman, E. 2005, ApJ, 630, 740
9. Chandran, B. D. G. 2005, ApJ, 632, 809
10. Cho, J. et al. 2003, ApJL, 589, L77
11. Churazov, E. et al. 2001, ApJ, 554, 261
12. Churazov, E. et al., 2002, MNRAS, 332, 729
13. Croston, J. H., et al. 2005, MNRAS, 357, 279
14. Croton, D. J., et al. 2006, MNRAS, 365, 11
15. Dalla Bonta', E. et al., astro-ph/0610702
16. Dalla Vecchia et al., 2004, MNRAS, 355, 995
17. Enßlin, T. A., & Brüggen, M. 2002, MNRAS, 331, 1011

18. El-Zant, A. A., Kim, W.-T., & Kamionkowski, M. 2004, MNRAS, 354, 169
19. Fabian, A. C., et al., 2003, MNRAS, 344, L43
20. Fabian, A. C. et al. 2003, MNRAS, 344, L48
21. Fabian, A. C. et al. 2005, MNRAS, 363, 891
22. Fabian, A. C. et al. 2006, MNRAS, 366, 417
23. Faltenbacher, A. et al. 2005, MNRAS, 358, 139
24. Forman, W., et al. 2005, ApJ, 635, 894
25. Fujita, Y., Matsumoto, T., & Wada, K. 2004, ApJL, 612, L9
26. Fujita, Y., & Reiprich, T. H. 2004, ApJ, 612, 797
27. Fujita, Y., & Suzuki, T. K. 2005, ApJL, 630, L1
28. Gardini, A., & Ricker, P. M. 2004, Modern Physics Letters A, 19, 2317
29. Ghizzardi, S. et al., 2004, ApJ, 609, 638
30. Goldreich, P., & Sridhar, S. 1995, ApJ, 438, 763
31. Graham, J. et al., 2006, MNRAS, 368, 1369
32. Heinz, S., & Churazov, E. 2005, ApJL, 634, L141
33. Heinz, S. et al. 2006, astro-ph/0606664
34. Jones, T. W., & De Young, D. S. 2005, ApJ, 624, 586
35. Jubelgas, M., Springel, V., & Dolag, K. 2004, MNRAS, 351, 423
36. Kaiser, C. R., & Binney, J. 2003, MNRAS, 338, 837
37. Kaiser, C. R. et al. 2005, MNRAS, 359, 493
38. Kim, W.-T., El-Zant, A. A., & Kamionkowski, M. 2005, ApJ, 632, 157
39. Lauer, T. R. et al., 2006, astro-ph/0606739
40. Lazarian, A. 2006, ApJL, 645, L25
41. Mathews, W. G., Faltenbacher, A., & Brighenti, F. 2006, ApJ, 638, 659
42. McNamara, B. R. et al. 2005, Nature, 433, 45
43. Motl, P. M. et al., 2004, ApJ, 606, 635
44. Narayan, R., & Medvedev, M. V. 2001, ApJL, 562, L129
45. Nath, B. B., Madau, P., & Silk, J. 2006, MNRAS, 366, L35
46. Nulsen, P. E. J. et al., 2005, ApJ, 628, 629
47. Omma, H., & Binney, J. 2004, MNRAS, 350, L13
48. Omma, H., Binney, J., Bryan, G., & Slyz, A. 2004, MNRAS, 348, 1105
49. Parrish, I. J., & Stone, J. M. 2005, ApJ, 633, 334
50. Piffaretti, R., & Kaastra, J. S. 2006, A&A, 453, 423
51. Pizzolato, F., & Soker, N. 2006, MNRAS, 371, 1835
52. Pringle, J. E. 1989, MNRAS, 239, 479
53. Rafferty, D. A. et al., 2006, astro-ph/0605323
54. Rebusco, P., et al. 2006, MNRAS, 372, 1840
55. Reynolds, C. S., Heinz, S., & Begelman, M. C. 2002, MNRAS, 332, 271
56. Reynolds, C. S. et al. 2005, MNRAS, 357, 242
57. Ricker, P. M., & Sarazin, C. L. 2001, ApJ, 561, 621
58. Robinson, K., et al., 2004, ApJ, 601, 621
59. Roychowdhury, S., Ruszkowski, M., & Nath, B. B. 2005, ApJ, 634, 90
60. Ruszkowski, M., & Begelman, M. C. 2002, ApJ, 581, 223
61. Ruszkowski, M., Brüggen, M., & Begelman, M. C. 2004, ApJ, 611, 158
62. Ruszkowski, M., Brüggen, M., & Begelman, M. C. 2004, ApJ, 615, 675
63. Ruszkowski M., Ensslin, T., Bruggen, M., Heinz, S., Pfrommer, C. 2006, in prep.
64. Sanders, J. S., & Fabian, A. C. 2006, MNRAS, 371, L65
65. Sijacki, D., & Springel, V. 2006, MNRAS, 371, 1025
66. Vernaleo, J. C., & Reynolds, C. S. 2006, ApJ, 645, 83
67. Vogt, C., & Enßlin, T. A. 2005, A&A, 434, 67
68. von der Linden et al. 2006, MNRAS, submitted
69. Wyithe, J. S. B. 2006, MNRAS, 365, 1082

Simulating the AGN-ICM Interaction

M. Brüggen[1] and S. Heinz[2]

[1] International University Bremen, Campus Ring 1, 28759 Bremen, Germany
m.brueggen@iu-bremen.de
[2] Department of Astronomy, University of Wisconsin, Madison, WI, USA

Summary. In recent years, simulations of the interaction between active galactic nuclei (AGN) and the intracluster medium (ICM) have become increasingly realistic and sophisticated. Most hydrodynamical simulations are now three-dimensional, some simulations have started to explore the effects of conductivity, viscosity and magnetic fields. The latest generation of simulations start from realistic cluster environments taken out of cosmological simulations. In these proceedings, we wish to present two recent advances in our understanding of how the AGN interacts with the ICM. The first concerns the metal distributions in cool core clusters and the second concerns the morphology of AGN jets and their energy deposition subject to the ambient motions in the cluster.

1 Metal transport by AGN

The ICM has a metallicity of about 1/3 solar. However, cooling core clusters, i.e. clusters with a centrally peaked X-ray brightness, show peaked abundance profiles. A number of observations indicates that SNe Ia in the central galaxy are mainly responsible for the metal enrichment in the central part of clusters. However, the observed metallicity profiles are much broader than the light profiles of the central cluster galaxy. Hence, the difference in the light and the metal distributions are interpreted as the result of transport processes that have mixed metals into the ICM. [14] have assumed that the spreading is due to local turbulent motions in the ICM, and that the evolution of the metal density profile can be modelled as a diffusion process. Using the abundance profile from [8] for Perseus, [14] estimated that a diffusion coefficient of the order of 2×10^{29} cm^2 s^{-1} is needed to explain the width of the observed abundance profiles. Based on a model of iron enrichment from stellar mass loss in the central cluster galaxies of a few nearby clusters, [2] have used the total amount of iron observed in the ICM to obtain constraints on the age of the cooling cores (i.e. the time for which the central region have remained relatively undisturbed). They find very large ages of more than 7 Gyr ([2]) that provide new constraints for the modelling of the interaction regions and have consequences for the turbulent transport of metals. While it appears to be established that the metals produced by the central galaxy are dispersed into the ICM to form the broad abundance peaks, it remains unclear what the mechanism is via which the metals are transported.

As one likely mechanism, [17] studied the effect of AGN-inflated bubbles that rise buoyantly and redistribute the metals. Radio-loud active galactic nuclei (AGN) drive strong outflows in the form of jets that inflate bubbles or lobes. The lobes are filled with hot plasma and can heat the cluster gas in a number of ways (e.g. [4, 3, 7, 9, 13, 1, 16, 18, 6, 5]). Additionally, they transport matter and thus metal-rich gas outwards.

In our simulations we focussed on the following questions:

- How far out can bubbles carry metals?
- How efficient is transport by buoyant bubbles?
- Does this process cause characteristic features, e.g. anisotropies?
- How does the resulting metal distribution depend on bubble and cluster parameters?
- What effective diffusion coefficients does this form of transport yield?
- How do these coefficients compare with those found by [14]?

1.1 Method

The broad abundance peaks in clusters have been produced over a time span of several gigayears. If the peaks have been broadened by AGN, this process has taken a large number of activity cycles. In order to simulate the transport of metals by AGN-induced flows and to capture the hydrodynamical details important for the interaction between the AGN and the ICM, a fair resolution of the computational mesh is necessary. This prevents us from simulating the bubble transport over cosmological times. However, we can still study the efficiency of the metal transport by simulating a small number of AGN cycles and parametrise the transport efficiency. A convenient parametrisation that has been applied previously to observations ([14]) is based on a diffusion description. Even though we do not propose that the metal transport occurs via microscopic diffusion, transport processes like these may be described by a diffusion parameter. This allows a direct comparison to observations without having to simulate the entire life time of the cluster.

We model the hydrodynamical evolution of the ICM in a galaxy cluster. Initially, the ICM is set in hydrostatic equilibrium in a static cluster potential. We assume that a central cluster galaxy injects metals into the ICM with a rate proportional to its light distribution. We now trace the distribution of the metals by injecting a tracer fluid into the ICM at a rate that is proportional to the light distribution of the central galaxy, which is modelled with a Hernquist profile. Note that the injection rate could also be time-dependent. [14] have used a time-dependent metal injection rate that accounts for the higher supernova rate in the past and the evolution of the stellar population. However, we follow the evolution of the cluster for only about Gyr. For typical cases studied in [14], the metal injection rate does not change significantly over this time. Hence in our simulations we use a temporally constant metal injection rate.

Finally, we model the AGN activity by inflating ambipolar pairs of underdense bubbles in the ICM that rise buoyantly and thus stir the ICM. The simulations were performed with the FLASH code ([10]), a multidimensional adaptive mesh refinement code that solves the Riemann problem on a Cartesian grid using the Piecewise-Parabolic Method.

1.2 Results

Our simulations show that buoyantly rising bubbles have a significant impact on the metal distribution in a cluster. They transport metals to larger distances from the cluster centre. The most efficient transport occurs along the bubbles' direction of motion, leading to an asymmetrical metal distribution (see Fig. 1).

To infer the global impact of this metal transport, we have calculated cumulative metal mass profiles and averaged metal density profiles. In order to compare profiles of different timesteps, we have normalised all profiles to $\dot{M}_{\mathrm{Metal}}t$, where $\dot{M}_{\mathrm{Metal}}$ is the total metal mass injection rate and t is the elapsed time. Figure 1 shows that the bubbles also change the global metal distribution significantly.

The diffusion coefficients inferred from our simple experiments lie at values of around $\sim 10^{29}$ cm^2 s^{-1} at a radius of 10 kpc (see Fig. 2). Right at the centre, they rise to about $\sim 10^{31}$ cm^2 s^{-1} and fall to about $\sim 10^{25}$ cm^2 s^{-1} at a radius of 100 kpc. Modulo a factor of a few, the coefficients for all runs lie around these values. Differences are induced by the bubble sizes, the initial positions, the recurrence times and the pressure profile in the cluster. Interestingly, the runs modelled on the Perseus cluster yield diffusion coefficients that agree very well with those estimated by assuming a simple diffusion model (see [14]). As we do not take into account motions that may have been induced by mergers etc., the values for the diffusion coefficients constitute lower limits on the real values. It is also striking that the diffusion coefficients fall so steeply with radial distance from the centre. Over two decades in radius they decrease by almost 10 orders of magnitude. Very approximately, they fall like r^{-5}. This strong radial dependence of the transport efficiency is, partially, a result of the three-dimensional nature of the bubble-induced motions, i.e. uplifted metals are diluted over the entire radial shell. A second effect is the decaying "lift" power of the bubbles as they rise.

Summarising, our investigations led to the following results:

-   Larger bubbles lead to mixing to larger radii. For realistic parameters and in the absence of stabilising forces, bubbles hardly move beyond distances of 150 kpc from the centre.
-   Mixing scales roughly with bubble frequency.
-   The metal distribution is not sensitive to the way of how bubbles are inflated.

- In hydrostatic cluster models, the resulting metal distribution is very elongated along the direction of the bubbles. Anisotropies in the cluster and/or ambient motions are needed if the metal distribution is to be spherical.
- The diffusion coefficients inferred from our simple experiments lie at values of around $\sim 10^{29}$ cm^{2}s^{-1} at a radius of 10 kpc.
- The runs modelled on the Perseus cluster yield diffusion coefficients that agree very well with those inferred from observations (see [14]).

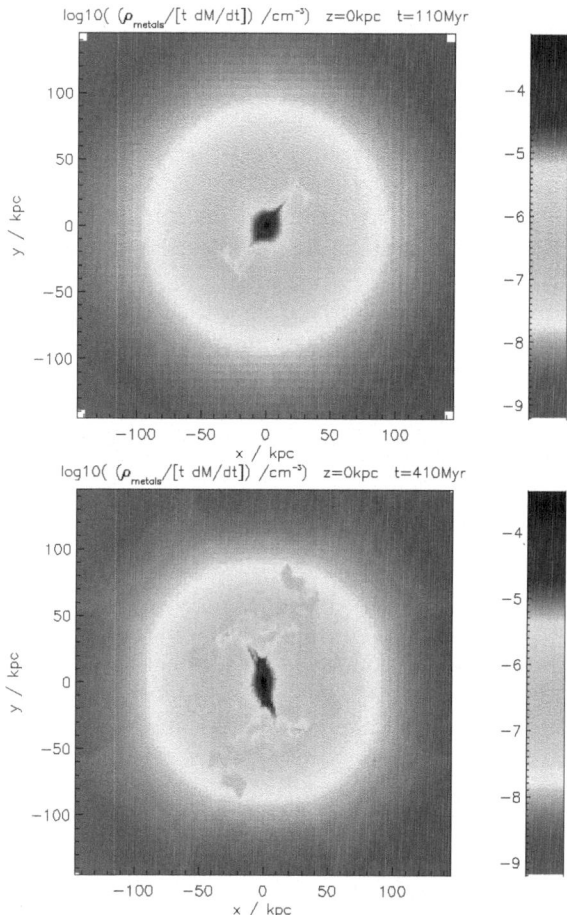

Fig. 1. Slices of the metal distribution (for details see Roediger et al. 2006).

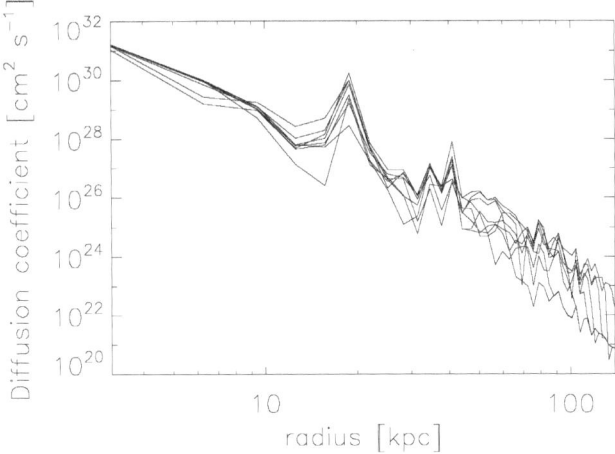

Fig. 2. Diffusion coefficients as a function of radius inferred from simulations presented in Roediger et al. (2006). The different curves denote different times.

2 Simulating the interaction of jets with dynamic cluster atmospheres

Numerical simulations of hot, underdense bubbles in clusters of galaxies have been performed by a number of authors [7, 4, 3, 15, 18, 9, 13]. Common to all of these simulations is that they use a spherically symmetric, analytical profile for the ICM. However, in reality the dynamics of the jet may be strongly affected by the ambient medium. Bulk velocities may advect and distort the jet and the radio lobe. Density inhomogeneities can affect both the jet propagation and the deposition of entropy in the ICM by the jet. For the purpose of studying the extent of cluster heating by AGN, the crucial question is *exactly where and how much* energy is deposited by the jet in the ICM. The primary aim of this work is therefore to study the dynamics of the jet and the extent of ICM heating subject to the motions and inhomogeneities of the ICM.

In a letter by [11], we present first results from hydrodynamical simulations of AGN heating in a realistic galaxy cluster. For the first time taking into account the dynamic nature of the cluster gas and detailed cluster physics. The simulations successfully reproduce the observed morphologies of radio sources in clusters. We find that cluster inhomogeneities and large scale flows have significant impact on the morphology of the radio source. Shear and rotation in the intra-cluster medium move large amounts of cold material back into the path of the jet, ensuring that subsequent jet outbursts encounter a sufficient column density of gas to couple with the inner cluster gas, thus alleviating the problem of evacuated channels discussed in the recent literature. The same effects redistribute the excess energy ΔE deposited the jet, making the distribution of ΔE at late times consistent with being isotropic.

2.1 Morphology of the jet

The initial conditions of our simulation are based on a rerun of the S2 cluster from [19], whose properties are sufficiently close to a typical, massive, X-ray bright cluster with a mass of $M \sim 7 \times 10^{14} M_\odot$ and a central temperature of 6 keV. The output of the GADGET SPH simulation serves as the initial conditions for our simulation. Further details about the setup are given in the contribution by Heinz et al. in these proceedings.

While the jet is active, the jet-inflated cocoon and the surrounding swept up shell generally follow the morphological evolution discussed previously in the literature: The jets inflate two oblong cocoons, which push thermal gas aside. The early expansion is supersonic, shocking the surrounding gas directly, while at later times the expansion slows down and eventually becomes sub-sonic. Fig. 3 shows a time series of density and entropy cuts as well as simulated X-ray maps (using the Chandra ACIS-I response and assuming an APEC thermal model) and low-frequency radio maps (assuming equipartition and neglecting radiative cooling). The aspect ratio of the lobes is roughly 3 and decreases with time, indicating the lateral spreading and shear of the radio plasma.

Regarding the morphological evolution of the source, we can draw two conclusions:

First, it is clear from the series of snapshots shown in Fig. 3 that the non-sphericity of the atmosphere induces significant asymmetries in the two sides of the cocoon. Given that the difference is so strong, it is clear that the impact on the morphology as a whole must be significant. This justifies our initial motivation for this study: in order to investigate the evolution of radio sources and their impact on galaxy cluster atmospheres, one cannot neglect the dynamic nature of the atmospheres prior to jet injection.

Second, the dynamic nature of the cluster atmosphere significantly alters the late stage evolution, after the jet switches off (between the fourth and the fifth frame at 3.3×10^6 yrs) and the source evolution becomes sub-sonic. During the sub-sonic stage, the impact of the pressure and density of the surrounding gas become more and more important. In particular, the rotation that is present in the cluster has clearly sheared the plasma away from the jet axis. We will quantify these results below. It is already clear, however, that this effect can solve the dilemma of launching multiple jet episodes into the same cluster: any evacuated channel created by previous outbursts will have been buffeted around sufficiently to move enough cluster gas back into the path of the new jet to provide a sufficient cross section for interaction and to couple the jet to the inner cluster gas.

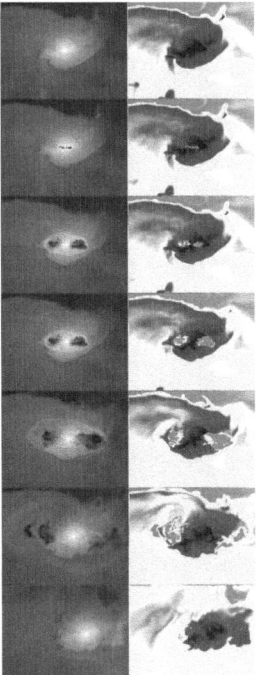

Fig. 3. Left panels: Time series of unsharp masked X-ray images at 0, 5, 10, 20, 40, 80 and 160 Myrs after jet onset. Right panels: entropy cut through the cluster centre for the same epochs. The images are 450 kpc in width and 335 kpc in height.

2.2 Iron-line studies

One way to probe the dynamics of the ICM is to look at the emission lines of heavy ions. The thermal broadening of emission lines of heavy ions is small enough such that Doppler shifts due to bulk motions may be detected with the next generation of X-ray observatories. Doppler shifts of emission lines may be detected as long as the ICM is optically thin with respect to the line considered (see also [12]). We have investigated the spectral signatures of AGN-induced motions at the example of the complex near the FeXXV K_α line as seen with Constellation-X.

Besides inspecting spectra along individual lines of sight one can attempt to quantify the effects caused by bubble motions by reducing the line shifts to a single parameter. To demonstrate this we have produced a synthetic Doppler map of our simulation. Here, red denotes velocities between -1400 to -425 km/s, green from -425 km/s to 425 km/s and blue from 425-1400 km/s. Fig. 4 shows such a map which reveals an imprint from the blue- and redshift produced by the gas pushed out radially by the bubbles.

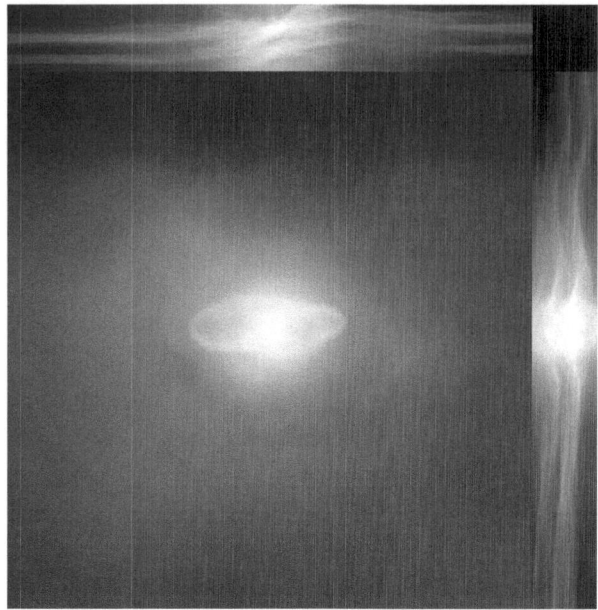

Fig. 4. Predicted iron Fe XXVI-Kα line image seen by Constellation X (red: -1400 to -425 km/s, green: -425 km/s to 425 km/s, blue: 425-1400 km/s)

2.3 Heating the cluster

At late times ($t \gtrsim 80\,\mathrm{Myrs}$) the cooling time for the inner few$\times 10^{12}\,\mathrm{M}_\odot$ has been increased by about 50%, more at earlier times. This indicates that, on average, the jet can halt cooling for a time significantly longer than the time the jet is active. Future simulations including radiative cooling will be necessary to investigate whether this increase is sufficient to offset cooling entirely and whether this result is consistent with *XMM*-Newton and *Chandra* constraints on cluster cooling rates.

One of the criticisms of models of cluster heating by jets is the non-spherical nature of the energy input. The problem is twofold: Firstly, if the energy is predominantly injected into the axial direction, the equatorial cluster gas does not benefit from the energy injection and can cool unimpeded, while the axial gas is super-heated, making AGN heating very inefficient. Secondly, the channel of jet exhaust and super-heated gas along the jet axis would allow subsequent episodes of jet activity to punch easily through the inner cluster and deposit their energy well outside the central region, thus allowing the inner regions to cool unimpeded.

If cluster mass can be moved around to re-fill the jet channel and to buffet the fossil radio plasma around, it is plausible to assume that the energy input of the jet might be isotropised more rapidly as well. The entropy panel of Fig. 3 shows this qualitatively: there is no clear high entropy channel along the axis of the jet. In fact, the draft of the rising bubbles draws low entropy material out of the central cluster. This is borne out by a quantitative analysis: At early times, the excess energy is clearly peaked around the jet axis (within about 30°),but at late times (after about 160 Myrs), the distribution is consistent with being isotropic. This indicates that individual episodes of jet activity can, in fact, distribute their energy rather isotropically into the cluster. More detailed simulations including radiative cooling and multiple jet episodes will be necessary to develop a detailed picture of how heating and cooling in realistic cluster atmospheres proceed.

The simulations shown here model a jet at the upper end of the power scale of typical radio-loud AGN. The results shown above resemble the source Cygnus A which is an extreme and rare source. Most AGN have mechanical powers that are lower by two to three orders of magnitude. Even though these sources are much weaker, they are also much more common and are likely to have a larger impact on the thermal evolution of the ICM than the very powerful sources. The simulation of weaker jets requires a finer resolution at the base of the jet and, therefore, is more expensive. This work is currently under way.

3 Main open questions

The two examples described above are just two examples taken from ourown work that show how simulations have become mature enough to address the physical challenges posed by AGN feedback.

However, while simulations have made significant progress, we would like to end with a list of open questions that delineate the problems that still need to be solved in this field:

- Why do AGN-blown cavities remain intact and appear spherical while simulations suggest that the cavities fragment?
- How does feedback operate?
- How are the bubbles inflated?
- Where and how is energy dissipated?
- How long do bubbles live? Are they stabilised, for example by magnetic fields?
- What is the equation of state of the bubbles?
- How important are microphysical processes such as viscosity and conduction?
- Is multiphase physics important?
- What are the relative contributions to the kinetic energy input of the central AGN and other processes such as mergers etc.?

Acknowledgement. We acknowledge the support by the DFG grant BR 2026/3 within the Priority Programme "Witnesses of Cosmic History" and the supercomputing grants NIC 1927 and 1658 at the John-Neumann Institut at the Forschungszentrum Jülich. Some of the simulations were produced with STELLA, the LOFAR BlueGene/L System. The results presented were produced using the FLASH code, a product of the DOE ASC/Alliances-funded Center for Astrophysical Thermonuclear Flashes at the University of Chicago.

References

1. J. F. Basson & P. Alexander, MNRAS, 2003, 339, 353
2. H. Böhringer, K. Matsushita, E. Churazov, A. Finoguenov, & Y. Ikebe, A&A, 2004, 416, L21
3. M. Brüggen & C. R. Kaiser, Nature, 2002, 418, 301
4. M. Brüggen, C. R. Kaiser, E. Churazov, & T. A. Enßlin, MNRAS, 2002, 331, 545
5. F. Brighenti & W. G. Mathews, ApJ, 2006, 643, 120
6. M. Brüggen, M. Ruszkowski, & E. Hallman, ApJ, 2005, 630, 740
7. E. Churazov, M. Brüggen, C. R. Kaiser, H. Böhringer, & W. Forman, ApJ, 2001, 554, 261
8. E. Churazov, W. Forman, C. Jones, & H. Böhringer, ApJ, 2003, 590, 225
9. C. Dalla Vecchia, R. G. Bower, T. Theuns, M. L. Balogh, P. Mazzotta, & C. S. Frenk, MNRAS, 2004, 355, 995
10. B. Fryxell, K. Olson, P. Ricker, F. X. Timmes, M. Zingale, D. Q. Lamb, P. Mac-Neice, R. Rosner, J. W. Truran, & H. Tufo, ApJS, 2000, 131, 273
11. S. Heinz, M. Brüggen, A. Young, & E. Levesque, ArXiv Astrophysics e-prints, astro-ph/0606664
12. M. Hoeft, M. Brüggen, G. Yepes, 2005, AN, 326, 613
13. H. Omma, J. Binney, G. Bryan, & A. Slyz, MNRAS, 2004, 348, 1105
14. P. Rebusco, E. Churazov, H. Böhringer, & W. Forman, MNRAS, 2005, 359, 1041
15. C. S. Reynolds, S. Heinz, & M. C. Begelman, ApJ, 2001, 549, L179
16. C. S. Reynolds, S. Heinz, & M. C. Begelman, MNRAS, 2002, 332, 271
17. E. Roediger, M. Brueggen, H. Böhringer, E. Churazov,, & P. Rebusco, MNRAS submitted.
18. M. Ruszkowski, M. Brüggen, & M. C. Begelman, ApJ, 2004, 611, 158
19. V. Springel, N. Yoshida, & S. D. M. White, New Astronomy, 2001, 6, 79

Simulations of Jets in Dynamic Galaxy Cluster Atmospheres

S. Heinz[1,2], M. Brüggen[3], A. Young[2] and E. Levesque[2]

[1] University of Wisconsin, Madison, 6508 Sterling Hall, 475 N. Charter St.,
   Madison, WI 53706, heinzs@astro.wisc.edu
[2] Kavli Institute for Astrophysics and Space Research, Massachusetts Institute of
   Technology, 77Massachusetts Avenue, Cambridge, MA 02139
[3] International University Bremen, Campus Ring 1, 28759 Bremen, Germany

Summary. In light of the suggestion that radio galaxies are responsible for regulating cooling flows, we present the first generation of galaxy cluster simulations that include powerful jets from black holes. We find that cluster inhomogeneities and large scale flows have significant impact on the morphology of the radio source and cannot be ignored a-priori when investigating radio source dynamics. Shear and rotation in the intra-cluster medium move large amounts of cold material back into the path of the jet, ensuring that subsequent jet outbursts encounter a sufficient column density of gas to couple with the inner cluster gas. The same effects redistribute the excess energy ΔE deposited by the jet, making the distribution of ΔE at late times consistent with being isotropic.

1 Introduction

The central galaxy in almost every strong cooling core contains an active nucleus and a jet–driven radio galaxy. The recently discovered correlation between the Bondi accretion rates and the jet power in nearby, X-ray luminous elliptical galaxies [1] shows that the AGN driven radio galaxies feed back enough energy to quench cooling and star formation, thus providing a possible explanation for the observed cut-off at the bright end of the galaxy luminosity function [2, 8, 3] and for the entropy floor in cool core clusters.

Numerical simulations of radio galaxy drive relativistic bubbles in galaxy clusters have been performed by a number of authors e.g., [7, 17, 5, 4, 15, 16, 9, 13, 12]. Common to all of these simulations is that they use a spherically symmetric, analytic profile for the ICM. [14] and [6] simulated AGN feedback in a cosmologically evolved clusters but did not model the jets directly. Most recently, [20] performed 3D simulations of a jet in a hydrostatic, spherically symmetric cluster model. In their simulation, the jet power was modulated by the mass accretion rate across the inner boundary. In all of their models, jet heating failed to prevent catastrophic cooling of gas at the center because the jet preferentially heated gas along the jet axis rather than heating matter in the equatorial plane. However, the dynamics of the jet may be strongly affected by the ambient medium. Bulk velocities may advect and distort the jet and the radio lobe. Density inhomogeneities can affect both the jet propagation and the deposition of entropy in the ICM by the jets. For the purpose of studying cluster heating by AGN, the crucial question is *exactly where and how much* energy is deposited by the jet in the ICM. We have undertaken the first hydrodynamical simulations of AGN heating in a realistic galaxy cluster, which we will briefly present below.

2 Simulating Jets in Clusters

2.1 Technical Setup

The initial conditions of our simulation are based on a rerun of the S2 cluster from [19], whose properties are sufficiently close to a typical, massive, X-ray bright cluster with a mass of $M \sim 7 \times 10^{14} M_\odot$ and a central temperature of 6 keV. The output of the GADGET SPH simulation serves as the initial conditions for our simulation. We use the FLASH code [10] which is a modular block-structured adaptive mesh refinement code, parallelized using the Message Passing Interface, including dark matter and self gravity of the gas. For the relatively short physical time of the jet simulation (160 Myrs), radiative cooling and star formation are neglected, though they were included in the constitutive SPH simulation. The computational domain is a $2.8 \mathrm{Mpc}^3$ box around the cluster's center of mass. The maximum resolution at the grid center corresponds to a cell size of 174pc.

The jet material is injected equally in opposite directions with velocity $v_\mathrm{jet} = 3 \times 10^9 \, \mathrm{cm \, s}^{-1}$ and an internal Mach number of 32. The total jet power of the run presented here was $W_\mathrm{jet} = 10^{46} \, \mathrm{ergs \, s}^{-1}$, corresponding to a rather powerful source, comparable to Cyg A, and in line with the powers now implied by the large scale moderate shocks found around numerous radio sources. We included the so-called dentist drill effect [18] by letting the jet axis random walk within an opening angle of $20°$.

2.2 Results

Source evolution follows the well known picture for powerful sources e.g., [15]: While the jet is active, the jet-inflated cocoon and the surrounding swept up shell generally follow the morphological evolution discussed previously in the literature: The jets inflate two oblong cocoons which push thermal gas aside. The early expansion is supersonic, shocking the surrounding gas directly, while at later times the expansion slows down and eventually becomes subsonic. Fig. 1 shows a time series of entropy cuts as well as simulated Chandra X-ray maps and low-frequency radio maps (assuming equipartition). The aspect ratio of the lobes is roughly 3 and decreases with time, indicating the lateral spreading and shear of the radio plasma.

The X-ray maps show a clear increase in the surface brightness in the outgoing shock/compression wave and a wispy, turbulent wake behind the

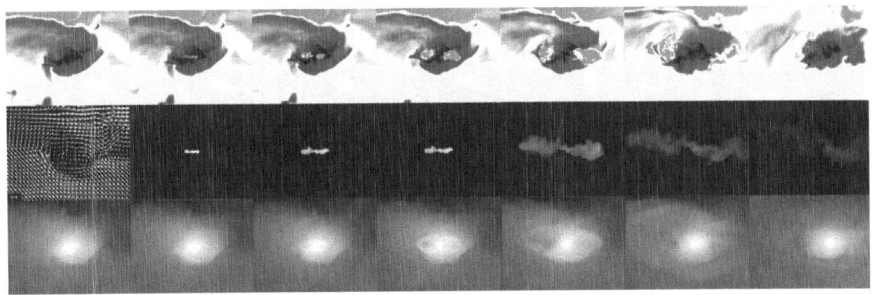

Fig. 1. Left-to-right: Time series of snapshots at 0Myrs, 5Myrs, 10Myrs, 20Myrs, 40Myrs, 80Myrs, and 160Myrs after jet onset. *Top panels*: entropy cut through cluster center. *Middle-left panel*: velocity map across the central plane of the images (arrow inset indicates 1000 km/s scale). *Middle six panels*: low-frequency radio synchrotron map. *Bottom panels*: X-ray map (red: 0.3-2 keV, green: 2-5 keV, blue: 5-10 keV). The images are 450 kpc in width and 335 kpc in height.

wave (Fig. 1). Two distinctions to previously published results are important to point out regarding the morphological evolution of the source:

First, it is clear from the series of snapshots shown in Fig. 1 that the non-sphericity of the atmosphere induces significant asymmetries in the two sides of the cocoon. This justifies our initial motivation for this study: In order to investigate the evolution of radio sources and their impact on galaxy cluster atmospheres, one cannot neglect the dynamic nature of the atmospheres prior to jet injection.

Second, the dynamic nature of the cluster atmosphere significantly alters the late stage evolution, after the jet switches off (between the fourth and the fifth frame at 3.3×10^6 yrs) and the source evolution becomes sub-sonic. During the sub-sonic stage, the impact of the pressure and density of the surrounding gas become more and more important. In particular, the rotation that is present in the cluster (see Fig. 1) has clearly sheared the plasma away from the jet axis. Qualitatively, it is clear that any evacuated channel created by previous outbursts will have been buffeted around sufficiently to move enough cluster gas back into the path of the new jet to provide a sufficient cross section for interaction and to couple the jet to the inner cluster gas.

The spiral-like morphology of the radio source at late stages is very reminiscent of the large scale (> 30 kpc) morphology of M87 and Abell 4059 [11], where the large scale radio bubbles are clearly mis-aligned with the inner jet. This leads us to suggest that there might be significant rotation or shear present in the gas of these clusters.

As found in earlier investigations of jet-cluster interactions, the impact of the jet on the cluster leads to an increase in the central entropy and a decrease in density, i.e, a net heat input. The bottom panel of Fig. 2 shows the cumulative radial mass, i.e., the gas mass $M(<r)$ contained within radius

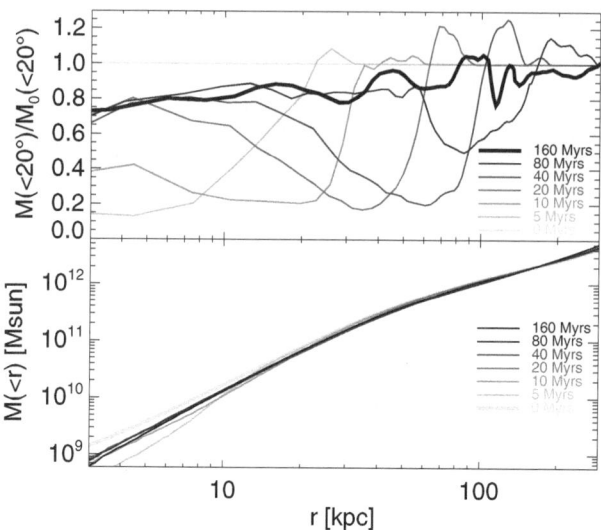

Fig. 2. *Bottom panel:* cumulative gas mass M_R within radius r as a function of r at different times (grey scales); note the general inflation of the cluster atmosphere by the energy injection into the center; *Top panel:* ratio of target mass M_{20} (mass contained within $20°$ of jet axis) over M_{20} for the control simulation, indicating the relative gas depletion around the jet axis. The residual $\sim 20\%$ reduction after 160 Myrs is consistent with the general reduction in density seen in the M_R.

r from the cluster center. The initial time step lies above all other curves out to a radius of about 60 kpc, indicating that the cluster has been inflated (i.e., mass moved out).

One of the criticisms of models of cluster heating by jets is the non-spherical nature of the energy input. The problem is twofold: Firstly, if the energy is predominantly injected into the axial direction, the equatorial cluster gas does not benefit from the energy injection and can cool unimpeded, while the axial gas is super-heated, making AGN heating very inefficient. Secondly, the channel of jet exhaust and super-heated gas along the jet axis would allow subsequent episodes of jet activity to punch easily through the inner cluster and deposit their energy well outside the central region, thus allowing the inner regions to cool unimpeded.

Shear and rotation in the central cluster can rectify the second problem: The top panel of Fig. 2 shows the radial profile of the target mass M_{20} contained in a $20°$ angle from the jet axis, relative to that of the control simulation. The amount of gas mass is significantly reduced during the early stages, however, at late times, M_{20} increases back up to 80% of the unperturbed value, consistent with the general reduction of density in the central cluster by the entropy input. Such a small reduction is dynamically insignif-

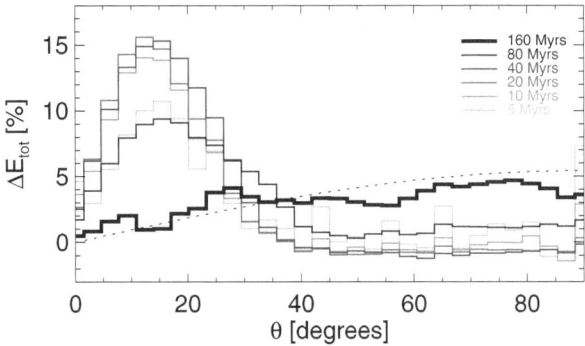

Fig. 3. Angular distribution (relative to mean jet axis) of excess energy ΔE injected by jet into thermal gas phase as percentage relative to total injected energy. For comparison, the dotted line shows the isotropic case.

icant, i.e., the amount of target mass is more than sufficient for subsequent jet episodes to couple to the cluster at *any* radius. Thus, cluster dynamics can rectify the problem of evacuated channels raised by [20].

Cluster dynamics can also solve the first problem: Fig. 3 shows the excess energy in the thermal cluster gas as a function of polar angle θ away from the jet axis (measured around the cluster center). At early times, the excess energy is clearly peaked around the jet axis, but at late times (160 Myrs, solid black line), the distribution is consistent with being isotropic (dotted black line). This demonstrates that individual episodes of jet activity can, in fact, distribute their energy rather isotropically into the cluster.

3 Summary

We have presented simulations of jet-cluster feedback that integrate a correct representation of cluster dynamics, including dark matter, and the collimated input of energy and momentum from an AGN. The simulations accurately reproduce the observed morphological appearance of powerful radio sources like Cyg A. We find that the dynamic structure of the cluster in the form of rotation and shear in the velocity field and the inhomogeneity and anisotropy of the cluster have significant impact on the evolution of the jets and the radio lobes they inflate. At late times, the energy deposited into the cluster by the jets is distributed well away from the axis, consistent with roughly spherically symmetric energy input. The inner cluster has been sufficiently rearranged to erase the low density channel blasted out by the jet and move enough material into the way of subsequent jet outbursts to couple efficiently with the inner cluster. This solves the problem of low efficiency feedback found in simulations in spherically symmetric, static atmospheres.

Acknowledgement. We thank Volker Springel for providing us with a set of Gadget simulated clusters. We thank Mateusz Ruszkowski, Chris Reynolds, Mitch Begelman, and Paul Nulsen for helpful discussions. SH acknowledges support by the National Aeronautics and Space Administration through Chandra Postdoctoral Fellowship Award Number PF3-40026 issued by the Chandra X-ray Observatory Center, operated by the Smithsonian Astrophysical Observatory for and on behalf of the National Aeronautics Space Administration under contract NAS8-39073. MB acknowledges support by DFG grant BR 2026/2 and the supercomputing grant NIC 1658 at the John von-Neumann center for computing at the Forschungszentrum Jülich. The software used in this work was in part developed by the DOE-supported ASCI/Alliance Center for Astrophysical Thermonuclear Flashes at the University of Chicago.

References

1. Allen, S. W., Dunn, R. J. H., Fabian, A. C., Taylor, G. B., & Reynolds, C. S. 2006, MNRAS, 372, 21

2. Benson, A. J., Bower, R. G., Frenk, C. S., Lacey, C. G., Baugh, C. M., & Cole, S. 2003, ApJ, 599, 38

3. Bower, R. G., Benson, A. J., Malbon, R., Helly, J. C., Frenk, C. S., Baugh, C. M., Cole, S., & Lacey, C. G. 2006, MNRAS, 659

4. Brüggen, M. & Kaiser, C. R. 2002, Nature, 418, 301

5. Brüggen, M., Kaiser, C. R., Churazov, E., & Enßlin, T. A. 2002, MNRAS, 331, 545

6. Brüggen, M., Ruszkowski, M., & Hallman, E. 2005, ApJ, 630, 740

7. Churazov, E., Brüggen, M., Kaiser, C. R., Böhringer, H., & Forman, W. 2001, ApJ, 554, 261

8. Croton, D. J. e. a. 2005, MNRAS, 356, 1155

9. Dalla Vecchia, C., Bower, R. G., Theuns, T., Balogh, M. L., Mazzotta, P., & Frenk, C. S. 2004, MNRAS, 507

10. Fryxell, B. et al. 2000, ApJS, 131, 273

11. Heinz, S., Choi, Y., Reynolds, C. S., & Begelman, M. C. 2002, ApJ, 569, L79

12. Omma, H. & Binney, J. 2004, MNRAS, 350, L13

13. Omma, H., Binney, J., Bryan, G., & Slyz, A. 2004, MNRAS, 348, 1105

14. Quilis, V., Bower, R. G., & Balogh, M. L. 2001, MNRAS, 328, 1091

15. Reynolds, C. S., Heinz, S., & Begelman, M. C. 2001, ApJ, 549, L179, rHB

16. Ruszkowski, M., Brüggen, M., & Begelman, M. C. 2004, ApJ, 611, 158

17. Saxton, C. J., Sutherland, R. S., & Bicknell, G. V. 2001, ApJ, 563, 103

18. Scheuer, P. A. G. 1982, in IAU Symp. 97: Extragalactic Radio Sources, ed. D. S. Heeschen & C. M. Wade, 163–165

19. Springel, V., White, M., & Hernquist, L. 2001, ApJ, 549, 681

20. Vernaleo, J. C. & Reynolds, C. S. 2006, ApJ, 645, 83

Can AGN and Radio Sources Reheat the Intracluster Medium?

D.S. De Young[1], S. M. O'Neill[2] and T. W. Jones[2]

[1] National Optical Astronomy Observatory, USA deyoung@noao.edu
[2] Dept. of Astronomy, University of Minnesota, USA

1 Introduction

The so called "cooling flow problem" in rich clusters of galaxies has been known for many years. In such clusters the number density of the intracluster medium (ICM) in the inner regions is sufficiently large that the thermal bremsstrahlung cooling time is less than a Hubble time. Hence this gas should cool, and pressure support in the inner ICM should be lost, resulting in inflow of the ICM gas from the outer regions. The several observational signatures expected from such cooling flows are not generally seen, and thus the "cooling flow problem" arose. The ICM clearly cools, as is evidenced by the observed X-ray emission, but it does not seem to flow.

A way out of this dilemma would be to provide some means of reheating the cooled ICM gas, and natural candidates for provision of the energy needed for reheating are the active galaxies often found in rich clusters of galaxies. This is not a new idea, since it has long been known that the total energies estimated to be injected by AGNs and radio sources in rich cluster are sufficient to provide the necessary reheating energy. However, this is an integral argument and is overly simplistic. Just because the AGN can provide enough total energy to reheat the ICM does not in any way guarantee that that energy will be transformed into heat in the ICM, nor is it clear that this energy will be deposited on the right time scale to overcome cooling within the gas cooling time. And finally, it is not at all clear that the localized energy injection from the AGN can become distributed throughout the volume of the ICM in the inner regions in a timely manner that will prevent inflow. Localized heating could cause hot ICM bubbles to rise, while the major part of the inner ICM could continue to flow inward.

Hence this simple concept requires further development to see if it is at all feasible. Recent high resolution X-ray images of cavities created in the ICM by radio emitting plasma ejected from AGN have renewed general interest in the AGN reheating picture. A combination of radio and X-ray data clearly shows that the radio emitting plasma has displaced the ambient ICM and has inflated a cavity that appears to be in pressure equilibrium with the ambient medium. In general, the data indicate that the cavities are expanding subsonically, if at all, that the internal density is lower than the ambient density, and that the internal energy of the cavities exceeds the energy given by equipartition arguments by large factors. This latter calorimetry from pdV work done in inflating the bubble can provide for the first time constraints on the particle content of jets emanating from active galactic nuclei (e.g., [5]). In terms of interaction with the ICM, these observations imply that the radio emitting cavities are buoyantly rising through the ambient ICM. This result has given rise to a number of speculations about how such cavities and bubbles could reheat the cooling ICM. These are: 1) Mixing of the energetic radio plasma and the ambient ICM; 2) Lifting of the ICM in the wake of rising bubbles; 3) Entrainment of the ICM along the surface of the bubbles via surface instabilities; and 4) creation of sound waves and dissipation of that energy in the ICM. To date no self consistent mixing calculations have been done that treat the overall reheating of the ICM throughout a large volume in the cores of clusters. However, examination of the evolution of rising bubbles through the ICM has been done in order to evaluate the bubble evolution and the mixing of bubble material with the ambient medium, together with lifting and mixing of ICM layers that are at different temperatures.

Both two and three dimensional hydrodynamic simulations of bubble evolution have been performed with high resolution (e.g., [2, 9]). These hydrodynamic simulation results are very encouraging for the bubble reheating picture because they clearly show a great deal of mixing between the bubble material and the ambient gas. In addition, as the bubble is fragmented and disperses, the volume of the ICM that is affected by lifting and mixing is much larger than the original dimensions of the bubble. A problem with this picture appears when observations were obtained that show the presence of "relic" radio bubbles which appear to be unattached from the parent AGN and that are much older than the smaller bubbles near the nucleus. A well known example of this are the active and relic radio bubbles in the core of the Perseus cluster (e.g., [6]). The surprising aspect of these relic radio bubbles is that they are intact. Their ages, estimated either from buoyant rise times or from synchrotron lifetime arguments, are of order 10^8 years. These ages are longer than the instability development times for either the Rayleigh-Taylor or Kelvin-Helmholtz instabilities that fragment the bubble as deduced from the numerical simulations and from analytic estimates (e.g., [4]). Something appears to be stabilizing the rising bubbles against fragmentation, and a natural candidate is the magnetic field in the intracluster medium.

2 Effects of Ambient Magnetic Fields

If a cavity is inflated in a magnetized gas, then in the MHD approximation, for any initial field configuration the field on the surface of the inflated cavity will have a significant tangential component. If the initial field is random on scales comparable to the cavity size, the resultant field on the surface of the inflated cavity will have an appearance similar to the threads on the surface of a ball of twine. This tangential field component will inhibit or delay the on-set and development of Rayleigh-Taylor (R-T) and Kelvin-Helmholtz (K-H) instabilities that lead to fragmentation of the bubble and subsequent mixing with the ambient ICM. Magnetic fields in the ICM of rich clusters of galaxies have been measured by a number of methods, and all of these result in field strengths of $1 - 10$ microGauss [3]. A simple linear stability analysis of the R-T and K-H instabilities with fields of this strength in average ICM condi-tions shows ([4]) that the fields inhibit the growth of these instabilities for times comparable to the estimated radio bubble lifetimes. However, a linear stability analysis is only indicative, and a fully nonlinear MHD calculation of the evolution of radio emitting cavities in the ICM is needed to accurately treat this problem.

2.1 MHD Simulations of Bubble Evolution

Early 2D MHD simulations of the evolution of a buoyant bubble in a mag-netized medium were carried out by [1]. In this case the magnetic field was inserted "by hand" instead of calculated self consistently, and the value of the field used was larger than that found in most clusters. However, these results were indicative, and they showed that under the assumptions of the calculation, the cavity remained intact as it rose through the ambient gas. A series of self consistent 2D MHD calculations was carried out by [8], and these again showed the rising cavities to be stable against destruction and mixing via R-T and K-H instabilities. A more complex set of 2D MHD simu-lations has been completed by [7], wherein the bubble inflation was included, together with the gravitational potential of both the parent galaxy and the cluster, including dark matter. One result of these simulations is shown in Figure 1, where in can be seen that once again the rising bubble is stabi-lized against disruption by the magnetic fields in the ICM. All runs used an isothermal ICM with a temperature of 3.3×10^7 K, and an ambient density at 5 kpc from the center of the gravitational potential of 0.1 particles cm^{-3}. All of the magnetic fields used in these simulations had initial values that were much less than the energy density of the ICM gas, i.e., the magnetic fields were initially completely unimportant in influencing the gasdynamics. The initial values of beta ranged from 120 to 75000, and most of the simulations were carried out with a constant beta, rather than a constant B, throughout the initial ICM volume. However, as the bubble is inflated, the value of the magnetic field on the surface increases, as expected from flux conservation, and this is true for arbitrary B field orientations, again as expected. In some cases the buoyant rise of the bubble also results in an enhancement of the field on the surface of the bubble.

In all cases where the bubble inflation ceases before the end of the calcu-lation, they first show a rapid initial rise, followed by a deceleration in their upward motion. This is an expected result, since the horizontal component of the ambient magnetic field provides a strong coupling to the ambient ICM

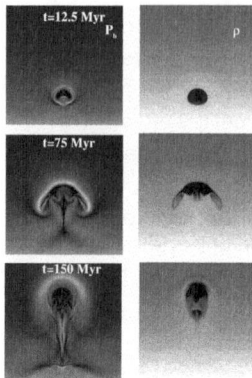

Fig. 1. Time evolution of a radio source bubble in a 2D MHD simulation for $\beta = 120$. Magnetic pressure is on the left, and density is shown on the right. (From Jones & De Young 2005.)

via mass loading of the field lines. Hence the previous estimates of bubble lifetimes obtained from analytic approximations to buoyant rise times need to be modified. In some cases these times may be modified by factors of two or more.

2.2 Three Dimensional MHD Simulations

The two dimensional calculations described above strongly suggest that the relic radio bubbles are maintained in an intact state due to suppression of the R-T and K-H instabilities by the ambient magnetic fields of the ICM. If so, then the suggested mixing and heating of the ICM by these objects may be much less effective than has been previously proposed. However, a definitive answer to this issue requires that a fully nonlinear simulation be performed in three dimensions so that no important MHD phenomena are suppressed. It is now well known that early 2D simulations of both MHD and hydrodynamic jet propagation resulted in spurious phenomena that were for some time believed to be real, and thus it is essential that 3D MHD calculations of relic radio source evolution be carried out.

We have generalized the above two dimensional MHD calculations to the three dimensional case, and 3D calculations with the same initial and boundary conditions as in the 2D case are being carried out. Here we present some early results of these calculations with varying values of β. In these cases the initial magnetic field in the bubble interior is set equal to zero, which was not the case for the 2D calculations. Figure 2 shows some results from a calculation with $\beta = 3000$. What is shown is a volume rendering (integration along the line of sight as opposed to a 2D slice through a volume) of the bubble material alone as it rises through the ICM. The ambient magnetic field in this case is perpendicular to the direction of motion of the rising bubble, and

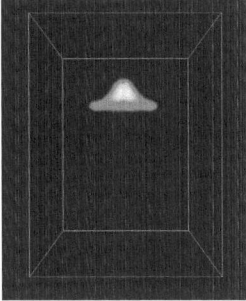

Fig. 2. Evolution of a radio source bubble in a 3D MHD simulation for $\beta = 3000$. Only the bubble material is shown in this volume rendered figure. The age of the bubble is 1.5×10^8 yr.

in this view the field lines lie in the plane of the figure. Figure 2 clearly shows once again that the bubble material is confined to a well defined volume and that the ICM and bubble material have not become mixed by surface instabilities at a time of 1.5×10^8 yr, which is comparable to estimates of the ages of currently observed relic radio bubbles and is much greater than the instability growth times for purely hydrodynamic simulations.

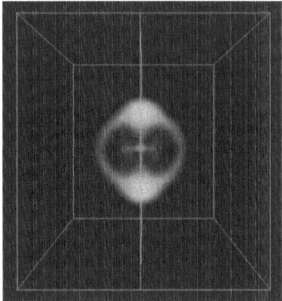

Fig. 3. Same as Figure 2 except the viewing angle is looking "down" on the bubble from "above".

Figure 3 shows the same bubble material at the same time, but in this case the volume rendering is from a view looking "down" on the bubble; i.e., antiparallel to the bubble direction of motion. Figure 3 clearly shows the need to perform these calculations in three dimensions. The effects of the field line tensions arising from distortions during the expansion of the bubble material into the ICM can be readily seen, and the expansion is definitely anisotropic. Figure 4 shows the same "look down" view as Figure 3, except in this case the β of the ambient ICM was set equal to 75000. Even in this very weak

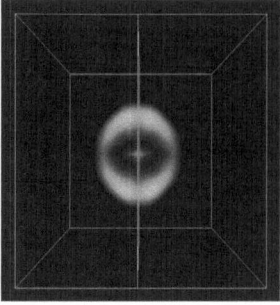

Fig. 4. Same as Figure 3 except that the ambient field is much weaker. Here $\beta = 75000$.

field case, the anisotropic effects of the ambient magnetic field on the rising bubble material are very evident.

3 Conclusions

Nonlinear MHD simulations in two and three dimensions of the evolution of radio source bubbles in the cluster ICM verify the preliminary conclusions of previous analytic studies. These simulations show that the ambient magnetic fields in the cluster ICM stabilize the rising bubbles against surface instabilities. This suppression of R-T and K-H instabilities allows the bubbles to remain intact for times of order 10^8 years, which is much longer than that predicted from purely hydrodynamic simulations. The bubble stability inhibits mixing of the bubble material with the ICM, and it also results in confinement of mixing and lifting of different elements of the ICM to a narrow column directly beneath the rising bubble. Both of these effects reduce the effectiveness of AGN created radio sources in reheating the ICM. Hence the role of AGN in solving the cluster ICM "cooling problem" is not as prominent as had been previously suggested, and a more detailed analysis of the overall energy input from radio sources into the ICM needs to be carried out. In particular, the global distribution of this energy and the timescale of its deposition should be calculated in a self consistent manner. The present calculations indicate that any energy deposition may be very localized and thus not reheat a significant volume of the ICM, and in addition the timescale for this process may be long and comparable to or greater than the cooling times in the inner regions of the ICM.

References

1. M. Brüggen & C. Kaiser, 2001, MNRAS, 325, 676
2. M. Brüggen & C. Kaiser, 2002, Nature, 418, 301

3. C. Carilli & G. Taylor, 2002, ARA&A, 40, 319
4. D. De Young, 2003, MNRAS, 343, 719
5. D. De Young, 2006, ApJ, 648, 200
6. A. Fabian et al. , 2000, MNRAS, 318, L65
7. T. Jones & D. De Young, 2005, ApJ, 624, 586
8. K. Robinson et al., 2004, ApJ, 601, 621
9. M. Ruzkowski, M. Brüggen, & M. Begelman, 2004, ApJ, 615, 675

AGN Heating Through Cavities and Shocks

P.E.J. Nulsen[1], C. Jones[1], W.R. Forman[1], L.P. David[1], B.R. McNamara[2,3], D.A. Rafferty[3], L. Bîrzan[3] and M.W. Wise[4]

[1] Harvard Smithsonian Center for Astrophysics, 60 Garden St MS6, Cambridge, MA 02138, USA pnulsen@cfa.harvard.edu
[2] Department of Physics & Astronomy, University of Waterloo, Ontario, Canada
[3] Astrophysical Institute and Department of Physics and Astronomy, Ohio University, Athens, OH 45701, USA
[4] Astronomical Institute "Anton Pannakoek," University of Amsterdam, Kruislaan 403, 1098 SJ Amsterdam, The Netherlands

1 Introduction

The X-ray emitting gas with cooling times much shorter than the Hubble time in elliptical galaxies, and at the centers of groups and clusters of galaxies is in an unstable state. If it is heated at a mean rate much less than the power it radiates, much of it would cool quickly to low temperatures, forming reservoirs of cold gas and young stars well in excess of those that are observed, e.g. [1, 2, 3]. If the gas is heated at a rate significantly exceeding the power it radiates, its cooling time would increase until it became comparable to the Hubble time or longer. Thus, the relatively high incidence of short central cooling times requires that the gas is heated at a mean rate closely matching the power it radiates. This is difficult to explain, unless heating rates are coupled to cooling rates. AGN heating is a natural vehicle to provide the coupling [4].

It has been established that the energy released by AGN at the centers of cooling flows is sufficient to have a significant impact on the cooling gas [5, 2]. Not only are the energies from AGN outbursts comparable to those needed to stop the gas from cooling, but the mean powers of the outbursts are well correlated with the powers radiated. While the heating process is not well understood, this would be a remarkable coincidence if AGN heating does not play a significant role in preventing the gas from cooling.

Three comments on AGN heating of cooling flows are made here. First, a simple physical argument is used to show that the enthalpy of a buoyant radio lobe is converted to heat in its wake. Thus, a significant part of "cavity" enthalpy is likely to end up as heat. Second, the properties of the repeated weak shocks in M87 are used to show that they can plausibly prevent gas close to the AGN from cooling. As the most significant heating mechanism at work closest to the AGN, shock heating probably plays a critical role in the feedback mechanism. Finally, results are presented from a survey of AGN heating rates in nearby giant elliptical galaxies. With inactive systems included, the overall AGN heating rate is reasonably well matched to the total cooling rate for the sample. Thus, intermittent AGN outbursts are energetically capable of preventing the hot atmospheres of these galaxies from cooling and forming stars.

2 Cavity Heating

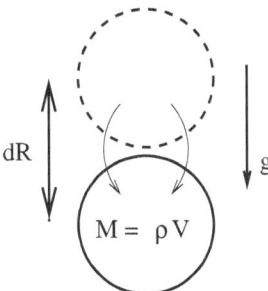

Fig. 1. Schematic of a buoyantly rising cavity

X-ray decrements over radio lobes are generally consistent with the lobes being devoid of hot gas, *e.g.* [6], so we treat them as massless. From the perspective of the ICM, a cavity (lobe) rises because ICM falls inward around it to fill the space it occupies (Fig. 1). This converts gravitational potential energy to kinetic energy in the gas flow around the rising cavity. Details of the flow depend on the viscosity, which is poorly determined. If it is high, the flow is laminar and the kinetic energy is dissipated as heat over a region comparable in size to the cavity (akin to Stoke's flow around a sphere [7]). If the viscosity is very low, the Reynolds number would be high and the flow turbulent. The turbulent region near the cavity would have a similar size to it. Turbulent kinetic energy is dissipated in the turn over time of the largest eddies [7], so that the dissipation time $t_d \sim r_{\rm cav}/v_{\rm cav}$, where $r_{\rm cav}$ is the radius of the cavity and $v_{\rm cav}$ is the speed of the eddy, which is comparable to the speed of the cavity. Since $t_d v_{\rm cav} \simeq r_{\rm cav}$, much of the turbulent kinetic energy is dissipated in a region of comparable size to the cavity. Thus, regardless of the viscosity, the kinetic energy created by cavity motion is dissipated locally as heat in a wake of similar size to the cavity.

The gravitational potential energy that is released as the cavity rises a small distance dR, subject to the gravitational acceleration g is (Fig. 1)

$$Mg\,dR = V\rho g\,dR \simeq -V\frac{dp}{dR}dR = -V\,dp, \qquad (1)$$

where $M = \rho V$ is the mass of gas displaced by the cavity, V is the volume of the cavity and ρ is the density of the external gas. Cavity motion is generally subsonic, so that the gas around the cavity is approximately hydrostatic, *i.e.*,

$\rho g \simeq -dp/dR$. The final dp here is the pressure change in the external gas over the distance dR, but a subsonic cavity maintains approximate local pressure equilibrium, so that dp may also be regarded as the change in pressure of the cavity. In terms of the enthalpy, $H = E + pV$, the first law of thermodynamics, $dE = T\,dS - p\,dV$, becomes $dH = T\,dS + V\,dp$. In an adiabatic process, the heat exchange, $T\,dS$, is zero and $dH = V\,dp$. Thus, equation (1) states that the potential energy released as the cavity rises is equal to the decrease in its enthalpy. No exotic process is required to explain how cavity enthalpy is converted to heat. This argument is less accurate for the largest cavities (with $r_{\mathrm{cav}} \simeq R$), but the corrections are of order unity. Fragmentation of cavities does not change the result, unless cavity contents dissipate. We conclude that enthalpy lost by rising cavities is dissipated as heat locally in their wakes [5].

3 Weak Shock Heating

Weak shocks associated with AGN outbursts are generally only detected in deep X-ray observations, so that relatively few are known, *e.g.* [8, 9, 10, 11, 12], although they are probably produced in most AGN outbursts. The energies required to drive the shocks are comparable to cavity enthalpies, but much of this energy will end up as potential energy in the ICM. The main requirement for stopping gas from cooling is to replace the entropy lost by radiation and, since the entropy jump, ΔS, varies as the cube of the pressure jump, weak shocks are not very effective at this [13]. The equivalent heat input per unit mass can be evaluated as

$$\Delta Q \simeq T\Delta S = E\Delta \ln \frac{p}{\rho^\gamma}, \tag{2}$$

where E is the specific thermal energy, p the pressure and ρ the density of the gas. Thus, the fractional heat input, $\Delta Q/E$, is given by the jump of $\ln p/\rho^\gamma$ in the shock.

Three weak shocks are visible in the X-ray image of M87 [14]. Fitting a simple model to the surface brightness profile of the innermost shock at 0.8 arcmin ($\simeq 3.7$ kpc) gives a Mach number of $\simeq 1.4$ and a shock age of $\simeq 2.4 \times 10^6$ y. Its equivalent heat input (equation 2), determined from the Mach number, is $\Delta Q/E \simeq 0.022$. Clearly, this single shock does very little to heat the gas. However, there is a second shock at about twice the radius, and the third shock, at $\simeq 3$ arcmin, required several times more energy [8]. Thus, a shock of comparable strength to the 0.8 arcmin shock may well occur every $\sim 2.5 \times 10^6$ y. The cooling time of the gas at 0.8 arcmin $\simeq 2.5 \times 10^8$ y, so that there is time for ~ 100 such shocks during the cooling time. Therefore, the combined heat input of the shocks ($100 \times 0.022 = 2.2$) is more than enough to make up for radiative losses from the gas.

These numbers are not to be taken literally, but they make a plausible case that the modest heating of repeated weak shocks can stop the gas at 0.8

arcmin from the AGN in M87 from cooling. The cavity heating discussed in section 2 is only effective beyond the radius where the radio lobes form. Some other mechanism is required to prevent gas closer to the AGN from cooling. Since the gas closest to the AGN is likely to be its most significant source of fuel, the heating process that affects this gas probably plays a critical role in the feedback cycle that prevents gas from cooling. Thus, while weak shock heating is unlikely to be the dominant mode [15], it may well play a major role in the AGN feedback process.

4 AGN Heating in Nearby Elliptical Galaxies

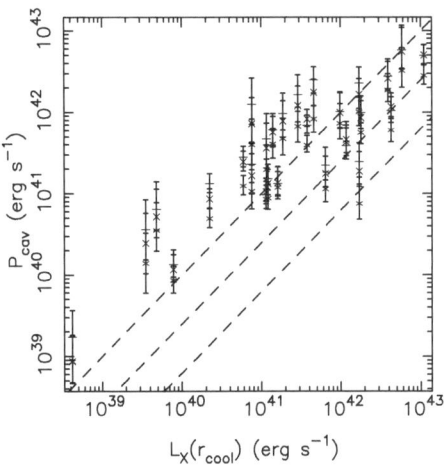

Fig. 2. Cavity heating power vs cooling power for nearby elliptical galaxies. Heating powers, P_{cav}, are estimated as pV/t_a, using three different estimates of the age, t_a, for the 27 galaxies with cavities in the sample of Jones *et al.*. Cooling power is the X-ray luminosity from within the projected radius where the cooling time equals 7.7×10^9 y. The dashed lines show where heating power equals cooling power, for heat inputs per cavity of pV, $4pV$ and $16pV$, from top to bottom.

To date, deeper X-ray observations of clusters have almost invariably revealed more cavities, *e.g.* [14], so that heating powers are likely to be underestimated from existing data [5, 2]. Progress in this areas is also hampered by relatively poor understanding of the selection effects for finding cavities [5]. A deep survey of a complete sample of cooling flow clusters is required to assess the overall significance of AGN heating in clusters. In the mean time, Jones *et al.* (in preparation and this proceedings) have assembled X-ray observations of a nearly complete sample of nearby giant elliptical galaxies. This includes $\simeq$ 160 galaxies, 109 of which show significant diffuse emission from hot gas after removal of resolved and unresolved point sources. Of those, 27 have significant AGN cavities. The sample is used here to assess the overall significance of AGN heating.

Our determinations of AGN heating rate parallel those of Bîrzan *et al.* [5] and Rafferty *et al.* [2], with minor modifications. Where gas temperatures are not available from the literature, they are estimated from the velocity dispersion, σ, of a galaxy as $kT = 1.5\mu m_H\sigma^2$ (consistent with the median of $\mu m_H\sigma^2/(kT)$ for the remaining galaxies). For the three galaxies without a temperature or velocity dispersion, the gas temperature is set to the median value of 0.7 keV. Abundances are assumed to be 0.5 solar. Electron densities are determined from beta model fits. Heating powers are estimated as pV per cavity, divided by one of the three estimates of cavity age, the sound crossing time, the buoyant rise time, or the refill time [5]. The cooling power is taken as the X-ray luminosity from within the projected radius where the cooling time equals 7.7×10^9 y. Our cooling powers are not corrected for mass deposition, *cf.* [5, 2].

The three resulting estimates of heating power are plotted against the cooling power in Fig. 2. Note that for UGC 408, the AGN appears to lie near the center of a single cavity, suggesting that the system is viewed almost along its radio axis, with its cavities and AGN projected on top of one another. The apparent distance from the AGN to the center of the cavity is then much smaller than the real distance, causing the age of the outburst to be underestimated and the heating power to be overestimated. Consistent with this, the naive estimates of its heating power are exceptionally high. For UGC 408 alone, we have therefore assumed that the distance from the center of the cavity to the AGN is equal to the semimajor axis of its cavity. The corrected heating powers agree with those for other systems with similar cooling powers. This correction reduces our estimate of the total heating power for all systems by a factor of ~ 2.

Approximately half of the outbursts in cooling flow clusters have heating powers that match or exceed their cooling powers for a heat input of $4pV$ per cavity [2]. By contrast, this is true for all of the nearby giant ellipticals in our sample (Fig. 2), apart from NGC 1553 [17]. Conversely, only about one quarter of the giant ellipticals with significant emission from hot gas have cavities, whereas the fraction of cluster cooling flows with outbursts is closer to 70% [16]. Consistent with the AGN feedback model, this suggests that the duty cycles of outbursts in larger systems have to be greater to keep the mean heating power at the level needed to stop cooling.

Allowing $1pV$ per cavity, the total heating powers for the 27 systems are 2.6×10^{43}, 2.9×10^{43} and 1.5×10^{43} erg s^{-1}, corresponding to ages of the sound crossing times, the buoyant rise times and the refill times, respectively. The total cooling power for all 109 galaxies with significant emission from diffuse hot gas is 1.06×10^{44} erg s^{-1}, so that the ratios of total cooling power to total heating power for the three age estimates are 4.1, 3.6 and 6.9, in turn.

The enthalpy of a cavity dominated by relativistic gas is $4pV$ and this value might, typically, be doubled by the "shock energy" (more precisely, the $p\,dV$ done as the cavity was inflated). While there is still significant systematic uncertainty in these numbers, they make a good case that AGN outbursts in nearby giant elliptical galaxies can prevent their X-ray emitting gas from cooling and forming stars. The outbursts are intermittent, with the AGN active $\sim 25\%$ of the time.

Acknowledgement. This work was partly supported by NASA grant NAS8-01130.

References

1. B.R. McNamara, these proceedings
2. D.A. Rafferty, B.R. McNamara, P.E.J. Nulsen, M.W. Wise, ApJ, in press (astro-ph/0605323)
3. A.C. Edge, 2001, MNRAS, 328, 762
4. E. Churazov, R. Sunyaev, W. Forman, H. Böhringer, 2002, MNRAS, 332, 729
5. L. Bîrzan, D.A. Rafferty, B.R. McNamara, M.W. Wise, P.E.J. Nulsen, 2004, ApJ, 607, 800
6. M.W. Wise, B.R. McNamara, P.E.J. Nulsen, J.C. Houck, L.P. David, 2006, ApJ, submitted
7. L.D. Landau, E.M. Lifshitz, 1959, *Course of Theoretical Physics, Volume 6, Fluid Mechanics*, (Pergamon, Oxford)
8. W. Forman, P. Nulsen, S. Heinz, F. Owen, J. Eilek, A. Vikhlinin, M. Markevitch, R. Kraft, E. Churazov, C. Jones, 2005, ApJ, 635, 894
9. B.R. McNamara, P.E.J. Nulsen, M.W. Wise, D.A. Rafferty, C. Carilli, C.L. Sarazin, E.L. Blanton, 2005, Nature, 433, 45
10. P.E.J. Nulsen, B.R. McNamara, M.W. Wise, L.P. David, 2005, ApJ, 628, 629
11. A.C. Fabian, J.S. Sanders, G.B. Taylor, S.W. Allen, C.S. Crawford, R.M. Johnstone, K. Iwasawa, 2006, MNRAS, 366, 417
12. M. Machacek, P.E.J. Nulsen, C. Jones, W.R. Forman, 2006, ApJ, 648,947
13. L.P. David, P.E.J. Nulsen, B.R. McNamara, W. Forman, C. Jones, T. Ponman, B. Robertson, M. Wise, 2001, ApJ, 557, 546
14. W. Forman, E. Churazov, C. Jones, M. Markevitch, P. Nulsen, A Vikhlinin, M. Begelman, H. Böhringer, J. Eilek, S. Heinz, R. Kraft, F. Owen, 2006, ApJ, submitted (astro-ph/0604583)
15. A.C. Fabian, C.S. Reynolds, G.B. Taylor, R.J.H. Dunn, 2005, MNRAS, 363, 891
16. R.J.H. Dunn, A.C. Fabian, G.B. Taylor, 2005, MNRAS, 364, 1343
17. E.L. Blanton, C.L. Sarazin, J.A. Irwin, 2001, ApJ, 552,106

The Difficulty of the Heating of Cluster Cooling Flows by Sound Waves and Weak Shocks

Y. Fujita[1] and T. Ken Suzuki[2]

[1] Department of Earth and Space Science, Graduate School of Science, Osaka University, Toyonaka, Osaka 560-0043, Japan
fujita@vega.ess.sci.osaka-u.ac.jp
[2] Graduate School of Arts and Sciences, University of Tokyo, Komaba, Meguro, Tokyo 153-8902, Japan
stakeru@provence.c.u-tokyo.ac.jp

Summary. We investigate heating of the cool core of a galaxy cluster through the dissipation of sound waves and weak shocks excited by the activities of the central active galactic nucleus (AGN). Using a weak shock theory, we show that this heating mechanism alone cannot reproduce the observed temperature and density profiles of a cluster, because the dissipation length of the waves is much smaller than the size of the core and thus the wave energy is not distributed to the whole core.

1 Introduction

The failure of standard cooling flow models indicates that the gas is prevented from cooling by some heating source(s). At present, the most popular candidate for the heating source is the active galactic nucleus (AGN) at the cluster center. However, it is not understood how the energy ejected by the AGN is transfered into the surrounding ICM. One idea is that bubbles inflated by AGN jets move outward in a cluster by buoyancy and mix with the surrounding ICM [1, 2, 3]. As a result of the mixing, hot ICM in the outer region of the cluster is brought into, and subsequently heats, the cluster center. The other idea is that the dissipation of sound waves created through the AGN activities. In fact, sound waves or weak shocks that may have evolved from sound waves are observed in the Perseus and the Virgo clusters [4, 5]. It was argued that the viscous dissipation of the sound waves is responsible for the heating of a cool core [4, 6]. They estimated the dissipation rate assuming that the waves are linear. However, when the amplitude of sound waves is large, the waves rapidly evolve into non-linear weak shocks [7], and their dissipation can be faster than the viscous dissipation of linear waves. Although [4] argued that weak shocks are present, their evolution from sound waves was not considered. Numerical simulations of dissipation of sound waves created by AGN activities were also performed [8]. Their results actually showed that the sound waves became weak shocks. However, their simulations were finished before radiative cooling became effective. Thus, the long-term balance between heating and cooling is still unknown. In this paper, we consider the evolution of sound waves to weak shocks, and analytically estimate the 'time-averaged' energy flux of the propagating waves as a function of distance, explicitly taking account of the dissipation at weak shock fronts and its global balance with radiative cooling. We assume a Hubble constant of $H_0 = 70$ km s^{-1} Mpc^{-1}. The details of the models and results are given in [9].

2 Models

We assume that sound waves are created by central AGN activity. The waves propagate in the ICM outwards. These waves, having a relatively large but finite amplitude, eventually form shocks to shape sawtooth waves. If the velocity amplitude is larger than ~ 0.1 times the sound velocity (the Mach number is $\gtrsim 1.1$), those waves steepen and become weak shocks after propagating less than a few wavelengths [10]. These shock waves directly heat the surrounding ICM by dissipating their wave energy. We adopt a heating model for the solar corona based on a weak shock theory ([10, 7], see also [11]). We assume that a cluster is spherically symmetric and steady.

The equation of continuity is

$$\dot{M} = -4\pi r^2 \rho v \, , \tag{1}$$

where $\dot{M}$ is the mass accretion rate, r is the distance from the cluster center, ρ is the ICM density, and v is the ICM velocity. The equation of momentum conservation is

$$v\frac{dv}{dr} = -\frac{GM(r)}{r^2} - \frac{1}{\rho}\frac{dp}{dr} - \frac{1}{\rho c_s\{1 + [(\gamma+1)/2]\alpha_w\}}\frac{1}{r^2}\frac{d}{dr}(r^2 F_w) \tag{2}$$

where G is the gravitational constant, $M(r)$ is the mass within radius r, p is the ICM pressure, c_s is the sound velocity, $\gamma(= 5/3)$ is the adiabatic constant, and α_w is the wave velocity amplitude normalized by the ambient sound velocity ($\alpha_w = \delta v_w/c_s$). The wave energy flux, F_w, is given by

$$F_w = \frac{1}{3}\rho c_s^3 \alpha_w^2 \left(1 + \frac{\gamma+1}{2}\alpha_w\right) \, . \tag{3}$$

The energy equation is

$$\rho v \frac{d}{dr}\left(\frac{1}{2}v^2 + \frac{\gamma}{\gamma-1}\frac{k_B T}{\mu m_H}\right) + \rho v \frac{GM(r)}{r^2} + \frac{1}{r^2}\frac{d}{dr}[r^2(F_w + F_c)] + n_e^2 \Lambda(T) = 0 \, , \tag{4}$$

where k_B is the Boltzmann constant, T is the ICM temperature, $\mu(= 0.6)$ is the mean molecular weight, m_H is the hydrogen mass, F_c is the conductive flux, n_e is the electron number density, and Λ is the cooling function. The term

$\nabla \cdot \boldsymbol{F}_w$ indicates the heating by the dissipation of the waves. The equation for the time-averaged amplitude of the shock waves is given by

$$\frac{d\alpha_w}{dr} = \frac{\alpha_w}{2}\left[-\frac{1}{p}\frac{dp}{dr} - \frac{2(\gamma+1)\alpha_w}{c_s\tau} - \frac{2}{r} - \frac{1}{c_s}\frac{dc_s}{dr}\right], \qquad (5)$$

where τ is the period of waves, which we assume to be constant [7, 10]. The second term of the right side of equation (5) denotes dissipation at each shock front.

3 Results

For parameters of our model cluster, we adopt the observational data of the Perseus cluster [12]. We assume that $r_s = 280$ kpc, $M(r_{1000}) = 3.39 \times 10^{14} M_\odot$, and $r_{1000} = 826$ kpc, where the mean density within r_Δ is Δ times the critical density of the Universe. Waves are injected at the inner boundary $r = r_0$, which should be close to the size of bubbles observed at cluster centers. We assume that $\lambda_0 = r_0$, where λ_0 is the initial wavelength. If the waves are injected in a form of sound waves with amplitude $0.1 \lesssim \alpha_w < 1$, waves travel about λ_0 before they become shock waves [10]. Therefore, for $r_0 \le r \le r_0 + \lambda_0 = 2\lambda_0$, we assume that $\nabla \cdot \boldsymbol{F}_w = 0$ (eqs. [2] and [4]), and that the second term of the right-hand side of equation (5) is zero. The temperature, electron density, and wave amplitude at $r = r_0$ are T_0 n_{e0}, and α_{w0}, respectively. Unless otherwise mentioned, the first two are fixed at $T_0 = 3$ keV and $n_{e0} = 0.08$ cm^{-3}, respectively, based on the observational results of the Perseus cluster [13].

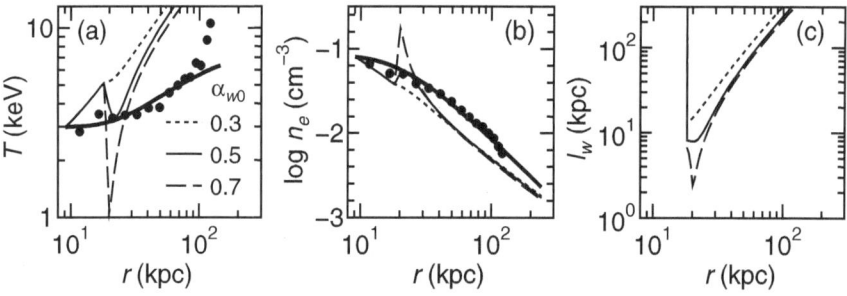

Fig. 1. (a) Temperature, (b) density, and (c) dissipation length profiles for $\alpha_{w0} = 0.3$ (dotted lines), 0.5 (thin solid lines), and 0.7 (dashed lines). Other parameters are $\tau = 1 \times 10^7$ yr, $\dot{M} = 50\,M_\odot\,\mathrm{yr}^{-1}$, and $f_c = 0$. Filled circles are *Chandra* observations of the Perseus cluster [13]. The bold solid lines correspond to a genuine cooling flow model of $\dot{M} = 500\,M_\odot\,\mathrm{yr}^{-1}$.

In Figure 1, we show the results when τ is fixed at 1×10^7 yr, and α_{w0} is changed. For the Perseus cluster, it was estimated that $\alpha_{w0} \sim 0.5$ [4]. The dissipation length is defined as $l_w = |F_w/\nabla \cdot \boldsymbol{F}_w|$. For these parameters, the initial wavelength is $\lambda_0 = 9$ kpc, which is roughly consistent with the *Chandra* observations [4]. Other parameters are $\dot{M} = 50\,M_\odot\,\mathrm{yr}^{-1}$, and $f_c = 0$, where f_c is the ratio of actual thermal conductivity to the classical Spitzer conductivity. In general, larger $\dot{M}$ reproduces observed temperature and density profiles better. However, large $\dot{M}$ is inconsistent with recent X-ray observations as was mentioned in § 1. For comparison, we show the results of a genuine cooling flow model ($\dot{M} = 500\,M_\odot\,\mathrm{yr}^{-1}$, $\alpha_{w0} = 0$, and $f_c = 0$) and the *Chandra* observations of the Perseus cluster [13]. Figures 1a and 1b show that only a small region is heated. The jumps of T and n_e at $r = 2\lambda_0 = 18$ kpc are produced by weak shock waves that start to dissipate there. The energy of the sound waves rapidly dissipates at the shocks, which is clearly illustrated in short dissipation lengths, $l_w \sim 2$–15 kpc (Fig. 1c). These dissipation lengths are smaller than those of viscous dissipation for linear waves, which can be represented by $l_v = 420\,\lambda_9^2\,n_{0.08}\,T_3^{-2}$ kpc, where the wavelength $\lambda = 9\,\lambda_9$ kpc, the density $n = 0.08\,n_{0.08}$ cm^{-3}, and the temperature $T = 3\,T_3$ keV [4]. In Figure 1, the ICM density becomes large and the temperature becomes small at $r \gtrsim 2\lambda_0$ so that the rapid shock dissipation is balanced with radiative cooling. Because of this, waves cannot reproduce the observed temperature and density profiles that gradually change on a scale of ~ 100 kpc.

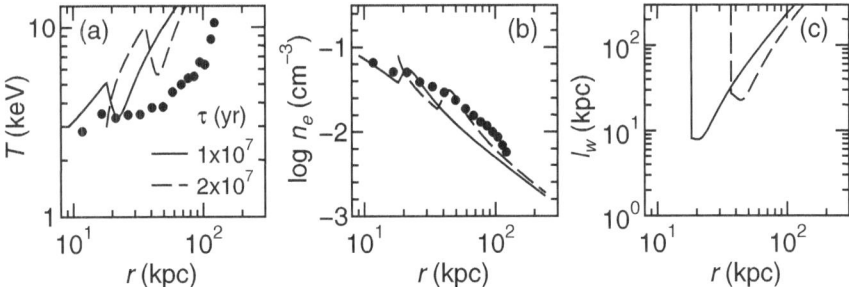

Fig. 2. Same as Fig. 1 but for $\tau = 1 \times 10^7$ yr (solid lines), and 2×10^7 yr (dashed lines). Other parameters are $\alpha_{w0} = 0.5$, $\dot{M} = 50\,M_\odot\,\mathrm{yr}^{-1}$, and $f_c = 0$.

In Figure 2, we present the results when $\tau = 2 \times 10^7$ yr. Compared with the case of $\tau = 1 \times 10^7$ yr, the wave energy dissipates in outer regions. However, the dissipation lengths are still smaller than the cluster core size (~ 100 kpc). Note that larger τ (or λ_0) means formation of larger bubbles. As indicated in [14], it is unlikely that the size of the bubbles becomes much larger than 20 kpc; the bubbles start rising through buoyancy before they become larger. On the other hand, when $\tau < 10^7$ yr ($\lambda_0 < 9$ kpc), the waves heat only the ICM around the cluster center. The predicted temperature and density profiles are obviously inconsistent with the observations.

The inclusion of thermal conduction changes the situation dramatically. Figure 3 shows the results when $f_c = 0.2$. The models including both wave heating and thermal conduction can well reproduce the observed temperature and density profiles. Figure 3c shows the contribution of the wave heating $(-\nabla \cdot \boldsymbol{F}_w)$ to compensating radiative cooling $(n_e^2 \Lambda)$. Since $-\nabla \cdot \boldsymbol{F}_w / n_e^2 \Lambda > 1/2$ for $r \sim 20$–30 kpc, the wave heating is more effective than the thermal conduction in that region.

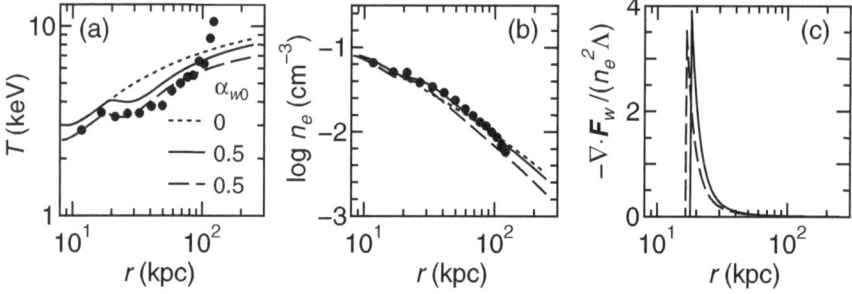

Fig. 3. (a) Temperature, (b) density, and (c) dissipation strength profiles for $\alpha_{w0} = 0$ and $T_0 = 3$ keV (dotted lines), $\alpha_{w0} = 0.5$ and $T_0 = 3$ keV (solid lines), and $\alpha_{w0} = 0.5$ and $T_0 = 2.5$ keV (dashed lines). Other parameters are $\tau = 1 \times 10^7$ yr, $\dot{M} = 50\ M_\odot\ \mathrm{yr}^{-1}$, and $f_c = 0.2$. Filled circles are *Chandra* observations of the Perseus cluster [13].

4 Discussion

We showed that sound waves created by the central AGN alone cannot reproduce the observed temperature and density profiles of a cluster, because the dissipation length of the waves is much smaller than the size of a cluster core and the waves cannot heat the whole core. The results have been confirmed by numerical simulations [15]. On the other hand, we found that if we include thermal conduction from the hot outer layer of a cluster with the conductivity of 20% of the Spitzer value, the observed temperature and density profiles can be reproduced. The idea of the "double heating" (AGN plus thermal conduction) was proposed in [16].

However, the fine structures observed in cluster cores may show that the actual conductivity is much smaller than what we assumed [17]; the structures would soon be erased if the conductivity is that large. If the conductivity is small, we need to consider other possibilities. While we considered successive minor AGN activities, some authors consider that rare major AGN activity should be responsible for the heating of cool cores [18]. In this scenario, powerful bursts of the central AGN excite strong shocks and heat the surrounding gas in the inner region of a cluster on a timescale of $\gtrsim 10^9$ yr. Moreover, in this scenario, heating and cooling are not necessarily balanced at a given time, although they must be balanced on a very long-term average. This is consistent with the fact that there is no correlation between the masses of black holes in the central AGNs and the X-ray luminosities of the central regions of the clusters [19]. Alternative idea is that cluster mergers are responsible for heating of cool cores [20, 21]. In this "tsunami" model, bulk gas motions excited by cluster mergers produce turbulence in and around a core, because the cooling heavy core cannot be moved by the bulk gas motions, and the resultant relative gas motion between the core and the surrounding gas induces hydrodynamic instabilities. The core is heated by turbulent diffusion from the hot outer region of the cluster. Since the turbulence is produced and the heating is effective only when the core is cooling and dense, fine-tuning of balance between cooling and heating is alleviated for this model.

References

1. E. Churazov, M. Brüggen, C. R. Kaiser et al., 2001, ApJ, 554, 261
2. V. Quilis, R. G. Bower, M. L. Balogh, MNRAS, 2001, 328, 1091
3. C. J. Saxton, R. S. Sutherland, G. V. Bicknell, 2001, ApJ, 563, 103,
4. A. C. Fabian, J. S. Sanders, S. W. Allen et al, MNRAS, 2003, 344, L43
5. W. Forman, P. Nulsen, S. Heinz et al., 2005, ApJ, 635, 894
6. A. C. Fabian, C. S. Reynolds, G. B. Taylor et al.: 2005, MNRAS, 363, 891
7. R. F. Stein, R. A. Schwartz, 1972, ApJ, 177, 807
8. M. Ruszkowski, M. Brüggen, M. C. Begelman, 2004, ApJ, 615, 675
9. Y. Fujita, T. K. Suzuki, 2005, ApJL, 630, L1
10. T. K. Suzuki, ApJ, 2002, 578, 598
11. Y. Fujita, T. K. Suzuki, K. Wada, 2004, ApJ, 600, 650
12. S. Ettori, S. De Grandi, S. Molendi, 2002, A&A, 391, 841
13. J. S. Sanders, A. C. Fabian, S. W. Allen et al, 2004, MNRAS, 349, 952
14. E. Churazov, W. Forman, C. Jones et al., 2000, A&A, 356, 788
15. W. G. Mathews, A. Faltenbacher, F. Brighenti, 2006, ApJ, 638, 659
16. M. Ruszkowski, M. C. Begelman, 2002, ApJ, 581, 223
17. Y. Fujita, C. L. Sarazin, J. C. Kempner et al, 2002, ApJ, 575, 764
18. N. Soker, R. E. White, L. P. David et al., 2001, ApJ, 549, 832
19. Y. Fujita, T. H. Reiprich, 2004, ApJ, 612, 797
20. Y. Fujita, T. Matsumoto, K. Wada, 2004, ApJL, 612, L9
21. Y. Fujita, T. Matsumoto, K. Wada et al, 2005, ApJL, 619, L139

The Lifecycle of Powerful AGN Outflows

C. R. Kaiser[1] and P. N. Best[2]

[1] School of Physics & Astronomy, University of Southampton, Southampton,
   SO17 1BJ, UK crk@soton.ac.uk
[2] Institute for Astronomy, Royal Observatory Edinburgh, Blackford Hill,
   Edinburgh, EH9 3HJ, UK pnb@roe.ac.uk

Summary. During the course of this conference, much evidence was presented that points to an intimate connection between the energetic outflows driven by AGN and the energy budget and quite possibly also the evolution of their gaseous environments. However, it is still not clear if and how the AGN activity is triggered by the cooling gas, how long the activity lasts for and how these effects give rise to the observed distribution of morphologies of the outflows.

In this contribution we concentrate on the high radio luminosity end of the AGN population. While most of the heating of the environmental gas may be due to less luminous and energetic outflows [1], these more powerful objects have a very profound influence on their surroundings [2]. We will describe a simple model for powerful radio galaxies and radio-loud quasars that explains the dichotomy of their large-scale radio morphologies as well as their radio luminosity function.

1 The FR dichotomy

In their original paper Fanaroff and Riley [3] pointed out that the large-scale radio structure of radio-loud AGN is either edge-darkened (FRI) or edge-brightened (FRII) and that there is a reasonably sharp transition between these two morphologies at around a luminosity of $10^{26}\,\mathrm{W\,Hz^{-1}}$ at an observing frequency of $151\,\mathrm{MHz}$. High-resolution radio observations show thin, presumably laminar jet flows ending in strong shocks for the more luminous FRII-type objects. After passing through the shock, visible as luminous 'hotspots' in radio observations, the jet material inflates radio lobes around the jets. Objects classed as FRI-types show a more varied radio morphology. While some show radio lobes reminiscent of those of FRII-type object without radio hotspots, other FRI-type objects contain turbulent jets. The appearance of these turbulent jets is similar to that of smoke rising out of a chimney. Here we concentrate on the latter class of objects.

There are also several other observational differences between the AGN giving rise to the jet flows in the two FR classes. Usually FRI-type objects show weaker optical continuum [4] and line emission [5]. Their X-ray emission is also less luminous [6]. All this evidence points to weaker emission from the accretion disc in the AGN in FRI-type sources, which is also consistent with less radiative heating of the dust in their host galaxies [7]. However, the so-called Low-Excitation Radio Galaxies (LERG) with FRII morphology and radio luminosities close to the value separating the FR classes appear to contain AGN more similar in their emission properties to those usually found in FRI-type objects [4, 8, 6]. The opposite combination of a bright AGN with an FRI-type radio morphology appears to be rare, but some detections have been claimed [9].

Finally, the dividing line between the FR classes in terms of radio luminosity is related to the optical luminosity of the host galaxy as $L_{\mathrm{B}}^{1.8}$ in the B-band [10].

2 The Radio Luminosity Function (RLF)

The RLF at low cosmological redshifts is well approximated by a broken power-law [11]. The cosmological evolution of the radio luminosity function seems to imply that this shape is caused by two distinct populations, one of which dominates below the break and the other contributing mainly at luminosities above the break [11, 12]. The break luminosity of the RLF is suspiciously close to the dividing line between the two FR classes. However, current samples of extragalactic radio sources do not contain information on radio morphology for enough sources or are too small to allow the determination of separate RLFs for the two classes.

In the following we will use a break luminosity of $P_{\mathrm{break}} \sim 5 \times 10^{25}\,\mathrm{W\,Hz^{-1}}$ at an observing frequency of 1.4 GHz. We also use the power-law slopes of the RLF as -0.6 below P_{break} and -2.4 above the break [12]. Note here, that observationally the RLF is determined in units of density per logarithm of radio luminosity. If we work in units of density per radio luminosity, then the slopes of the RLF are -1.6 and -3.4, respectively.

3 A unified model for both FR classes and the RLF

3.1 Modelling FRI-type sources

Due to their turbulent jets, it is notoriously difficult to construct models for the evolution of FRI-type sources. The flow structure is complicated and virtually impossible to model accurately. This implies that the strength and structure of the magnetic field is not predictable. It is also not clear how and where the relativistic electrons giving rise to the observed synchrotron radiation are accelerated. Hence it is also not possible to estimate how much of the energy transported by the jets is dissipated to these electrons.

Some insight can be gained from arguments based on conservation laws [13], but this does not lead to a straightforward model describing the evolution of the jet flow and the emitted radiation. Models of the velocity field of the jet and the underlying magnetic fields and distributions of relativistic electrons can be crafted on high resolution radio images of FRI-type sources [14], but these models do not make predictions for the emission properties of these objects in general. However, both approaches can be used as guides towards some rough principles determining the radio luminosity and its evolution.

Most FRI-type source with turbulent jets show a low luminosity, confined jet similar to the jets seen in FRII-type morphologies extending for a short distance from the radio core. At the end of this possibly laminar jet the flow suddenly widens and brightens considerably in the 'flare point'. After this the jet appears to be completely turbulent, it spreads further and fades into the surroundings as it travels outwards. The sudden increase in the luminosity of the jet in the flare point and the fading of the emission further out may indicate that most of the emitting particles are accelerated in the flare point with little or no additional acceleration at larger distances. If the energy transport rate of the jet or 'jet power' and the flare point are not evolving, then a zeroth-order model for FRI sources would predict their radio luminosity to remain constant throughout their lifetime. Furthermore, if the flare point does not evolve, then it must be in pressure equilibrium with the surrounding gas. It is then very likely that the radio luminosity of the flare point and that of the turbulent jet are linearly proportional to the jet power.

In the following we adopt this simple model for FRI-type sources. If we now assume that the lifetime of the jet flows does not depend on the jet power, Q, and also that the RLF is dominated by FRI-type objects below P_{break}, then we can infer the 'birth rate' of radio-loud AGN directly from the RLF to be proportional to $Q^{-1.6}$.

3.2 Modelling FRII-type sources

It is possible to formulate analytical models for the radio luminosity evolution of FRII-type objects. The reason is that in this class the underlying fluid flow is laminar and hence more amenable to modelling. Also, high resolution radio observations imply little or no acceleration of relativistic particles away from the radio hotspots at the ends of the laminar jets [15].

Most current models are based on the fundamental picture of a jet in pressure equilibrium with its own lobe [16]. Once the dynamics of the jet-lobe system are established [17], we can calculate the resulting radio emission taking into account the cumulative energy losses of the radiating electrons [18]. The gas density in the environment of the source is modelled as a modified β-profile, i.e.

$$\rho = \frac{\rho_0}{\left[1 + (r/a_0)^2\right]^{\beta/2}}. \tag{1}$$

This profile is well approximated by a constant density ρ_0 inside the core radius a_0 and by a simple power-law $\rho = \rho_0 (r/a_0)^{-\beta}$ further out. The luminosity evolution in the power-law regime is given by $P \propto t^{-1/2}$ for $\beta = 2$. Here we neglect radiative losses of the relativistic electrons due to the emitted synchrotron radiation as these mainly affect young objects [18]. We also ignore the energy losses due to Inverse Compton (IC) scattering of cosmic microwave background (CMB) photons off the electrons because these affect mainly old sources (see below).

If we assume that the RLF above P_{break} is dominated by FRII-type objects, then we can determine the slope of the RLF from the birth function of sources with a given jet power and the luminosity evolution determined above. The model predicts an RLF proportional to $P^{-3.4}$ in excellent agreement with the observations.

3.3 The connection between the FR classes

What decides the radio morphology a given sources develops? This must clearly be connected to the onset of turbulent disruption of the jet flow. In FRII-type sources the jets are located inside the radio lobes. These contain mainly the material that has been transported along the jet. From geometric considerations it follows that the gas density in the lobe is lower than that inside the jet. Therefore the jets in FRII sources are very well protected against turbulent disruption which would require the jets to be in contact with much denser gas.

The inflation of lobes depends crucially on the formation of a bow shock around the lobes. In other words, the lobes must be overpressured with respect to the ambient gas. If they are not, then Rayleigh-Taylor instabilities develop on the lobe surface [16]. In this case, the denser gas of the source environment will start to buoyantly replace the lobe material until it reaches the jet itself. At this point the interaction of the jet with the dense material will lead to turbulent disruption of the jet. So, the crucial condition for the jet to remain laminar is the overpressured lobe.

At large distances the pressure in the source environment will decrease proportional to r^{-2}. The dynamical model of the jet-lobe system predicts that the pressure inside the lobe also decreases with r^{-2} in this regime [17]. Hence the lobe remains overpressured with respect to its surroundings at all times provided it was overpressured in the beginning. The situation is quite different in the core region, where the external pressure is essentially constant, but the pressure inside the lobe decreases as $r^{-4/3}$. Since the pressure inside the lobe also depends on the jet power, we can summarise this result as follows. All sources start out with an FRII-type morphology. The lobe pressure in sources with weaker jets falls so rapidly that they come into pressure equilibrium with their surroundings while they are still located in the core region. Their jets develop turbulent flows with an FRI-type radio morphology. These sources dominate the RLF at low radio luminosities. More powerful jets grow beyond the core region before the pressure in their lobes decreases too much. They retain their FRII-type morphology and form the bulk of the population at the high luminosity end of the RLF.

For a given environment density distribution we can calculate from the model the jet power at which the divide between the FR classes occurs.

For reasonable values of the necessary parameters we obtain 10^{37} W, close to the jet power suggested for the divide from observations of emission line strengths [19]. We can also determine the radio luminosity associated with the divide. This depends on the density of the source environment which also determines the X-ray luminosity of this gas. Using the observed correlation of X-ray luminosity and luminosity in the B-band of giant elliptical galaxies [20], we predict that the dividing line between the FR classes should occur at around 10^{26} W Hz^{-1} with a dependence on the B-band luminosity of the host galaxy as $L_{\mathrm{B}}^{1.5\rightarrow 2}$. Both predictions are again in very good agreement with observations [12, 10].

4 A second switch

The above model establishes a connection between the break in the RLF and the division between the FR classes in terms of radio morphology on the one side and the jet power and the density of the source environment on the other. Therefore in its current form it cannot explain the differences between the FR classes observed outside the radio band.

The observational evidence suggests that the AGN predominantly associated with large-scale radio structure of type FRI are comparatively inefficient at producing observable radiation. For jet producing X-ray binaries the radiatively inefficient accretion state is associated with the production of weak jets. Powerful jet ejections are connected to the transition to the radiatively efficient accretion state [21]. If the same connection applies to the accretion discs of AGN and their jets, then this would explain the observations as less powerful jets arising from radiatively inefficient accretion discs are more susceptible to turbulent disruption and hence preferentially develop FRI-type morphologies.

As we have seen above, the division between the FR classes depends not only on the jet power but also on the properties of the source environment. Hence we may expect to observe some hybrid objects, i.e. radiatively inefficient AGN with an FRII-type radio morphology or bright AGN with FRI-type radio structures. The former can be identified as the LERGs. Their radio luminosities are always close to the dividing line between the FR classes [22] and they also often show weak radio hotspots and bright jet flows [23]. Both properties may indicate relatively weak jets that are affected by turbulence.

Radiatively efficient AGN with turbulent jets appear to be rare. This may indicate that the jet power associated with the division of radiatively efficient and inefficient accretion is below 10^{37} W; the jet power at which the FR divide occurs in typical source environments. It will be interesting to further pursue this connection between the jets emerging from AGN and those of X-ray binary stars. In particular the evidence for the X-ray binaries shows that individual objects cycle through accretion states and produce jets of very different power along the way. Is this also true for AGN?

References

1. P. N. Best, C. R. Kaiser, T. M. Heckman, G. Kauffmann, 2006, MNRAS, 368, L67
2. J. F. Basson, P. Alexander, 2003, MNRAS, 339, 353
3. B. L. Fanaroff, J. M. Riley, 1974, MNRAS, 167, 31
4. M. Chiaberge, A. Capetti, A. Celotti, 2000, A&A, 355, 873
5. R. G. Hine, M. S. Longair, 1979, MNRAS, 188, 111
6. M. J. Hardcastle, D. A. Evans, J. H. Croston, 2006, MNRAS, 370, 1893
7. S. A. H. Müller, et al., 2004, A&A, 426, L29
8. S. A. Baum, E. L. Zirbel, C. P. O'Dea, 1995, ApJ, 451, 88
9. K. M. Blundell, S. Rawlings, 2001, ApJ, 562, L5
10. M. J. Ledlow, F. N. Owen, 1996, AJ, 112, 9
11. J. S. Dunlop, J. A. Peacock, 1990, MNRAS, 247, 19
12. C. J. Willott, S. Rawlings, K. M. Blundell, M. Lacy, S. A. Eales, 2001, MNRAS, 322, 536
13. G. V. Bicknell, 1995, ApJ Supp., 101, 29
14. R. A. Laing, A. H. Bridle, 2004, MNRAS, 348, 1459
15. C. L. Carilli, R. A. Perley, J. W. Dreher, J. P. Leahy, 1991, ApJ, 383, 554
16. P. A. G. Scheuer, 1974, MNRAS, 166, 513
17. C. R. Kaiser, P. Alexander, 1997, MNRAS, 286, 215
18. C. R. Kaiser, J. Dennett-Thorpe, P. Alexander, 1997, MNRAS, 292, 723
19. S. Rawlings, R. Saunders, 1991, Nat., 349, 138
20. E. O'Sullivan, D. A. Forbes, T. J. Ponman, 2001, MNRAS, 328, 461
21. R. P. Fender, T. M. Belloni, E. Gallo, 2004, MNRAS, 355, 1105
22. P. D. Barthel, 1989, ApJ, 336, 606
23. M. J. Hardcastle, P. Alexander, G. G. Pooley, J. M. Riley, 1998, MNRAS, 296, 445

MHD Simulations of Clusters of Galaxies Including Radiative Cooling

N. Asai[1], N. Fukuda[2] and R. Matsumoto[3]

[1] Graduate School of Science and Technology, Chiba University, 1-33 Yayoi-cho, Inage-ku, Chiba 263-8522, Japan `asai@astro.s.chiba-u.ac.jp`

[2] Department of Computer Simulation, Faculty of Informatics, Okayama University of Science, 1-1 Ridai-cho, Okayama 700-0005, Japan `fukudany@sp.ous.ac.jp`

[3] Department of Physics, Faculty of Science, Chiba University, 1-33 Yayoi-cho, Inage-ku, Chiba 263-8522, Japan `matumoto@astro.s.chiba-u.ac.jp`

Summary. We carried out 2D and 3D magnetohydrodynamic simulations including radiative cooling and anisotropic thermal conduction of a magnetically turbulent intracluster medium. In 3D simulations, we found that magnetic pressure increases by a factor of 2 at 1.4 Gyr because magnetic fields accumulate in the center along with the plasma contracting by the cooling instability. In 2D simulations including thermal conduction, we found that a low temperature region exists along the loop-shaped magnetic field lines. Thus, the restriction of thermal conduction across magnetic field lines enables the coexistence of hot and cool plasmas in cluster cores.

1 Introduction

Observations by *ASCA* and *XMM-Newton* of cluster cores suggest that heat sources are required to compensate for the radiative cooling in the cluster cores [1, 2]. In this paper, we focus on the effects of conductive heating from the ambient hot intracluster medium (ICM) and magnetic heating.

When a magnetized plasma contracts spherically by cooling, the radius of magnetic flux tubes embedded in the plasma will decrease. The strength of magnetic fields frozen to the plasma will increase according to $B \propto R^{-2}$, where R is the radius of the flux tube. The magnetic pressure is then enhanced as $P_{\mathrm{mag}} = B^2/8\,\pi \propto R^{-4}$. As the plasma contracts, since $\rho \propto R^{-3}$ by mass conservation, $P_{\mathrm{gas}} \propto \rho T$ increases slower than R^{-3} when the temperature decreases with radius due to cooling. Thus, $\beta = P_{\mathrm{gas}}/P_{\mathrm{mag}}$ decreases as the plasma contracts. If $\beta \leq 1$, the magnetic energy released by magnetic reconnections contributes to the plasma heating.

Since thermal conductivity strongly depends on the direction of the magnetic fields, conductive heating of cluster cores should be studied by considering the magnetic fields [3]. Heat is conducted from the outer hot region to the cool core along the magnetic channels connecting the cool core and the ambient ICM. Moreover, the anisotropic thermal conduction leads to the coexistence of hot and cool components in cluster cores as indicated by observations [4].

2 Simulation Model

We solved ideal 2D and 3D MHD equations in a Cartesian coordinate system by using a solver based on a modified Lax-Wendroff method with artificial viscosity implemented to the Coordinated Astronomical Numerical Software (CANS). We assume that the density distribution is given by an isothermal β-model profile and that the magnetic fields are turbulent. The size of the computational box is $150 \, \mathrm{kpc}^3$, with 192^3 grid points for a 3D model (model M3), and $900 \, \mathrm{kpc}^2$ with 1024^2 grid points for a 2D model (model MC2). For model M3, radiative cooling is included. The mean magnetic field strength B_0 is $0.03 \, \mu\mathrm{G}$ (plasma β is 7.5×10^4). For model MC2, both radiative cooling and anisotropic thermal conduction are included. The mean magnetic field strength B_0 in the central region ($300 \, \mathrm{kpc} \times 300 \, \mathrm{kpc}$) is $0.07 \, \mu\mathrm{G}$ (plasma β is 1.0×10^4). We assume that the conductivity parallel to the field lines has the Spitzer value, but the conductivity perpendicular to the field line is zero. We apply absorbing boundary conditions for both models. The units of length, velocity, density, and time are $r_0 = 15 \, \mathrm{kpc}$, $v_0 = 1000 \, \mathrm{km \, s^{-1}}$, $\rho_0 = 5 \times 10^{-27} \, \mathrm{g \, cm^{-3}}$, and $t_0 = 2 \times 10^7 \, \mathrm{yr}$, respectively.

3 Results

We present the results of 3D simulations showing the growth of magnetic fields. Around the dense central region where gas contracts, the radiative cooling leads to the cooling instability. Fig. 1 (a) shows a snap shot of the distribution of the magnetic field strength at $t = 1.4 \, \mathrm{Gyr}$. Magnetic fields accumulate and are enhanced around the center. Fig. 1 (b) shows the growth of magnetic pressure in the central region. The magnetic pressure grows exponentially and increases by a factor of 2 at $t = 1.4 \, \mathrm{Gyr}$. In this simulation, however, the enhanced magnetic fields are still too small to contribute to the heating of the cooling core.

Next, we present the result of 2D simulations showing the effects of magnetic fields on thermal conduction. Fig. 2 shows the close-up snap shot of the temperature distribution at $t = 1.2 \, \mathrm{Gyr}$. Loop-shaped structures of low temperature plasma appear around the coolest region along the direction of magnetic fields because thermal conduction is restricted by magnetic fields. Heat is conducted only in the direction parallel to the magnetic field lines. In this simulation, however, since the magnetic field lines do not connect the core and the ambient plasma, the core cannot be heated by the heat conduction from the ambient plasma. The loop structures are similar to those expected from the *cD-corona* model [5] and may explain the coexistence of hot and cool components in cluster cores indicated by observations [4].

Acknowledgement. We thank T. Yokoyama for developments of the coordinated astronomical numerical software (CANS). This work is supported by JSPS Re-

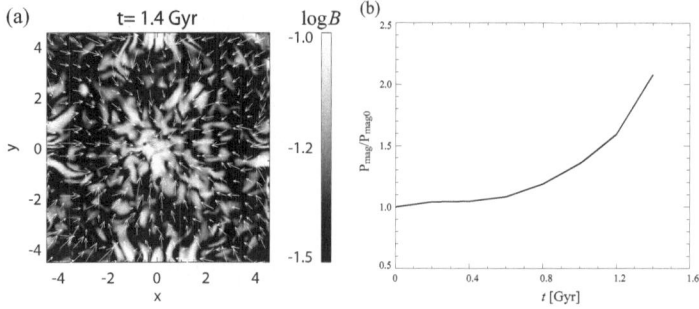

Fig. 1. (a) Distribution of magnetic field strength in the $z = 0$ plane at $t = 1.4\,\mathrm{Gyr}$ for model M3. Arrows show the velocity vectors. (b) The growth of the mean magnetic pressure normalized by the initial value in the central region ($[30\,\mathrm{kpc}]^3$)

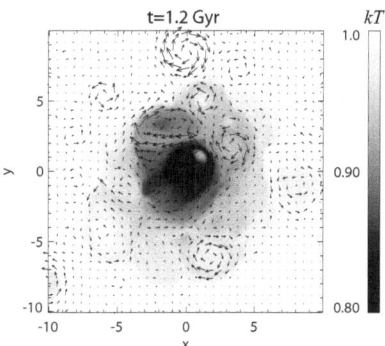

Fig. 2. Close-up ($300\,\mathrm{kpc} \times 300\,\mathrm{kpc}$) snap shot of temperature distribution at $t = 1.2\,\mathrm{Gyr}$ for model MC2. Arrows show the magnetic field vectors

search Fellowships for Young Scientists. Numerical computations were carried out on VPP5000 at the Center for Computational Astrophysics, CfCA, of the National Astronomical Observatory of Japan and joint research program of IMIT, Chiba University.

References

1. Y. Ikebe, K. Makishima, H. Ezawa et al: Astrophys. J. **481**, 660 (1997)
2. J. R. Peterson, F. B. S. Paerels, J. S. Kaastra et al: Astron. Astrophys. **365**, L107 (2001)
3. E. Bertschinger, A. Meiksin: Astrophys. J. **306**, L1 (1986)
4. Y. Fukazawa, T. Ohashi, A. C. Fabian et al: Publ. Astron. Soc. Japan **46**, L55 (1994)
5. K. Makishima, H. Ezawa, Y. Fukazawa et al: Publ. Astron. Soc. Japan **53**, 401 (2001)

Quantifying "Feedback" in Cool Core and Non-Cool Core Clusters

I. G. McCarthy[1], A. Babul[2], R. G. Bower[1] and M. L. Balogh[3]

[1] Department of Physics, University of Durham, Durham, DH1 3L3, United Kingdom i.g.mccarthy@durham.ac.uk
[2] Department of Physics and Astronomy, University of Victoria, Victoria, BC, Canada, V8P 5C2
[3] Department of Physics and Astronomy, University of Waterloo, Waterloo, ON, Canada, N2L 5L4

1 Introduction

The lack of cool gas with $kT \leq 1 - 2$ keV in "cool core" (CC) clusters implies that some form of heating is offsetting radiative cooling in these systems. There is mounting evidence for a self-regulatory feedback scenario in which cooling gas feeds a supermassive black hole, following which an AGN 'turns on' and heats the gas [1]. Most compelling are recent X-ray images with radio overlays that reveal bubbles of hot plasma rising buoyantly through the ICM [2]. The estimated energy needed to inflate these bubbles can be quite large, ranging from $\approx 10^{58-61}$ ergs [3]. If this energy can be dissipated in the ICM (e.g., via thermalisation of PdV work done on the ICM by the rising bubbles), it can possibly completely offset radiative losses in many/most CC clusters.

While the observed bubbles seem able to *maintain* the present structure of CC clusters, how these systems got into this self-regulatory loop in the first place is unclear. Presumably, the properties of the ICM before the first AGN burst were quite different than they are today (see §2). In addition, we still lack an adequate understanding of non-cool core (NCC) clusters, systems that likely experienced some form of extreme heating in their past. Remarkably, recent estimates suggest that NCC clusters make up perhaps 50% of the cluster population [4, 5], yet there is still a surprising dearth of high quality observations of these systems. As part of a study that we will soon submit for publication [6], we have sought to shed light on these two important issues through detailed comparisons of simple models with the observed properties of CC and NCC systems.

2 Heating in CC and NCC Clusters

2.1 Analytic Models & Comparison to Observations

If the heating requirements necessary to explain the observed structure of clusters could be calculated, this would yield valuable clues to the nature of the heating sources once present in these systems. Fortunately, in the absence of feedback, radiative cooling and gravity establish a unique (steady-state) distribution for the ICM in a cluster of mass M. To explain, within r_{cool} the shapes of the ICM temperature and density profiles are dictated by the shapes of the gravitational potential and the cooling function [7]. Both of these are known quantities, with cosmological simulations providing the form of the potential well [8]. The normalisations of these profiles within r_{cool} are dictated by the properties of these profiles at larger radii. These, in turn, are set by gravitational processes alone and can be calculated reliably using non-radiative cosmological simulations [9]. We refer to this unique steady state condition as the "pure cooling" model.

To deduce the required amount of heating, one need only compare the observed structural properties of clusters with those predicted by the pure cooling model. We have made detailed comparisons of the model with the gas density, entropy, and temperature distributions of CC and NCC systems from *Chandra*, *XMM-Newton*, and *ROSAT* observations. As an example, in Figure 1 we compare the *ROSAT* gas density profiles of a flux-limited cluster sample with that predicted by the pure cooling model.

The minimum energy needed to transform the pure cooling state into some other state can be obtained by simply subtracting the total energy of the pure cooling state from the total energy of the final state (where the total energy is just the summation of the total thermal and potential energies of the gas). In this way, we calculate the energy required to explain the structure of CC and NCC systems, with these representing the "final" states.

2.2 Implications for Heating

In the case of CC systems (with $t_{cool} < 3$ Gyr), we find that approximately 10^{62} ergs must be injected into the ICM. This is huge, being roughly equivalent to half the total energy a CC cluster would radiate over its lifetime or roughly a factor of two larger than the most powerful AGN outburst known [11]. Furthermore, *every* CC cluster must have experienced a blast this large at least once during their life. NCC systems, on the other hand, require even more energy(!) — as much as a few times 10^{63} ergs is required. However, once the CC and NCC clusters are heated to this level, maintaining their structure is energetically much easier (see §1).

As discussed in [6] the only plausible way to lower the energy requirements of the initial burst is if the gas was heated prior to cluster formation (i.e., "preheating"). We calculate that if preheating occurred recently ($1 < z < 3$), the energy requirements can be lowered by up to two orders of magnitude. Given that there are no known sources of heating powerful enough to explain NCC systems if heated from within, this seems a real possibility. We envisage a scenario in which *all* clusters are preheated to varying degrees. Those preheated to a large degree, such that cooling was never subsequently important, evolved passively into NCC systems. Those preheated to a lesser degree, the

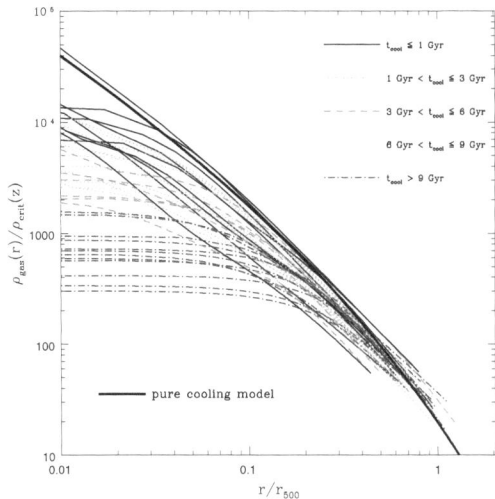

Fig. 1. Comparison of the pure cooling model with the *ROSAT* gas density profiles of a flux-limited cluster sample [10]. At large radii, the observed systems (both CC and NCC) and the theoretical model converge, implying that feedback has a minimal role compared to gravitational shock heating there. However, at small radii ($r < 0.3r_{500}$ or so), the observed systems are under-dense compared to the pure cooling model (particularly the NCC systems with $t_{\rm cool} > 3$ Gyr) — a signature of non-gravitational heating.

CC systems, cooled down into the self-regulatory loop (as opposed to being 'heated up' into the loop). Additional work is required to test this hypothesis.

References

1. C. R. Kaiser, J. Binney, 2003, MNRAS, 338, 837
2. A. C. Fabian et al., 2006, MNRAS, 366, 417
3. L. Birzan et al., 2004, ApJ, 607, 800
4. C. B. Peres et al., 1998, MNRAS, 298, 416
5. D. Hudson, T. Reiprich et al., this volume.
6. I. G. McCarthy, A. Babul, R. G. Bower, M. L. Balogh, 2006, MNRAS, to be submitted
7. E. Bertschinger, 1989, ApJ,, 340, 666
8. J. F. Navarro, C. S. Frenk, S. D. M. White, 1997, ApJ, 490, 493
9. G. M. Voit, S. T. Kay, G. L. Bryan, 2005, MNRAS, 364, 909
10. J. J. Mohr, B. Mathiesen, A. E. Evrard, 1999, ApJ, 517, 627
11. B. R. McNamara et al., 2005, Nature, 433, 45

Buoyant Bubbles in the Virgo Cluster

G. Pavlovski[1], C. R. Kaiser[1] , E. C. D. Pope[1,2,3] and H. Fangohr[3]

[1] School of Physics and Astronomy, University of Southampton, Southampton, SO17 1BJ, U.K. gbp@phys.soton.ac.uk (GP), gbp@phys.soton.ac.uk (CRK)
[2] School of Physics and Astronomy, University of Leeds, LS2 9JT, U.K. edpope@astro.soton.ac.uk (ECDP)
[3] School of Engineering Sciences, University of Southampton, Southampton, SO17 1BJ, U.K. fangohr@soton.ac.uk (HF)

1 Introduction

In this study we have constructed and tested 3D numerical model of the Virgo cluster, the core of which is thermally supported by the bubbles produced by the active galactic nucleus of the central galaxy.

The physical properties of the plasma (intracluster medium, ICM) are not precisely known. Usually the ICM models use the values of thermal diffusivity and kinematic viscosity calculated in the classical work of Spitzer [10] (see also review by M. Ruszkowski in this volume). It is accepted, however, that magnetic fields decrease the diffusivity, since they suppress movements of charged particles across the field lines. The same is believed to be true for the value of the viscosity as well. Usually a fraction of the Spitzer value is used to approximate thermal diffusivity of the ICM [3, 6], which assumes chaotic orientation of magnetic field lines (i.e., magnetic field lines are bent on scales smaller than the mean free path of an electron in the plasma).

2 The Model

In our model we do not explicitly include magnetic fields (and, therefore, possible effects of the anisotropy). The purely hydrodynamical setup includes thermal conduction and viscosity at a fraction of the Sptizer value to account for the suppression effects by the chaotic magnetic fields.

The effects of AGN heating are modelled by introduction of hot, low density regions (bubbles) close to the cluster core. We do not simulate inflation of the bubbles by AGN outflows, instead we introduce them as an isobaric perturbation of the temperature and density fields of the cluster, when the temperature of any computational cell in the vicinity of the centre falls below 1.5 keV, but not more frequently than once per 10^7 yr.

The perturbations (bubbles) are introduced with a fixed constant temperature, $T = 5 \times 10^{10}$ K, and a flat (constant) density profile, which is adjusted to ensure that the bubble stays in overall pressure equilibrium with the ambient medium. The initial radius of the bubbles was taken to be $r = 10^{22}$ cm = 3.2 kpc.

The gravitational potential of the cluster is modelled using the observational temperature and density profiles for Virgo determined in [4]. The observational data allows for more accurate modelling of the gravitational potential then a fitted theoretical profile (e.g., NFW profile [7]), as the gravity of the central galaxy can become dominant at the cluster core.

The cooling function in our model is based on the tabulated values from [11], calculated for the case of a fully ionised, optically thin plasma in collisional ionisation equilibrium, with half solar metallicity ([Fe/H]=-0.5).

2.1 Simulations

The simulations were performed using the Flash 2.3 hydrocode using an adaptive mesh (AMR), constructed with 16^3 nested blocks. The minimum level of refinement was set equal to 3, and the maximum to 7. This adaptive mesh is equivalent to a 2048^3 uniform grid, which provides a physical resolution of 4.89×10^{20} cm (158 pc), on the $10^{24} \times 10^{24} \times 10^{24}$ cm^3 grid. Boundary conditions were set to periodic in all directions. Initial conditions for the temperature were selected to be uniform, $T = 3 \times 10^7$ K, and the initial density distribution was determined from the requirement of hydrostatic equilibrium with fixed peak density, $\rho_0 = 5.15 \times 10^{-26}$ g cm^{-3} (the values were based on a number of 1D and 3D test simulations with no AGN heating, see also [8, 9], and E.C.D.Pope et al. in this volume).

We performed simulations with four values of thermal diffusivity and viscosity: 0, 10, 30, and 60 per cent of the Spitzer value. Particular emphasis in analysis of the simulations was placed on the parameters of the bubbles, their evolution, and dynamics.

3 Overview of the Results

The simulations showed that the energy of the bubbles is dispersed via two main channels. The thermal energy of the bubble is transferred to the ambient medium through the mechanism of thermal diffusion, which in case of the simulation with no modelled physical thermal diffusivity is a numerical artifact. In reality such diffusion can be turbulent in nature. However, to model diffusivity in the non-diffusive and non-viscous case would either require sufficiently high numerical resolution to resolve the dissipative scale (which is prohibitively computationally expensive), or call for an alternative numerical scheme such as the large eddy simulation (LES) paradigm. When physical diffusivity is present, the heat transfer occurs through the thermally conductive boundary of the bubble, and it is regulated by the model parameters for suppression of the Spitzer diffusivity.

The gravitational energy of the bubble is dispersed via "aerodynamic" forces: Kutta-Zhukovsky [1, 2, 5], and drag force. The buoyancy force pushes the bubble away from the centre, and the motion result in spin up of the vortex ring inside the bubble. This vortex motion and the motion of the bubble through the cluster atmosphere produce the aerodynamic Kutta-Zhukovsky forces that expand the bubble into a spheroid (and later into a torus), and counteract buoyancy.

The resulting temperature and density profiles suggest that the plasma of the ICM is likely to be diffusive and viscous, but all simulated profiles showed a stark difference from their observational counterparts, which suggests that other physical mechanisms (magnetic fields, large scale turbulence, etc) might be important elements in determining the exact thermal balance of the cluster.

Acknowledgement. The software used in this work was in part developed by the DOE-supported ASC / Alliance Center for Astrophysical Thermonuclear Flashes at the University of Chicago.

References

1. Y. D. Afanasyev, Phys. Fluids, 2006, 18, 037103
2. Y. D. Afanasyev and N. V. Korabel, Phys. Fluids, 2004, 16, 3850
3. B. D. G. Chandran and S. C. Cowley, Phys. Rev. Lett., 1998, 80, 3077
4. S. Ghizzardi, S. Molendi, F. Pizzolato, and S. De Grandi, 2004, ApJ, 609, 638
5. L. D. Landau and E. M. Lifshitz, 1987, Fluid Mechanics, volume 4, (Pergamon Press, second edition)
6. R. Narayan and M. V. Medvedev, 2001, ApJL, 562, L129
7. J. F. Navarro, C. S. Frenk, and S. D. M. White, 1997, ApJ, 490, 493
8. E. C. D. Pope, G. Pavlovski, C. R. Kaiser, and H. Fangohr, 2005, MNRAS, 364, 13
9. E. C. D. Pope, G. Pavlovski, C. R. Kaiser, and H. Fangohr, 2006, MNRAS, 367, 1121
10. L. Spitzer, 1962, Physics of Fully Ionized Gases, (Interscience, New York)
11. R. S. Sutherland and M. A. Dopita, 1993, ApJS, 88, 253

The Role of AGN Feedback and Gas Viscosity in Hydrodynamical Simulations of Galaxy Clusters

D. Sijacki and V. Springel

Max-Planck-Institut für Astrophysik, Karl-Schwarzschild-Straße 1,
85740 Garching bei München, Germany deboras@mpa-garching.mpg.de

Summary. We study the imprints of AGN feedback and physical viscosity on the properties of galaxy clusters using hydrodynamical simulation models carried out with the TreeSPH code GADGET-2. Besides self-gravity of dark matter and baryons, our approach includes radiative cooling and heating processes of the gas component and a multiphase model for star formation and SNe feedback [1]. Additionally, we introduce a prescription for physical viscosity in GADGET-2, based on an SPH discretization of the Navier-Stokes and general heat transfer equations. Adopting the Braginskii parameterization for the shear viscosity coefficient, we explore how gas viscosity influences the properties of AGN-driven bubbles. We find that the morphology and dynamics of bubbles are significantly affected by the assumed level of physical viscosity in our simulations, with higher viscosity leading to longer survival times of bubbles against fluid instabilities. In our cosmological simulations of galaxy clusters, we find that the dynamics of mergers and the motion of substructures through the cluster atmosphere is significantly affected by viscosity. We also introduce a novel, self-consistent AGN feedback model where we simultaneously follow the growth and energy release of massive black holes embedded in a cluster environment. We assume that black holes accreting at low rates with respect to the Eddington limit are in a radiatively inefficient regime, and that most of the feedback energy will appear in a mechanical form. Thus, we introduce AGN-driven bubbles into the ICM with properties, such as radius and energy content, that are directly linked to the black hole physics. This model leads to a self-regulated mechanism for the black hole growth and overcomes the cooling flow problem in host halos, ranging from the scale of groups to that of massive clusters.

1 Physical viscosity in SPH simulations of galaxy clusters

There is growing observational evidence [2, 3, 4] that gas viscosity in massive, hot clusters might not be negligible, and that it could play an important role in dissipating energy generated by AGN-driven bubbles or during merger events.

1.1 Numerical implementation

Within the framework of the entropy conserving formulation of SPH [5] in GADGET-2 [6, 7], we have implemented a treatment of physical viscosity that accounts both for the shear and bulk part, as explained in detail in [8]. In particular, we have derived novel SPH formulations of the Navier-Stokes and general heat transfer equations, and for the shear viscosity coefficient we have adopted Braginskii's parameterization [9, 10]. We have tested our numerical scheme extensively on a number of hydrodynamical problems with known analytic solutions, recovering these solutions accurately.

1.2 AGN-driven bubbles in a viscous ICM

As a first application of our physical viscosity implementation in GADGET-2, we have analyzed AGN-induced bubbles in a viscous intracluster gas. We consider models of isolated galaxy clusters consisting of a static NFW dark mater halo with a gas component which is initially in hydrostatic equilibrium. AGN heating has been simulated following a phenomenological approach, outlined in [11], where the bubbles are recurrently injected in the central cluster region. In Fig. 1 we show temperature maps of a $10^{15}\,h^{-1}M_\odot$ galaxy cluster that is subject to AGN heating and has a certain level of physical viscosity. In the left-hand panel, the Braginskii viscosity has been suppressed by a factor of 0.3, while in the right-hand panel, the simulation has been evolved with the full Braginskii viscosity. It can be seen that the morphologies, maximum clustercentric distance reached and survival times depend strongly on the assumed level of physical viscosity. With unsuppressed Braginskii viscosity, bubbles rise up to $\sim 300h^{-1}$kpc in the cluster atmosphere without being disrupted, and up to $2-3$ bubble episodes can been detected, indicating that the bubbles survive as long as $\sim 2 \times 10^8$ yr. However, in the case of suppressed physical viscosity by a factor of 0.3 bubbles start to disintegrate at roughly $150h^{-1}$kpc.

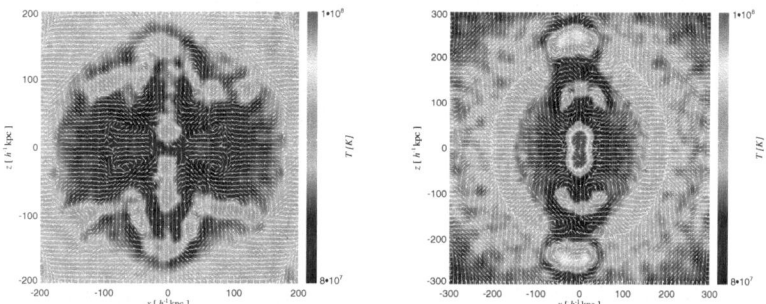

Fig. 1. Mass-weighted temperature maps of a $10^{15}\,h^{-1}M_\odot$ isolated halo, subject to AGN bubble heating. The velocity field of the gas is over-plotted with white arrows.

1.3 Cosmological simulations of viscous galaxy clusters

We have carried out fully self-consistent cosmological simulations of galaxy clusters with certain amounts of physical viscosity. We have performed both viscous non-radiative simulations and runs with additional cooling and star formation, in order to understand the complex interplay of these different physical ingredients. In Fig. XII we show density maps of a non-radiative galaxy cluster simulation at $z = 0.1$ without any physical viscosity (left-hand panel) and with 0.3 of Braginskii shear viscosity (right-hand panel). We find that the introduction of a modest level of physical viscosity has a significant impact on galaxy cluster properties. The dynamics of clusters during merging events is affected, with viscous dissipation processes generating an entropy excess in cluster peripheries. Also, due to the viscous dissipation, smaller structures entering more massive halos are more efficiently stripped of their gaseous content, which forms narrow and up to $100h^{-1}$ kpc long tails, as visible on the right-hand panel of Fig. XII. These features of viscous dissipation are very prominent, occurring already at quite early cosmic times and regardless of the presence of radiative cooling in the runs. However, even though viscous dissipation occurs in the central cluster region, it does not provide sufficient heating to prevent the formation of a central cooling flow at low redshifts, indicating that at least one other physical process is needed to reconcile observational findings with simulations.

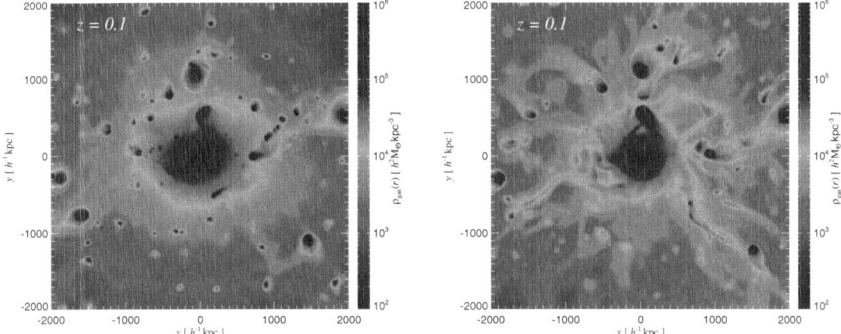

Fig. 2. Projected gas density maps of a galaxy cluster simulation at redshift $z = 0.1$, as indicated in the upper-left corner of the panels. The left-hand panel shows the gas density distribution in the case of a non-radiative run, while the right-hand panel gives the gas density distribution when Braginskii shear viscosity is "switched-on", using a suppression factor of 0.3.

2 Self-regulated AGN feedback in simulations of galaxy clusters

2.1 Methodology

We have developed a novel AGN feedback model that simultaneously tracks the growth of massive black holes in cluster environments, and provides heating in the form of AGN-driven bubbles [12]. In our simulations of clusters a self-regulated feedback loop is established, leading to an equilibrium where black hole growth is restricted and the central ICM is heated, overcoming the cooling flow problem.

We model the black hole growth according to the prescriptions outlined in [13, 14]. In these studies, the Bondi formula has been adopted for the accretion rate onto a black hole, and the Eddington limit has been imposed. Here, we link the black hole properties, namely its mass and accretion rate, with the physics of AGN-blown bubbles. We parameterize our scheme in terms of bubble energy and radius, and we consider recurrent episodes of bubble injection. Specifically, we introduce a threshold in black hole accreted mass above which a bubble event is triggered. We relate the bubble energy with the black hole properties as follows,

$$E_{\rm bub} = f \, \epsilon_r \, c^2 \, \delta M_{\rm BH} \,, \tag{1}$$

where f is the fraction of energy that goes into the bubbles[1], ϵ_r is the standard radiative efficiency that we assume to be 0.1, and $\delta M_{\rm BH}$ is the mass growth of a black hole between two successive bubble episodes. Moreover, we link the bubble radius both to $\delta M_{\rm BH}$ and to the density of the surrounding ICM, in the following way

$$R_{\rm bub} = R_{\rm bub,0} \left(\frac{E_{\rm bub}/E_{\rm bub,0}}{\rho_{\rm ICM}/\rho_{\rm ICM,0}} \right)^{1/5} \,, \tag{2}$$

where $R_{\rm bub,0}$, $E_{\rm bub,0}$, and $\rho_{\rm ICM,0}$ are normalization values for the bubble radius, energy content and ambient density. The relation for the bubble radius is motivated by the solutions for the radio cocoon expansion in a spherically symmetric case [15].

2.2 Black hole growth and feedback in isolated galaxy clusters

We have performed simulations of isolated galaxy clusters for a range of different masses, from $10^{13} \, h^{-1} {\rm M}_\odot$ to $10^{15} \, h^{-1} {\rm M}_\odot$, analyzing the black hole growth and feedback over a large time span. The initial conditions have been set up by imposing hydrostatic equilibrium for the gas within a static NFW dark mater halo, and by introducing a seed black hole sink particle with

---

[1] For low accretion rates, $\dot{M}_{\rm BH} < 10^{-2} \dot{M}_{\rm Edd}$, we assume that most of the energy is in mechanical form.

$10^5\,h^{-1}\mathrm{M}_\odot$. The simulations have been evolved for $0.25t_{\mathrm{Hubble}}$ with radiative cooling, star formation, and AGN feedback. In the left-hand panel of Fig. 3 we show how the black hole mass is growing for three halos of increasing mass. After the initial rapid growth, bubble feedback regulates the black hole mass, which remains practically constant for more than a Gyr of the simulated time. The mass accretion rate onto the black hole (see right-hand panel of Fig. 3) drops after the initial phase to very low values of order $10^{-3}M_{\mathrm{Edd}}$. It can been seen that the black hole accretion rate shows occasional bursts during short time intervals, reflecting the recurrent nature of the bubble feedback. However, these jumps in black hole accretion rate do not contribute significantly to the growth of the black hole itself. In Fig. 4 we plot entropy

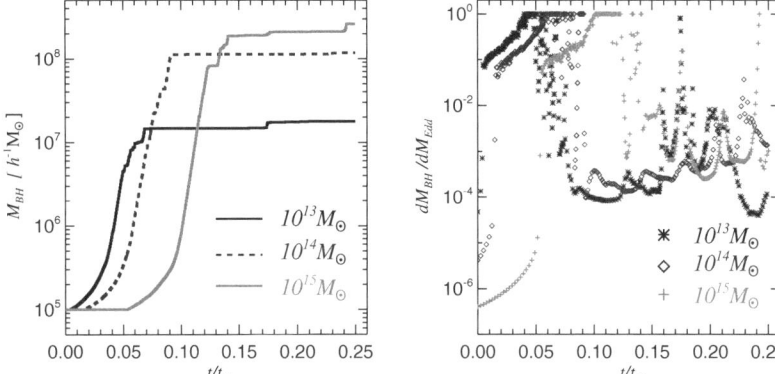

Fig. 3. Black hole mass growth and black hole accretion rate in Eddington units for three isolated galaxy clusters simulations. After the initial rapid growth from a $10^5\,h^{-1}\mathrm{M}_\odot$ seed, the black hole mass is regulated and the accretion rate drops to low values.

and temperature profiles of a $10^{14}\,h^{-1}\mathrm{M}_\odot$ cluster at $0.25t_{\mathrm{Hubble}}$. We compare the runs without AGN heating with the simulations with different thresholds δM_{BH} for the bubble triggering. In the case of highest δM_{BH}, the bubble frequency is lowest, but the energy injected into the bubbles is large. This model corresponds to a sporadic, but rather powerful AGN activity and heats the ICM very efficiently. On the other hand, low values of δM_{BH} imply more frequent and gentle bubble feedback that affects the ICM properties mildly, but that can still prevent the overcooling in the central regions of clusters.

3 Conclusions

Unless heavily suppressed by magnetic fields, physical gas viscosity in hot, massive clusters appears to be an important physical ingredient, changing significantly the properties of AGN-driven bubbles and influencing the dynamics of clusters during merging events. The ever more accurate X-ray data

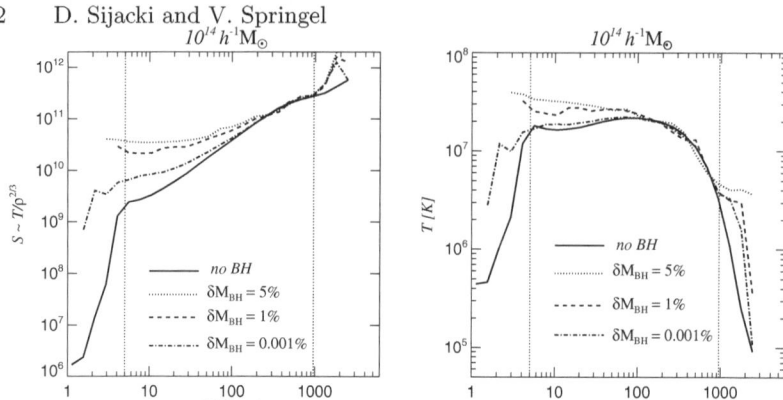

Fig. 4. Entropy (left-hand panel) and mass-weighted temperature (right-hand panel) radial profiles of a $10^{14}\,h^{-1}\mathrm{M_\odot}$ isolated cluster. With increasing δM_{BH} bubble feedback becomes more infrequent, but also more violent, leading to a substantial heating of the central gas.

on galaxy clusters will allow detailed comparisons with simulations, that can constrain the effective level of viscosity present in these systems. AGN heating is a very promising candidate to solve the cooling flow riddle in clusters. Adopting the theoretical model outlined here we discuss in forthcoming work [12] fully cosmological simulations of self-regulated AGN feedback, trying to understand this physical mechanism in more depth.

Acknowledgement. We are grateful to Simon White, Eugene Churazov and Andrea Merloni for many constructive discussions and comments.

References

1. V. Springel, L. Hernquist, 2003, MNRAS, 339, 289
2. M. Markevitch et al., 2002, ApJ, 567, 27
3. M. Sun et al., 2006, ApJ, 637, 81
4. A. C. Fabian et al., 2006, MNRAS, 366, 417
5. V. Springel, L. Hernquist, 2002, MNRAS, 333, 649
6. V. Springel, 2005, MNRAS, 364, 1105
7. V. Springel, N. Yoshida, S. D. M. White, 2001, New Astron., 6, 79
8. D. Sijacki, V. Springel, 2006, MNRAS, 371, 1025
9. S. I. Braginskii, JETP, 1958, 33, 459
10. S. I. Braginskii, 1965, Rev. Plasma Phys., I, 205
11. D. Sijacki, V. Springel, 2006, MNRAS, 366, 397
12. D. Sijacki et al., 2006, in prep.
13. T. Di Matteo, V. Springel, L. Hernquist, 2005, Nature, 433, 604
14. V. Springel, T. Di Matteo, L. Hernquist, 2005, MNRAS, 361, 776
15. M. C. Begelman, D. F. Cioffi, 1989, ApJ., 345, 21

Cold Feedback in Cooling–Flow Galaxy Clusters

F. Pizzolato

Department of Physics
Technion–Israel Institute of Technology
Haifa 32000
fabio@physics.technion.ac.il

Summary. We put forward an alternative view to the Bondi–driven feedback between heating and cooling of the intra-cluster medium (ICM) in cooling flow (CF) galaxies and clusters.

We adopt the popular view that the heating is due to an active galactic nucleus (AGN), i.e. a central black hole accreting mass and launching jets and/or winds. We propose that the feedback occurs with the *entire* cool inner region ($r \lesssim 5-30$ kpc). A moderate cooling flow *does* exist here, and non–linear over–dense blobs of gas cool fast and are removed from the ICM before experiencing the next major AGN heating event. Some of these blobs may not accrete on the central black hole, but may form stars and cold molecular clouds. We discuss the conditions under which the dense blobs may cool to low temperatures and feed the black hole.

1 Introduction

The observations of galaxy clusters with the last generation X-ray satellites *Chandra* and *XMM-Newton* failed to detect the large amounts of cool gas predicted by the old version [6] of the cooling flow (CF) model (see e.g. [16] for a recent review). The most straightforward explanation is that the intra-cluster medium (ICM) in CF clusters is heated by some mechanism, the currently most popular candidate being the active galactic nucleus (AGN) residing at the core of the cluster dominant galaxy.

Most models of AGN heating agree in that there is some feedback between the heating and the radiative cooling, possibly resulting in intermittent AGN activity. There are two approaches to the AGN/ICM feedback. In the *hot feedback* scenario (see e.g. [14]) the ICM never cools below X–ray emitting temperatures; the AGN accretion pattern is Bondi–like, and is determined by the ICM properties at the Bondi radius, which is typically a few tens of parsec.

In the second approach, dubbed *cold feedback*, (see [18] and [21] for more details), the black hole accretes *cold* gas, although the amount of cooling mass is much below that predicted by the old cooling flow model. In this case the feedback takes place within a region extending to a distance of $\approx 5-30$ kpc from the cluster centre.

2 The Cold Feedback Scenario

The cold feedback scenario entails a cycle in the cooling/accretion activity.

We suggest that this cycle starts with a major AGN outburst, which injects a huge amount of energy into the ICM. The AGN outburst interacts in a very complicated fashion with the ICM [1], e.g. it heats and inflates radio bubbles, and may also stir some turbulence. The ICM itself is displaced and thickened by the rising bubbles, as shown by their rims' enhanced X–ray brightness [2]. A non-homogeneous thickening may result in the formation of a multi-phase gas, consisting e.g. of pockets of cold gas with a wide spectrum of densities. Owing to their over-density, these blobs fall to the black hole.

If these these blobs have a significant amount of angular momentum, they cannot feed the black hole. On the other hand, low angular momentum blobs are allowed to accrete, igniting a fresh AGN outburst, which in turn disturbs the ICM, restarting the cycle with a new injection of blobs.

In this duty cycle some of the gas cools to low temperatures ($\lesssim 10^4$K) before the next major heating episode, while the rest is heated back to a relatively high temperature. The presence of a detectable amount of gas cooling below X-ray emitting temperatures is a specific prediction of this model. Indeed, in the CF cluster A 2597 both extreme-UV and X-ray observations indicate a mass cooling rate of $\sim 100 \ M_\odot \ \mathrm{yr}^{-1}$, which is ~ 0.2 of the value quoted in the past based on *ROSAT* X-ray observations (see the discussion in [11]). In the CF cluster A 2029, [5] find a substantial amount of gas at a temperature of $\approx 10^6$K; a CF model gives a mass cooling rate of $\sim 50 \ M_\odot \ \mathrm{yr}^{-1}$.

According to the results of [18] and [21], a key role in the cold bubbles' accretion *and feedback* is played by the ICM entropy profile.

Long after an outburst, following an extended period of cooling, the entropy profile is steep. In this case it is difficult for an infalling blob to accrete and feed the black hole: on its way it will reach an equilibrium radius where its density equals that of its surroundings; it then dissolves *before* accreting on the black hole. Only blobs which come from regions which have a small entropy difference with respect to the core can accrete. Therefore, the first accretion episodes are most likely to involve small blobs, stemming not far from the centre. The AGN activity induced by their accretion is rather weak, and only raises the entropy very close to the AGN. In the meantime, the ICM further out keeps cooling, reducing the cooling time of the dense blobs. In addition, the combined action of cooling the far regions and heating the central region flattens the cluster's entropy profile. By the same token, due to the flat entropy profile now even far blobs may accrete on the AGN, triggering a major outburst. This may therefore heat the cluster on large scales.

There are four hurdles the blobs must overcome before accreting.

1. Some pockets may be engulfed by the expanding radio lobes. As [18] (and references therein) argue, if the blobs are initially dense enough

(say, $10 - 100$ times the ambient medium) they can survive the shock, and accrete unhindered through the radio lobes.

2. These blobs must withstand the ICM thermal conduction: if it is too efficient, it would be able to evaporate these cold clouds *before* they can accrete on the AGN. We find from Figure 3 of [13] that the effective heat conduction should be $\lesssim 10^{-3}$ times the nominal [24] not to evaporate a blob of radius $a \sim 10 - 100$ pc. Such a strong suppression factor is supported by theoretical considerations [17, 12]. The possible existence of magnetic turbulence [19] may also strongly affect the transport coefficients, including thermal conductivity. A somewhat suppressed conduction is also consistent with some recent observations (M87: [10], NGC 5044: [3]).

3. As [15] pointed out, *in the absence of a cohesive force* a blob would be torn apart by the ram pressure in $\sim 10^7$ yr, i.e. a time considerably shorter than the time taken by the blob to fall to the centre. Some kind of cohesive force (like a magnetic tension) must then be at work to prevent the blob disruption.

4. If the blobs' angular momentum is too high, they are prevented from approaching the central black hole: the flux would merely stagnate, cool down and condense in filaments or stars, and the AGN fuelling is cut off altogether, thus making the feedback impossible. Indeed, the existence of a circumnuclear disc with radius $R_d \approx 10^2$ pc around M87 [8, 7] shows that the flow possesses an amount of angular momentum.

   However, the blobs are expected to form and accrete only in a region of the same extension as the inner gas entropy plateau ($\sim 5 - 30$ kpc). The circularisation radius of this flow is expected to be of the same order as the actual size of the circumnuclear disc of M87 [18], i.e. direct accretion is possible down to the immediate vicinity of the black hole. In addition, the blobs may stem directly from ICM disturbances driven by an early AGN activity, but also from galaxies mass-stripping [22]. Since the galaxies do not have an ordered bulk motion, also the blobs stripped from them are also unlikely to organise in an ordered flow with high net angular momentum.

So, to summarise, under some reasonable assumptions the cold blobs are able to survive long enough to accrete on the AGN and hence provide feedback.

The dense blobs that sink to the centre feed the AGN. The feedback is with the *entire* cool inner region, and not only with the gas close to the black hole. Any over-cooling taking place in the inner region, where the temperature profile is flat, will lead to many small and dense blobs, which feed the AGN.

A Bondi accretion radius as large as the disc around the black hole of M87 further suggests that the simple Bondi accretion flow [4, 14] does not hold; the accreted material has a larger angular momentum, and may come from much larger radii. Also for cluster A 1835 the Bondi accretion is unlikely to be the main engine powering the feedback [9].

3 Summary

We propose that the feedback occurs with the entire cool inner region, $r \lesssim 5 - 30$ kpc, in what we term a *cold-feedback model*. In the proposed scenario non-linear over-dense blobs of gas cool fast and are removed from the ICM before the next major AGN heating event in their region. It is important to note that an AGN burst can take place and heat other regions, since the jets and/or bubbles may expand in other directions as well. The typical interval between such heating events at a specific region is $\approx 10^8$ yr. Some of these blobs cool and sink toward the central black hole, while others may form stars and cold molecular clouds.

Four conditions should be met in the inner region participating in the feedback heating.

1. In order for the blob not to reach an equilibrium point before accreting, a shallow ICM entropy profile is required. The relevant dense blobs then must form within the cluster core, typically $\lesssim 5-30$ kpc from the centre. We note that the lower segment of magnetic flux loops can be prevented from reaching the stabilising point by the upward force of the magnetic tension inside the loop [20]. Therefore, some perturbations can be formed at large distances, where density profile is steep, and still cool to low temperature and feed the central black hole.
2. Non-linear perturbations are required. These presumably formed mainly by previous AGN activity, e.g. jets and radio lobes.
3. The cooling rate of these non-linear perturbations is short relative to few times the typical interval between successive AGN outbursts.
4. The blobs must not be evaporated by thermal conduction before they are delivered to the AGN. This requires a strong suppression of thermal conduction.

The first and the third condition, which are not completely independent of each other, require that the initial ICM cools by a factor of a few before the feedback starts operating, and the second condition requires that the inner region must be disturbed.

The cold-feedback model has the following implications and predictions

1. The optical filaments observed in many CF-clusters and the cooler molecular gas detected via CO observations come from cooling ICM (with some amount possibly from stripping from galaxies).
2. Some X-ray emission from gas at temperatures $\lesssim 10^7$K is expected, consistent with the moderate CF model. This is much more than in many other AGN heating models, but at least an order of magnitude below

what predicted by the old CF model, We stress that in the cold-feedback heating, cooling flows *do exist*. Such gas cooling to below X-ray emitting temperatures was found recently in two CF clusters (A2597: [11]; A2029: [5]).

3. The feeding of the central black hole with cold gas in the cold feedback models makes the process similar in some aspects to that of AGN in spiral galaxies. Therefore, the outflow can be similar [23].

4. It is possible that in the cold feedback model a substantial fraction of gas that cooled to low temperatures and was accreted to the accretion disc around the central black hole, is injected back to the ICM at non-relativistic velocities [23].

References

1. Begelman, M. C. 2004, in Coevolution of Black Holes and Galaxies, ed. L. C. Ho, pp374
2. Blanton, E. L., Sarazin, C. L., McNamara, B. R., & Wise, M. W. 2001, ApJ, 558, L15
3. Buote, D. A., Lewis, A. D., Brighenti, F., & Mathews, W. G. 2003, ApJ, 594, 741
4. Churazov, E., Sunyaev, R., Forman, W., & Böhringer, H. 2002, MNRAS, 332, 729
5. Clarke, T. E., Blanton, E. L., & Sarazin, C. L. 2004, ApJ, 616, 178
6. Fabian, A. C. 1994, ARA&A, 32, 277
7. Ford, H. C., Harms, R. J., Tsvetanov, Z. I., Hartig, G. F., Dressel, L. L., Kriss, G. A., Bohlin, R. C., Davidsen, A. F., Margon, B., & Kochhar, A. K. 1994, ApJ, 435, L27
8. Harms, R. J., Ford, H. C., Tsvetanov, Z. I., Hartig, G. F., Dressel, L. L., Kriss, G. A., Bohlin, R., Davidsen, A. F., Margon, B., & Kochhar, A. K. 1994, ApJ, 435, L35
9. McNamara, B. R., Rafferty, D. A., Bîrzan, L., Steiner, J., Wise, M. W., Nulsen, P. E. J., Carilli, C. L., Ryan, R., & Sharma, M. 2006, ApJ, 648, 164
10. Molendi, S. 2002, ApJ, 580, 815
11. Morris, R. G. & Fabian, A. C. 2005, MNRAS, 358, 585
12. Nath, B. B. 2003, MNRAS, 340, L1
13. Nipoti, C. & Binney, J. 2004, MNRAS, 349, 1509
14. Nulsen, P. 2004, in The Riddle of Cooling Flows in Galaxies and Clusters of galaxies, ed. T. Reiprich, J. Kempner, & N. Soker, 259–262
15. Nulsen, P. E. J. 1986, MNRAS, 221, 377
16. Peterson, J. R. & Fabian, A. C. 2006, Physics Reports, 427, 1
17. Pistinner, S., Levinson, A., & Eichler, D. 1996, ApJ, 467, 162
18. Pizzolato, F. & Soker, N. 2005, ApJ, 632, 821
19. Schekochihin, A. A., Cowley, S. C., & Dorland, W. 2006, astro-ph/0610810
20. Soker, N. 2004, MNRAS, 350, 1015
21. —. 2006, New Astronomy, 12, 38
22. Soker, N., Bregman, J. N., & Sarazin, C. L. 1991, ApJ, 368, 341
23. Soker, N. & Pizzolato, F. 2005, ApJ, 622, 847
24. Spitzer, L. 1956, Physics of Fully Ionized Gases, (Interscience, New York)

AGN Heating of Cooling Flow Clusters: Problems with 3D Hydrodynamic Models

J. C. Vernaleo[1,2] and C. Reynolds[1,3]

[1] Department of Astronomy, University of Maryland, College Park, MD 20742
[2] vernaleo@astro.umd.edu
[3] chris@astro.umd.edu

1 Introduction

Relaxed galaxy clusters have central cooling times less than the age of the cluster. However, there are observational limits to the amounts of cool gas present, and XMM-Newton spectroscopy shows nothing below $\sim \frac{1}{3}T_{virial}$ $(1-2 \text{ keV})$. This discrepancy is the heart of the classical cooling flow problem.

Viewed from another angle, the galaxy luminosity function is truncated at the high end [1]. So whatever offsets cooling probably also stops the formation of massive galaxies. This must occur on many mass and temperature scales: therefore some self regulation seems to be required. AGN (Active Galactic Nuclei) are often given as a possible solution. AGN inject energy on the same order as cooling luminosity ($\sim 10^{45}-10^{46} \text{ erg } s^{-1}$). They are fed by accretion, so self regulation may come naturally.

Numerous observations indicate that AGN can have an impact on large scale structure. We have performed a set of high resolution, three dimensional simulations of a jetted AGN embedded in a relaxed cooling cluster [5]. To the best of our knowledge, these are the first simulation to include both full jet dynamics and a feedback model. These ideal hydrodynamic simulations show that, although there is enough energy present to offset cooling on average, the jet heating is not spatially deposited in a way that can prevent catastrophic cooling of the cluster.

2 Models

For comparison, we start with a pure cooling cluster. All other models start with the same, with some additional effects added. The initial cluster is modelled after a rich, relaxed cluster with a β-law density profile: $r_{core} = 100 \text{ kpc}$, $n_0 = 0.01 \text{ cm}^{-3}$, and $c_s = 1000 \text{ km s}^{-1}$. The cooling is modelled as thermal bremsstrahlung emission, following [3]. Due to n^2 dependence of the ICM cooling, in the absence of any feedback, cooling runs away, showing a featureless increase with time It gets to a level we set as 'catastrophic' by around 250 Myrs.

We added several types of feedback to our models. The first type is a single jet outburst lasting 50 Myrs. To model actual feedback, the velocity of the jet (and hence the kinetic luminosity of the source) was varied based on $\dot{M}$ across the inner edge of the simulation. This was done with various time delays (up to 100 Myrs) and efficiencies ($\eta = 0.00001 - 0.1$).

For the single burst jet, with a kinetic luminosity of $L_{kin} = 9.3 \times 10^{45}$ erg s^{-1} and a Mach 10.5 jet (see also [2]), catastrophic cooling was delayed by about 50 Myrs. The results of this simulation (which show the bubble the jet inflates) can be seen in Figure 1.

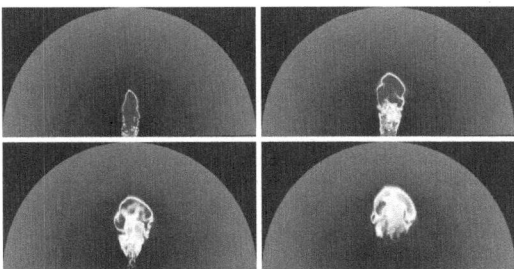

Fig. 1. Entropy for single jet simulation.

For feedback, mass flow across inner boundary was calculated and used to set a jet velocity assuming some efficiency η of the central blackhole using the formula $v_{jet} = \left(\frac{2\eta \dot{M}c^2}{A\rho}\right)^{\frac{1}{3}}$.

The most realistic model seems to be the low efficiency ($\eta = 10^{-4}$) model with a delay of 100 Myrs (close to the dynamical time for the galaxy). Even this only delays the cooling catastrophe (see Figure 2 for mass accretion rates).

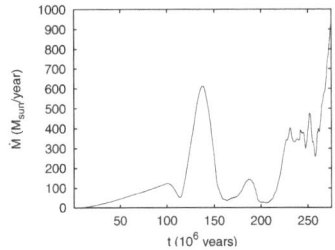

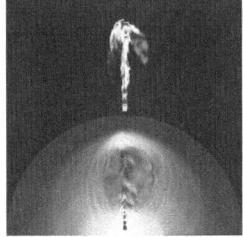

Fig. 2. Mass accretion for delayed feedback.

Fig. 3. Channel formation (Temperature (top) Pressure (bottom)).

The jet seems to cut a channel (see Figure 3) in the ICM which allows it to avoid heating the inner regions. This explains why simulations with bubbles placed in the center can do better at halting cooling than jets, but are less realistic.

3 Conclusions

We have preformed the first simulations that we are aware of to include both the full dynamics of a jet and an feedback model. When the full dynamics of the jet are included, ideal hydrodynamics interactions do not seem able to offset cooling on average, even though they are energetically capable of doing so. We conclude that either some physical process beyond that captured by our ideal hydrodynamic simulations (e.g., plasma transport processes, cosmic ray heating, dramatic jet precession, or ICM turbulence) is relevant for thermalizing the AGN energy output, or the role of AGN heating of cluster gas has been overestimated.

4 ZEUS-MP

All simulations were done using the ZEUS-MP 3D parallel hydrocode (a version of the code in [4]). We have updated and modified the NCSA release (v1.0). Our modifications (v1.5) and documentation are publicly available at: http://www.astro.umd.edu/~vernaleo/zeusmp.html
There is also a version 2 of ZEUS-MP which we are not affiliated with.

Acknowledgement. Simulations were performed on the Beowulf cluster ("The Borg") supported by the Center for Theory and Computation (CTC) in the Department of Astronomy, University of Maryland. This work was partly funded by the *Chandra* Cycle-5 Theory & Modelling program under grant TM4-5007X.

References

1. A. J. Benson, R. G. Bower, C. S. Frenk, C. G. Lacey, C. M. Baugh, & S. Cole. 2003, ApJ, 599, 38
2. C. S. Reynolds, S. Heinz, & M. C. Begelman. 2002, MNRAS, 332, 271
3. M. Ruszkowski & M. C. Begelman. 2002, ApJ, 581, 223
4. J. M. Stone & M. L. Norman. 1992, ApJS, 80, 753
5. J. C. Vernaleo & C. S. Reynolds. 2006, ApJ, 645, 83

Heating Rate Profiles in Galaxy Clusters

E. C. D. Pope[1,2,3], G. Pavlovski[2], C. R. Kaiser[2] and H. Fangohr[3]

[1] School of Physics and Astronomy, University of Leeds, Leeds, LS2 9JT,
  e.c.d.pope@leeds.ac.uk
[2] School of Physics and Astronomy, University of Southampton Highfield,
  Southampton, SO17 1BJ
[3] School of Engineering Sciences, University of Southampton Highfield,
  Southampton, SO17 1BJ

Summary. The results of hydrodynamic simulations of the Virgo and Perseus clusters suggest that thermal conduction is not responsible for the observed temperature and density profiles. As a result it seems that thermal conduction occurs at a much lower level than the Spitzer value. Comparing cavity enthalpies to the radiative losses within the cooling radius for seven clusters suggests that some clusters are probably heated by sporadic, but extremely powerful, AGN outflows interspersed between more frequent but lower power outflows.

1 Introduction

The two candidates for heating cluster atmospheres are Active Galactic Nuclei (AGNs) and thermal conduction. Heating by AGN is thought to occur through the dissipation of the internal energy of plasma bubbles inflated by the AGN at the centre of the cooling flow. Since these bubbles are less dense than the ambient gas, they are buoyant and rise through the intracluster medium (ICM) stirring and exciting sound waves in the surounding gas. This energy may be dissipated by means of a turbulent cascade, viscous processes, or aerodynamic forces. Deep in the central galaxy other processes such as supernovae and stellar winds will also have some impact on the ambient gas.

Thermal conduction may also play a significant role in transferring energy towards central regions of galaxy clusters given the temperature gradients which are observed in many clusters.

2 The Model

2.1 General heating rates

Starting from the assumption that the atmospheres of galaxy clusters are spherically symmetric, and in a quasi steady-state, it is possible, using the fluid energy equation, to derive what the radial time-averaged heating rate must be in order to maintain the observed temperature and density profiles. The flow of the gas is assumed to be subsonic meaning that the cluster atmosphere is in approximate hydrostatic equilibrium allowing the gravitational acceleration to be calculated from observations of the temperature and density profiles. To avoid anomalies when calculating spatial derivatives, continuous analytical functions are fitted through the density and temperature data. This ensures that there are not any large discontinuities which may result in extreme, and erroneous, heating rates later on in the calculations. This model is described in greater detail in [9].

2.2 Thermal conduction

Thermal conduction of energy from the cluster outskirts may provide the required heating of the central regions without an additional energy source, like an AGN. The thermal conductivity is assumed to be given by [11], but includes a suppression factor designed to take into account the possible effects of magnetic fields. For a steady-state to exist, the heating by thermal conduction must equal the heating rate. From this criterion, the radial suppression factor can be deduced.

2.3 Heating by AGNs

The time-averaged mechanical power of an AGN can be estimated by dividing the cavity enthalpy by a characteristic timescale, see for example [1]. We assume that the radio-emitting plasma that fills the cavities is relativistic and that half of the outburst energy is deposited in the ICM by shocks. An accurate estimate of the time-averaged jet power requires the average time between consecutive AGN outbursts to be known. However, since this parameter is rarely known, a typical choice would be the buoyant timescale required for the cavity to rise to its current location. An alternative method is to assume a particular value for the duty cycle of the AGN. In this study it is initially assumed that the duty cycle of each AGN is 10^8 yrs.

Note that this is simply an estimate of the rate at which energy is injected by the AGN and is not related to any particular physical process by which this energy is dissipated, e.g. the viscous dissipation of sound waves.

An estimate of the duty cycle required to balance the radiative losses within the cooling radius is obtained by calculating the volume integral of the heating rate within this region and comparing this with the bubble enthalpy.

3 Results:1

Radial suppression factors and AGN duty cycles are calculated for a sample of seven objects for which temperature and density were available, as well as the information about the X-ray cavities inflated by their central AGNs. These objects are the Virgo [5], Perseus [10] and Hydra A [3] clusters, A2597 [6], A2199 , A1795 [4] and A478 [12] with cavity parameters taken from [1].

3.1 Suppression factors

The results show that the suppression factors must be finely tuned if thermal conduction is to balance the radiative losses. Such a high degree of fine-tuning suggests that thermal conduction is unlikely to be a dominant heating mechanism in galaxy clusters. Furthermore, in many cases, the required suppression factors exceed the physical maxmimum of unity. This is most true for the Virgo, Hydra A and A2597 clusters. In contrast, it appears that thermal conduction could, in principle, balance the radiative losses in the Perseus, A2199, A478 and probably A1795. The effect of thermal conduction on a cluster from each of these two groups is investigated in more detail using numerical simulations discussed in the next section.

3.2 AGN Duty Cycles

The duty cycles for Virgo and A478 are of the order of 10^6 yrs which is very short compared to the predicted lifetimes of AGN [7]. In contrast, the Hydra A cluster requires recurrent outbursts of magnitude similar to the currently observed one only every 10^8 yrs, or so. The required duty cycles for the remaining AGNs are of the order of 10^7 yrs. From this, the obvious conclusion is that if thermal conduction is negligible and if this sample is representative of galaxy clusters in general, then many, if not all, clusters will probably be heated, at certain points in time, by extremely powerful AGN outbursts. Furthermore, it is worthwhile pointing out that roughly 71% of cD galaxies at the centres of clusters are radio-loud [2] which is larger than for galaxies not at the centres of clusters. This may suggest that the galaxies at the centres of clusters are indeed active more frequently than other galaxies.

4 The Simulations

Numerical simulations of mock Virgo and Perseus clusters were performed using the FLASH hydrodynamics code. Four simulations were performed for each cluster to investigate the effect of different values of the thermal conduction on the evolution of the cluster temperature and density profiles. The simulations of the Virgo cluster are described in more detail in [8].

5 Results:2

5.1 Temperature and Electron Number Density Profiles

For the Virgo cluster, the temperature and density data are presented as spherically averaged profiles, rather than 1-d slices through the cluster. The temperatures and densities for two cases are compared to the observations of

[5] in Figure 1. The two cases shown here are: zero thermal conduction and Spitzer thermal conduction.

The simulations of the Perseus cluster were 1-d, but spherically symmetric, meaning that the data did not require spherical averaging. Results are shown in Figure 2 for the same cases as the Virgo cluster.

The qualitative response of the two clusters to the presence of thermal conduction is rather similar: it seems that thermal conduction probably cannot indefinitely prevent a cooling catastrophe from occuring. This is characterised by a large dip and a peak in the temperature and density profiles, respectively.

In the case of the Virgo cluster, even full Spitzer thermal conduction can only postpone the cooling catastrophe for a few Gyrs. The result is roughly the same for simulations of the Perseus cluster where the thermal conduction is sub-Spitzer. The main difference is that in the Perseus cluster the time taken for a cooling catastrophe to develop is significantly longer than for the Virgo cluster. To some extent this is because the density of the gas is lower in Perseus but also because the energy transport by thermal conduction is greater, due to the higher gas temperatures.

In the Perseus cluster, the simulations show that including full Spitzer thermal conduction can avert a cooling catastrophe, for at least a Hubble time, by transfering energy at such a high rate that it essentially keeps the temperature profile flat. This also prevents the density profile from evolving significantly. However, the rapid transfer of energy by thermal conduction leads to an additional problem: thermal conduction 'heats' the central regions of galaxy clusters by transfering thermal energy from outskirts yet there is not an infinite supply. Thus, while energy is being transfered the average gas temperature drops. Eventually the entire ICM would cool to temperatures where it is no longer observed in the X-rays. Essentially the problem with thermal conduction is that it does not 'add' energy to a system, it merely transfers it from one region to another.

6 Conclusion

The results of simulations including thermal conduction qualitatively agree with predictions, based on energetic arguments, for determining which clusters will be most influenced by thermal conduction. However, for clusters such as Perseus, where thermal conduction can balance the radiative losses, the simulated temperature profile never converges with the observations. The results of these simulations suggest that thermal conduction must be drastically reduced.

Current techniques of estimating AGN power output are still relatively uncertain. Nevertheless, the values obtained using these methods sometimes require duty cycles as short as a few Myrs to balance the radiative losses.

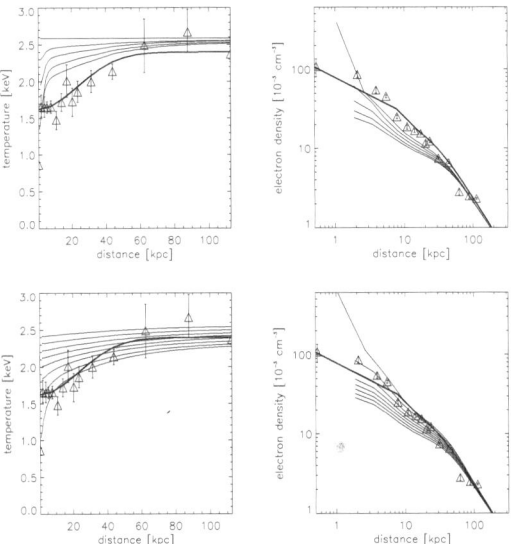

Fig. 1. Temperature and density profiles for the Virgo cluster evolving with time for zero thermal conduction (top) and Spitzer thermal conduction (bottom). The thick lines for both temperature and density are the functions fitted to the data points (triangles) by [5]. For zero thermal conduction the top line in the temperature plot shows the temperature profile after 3.17×10^8 yr and the bottom line at time of the end of the simulation. The intermediate lines represent the temperatures at intervals of 3.17×10^8 yr after the top temperature profile. The temporal sequence of the lines is reversed (bottom to top) in the density plot. For Spitzer thermal conduction the data are plotted every 6.34×10^8 yrs.

An alternative possibility is that the AGN power output is currently under-estimated, meaning that the duty cycle is actually longer. Another strong possibility is that extemely powerful outbursts occur between more frequent but less powerful outbursts.

References

1. Bîrzan, L., Rafferty, D. A., McNamara, B. R., Wise, M. W., Nulsen, P. E. J., 2004, ApJ, 607, 800
2. Burns, J. O., 1990, BAAS, 22, 821
3. David, L. P., Nulsen, P. E. J., McNamara, B. R., Forman, W., Jones, C., Ponman, T., Robertson, B., Wise, M., 2001, 557 546
4. Ettori, S., Fabian, A. C., Allen, S. W., Johnstone, R. M., 2002, MNRAS, 331, 635
5. Ghizzardi, S., Molendi, S., Pizzolato, F., De Grandi, S., 2004, ApJ, 609, 638

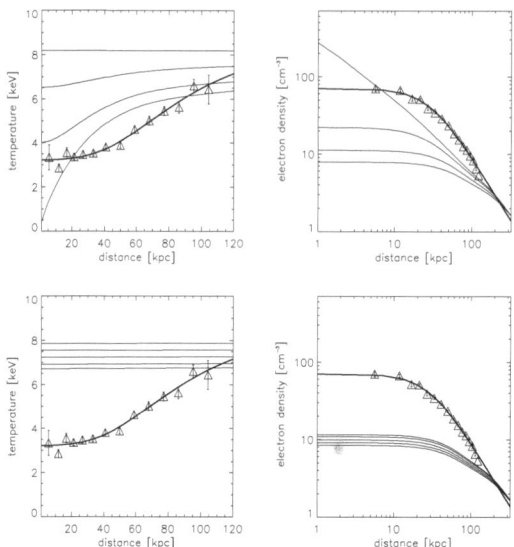

Fig. 2. Temperature and density profiles for the Perseus cluster evolving with time for zero thermal conduction (top) and Spitzer thermal conduction (bottom). The thick lines for both temperature and density are the functions fitted to the data points (triangles) by [10]. The top line in the temperature plot shows the temperature profile after 6×10^9 yr and the bottom line at time of the end of the simulation. The intermediate lines represent the temperatures at intervals of 3×10^9 yrs after the top temperature profile. The temporal sequence of the lines is reversed (bottom to top) in the density plot.

6. McNamara, B. R., Wise, M. W., Nulsen, P. E. J., David, L. P., Carilli, C. L., Sarazin, C. L., O'Dea, C. P., Houck, J., Donahue, M., Baum, S., Voit, M., O'Connell, R. W., Koekemoer, A., 2001, ApJ, 562, L149

7. Nipoti, C., Binney, J., 2005, MNRAS, 361, 428

8. Pope, E. C. D., Pavlovski, G., Kaiser, C. R., Fangohr, H., 2005, MNRAS, 364, 13

9. Pope, E. C. D., Pavlovski, G., Kaiser, C. R., Fangohr, H., 2006, MNRAS, 367, 1121

10. Sanders, J. S., Fabian, A. C., Allen, S. W. and Schmidt, R. W., 2004, MNRAS, 349, 952

11. Spitzer, L., Physics of Fully Ionized Gases, Wiley-Interscience, New York, 1962.

12. Sun, M., Jones, C., Murray, S. S., Allen, S. W., Fabian, A. C., Edge, A. C., 2003, ApJ, 547, 619

Part V

Entropy Structure

Heating, Cooling, and Intracluster Entropy

G. M. Voit

Department of Physics and Astronomy, Michigan State University, East Lansing, MI 48824, USA voit@pa.msu.edu

Summary. It is now generally acknowledged that gas at the centers of galaxy clusters with short central cooling times is not simply cooling, flowing toward the center, and condensing. Some sort of heat source must replenish much of the thermal energy radiated by the gas in the cores of those clusters, and the active galactic nucleus at the center of the brightest cluster galaxy seems like the likeliest source of energy. Whatever the energy source is, it must leave an imprint on the entropy structure of the intracluster medium. In search of that imprint we have measured the radial entropy profiles of some classic cooling-flow clusters and have found a remarkable degree of regularity among them. They asymptotically approach a pure-cooling profile at ~ 100 kpc but depart from the pure-cooling profile within 10-20 kpc of the cluster center, at an entropy scale of ~ 10 keV cm^2. Any model purporting to explain why the core gas does not simply cool and condense must explain these features of the core entropy profiles. A simple episodic feedback model can account for the observed profiles with outbursts of $\sim 10^{45}$ erg s^{-1} occurring on a ~ 100 Myr timescale. At radii < 30 kpc these outbursts would deposit heat primarily through shocks. Rising bubbles would dominate the heat deposition at larger radii.

1 Introduction

Many astronomers I know are not particularly comfortable with thinking about entropy, but it can be a very useful concept when one is trying to understand the roles of heating and cooling in clusters because entropy is the thermodynamic quantity most directly related to gains or losses of heat energy. Focusing on temperature alone can sometimes be misleading, particularly when gravity is involved. For example, consider what happens to an optically thick cloud of self-gravitating gas whose cooling time is less than the age of the universe. Its temperature *rises* as the cloud radiates away its thermal energy because the consequent gravitational contraction releases twice the amount of energy lost in the form of radiation.

The relationship between temperature and radiative losses in the intracluster medium is more subtle because the gas is generally not self-gravitating. If radiative losses are significant, the hot gas must contract within the potential well of the dark matter that binds it to the cluster. However, its temperature may go either up or down, depending on the shape of the potential well, the rate at which cooling occurs, and the amount of feedback that results. That is why focusing on entropy is often more fruitful than focusing on temperature when one is trying to gauge the impact of radiative cooling and feedback on the intracluster medium.

X-ray astronomers generally quantify entropy in terms of some version of the quantity $K = (k/\mu m_p)T\rho^{-2/3}$, which is the constant of proportionality in the polytropic equation of state for a monatomic ideal gas: $P = K\rho^{5/3}$. The quantity K is basically the logarithm of the thermodynamic entropy S, defined so that $\Delta S = \Delta(\text{heat})/T$, and is often expressed in units of keV cm^2. Be aware that entropy-like quantities given in these units in the cluster literature are sometimes evaluated using the electron density $(Tn_e^{-2/3})$ and other times using the total particle density $(Tn^{-2/3})$.

Radial entropy profiles have been determined for numerous clusters in which T and n_e have been measured as functions of radius [9, 10]. Surface-brightness deprojection under the assumption of spherical symmetry is the usual method employed to find the average values of T and n_e within spherical shells, meaning that these radial entropy profiles are only approximate (but see reference [6] for measurements of the spatial distribution of entropy that do not assume spherical symmetry). If the cluster is not totally relaxed, the gas within a spherical shell could be multiphase, containing gas of substantially different entropy levels. However, assuming in the deprojection analysis that the gas is single-phase still leads to a reasonably accurate estimate of the mean entropy, unless the differences in entropy are so extreme that the majority of the emission is coming from gas with a very small filling factor (see the Appendix of [4]).

This article is a short guide to what can be learned about heating and cooling in clusters from analyses of those intracluster entropy profiles. Section 2 discusses how structure formation generates entropy in the intracluster medium. Section 3 explains how cooling is thought to modify that entropy structure. Section 4 looks at the feedback response to cooling and the imprint of feedback on the entropy profiles of clusters. Section 5 considers the balance between heating and cooling in clusters, and § 6 briefly sketches how we might construct a more complete observational picture of the roles of cooling and feedback in clusters.

2 Entropy and Structure Formation

Merger shocks generate most of the entropy observed in the intracluster medium of a large galaxy cluster. These shocks produce entropy by converting the kinetic energy of infalling gas into heat. Thermalization of the kinetic energy of gas with pre-shock entropy K_1 and pre-shock density ρ raises the entropy to a post-shock value K_2 approximately given by

$$K_2 \approx \frac{v^2}{3(4\rho)^{2/3}} + 0.84K_1 \; , \tag{1}$$

where v is the velocity difference between the preshock gas and the postshock gas ([17]) The entropy gain

$$\Delta K = K_2 - K_1 \approx \frac{v^2}{3(4\rho)^{2/3}} - 0.16K_1 \ , \tag{2}$$

therefore depends on both the shock velocity and the density of the preshock gas.

Simulations show that gravitationally-driven shocks during the process of hierarchical structure formation produce a power-law entropy profile of the form $K \propto r^{\alpha}$, with $\alpha = 1.1-1.2$ [19]. The normalization of this power-law profile is determined by the cluster's mass. A halo of mass M_{200} within the radius r_{200}, defined so that mean mass density is $200\rho_{cr}$, has a characteristic temperature $T_{200} = GM_{200}/2r_{200}$. One therefore expects the intracluster medium in such a halo to have an entropy of approximately $K_{200} = T_{200}(200f_b\rho_{cr})^{-2/3}$. This is indeed the case in clusters simulated without radiative losses or feedback, which obey the self-similar relationship $K(r) \approx 1.3K_{200}(r/r_{200})^{1.1}$ outside of the cluster core ($r > 0.2r_{200}$) over a wide range of mass scales ([19]). Simulated entropy profiles within the cluster core ($r < 0.1r_{200}$) exhibit more diversity, but in this region cooling and feedback are expected to modify and regulate the entropy structure.

Recent numerical studies of idealized mergers show that mergers generate entropy through a two-step process [8]. In a head-on collision, the initial strong shock generates a large amount of entropy as the cores of the two merging objects collide. However, only about half the entropy produced in the merger comes from the initial shock. After this first shock passes through the cluster, the energy it deposits causes the intracluster medium to expand, thereby lowering the gas density. The remaining kinetic energy in the intracluster gas therefore dissipates in a relatively low-density medium, which increases the efficiency of entropy generation. Both episodes of entropy generation are essential to maintaining the self-similar entropy structure seen in simulated clusters.

Observations with *XMM-Newton* show that massive clusters with temperatures above 5 keV have self-similar entropy profiles with the same normalization and slope as those seen in non-radiative simulations, in which purely gravitational processes generate the entropy [9]. However, cooler, lower-mass clusters have more entropy than predicted by non-radiative simulations. In the most extreme cases, the entropy excess is more than a factor of 2 above the expectation from gravitational structure formation alone.

To first approximation, the deviation of cooler clusters from the predicted self-similar structure appears to preserve the shapes of the density and entropy profiles, while raising the normalization of the entropy profile and lowering the normalization of the density profile. However, *Chandra* measurements of temperature profiles suggest that the shapes of these profiles are not exactly preserved. Temperatures in objects < 3 keV peak at around $0.1r_{500}$

[13], while in hotter objects the peak temperature is at a larger scale radius $\sim 0.2 r_{200}$ [14]. This finding suggests that the entropy enhancement near the core radii of low-temperature objects is greater than at larger radii.

3 Entropy and Cooling

Some combination of radiative cooling and non-gravitational heating is presumably responsible for these deviations from self-similarity. Much progress has been made in understanding how excess entropy resulting from cooling and feedback affects the global scaling relations of clusters [15]. High-resolution simulations that include radiative cooling and supernova feedback generally do a good job of predicting the global properties of hot clusters [7]. However, those simulations do a poorer job of accounting for the scaling relations at < 2 keV, they vastly underpredict the amount of scatter in those relations, and they do not adequately explain why entropy in systems at < 5 keV is enhanced at all radii. Even when supernova-driven galactic winds are pushed to maximum efficiency, they still do not reproduce the observed entropy excesses [2], raising the possibility that AGN feedback might be needed to explain the observed entropy structure of the intracluster medium.

Here is where the entropy problem in clusters makes contact with the cooling-flow problem in cluster cores. There is currently great optimism about the hypothesis that AGN heating compensates for radiative cooling in cluster cores. If AGN heating is indeed the solution to the cooling-flow problem, then AGN heating integrated over the history of the universe should also have an impact on the entropy structure of the intracluster medium.

With these issues in mind, we have been studying the core entropy profiles in cooling-flow clusters with *Chandra* [5, 4]. In the absence of any type of feedback, one can construct a simple model for how radiative cooling alters the entropy profile of a cluster. Figure 1 shows how radiative losses affect the entropy profile in the pure-cooling model of [16]. The right-hand panel shows that uncompensated cooling tends to alter the entropy profiles of clusters in the 2-8 keV range so that they lie virtually on top of one another when plotted in physical K and r coordinates, with $K \propto r^{1.2}$ [18].

Comparing this universal pure-cooling profile to observations shows that the simple pure-cooling model acts as a lower boundary to the entropy profiles of real clusters. Figure 2 presents the entropy profiles of nine classic cooling-flow clusters observed by [4]. Although these clusters span a factor of three in temperature, their entropy profiles look very similar when plotted in physical coordinates. At large radii (> 100 kpc) they converge to the pure cooling model shown by the solid line. At small radii (< 30 kpc) they exceed the pure-cooling model by approximately 10 keV cm^2. The dashed line in the figure shows the pure-cooling model with 10 keV cm^2 added to it.

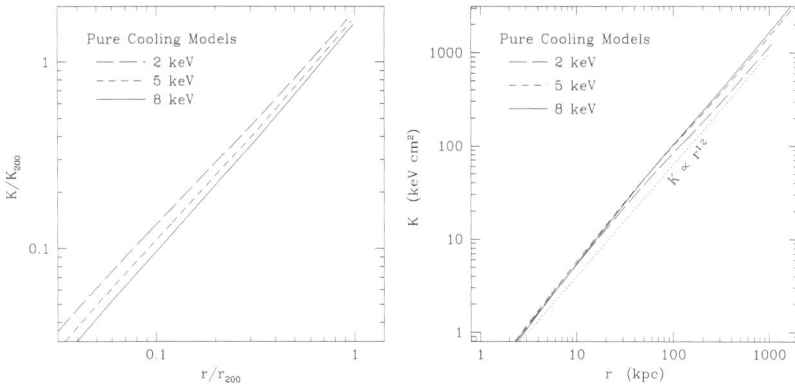

Fig. 1. Pure cooling models for the entropy profiles of clusters with three different temperatures: 2 keV (top), 5 keV (middle), and 8 keV (bottom). The pure-cooling models were calculated as described in [16]. The left-hand panel shows modified entropy profiles in scaled coordinates. Cooling shifts the entropy profiles of clusters to the left in this plot because condensing gas is at $r = 0$, and the effects of cooling are greater for cooler clusters. In physical coordinates, shown in the right-hand panel, these modified profiles lie virtually on top of one another, approximately following a $K \propto r^{1.2}$ law over the whole range in radius.

4 Core Entropy and Feedback

The shapes of the observed entropy profiles of cooling-flow clusters in Figure 2 suggest that radiative cooling is driving them all toward the pure-cooling state. However, some form of feedback is preventing convergence to that pure-cooling state. In some cases the the entropy profiles appear to flatten out at the 10 keV cm^2 level within 10 kpc. In other cases the departure from a pure-cooling model is more subtle, manifesting as a shallower entropy profile slope within ~ 10 kpc.

Whatever the feedback mechanism is, it must be acting quasi-steadily to maintain these entropy profiles in a monotonically increasing state. Large bursts of energy input at the center would produce large amounts of central entropy and temporary inversions in the central entropy gradient. Such inversions are generally not seen. However, X-ray cavities are common. These cavities are, in essence, high-entropy regions, but they contain a very small fraction of the mass. Thus, they have only a modest effect on the mean entropy of the region.

In order to gain some insight into the feedback process, we can look more closely at the ~ 10 keV cm^2 entropy gap between the observed central entropy profiles and the pure-cooling state. The cooling time of gas at this entropy level is

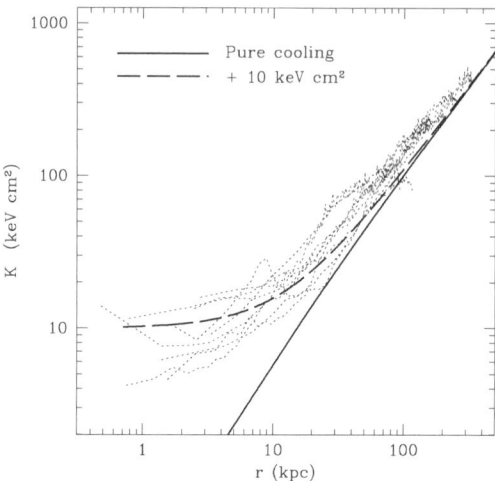

Fig. 2. Entropy profiles of nine classic cooling-flow clusters and the pure-cooling model. Dotted lines show the entropy profiles of nine cooling-flow clusters measured by [4]. The solid line shows the pure-cooling profile of a 5 keV cluster, which is virtually temperature-independent. The dashed line showing a profile created by adding a 10 keV cm^2 pedestal to the pure-cooling model is quite similar to the observed profiles.

$$t_c \approx 10^8 \, \text{yr} \left(\frac{K}{10 \, \text{keV cm}^2} \right)^{3/2} \left(\frac{T}{5 \, \text{keV}} \right)^{3/2} \tag{3}$$

Maintaining an entropy level of ~ 10 keV cm^2 therefore requires feedback to be periodic on a time scale of no greater than 100 Myr. Furthermore, feedback episodes cannot often raise the mean central entropy (not including cavities) to greater than ~ 10 keV cm^2. Otherwise, the central gas would linger for a long period of time at this higher entropy level, because of the longer cooling time, and many more cooling-flow clusters would be observed to have central entropy levels in excess of 10 keV cm^2.

Feedback with a 100 Myr duty cycle is a plausible solution. In such a cycle, a feedback episode would happen each time the gas at small radii cooled to an entropy $\ll 10$ keV cm^2, and the outburst would have to restore the central entropy to ~ 10 keV cm^2. The outburst could not simply deposit all the energy instantaneously at small radii, or else it would produce a large entropy spike at the center, which is generally not observed. The AGN energy must be introduced gradually and then propagated to larger radii without disrupting the power-law shape of the entropy profile.

In this context, it is enlightening to consider how a continuous outflow of energy from an AGN produces entropy in the intracluster medium [18]. First, let us assume that, prior to the outburst, the cluster core has a power-law structure with

$$K(r) = 150 \, \text{keV cm}^2 \left(\frac{r}{100 \, \text{kpc}} \right) \tag{4}$$

$$T(r) = T_{5,100} \left(\frac{r}{100 \, \text{kpc}} \right)^{1/3} \tag{5}$$

$$\rho(r) = 0.006 \, \text{cm}^{-3} \, T_{5,100}^{3/2} \left(\frac{r}{100 \, \text{kpc}} \right)^{-1} . \tag{6}$$

This is a reasonable approximation to the cluster cores shown in Figure 2 in the 10-100 kpc range. Then let us assume that the outburst drives a spherically symmetric shock front into the intracluster medium. This assumption may seem unwarranted, given that the observed AGN interactions with the ICM are often highly assymmetric, but it is still useful for establishing an order-of-magnitude connection between outflow power and the necessary amount of entropy deposition.

Given these assumptions, the first stage of entropy generation is an outflow-driven shock. The entropy generated at each radius by that shock can be estimated with a Sedov-like solution. For a kinetic power output L, the shock radius r and velocity v obey the condition $L \sim \rho r^2 v^3$, meaning that the shock velocity is $v \sim (L/\rho r^2)^{1/3}$. Plugging this expression into the shock jump condition gives $\Delta K \sim v^2 \rho^{-2/3} \sim$ constant. In other words, an outflow-driven shock produces an entropy jump that is constant with radius, in accord with observations. Doing things a little more quantitatively [18], one finds that

$$\Delta K \approx 23 \, \text{keV cm}^2 \, L_{45} \, T_{5,100}^{-2} \, f_P^2 \ , \tag{7}$$

where L_{45} is the kinetic power in units of 10^{45} erg s^{-1} and f_P is the fraction of the total energy output in the form of kinetic energy in the shock front. From this equation, it is evident that a kinetic power $\sim 10^{45}$ erg s^{-1} is needed to produce an entropy boost of $\sim 10 \, \text{keV cm}^2$ in the center of a cooling-flow cluster.

If the kinetic output continues at a constant rate of $\sim 10^{45}$ erg s^{-1}, the shock front will eventually become subsonic at a radius ~ 30 kpc. At that point, the hot bubble of AGN ejecta that was driving the shock front can begin to rise through buoyancy. If one assumes that the gravitational potential energy released as the bubble rises is transformed into heat energy at the same radius it was released [3, 11], one finds that the entropy jump at each radius due to bubble heating is

$$\Delta K_{\text{bubble}} \propto r^{-3/2} \tag{8}$$

for a relativistic equation of state within the bubble and

$$\Delta K_{\text{bubble}} \propto r^{-8/5} \tag{9}$$

for a non-relativistic equation of state. What this means is that repeated episodes of bubble heating will inevitably produce an entropy inversion, if the bubbles begin buoyantly rising at too small a radius. Entropy inversions can be avoided if the bubbles are injected at ~ 30 kpc, because the entropy jump per cycle is then small compared with the entropy at that radius, but then some other mechanism is needed to heat gas within that radius. A certain amount of outflow-driven shock heating at < 30 kpc would therefore seem to be inescapable.

5 Balancing Heating and Cooling

Feedback is a natural mechanism for balancing heating and cooling in the vicinity of the active nucleus, but it is not so clear how AGN heating can balance cooling over the entire region where the cooling time is less than a Hubble time. Taking the power-law representation of core structure from the previous section, we can write an expression for the entropy decrease due to free-free cooling at each radius over a time period t:

$$\Delta K_{\text{cool}} \approx -9.3 \, \text{keV} \, \text{cm}^2 \, T_{5,100} \left(\frac{r}{10 \, \text{kpc}} \right)^{-1/6} \left(\frac{t}{10^8 \, \text{yr}} \right) \tag{10}$$

Notice that the entropy drop is nearly a constant function of radius in the power-law core model. Thus, over the course of one cooling and feedback cycle, entropy production by an outflow-driven shock is nearly the same as the entropy losses at all radii. However, the same is not true of bubble heating. If bubble heating balances cooling at small radii, then it falls short of cooling at larger radii, at least in the simple power-law core model we have adopted here.

Perhaps heating need not balance cooling at all radii. Over the course of a cooling-feedback cycle, heating and cooling must be approximately equal within 10 kpc, because if heating exceeds cooling, then the net gain of heat over each cycle would lead to a buildup in central entropy, eventually resulting in an isentropic core larger than 10 kpc. However, heating can exceed cooling at larger distances without contradicting the observations. A net entropy gain per cycle at $\gg 10$ kpc would simply build up over time, leading to an entropy excess at the core radius.

Detailed numerical modeling is needed to determine whether central AGN heating, integrated over time, can produce the entropy excesses seen at all radii in low-temperature clusters. Because merger shocks can amplify entropy that is injected early in time, a complete calculation must involve both the effects of central heating and hierarchical merging. Early efforts along these lines look promising [12]. However, the AGN feedback mechanisms that have so far been implemented in cosmological simulations remain primitive.

6 Observational Prospects

The core entropy survey we have undertaken at Michigan State is not yet complete. It is based on the *Chandra* archive, and the first objects we studied tended to be famous objects of high central surface brightness. Those turned out to have strikingly similar core entropy profiles, but that sample is obviously biased toward classic cooling-flow clusters. Following up a few other less famous cooling clusters without significant AGN activity, we found that objects showing little evidence for recent feedback have longer central cooling times and higher central entropy, suggesting that very large AGN outbursts might sometimes shut off the feedback loop for time intervals exceeding 1 Gyr [5].

If we are to understand how AGN feedback couples to the intracluster medium in the context of hierarchical structure formation, we will need to study the entropy structure of cluster cores in a less biased cluster sample more representative of the entire cluster population. One sample that could be very useful is the *REXCESS* sample for which there is now a full set of deep *XMM-Newton* observations [1]. Followup of the *REXCESS* survey with *Chandra* would give us a more complete and statistically less biased picture of how the entropy profiles of clusters depend on cluster mass and morphology.

References

1. H. Böhringer et al., A&A, in press (2007)
2. S. Borgani et al.: MNRAS, **361**, 233 (2005)
3. E. Churazov, R. Sunyaev, W. Forman, H. Böhringer: MNRAS **332** 729 (2002)
4. M. Donahue, D. J. Horner, K. W. Cavagnolo, G. M. Voit: ApJ **643**, 730 (2006)
5. M. Donahue, G. M. Voit, C. P. O'Dea, S. Baum, W. B. Sparks: ApJ **630**, L113 (2005)
6. A. Finoguenov, D. S. Davis, M. Zimer, J. S. Mulchaey: ApJ **646**, 143 (2006)
7. A. V. Kravtsov, A. Vikhlinin, D. Nagai: ApJ **650**, 128 (2006)
8. I. G. McCarthy et al.: MNRAS, submitted (2006)
9. G. W. Pratt, M. Arnaud, E. Pointecouteau: A&A **446**, 429 (2006)
10. R. Piffaretti et al.: A&A **433**, 101 (2005)
11. M. Ruszkowski, M. Begelman: ApJ **581**, 223 (2002)
12. D. Sijacki, V. Springel: A&A **366**, 397 (2006)
13. M. Sun et al.: in preparation
14. A. Vikhlinin et al.: ApJ **640**, 691 (2006)
15. G. M. Voit: Rev. Mod. Phys. **77**, 207 (2005)
16. G. M. Voit, G. L. Bryan, M. L. Balogh, R. G. Bower: ApJ **576**, 601 (2002)
17. G. M. Voit, M. L. Balogh, R. G. Bower, C. G. Lacey, G. L. Bryan: ApJ **593**, 272 (2002)
18. G. M. Voit, M. Donahue: ApJ **634**, 955 (2005)
19. G. M. Voit, S. T. Kay, G. L. Bryan: MNRAS **364**, 909 (2005)

Entropy Generation in Merging Galaxy Clusters

M. Balogh[1], I. G. McCarthy[2], R. Bower[2], and G. M. Voit[3]

[1] Department of Physics and Astronomy, University of Waterloo, Waterloo, ON, Canada, N2L 5L4 mbalogh@uwaterloo.ca
[2] Department of Physics, University of Durham, South Road, Durham UK, DH1 3LE
[3] Department of Physics and Astronomy, MSU, East Lansing, MI 48824

1 Introduction

This conference has seen much discussion about non-gravitational heating of the intracluster medium, as required to reduce cooling rates and central densities sufficiently to explain the observed properties of galaxy clusters. The amount of energy input required is often computed by comparing to the entropy profile that would be expected from gravitational processes alone, as determined, for example, from cosmological simulations. Indeed, observations of relaxed clusters show that, except for the innermost regions, their entropy profiles follow a power-law profile that scales self-similarly with the cluster mass e.g., [7, 5, 13].

However, the origin of this default entropy profile and apparent self-similarity is not really understood in detail. Models of smooth, spherical accretion are able to reproduce the power-law slope of the entropy gradient e.g., [2, 1, 11, 3], but the normalization is very sensitive to the initial gas density distribution. In particular, if accretion is lumpy rather than smooth (as expected in a universe dominated by cold dark matter), insufficient entropy is generated to explain the observations [12]. If there is such a strong dependence on initial density, it is unclear why (or if) numerical simulations with different resolutions (and hence smoothing scales) and implementations (e.g. Eularian or Lagrangian) are able to produce self-similar clusters.

The source of this puzzle is illuminated by writing the equation of hydrostatic equilibrium as

$$\frac{1}{\rho}\frac{d\left(\rho T\right)}{dr} = -\frac{\mu m_H}{k}\frac{GM}{r^2},$$

$$(1)$$

which describes how the gas density and temperature (ρ and T) depend on distance from the cluster centre, r. If the mass M is dominated by dark matter, then the right hand side is constant. The temperature of the gas must be close to the virial temperature T_{vir}, again because the potential is dominated by an external field (the dark matter). We see, therefore, that for any solution to this equation, an equally valid solution can be found by scaling the density (and the corresponding boundary conditions) by an arbitrary factor. If we define the "entropy" as

$$K = \frac{T}{\rho^{2/3}},\tag{2}$$

we see that the entropy profile of an isothermal cluster with $T \sim T_{\mathrm{vir}}$ can also be arbitrarily normalized and still yield a valid solution to the hydrostatic equilibrium equation. Merger shocks, as seen in both observations e.g., [4] and simulations e.g., [8] have a very complex geometry, and entropy is clearly not generated in a single, strong shock. Why, then, do observed clusters show such a striking uniformity in their entropy normalization? If we are to understand how the effects of early energy injection propagate through the mass assembly of clusters, and if we are to implement self-consistent cooling/feedback processes in semi-analytic models of galaxy formation, we must first understand how entropy is generated in purely gravitational processes. This is the purpose of our work, recently submitted for publication [6], which we summarize in these proceedings.

2 Simulations

We have executed a number of idealized simulations of two-body cluster mergers, using the Tree-SPH code GADGET-2 [9], as described in detail in [6]. By default the code implements the entropy-conserving SPH scheme of [10], which ensures that the entropy of a gas particle will be conserved during any adiabatic process. The simulations span a range of mass ratios, from 10:1 to 1:1, with the mass of the primary fixed at $M_{200} = 10^{15} M_{\odot}$. Initial conditions and orbital parameters were chosen to match the typical conditions seen in cosmological simulations. Initially, the gas is assumed to be in hydrostatic equilibrium with the dark matter, with an entropy profile similar to that seen in observations and cosmological simulations.

In these proceedings we discuss three interesting results from these experiments. These results are fairly general to all our simulations, although here we just focus on those derived from the head-on collisions[4].

2.1 Two shocks, not one

All of our simulations show that there are *two* main episodes, during which the gas entropy sharply increases. The first occurs when the cores of the clusters collide. It is only then, and not before, that there is sufficient kinetic energy to drive strong shocks through the gas. Therefore, the energy is not deposited primarily at the outskirts, as assumed by spherical accretion ("onion-skin") models [11, 12] but rather the cluster is heated from the inside-out.

---

[4] We have also run simulations without dark matter, which help to identify the role played by the collisionless component. We will not discuss those results here, but note that our interpretation of the simulations including dark matter often depends on what we have learned from the gas-only runs.

Following this event is a more extended, gradual increase in entropy, due primarily to the reaccretion of gas that was driven away from the cluster by the shock wave (and overpressurized gas at the centre) produced by the collision of the cores. This phase generates approximately the same amount of entropy as the initial shock, over a longer period of time. This phase is what we refer to as the second shock, although in fact the entropy is being generated in numerous weak shocks (and perhaps turbulent mixing) in the outer regions of the cluster[5].

2.2 Distributed heating

If the gas entropy is to scale in a self-similar way following a merger, then $K \propto T \propto M^{2/3}$. For example, a binary merger should result in a ~ 60 per cent increase in entropy. One remarkable result is that for all our simulations (with a range of impact parameters and mass ratios, and even the gas-only simulations), most of the gas satisfies this scaling after ~ 10 Gyr. The exceptions are at the outer boundary, where there is an excess of entropy generated due to artificial boundary effects, and in the inner ~ 10 per cent of gas, which is physically heated to $\sim$twice the self-similar value.

A very interesting result emerges when we simulate the merger of clusters with unequal masses. In this case, the final entropy profile again agrees with the self-similar scaling law, so a 10:1 merger ends up with an entropy that is ~ 6 per cent larger than that of the primary, initial halo. One might naively have expected the gas in each component to independently scale in this way; that is, for the gas in the small component to be more strongly shocked than in the primary. However, this is not what happens. Instead, the primary halo is *overheated*, relative to the self-similar expectation, while the secondary is *underheated*, as shown in Figure 1. In other words, much of the infall energy associated with the secondary goes into thermalising the gas in the primary, and heating is a distributed, rather than local, process. We find a remarkably robust relation between the energy thermalized in both components (primary and secondary):

$$\frac{E_{T,p}}{E_{T,s}} \approx \left(\frac{M_p}{M_s}\right)^{5/4}. \tag{3}$$

As the mass of the primary is increased, the fraction of energy thermalized within its gas also increases. We do not understand why this simple relation arises, but it seems to hold for a wide range of mass ratios and orbital parameters.

---

[5] This is not unreasonable, as the Rankine-Hugoniot equations used to determine the post-shock conditions from the pre-shock gas and the Mach number are just a consequence of energy and momentum conservation, and are independent of the path taken between the initial and final state.

2.3 Energy requirements

One would hope to be able to capture most of the relevant physics from these simulations in a simple analytic prescription. The first attempts e.g., [12] have assumed all the entropy is generated in a single strong shock. Voit et al. showed that, in a simple, spherical accretion model, this fails to produce sufficient entropy if the infalling matter is clumpy. Our simulations show that this conclusion still holds when we consider more realistic merger models. In particular, for a given mass ratio, we can calculate the maximum energy available to be thermalized, when the cores collide, as shown in Figure 2. If we then assume that all of this energy is thermalized in a single, strong shock, we are unable to produce enough entropy to explain the results of 3:1 or 10:1 mergers.

This is a puzzling result, as our naive expectation had been that a single, strong shock thermalizing all the available energy would yield the maximum entropy. This is true, in that thermalizing the energy in N weaker shocks does not produce as much entropy, if the system does not evolve between shocks. Where does the real system find the extra energy? The answer appears to be that the gas density actually decreases significantly between the first and second shock events. In [6], we show analytically that if the density drops 20-30% below its *pre-merger* value, then the second shock can generate enough entropy for the final cluster to attain its self-similar structure. From the simulations we directly measure that this is indeed what happens: the first shock actually drives gas outward, so that the density in the outer regions ends up lower than it was initially, by 20-30%.

3 Conclusions

Gravitational shock heating of clusters does not appear to be as simple a process as once envisaged. It is crucial that we understand this mechanism, if we are to improve semi-analytic models of galaxy formation and to understand the heating requirements of real clusters. Our simulations have shown that self-similarity is indeed achieved during cluster mergers, but that this does not happen in a single accretion shock, because there is insufficient energy available. Instead, entropy is generated in two major "shocks", that heat the gas from the inside-out, in a way that distributes most of the energy within the more massive clump.

Our ultimate goal is to construct an analytic model of cluster growth that self-consistently tracks the entropy of the gas as it is shocked or non-gravitationally heated. There is still much work to be done before we achieve this. Our next steps are to test our analytic, two-shock model against idealized simulations with perturbed initial conditions (e.g. preheating).

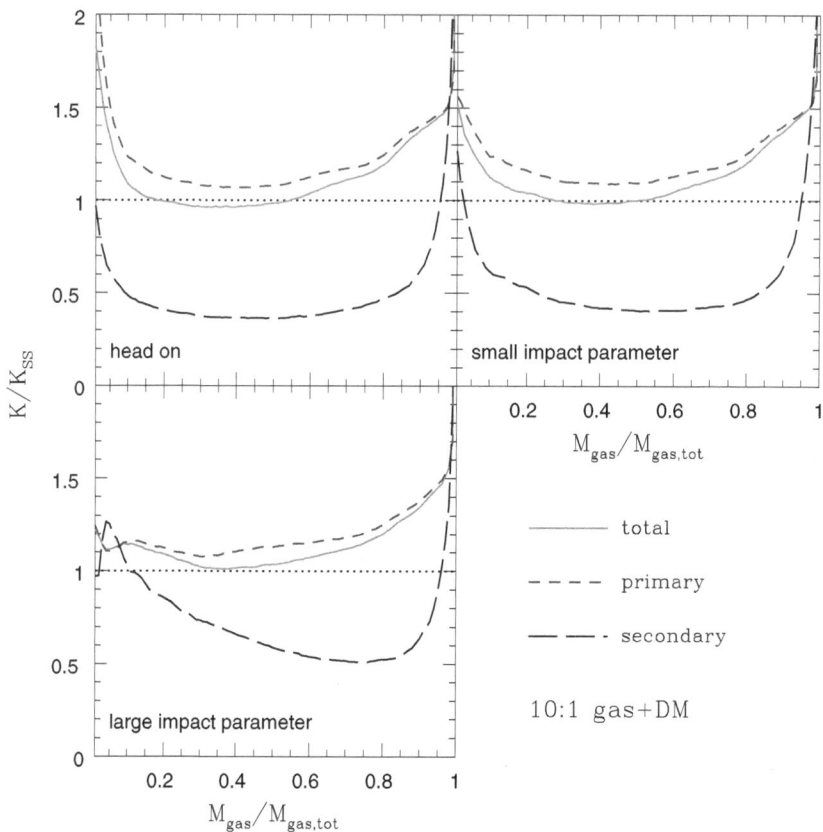

Fig. 1. The resulting $K(M_{\mathrm{gas}})$ distributions (entropy as a function of enclosed gas mass) for the 10:1 mass ratio mergers, normalized to K_{200} (the characteristic entropy of the halo). The dotted line, at $K/K_{200} = 1$, is the entropy distribution expected if the entropy growth is self-similar. We show the final entropy distributions for the primary (short-dashed), secondary (long-dashed), and total (solid) systems for three different orbital cases. Although most of the gas in the final system follows the self-similar expectation, this is achieved by overheating the primary and underheating the secondary.

Acknowledgement. We would like to thank our collaborators F. Pearce, T. Theuns, A. Babul, C. Lacey and C. Frenk, who are co-authors on this work as submitted in [6].

References

1. Abadi, M. G., Bower, R. G., & Navarro, J. F. 2000, MNRAS, 314, 759

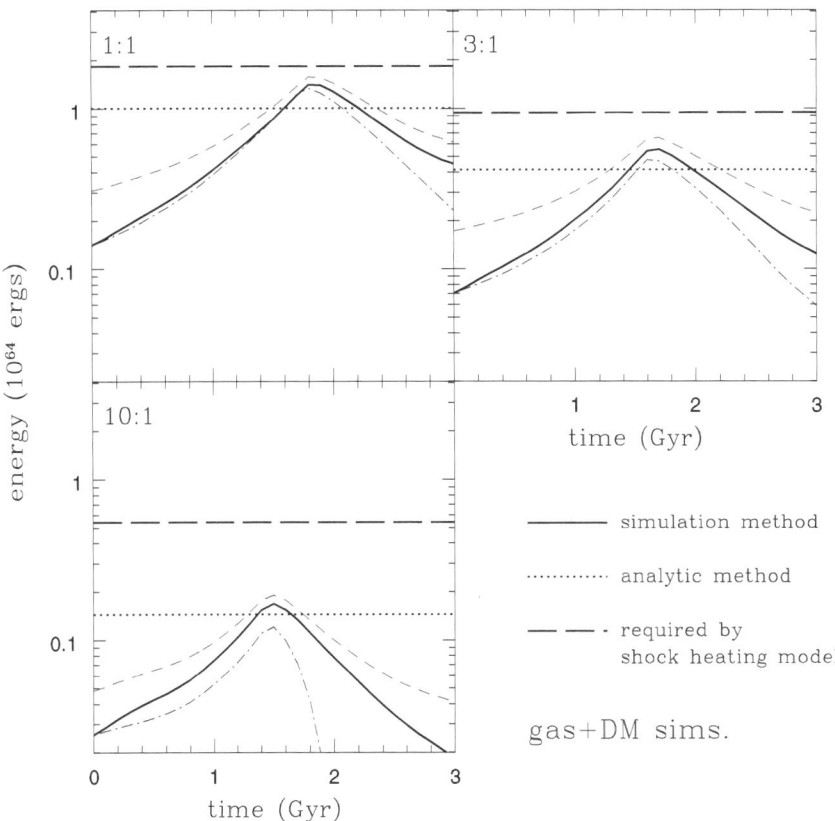

Fig. 2. Energy requirements in head-on mergers with three different mass ratios (as shown in the top-left corner of each panel). The horizontal, dashed line shows the amount of energy required to produce the final entropy distribution in the simulations, if all the entropy is generated in a single, strong shock. This is compared with the actual entropy available to be thermalized in the merger. The horizontal, dotted line shows the result of an analytic calculation of the maximum energy available, at the point where the two cores collide. The curved lines show the energy available as measured in the simulations, which reaches a maximum at the time when the cores collide. The different line styles correspond to different assumptions about the interaction between the dark matter and gas, as described in [6]. For the 3:1 and 10:1 simulations, there is not enough energy available to produce the observed entropy in a single shock.

2. Cavaliere, A., Menci, N., & Tozzi, P. 1998, ApJ, 501, 493
3. Dos Santos, S. & Doré, O. 2002, A&A, 383, 450
4. Markevitch, M., Govoni, F., Brunetti, G., & Jerius, D. 2005, ApJ, 627, 733
5. McCarthy, I., Fardal, M., & Babul, A. 2005, ApJ, submitted (astro-ph/050113)
6. McCarthy, I. et al. 2006, MNRAS, submitted
7. McCarthy, I. G., Balogh, M. L., Babul, A., Poole, G. B., & Horner, D. J. 2004, ApJ, 613, 811
8. Ryu, D., Kang, H., Hallman, E., & Jones, T. W. 2003, ApJ, 593, 599
9. Springel, V. 2005, MNRAS, 364, 1105
10. Springel, V. & Hernquist, L. 2003, MNRAS, 339, 289
11. Tozzi, P. & Norman, C. 2001, ApJ, 546, 63
12. Voit, G. M., Balogh, M. L., Bower, R. G., Lacey, C. G., & Bryan, G. L. 2003, ApJ, 593, 272
13. Voit, G. M., Kay, S. T., & Bryan, G. L. 2005, MNRAS, 364, 909

Entropy Profiles of Relaxed Galaxy Groups

F. Gastaldello[1], D. A. Buote[1], P. J. Humphrey[1], L. Zappacosta[1], F. Brighenti[2,3] and W. G. Mathews[2]

[1] University of California, Irvine, 4129 Frederick Reines Hall, Irvine, CA, 92697
gasta@uci.edu
[2] UCO/Lick Observatory, University of California, Santa Cruz, CA 95064
[3] Università di Bologna, via Ranzani 1, Bologna 40127, Italy

1 Introduction

X-ray observations have extensively shown that the structure of the Intra-cluster medium (ICM) departs from the self-similarity expected if only gravitational processes are shaping its properties [3, 7, 8].

Investigation of the entropy profiles is of fundamental importance because entropy records the thermodynamic history of the ICM, reflecting cooling and feedback heating from supernovae and AGN [10]. High quality XMM and Chandra cluster observations show that ICM entropy profiles have the $S(r) \propto r^{1.1}$ shape characteristic of gravitational structure formation outside of the core, but the overall normalization of these profiles scales as $T^{2/3}$ instead of T as in the self-similar prediction [6, 9, 1].

It is important to extend these studies to groups/poor clusters with temperatures less than 2 keV using higher quality *Chandra* and *XMM* data. At this scale the effects of feedback are more severe and should have a more dramatic impact on temperature and entropy profiles. We present here measurements of the entropy profiles of a sample of 16 groups with the best available XMM and Chandra data, originally selected for mass determination [2]. All results have been calculated using a concordance cosmological model with $\Omega_{\rm m} = 0.3$, $\Omega_\Lambda = 0.7$ and $H_0 = 70 \, {\rm km \, s^{-1} \, Mpc^{-1}}$.

2 Entropy profiles

The 16 objects in the sample are in the temperature range of 1-3 keV and in the mass range of $1 \times 10^{13} - 2 \times 10^{14}$ M$_\odot$. The entropy profiles are shown in Figure 1, scaled by the virial radius obtained by the NFW fit to the mass profile. Entropy profiles for groups (left panel of Figure 1, $10^{13} < M_{\rm vir} < 10^{14}$ M$_\odot$) show a large scatter and they are consistent with a broken power law with steeper inner slope (0.8-1.3) and flatter outer slope (0.4-0.6). A similar behavior has previously been noted for a sample of 8 groups [4].

NGC 5044, NGC 4325 and RGH 80 show a central flattening in the very inner core, a behavior seen in more massive clusters (e.g. Donahue et al. 2006). Its origin can be attributed to AGN heating or, in the case of RGH

80 (the only object in our sample which has two dominant elliptical galaxies in the center), to a merging event. NGC 2563 shows an entropy plateau at 0.1-0.2 $r_{\rm vir}$.

Entropy profiles for massive groups/poor clusters show less scatter similar to observations of more massive objects (right panel of Figure 1, $1 \times 10^{14} < M_{\rm vir} < 2 \times 10^{14}$ M$_\odot$). Local entropy modification, like the central flattening of AWM 4, is likely due to AGN heating, which has caused also the disappearance of the central cool core [5, 2].

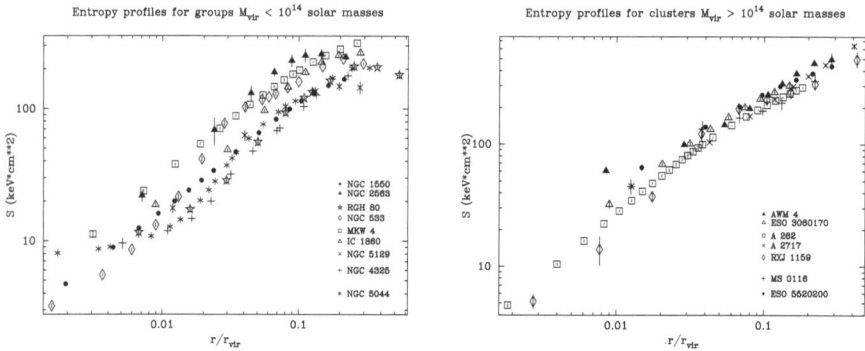

Fig. 1. *Left Panel:* Entropy profiles for group scale objects ($10^{13} < M_{\rm vir} < 10^{14}$ M$_\odot$). *Right Panel:* Entropy profiles for poor cluster scale objects ($1 \times 10^{14} < M_{\rm vir} < 2 \times 10^{14}$ M$_\odot$).

We compared the entropy profiles of the objects in our sample with the baseline entropy profile due to gravitational effects alone, derived from adiabatic numerical simulations [11]: $S(r) = 1.32\,S_{200}\,(r/r_{200})^{1.1}$ where S_{200} has been calculated as Eq. (65) of [10]:

$$S_{200} = \frac{1}{2} \left[\frac{2\pi}{15} \frac{G^2 M_{200}}{f_b H(z)} \right]^{2/3} , \tag{1}$$

where we assumed the baryon fraction to be $f_b = 0.14$, $H(z)^2 = H_0^2[\Omega_{\rm m}(1 + z)^3 + \Omega_\Lambda]$ and the mass M_{200}, enclosed within the radius r_{200} containing an overdensity of 200, calculated from our NFW fit to the gravitating mass. The scaled entropy profiles for all the objects in the sample are shown in the left panel of Figure 2 together with the baseline prediction, valid in the radial range $0.1 < r_{200} < 1.0$. In the right panel of Figure 2 we show the ratio of the observed profiles and the baseline prediction: this plot can be compared with the equivalent plot shown by [9] for a sample of 10 nearby clusters in the temperature range 2-9 keV: The trend of increasing deviation from the pure gravitational adiabatic prediction with decreasing mass(temperature) is confirmed by our data extending to the group scale.

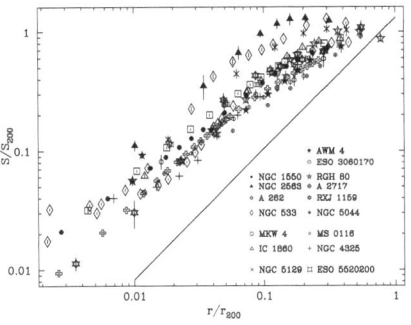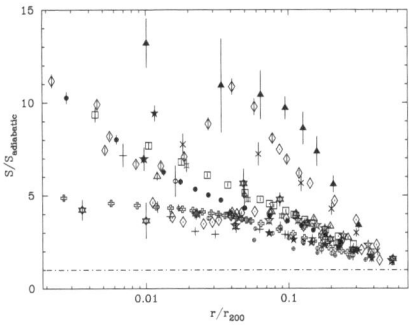

Fig. 2. Comparison of our observed entropy profiles with the baseline profile due to gravitational effects alone. *Left Panel:* The observed entropy profiles have been scaled to S_{200} using Eq. 1. The solid line represents the best fitting power law relation found from fitting SPH clusters in the radial range $0.1 < r_{200} < 1.0$ [11]. *Right Panel:* Ratio between the observed profiles and the best fitting power law baseline profile. The dot-dashed line shows a ratio of 1. Symbols are as in the left panel.

3 Conclusions

The large scatter in the entropy distribution of groups reflect their different and complex gasdynamical histories. The broken power law behavior challenges the models proposed so far and contrasts with the behavior of more massive clusters. The deviations from the baseline entropy profile due to gravitational effects are more severe and extend over a larger radial range compared to clusters. These results are consistent with the idea that AGN feedback is more important at the lower mass scale and aids in the breaking of the self-similar behavior.

References

1. M. Donahue et al. 2006, ApJ, 643, 730
2. F. Gastaldello et al. 2006, ApJ, submitted (astro-ph/0610134)
3. N. Kaiser. 1991, ApJ, 383, 104
4. A. Mahdavi, et al.. 2005, ApJ, 622, 187
5. E. O'Sullivan, et al. 2005, MNRAS, 357, 1134
6. R. Piffaretti, et al. 2005, A&A, 433, 101
7. T. J. Ponman, et al.. 1999, Nature, 397, 135
8. T. J. Ponman, et al.. 2003, MNRAS, 343, 331
9. G. W. Pratt, et al. 2006, A&A, 446, 429
10. G. M. Voit. 2005, Rev. Mod. Phys., 77, 207
11. G. M. Voit, et al. 2005, MNRAS, 364, 909

Observational Constraints on the ICM Temperature Enhancement by Cluster Mergers

N. Okabe[1] and K. Umetsu[2]

[1] Astronomical Institute, Tohoku University, Aramaki, Aoba-ku, Sendai, 980-8578, Japan okabe@astr.tohoku.ac.jp

[2] Academia Sinica Institute of Astronomy and Astrophysics (ASIAA) P.O. Box 23-141, Taipei 106, Taiwan, R.O.C. keiichi@asiaa.sinica.edu.tw

Summary. We present results from a combined weak lensing and X-ray analysis of the merging cluster A1914 at a redshift of $z = 0.1712$, based on R-band imaging data with Subaru/Suprime-Cam and archival Chandra X-ray data. Using the weak-lensing and X-ray data we explore the relationships between cluster global properties, namely the gravitational mass, the bolometric X-lay luminosity and temperature, the gas mass and the gas mass fraction, as a function of radius. We found that the gas mass fractions within r_{2500} and r_{vir} are consistent with the results of earlier X-ray cluster studies and with cosmic microwave background studies based on the WMAP observations, respectively. However, the observed X-ray temperature, $k_B T_{ave} = 9.6 \pm 0.3$keV, is significantly higher than the virial temperature, $k_B T_{vir} = 4.7 \pm 0.3$keV derived from the weak lensing distortion measurement. The X-ray bolometric luminosity-temperature ($L_X - T$) relation is consistent with the $L_X - T$ relation derived by previous statistical X-ray studies of galaxy clusters. Such correlations among the global cluster properties are invaluable observational tools for studying the cluster merger physics. Our results demonstrate that the combination of X-ray and weak-lensing observations is a promising, powerful probe of the physical processes associated with cluster mergers as well as of their mass properties.

1 Weak Lensing and X-ray Analyses

Weak Lensing Analysis

Weak lensing provides a direct measure of the projected mass distribution in the universe regardless of the physical/dynamical state of matter in the system. Therefore weak lensing enables the direct study of mass in clusters even when the clusters are in the process of (pre/mid/post) merging, where the assumptions of hydrostatic equilibrium or isothermality are no longer valid. We carried out a weak lensing analysis of the merging cluster A1914 with deep R_c-band data taken with Suprime-Cam on the Subaru telescope, covering the entire cluster region out to the cluster virial radius thanks to the wide field-of-view of $34' \times 27'$. The details of the analysis will be presented in [4] and [6]. We define a sample of background galaxies with magnitudes $21 \lesssim R_c \lesssim 26$ and half-light radii $r_h^* \lesssim r_h \lesssim 15$ pixels, yielding a mean galaxy number density of $n_g \simeq 48 \mathrm{arcmin}^{-2}$. We derived the radial profile of the tangential component of reduced gravitational shear, $g_+ = \gamma_+/(1 - \kappa)$, over the radial range of $3'$ to $17'$, where the irregularity in the mass distribution is less significant than that in the central region. The best-fitting NFW ([3]) profile to the Subaru distortion data (Figure 1, left panel) is obtained as follows: virial mass $M_{\mathrm{vir}} = (7.66 \pm 0.73) \times 10^{14} M_\odot h_{70}^{-1}$ (or viral radius $r_{\mathrm{vir}} = 12'.26 = 2.144 h_{70}^{-1}$ Mpc); concentration parameter $c = 5.07 \pm 1.75$. The best-fitting mass profile from weak lensing distortion measurements is in good agreement with the luminosity profile of cluster member galaxies, which will be presented in [6].

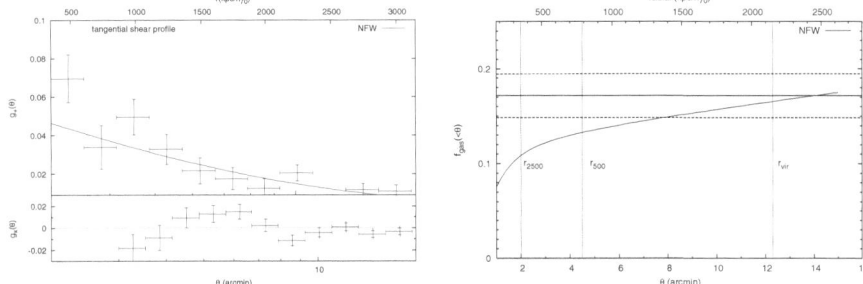

Fig. 1. *Left panel:* Radial profiles of the reduced tangential shear (upper panel) and the $45°$ rotated ($\times$) component (lower panel). The solid curve represents the best-fitting NFW shear profile. *Right panel:* Radial profile of the cluster gas mass fraction, f_{gas}, obtained by combining best fit results of weak lensing and X-ray analyses. Dashed horizontal lines indicate the 1σ confidence range of the cosmic baryon fraction.

X-ray Analysis

We used archival *Chandra* data to measure physical properties of the ICM in A1914. We performed a spectral fit with a single temperature model within a radius of $5'$. The average temperature and abundance were $k_B T_{\mathrm{ave}}(\theta < 5') = 9.6 \pm 0.3$ keV $A = 0.19 \pm 0.08$ at a 90% confidence level, respectively. The radial profile fit was performed on the observed X-ray surface brightness distribution ($\theta < 6'$) using a single β model, where we adopted the same center as for the weak lensing tangential shear measurement. The best fitting parameters were obtained as follows: cluster core radius, $r_c = 1.'03 \pm 0.'06$; slope parameter, $\beta = 0.727 \pm 0.014$; central electron density, $n_{e,0} = (1.46 \pm 0.26) \times 10^{-2} h_{70}^{-2}$ cm^{-3}.

2 Global Parameters of Merging Clusters

Gas Mass Fraction

We show in the right panel of Figure 1 the radial profile of the gas mass fraction, $f_{gas}(< r) = M_{gas}(< r)/M_{tot}(< r)$. We used the best-fitting models derived in §1 to calculate $f_{gas}(< r)$. Using the β model we extrapolated the gas mass profile outside $\theta = 6'$ where the weak lensing mass profile is available. The values within typical radii of r_{2500}, r_{500} and r_{vir}, at which the mean density is 2500, 500 and δ_{vir} times the critical density of the universe, are $f_{2500} = 0.108$, $f_{500} = 0.133$ and $f_{vir} = 0.165$, respectively. The central gas mass fraction, f_{2500}, is similar to those values for relaxed clusters derived by X-ray data alone, such as 0.091 ± 0.002 ([7]) and 0.117 ± 0.002 ([1]). We note the values of f_{500} deduced from X-ray studies show a large scatter among different clusters and different observations. We found a virial gas fraction, f_{vir}, of A1914, consistent with the cosmic mean baryon fraction Ω_b/Ω_m, constrained by the CMB observations ([5]). We emphasize that a combined weak-lensing and X-ray study will allow us to determine gas mass fractions even in merging clusters.

$L_X - T$ relation

We derived the X-ray bolometric luminosity $L_{X,bol}$ and temperature T with a single-temperature model. The resulting X-ray luminosity within r_{200} is $L_{X,bol} = 2.5 \times 10^{45} h_{70}^{-2}$ ergs s^{-1} with $k_B T_{vir} = 4.7^{+0.3}_{-0.3} (M/M_{vir})^{2/3}$ keV, which is consistent with the observed local $L_X - T$ relation from previous X-ray studies: $L_{X,bol}(< r_{200}) = 2.3^{+0.3}_{-0.3} \times 10^{45} (k_B T/9.6\text{keV})^{2.88\pm0.15} h_{70}^{-2}$ ergs s^{-1} ([2]). This indicates that the observed local $L_X - T$ relation holds even in the merging cluster A1914.

$M - T$ relation

The observed temperature $k_B T = 9.6 \pm 0.3$keV is significantly higher than the viral temperature, $k_B T_{vir} = 4.7^{+0.3}_{-0.3} (M/M_{vir})^{2/3}$ keV derived from the weak lensing analysis. If the cluster were virialized before the merging process, then the ICM temperature could be heated up by a factor of two. Based on this scenario, we constrain the heating energy of the ICM induced by the cluster merger as

$$\langle \Delta E_{ICM}(r < 5') \rangle = 4\pi(k_B T_{ave} - k_B T_{vir}) \int \sum_j n_j(r) r^2 dr \sim 2 \times 10^{62} \text{erg},$$

with $n_H = 0.82 n_e$, where we assumed the electron and ion temperatures are the same.

References

1. Allen, S. W. et al., 2004, MNRAS, 353, 457.
2. Arnaud, M. & Evrard, A. E., 1999, MNRAS, 305, 631.
3. Navarro, J. F., Frenk, C. S. & White, S. D. M. 1997, ApJ, 490, 493.

4. Okabe, N. & Umetsu, K., 2006, PASJ, submitted.
5. Spergel, D. N. et al., 2003, ApJS, 148, 175.
6. Umetsu, K. & Okabe, N., : in prep.
7. Vikhlinin, A. et al., 2006, ApJ, 640, 691.

Effects of AGN and Mergers on the Cores of Galaxy Groups

E. O'Sullivan[1], J. M. Vrtilek[1], L. P. David[1], T. J. Ponman[2], A. J. R. Sanderson[2] and J. Kempner[3]

[1] Harvard-Smithsonian Center for Astrophysics, 60 Garden Street, Cambridge, MA 02138, USA eosullivan@head.cfa.harvard.edu
[2] School of Physics and Astronomy, University of Birmingham, Edgbaston, B17 2TT, UK
[3] Department of Physics and Astronomy, Bowdoin College, 8800 College Station, Brunswick, ME 04011, USA

Summary. Using a sample of 12 galaxy groups observed by *XMM-Newton* and *Chandra*, we examine the effects of AGN heating and mergers on group properties and structure. While some systems possess clearly disturbed surface brightness features (e.g. cavities), in others high resolution temperature and abundance maps are required to reveal structure indicative of interactions. Comparison between the five systems with strongly disturbed cores and the more relaxed members of the sample suggests that central entropy and cooling time are only affected by the most powerful AGN activity. However, the mass-to-light ratio in the core of some systems drops below that expected for stars alone, suggesting that even in apparently relaxed groups AGN activity may push the gas out of hydrostatic equilibrium.

1 Introduction

Galaxy groups straddle the important mass range between large galaxy clusters and individual galaxy halos. This makes them a particularly interesting choice for studies of heating and cooling. The cores of group halos have densities comparable to those of massive clusters, but temperatures of only ~ 1 keV, leading to very short central cooling times. Most groups with extensive X-ray halos are dominated by giant elliptical or cD galaxies which host active nuclei. While groups are massive enough not to be disrupted by AGN outbursts (unlike individual galaxy halos), the energy input is likely to have a significant effect on the group gas. Studies of groups with radio-loud central galaxies show that they tend to fall below the L_X:T relation [1], suggesting that the AGN has either heated the halo or caused a lowering of the central density. Similarly, groups should be much more strongly affected by mergers and infall of individual galaxies than are larger clusters.

As part of a study of a sample of galaxy groups observed by *Chandra* and *XMM-Newton*, we have examined a number of systems which show evidence of disturbance and heating. In this paper we describe some of these groups and compare their properties to those of undisturbed systems, so as to examine the effects of heating and cooling on their structure.

2 Sample and Analysis

Our sample consists of 12 groups taken from the *XMM-Newton* and *Chandra* archives, selected to have high quality X-ray data and extended gaseous halos. All data were screened to remove periods of high background and images and spectra were extracted using standard techniques. Background spectra were created based on blank-sky data, with corrections for differences in soft flux. Annular spectra were extracted for all systems, and profiles of temperature and density were then used to estimate parameters such as gas entropy, cooling time, and total gravitational mass, under the standard assumptions of hydrostatic equilibrium and relaxation. Hardness maps were generated, and for systems with sufficient counts, adaptively binned spectral maps were used to examine the 2D temperature and abundance structure. The mapping software used [2] extracts spectra from regions centered on the pixels of the map, varying the extraction region to ensure the inclusion of a fixed number of source counts (typically 800-1000). The regions may be larger than the pixels and therefore can overlap in regions of low surface brightness. However, comparison between datasets and with more typical spectral extraction methods shows the resulting maps to be both accurate and reliable. Of the 12 systems examined, 5 show clear signs of disturbance. These are detailed below.

2.1 HCG 62

HCG 62 is the clearest case of an AGN interaction in our sample. Unusually for a galaxy group, a $\sim$49 ks *Chandra* observation revealed two large cavities to north and south of the group core. The energy required to create these structures is estimated to be a few 10^{56} erg [3]. The group does host a weak ($L_R \sim 2\times10^{38}$ erg s^{-1}) but extended radio source, which overlaps the southern cavity but does not have a clear jet and lobe structure. We estimate that the central AGN underwent a period of activity $\sim 2 \times 10^7$ yr ago, with a power output of $\sim 10^{42}$ erg s^{-1}, creating the cavities and heating the surrounding gas. However, it is notable that HCG 62 possesses a cool core, which was either not destroyed by the heating or has reformed over the past 20 Myr.

2.2 AWM4

AWM4 appears at first glance to be a completely relaxed system, having a regular mildly elliptical halo. An azimuthally averaged radial temperature profile shows the system to be isothermal to >250 kpc radius, and dynamical studies have shown the galaxy population to be relaxed with no substructure [4]. However, the dominant elliptical, NGC 6051, is an FR-I radio galaxy with jets extending $\sim$100 kpc into the surrounding intra-group medium (IGM). Spectral mapping reveals shock heated gas along the northern edge of the western jet, and possibly also to the south, and shows that the abundance distribution is also peaked north of the core, suggesting some enriched gas has been driven out of the core by the action of the AGN [2]. MKW4 (a system of similar mass with no central AGN) has a cool core and comparison suggests that the ongoing outburst in AWM4 has reheated (or prevented the formation of) a cool core in this group. The energy required to do so is very large by the standards of galaxy groups, $\sim 10^{59}$ erg. However, given the mechanical power output of the radio jets, $L_{mech} = 3 \times 10^{43}$ erg s^{-1}, this is not unreasonable.

2.3 NGC 4636

NGC 4636 is a nearby group on the edge of the Virgo cluster, dominated by a single large elliptical. *Chandra* imaging revealed a set of spiral arm–like weak shocks in the group core, probably produced by an AGN outburst beginning a few Myr ago [5]. VLA 1.4 GHz radio maps show small cavities being inflated in the very center of the galaxy [6]. Spectral mapping of the group finds cool, high abundance gas in the core and extending $\sim$30 kpc to the south-west. A cavity is embedded in the cool gas, visible as a hot region because of the reduction in low temperature emission along that line of sight [7]. Our interpretation of these features is that NGC 4636 has undergone a series of relatively small AGN outbursts with intervals of $\sim$10 Myr and energies of a few 10^{56}–10^{57} erg. The cavities created in these episodes rise buoyantly, drawing cool, enriched gas out into the plume we observe. The AGN is therefore countering cooling through both direct heating and gas mixing.

2.4 NGC 507

The NGC 507 group is currently undergoing a merger with the slightly smaller NGC 499 system. NGC 507 is perhaps the most disturbed system in our sample, with a number of sharp-edged surface brightness features in its core. It hosts an FR-I radio galaxy whose power output to date is estimated to be a few 10^{58} erg. Spectral mapping reveals that although there is a large region of cool gas, it is offset to the south of the galaxy core. The abundance distribution is clumpy and poorly correlated with the temperature structure. A strong surface brightness edge to the east of the core is produced by a combination of these features, with the decline in emission marking the boundaries of two separate regions, one cool, the other hotter but with high metallicity [8]. These structures suggest that the core has been seriously disrupted by the AGN activity, with gas motions on scales of $\sim$25 kpc. However, it is as yet unclear whether NGC 499 may have a role in the creation of this situation; if it has already passed through the NGC 507 group, tidal forces may have affected the group core, and possibly triggered the AGN outburst.

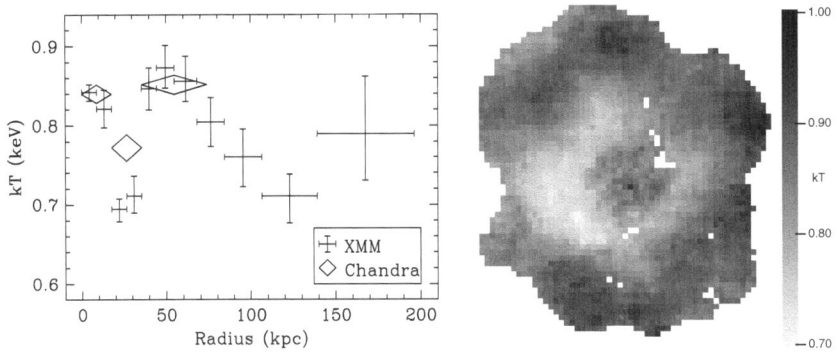

Fig. 1. *Left*: Temperature profiles of NGC 3411 based on *Chandra* and *XMM* data, showing the central temperature peak and surrounding cool region. *Right*: *XMM* temperature map of the group, showing the core and shell in two dimensions.

2.5 NGC3411

NGC 3411 is an apparently relaxed group with an almost circular halo, dominated by a large elliptical galaxy with several neighbouring disk galaxies. Radial spectral profiles show the halo to have a normal, centrally peaked abundance profile, but a highly unusual temperature profile; kT rises toward the center of the group, then drops as if there were a cool core, but then rises again to produce a central temperature peak (see Fig. 1). Both *Chandra* and *XMM-Newton* data show the same feature, which is 4.75σ significant. Temperature maps confirm that the group has a hot core surrounded by a shell of cooler gas.

One possible mechanism for the formation of this structure is the stripping of an infalling galaxy. A galaxy entering the group and passing through the dense core might be stripped by turbulent processes, losing a significant fraction of its gas, which would then sink through the group halo until it reached a region of equal entropy. It would then spread along this equal entropy surface to form a shell. However, while the mass of gas involved (3×10^9 $M_\odot$) is typical of an elliptical galaxy, there are no ellipticals of sufficient size within the group, and the larger disk galaxies are at sufficient distance to require the lifespan of the feature to be unfeasibly long ($> 10^9$ yr). Alternatively, NGC 3411 could be an example of a reheated cool core, with an AGN outburst heating the central cool gas to its current temperature. The AGN is currently not powerful enough to prevent cooling, and high resolution VLA imaging shows that there are no radio jets at present. An outburst with energy 2×10^{57} erg occurring ~ 30 Myr ago would have been sufficient to heat the core.

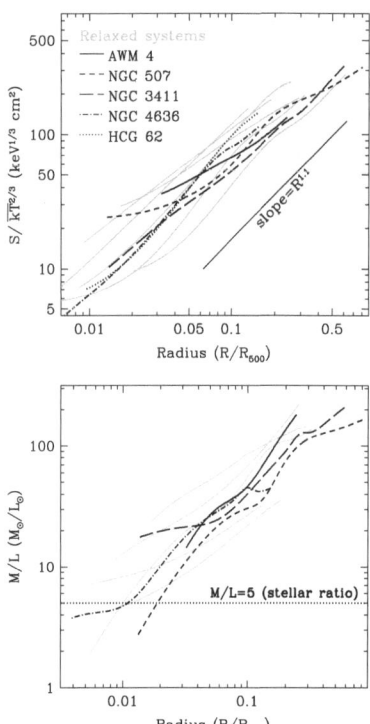
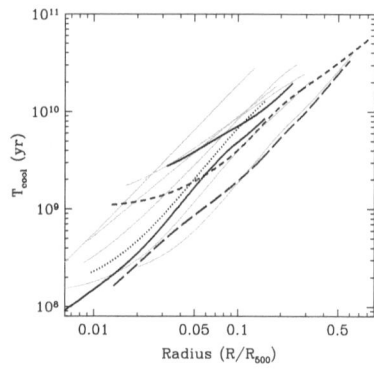

Fig. 2. Radial profiles of scaled entropy (upper left), cooling time (upper right) and mass-to-light ratio (lower left) for our sample of galaxy groups. The systems which show signs of strong disturbance are marked in black, the more relaxed systems in grey. A line of slope 1.1 is marked on the entropy plot to show the gradient expected from shock heating. The typical stellar B-band mass-to-light ratio of 5 $M_\odot/L_\odot$ is marked by a dotted horizontal line in the lower plot.

3 Entropy, Cooling Time and Mass-to-Light ratio

Having examined the five disturbed groups individually, we next compare their properties to those of the more relaxed systems. Figure 2 shows radial profiles of entropy, cooling time and mass-to-light ratio (M/L) for all the groups in the sample. Radius is plotted in units of the overdensity radius R_{500} so as to allow comparison of systems with differing physical sizes, and entropy is scaled by $\overline{kT}^{-2/3}$, where the characteristic group temperature $\overline{kT}$ is defined as the mean temperature between 0.1 and 0.3 R_{500}. This scaling corrects for the known dependence of entropy on system temperature in groups [9].

For systems in which the gas is mainly heated by shocks during formation, the entropy distribution is expected to have a gradient of 1.1 [10]. Most of our groups have profiles which match this expectation at least approximately. For the disturbed NGC 4636, NGC 3411 and HCG 62 groups, this suggests that AGN heating has occurred adiabatically, probably through weak shocks. However, there is some variation between systems, particularly in the cores. It is notable that both AWM 4 and NGC 507 deviate quite strongly from the general trend, having profiles which appear to flatten towards the center. If we extrapolate inwards, both systems seem likely to have central scaled entropies of $\sim$20-30 keV$^{1/3}$ cm^2, compared to $\leq$ 10 in the other groups. Similarly, both AWM 4 and NGC 507 have high central cooling times, $>10^9$ yr, compared to a few 10^8 yr for the other systems in the sample. These are also the groups hosting FR-I radio galaxies, which have the largest AGN power output in our sample ($\sim 10^{59}$ erg). It seems likely that the while the weaker outbursts seen in the other groups are capable of countering cooling, powerful outbursts with extended radio jets are required to reheat the cool core and alter the central entropy and cooling time significantly.

While the outer parts of all the groups have M/L ratios consistent with significant dark matter content, it is interesting to note that several systems have central M/L ratios lower than that expected from stars alone. NGC 507 is one of these, as is NGC 4636. One possibility is that the gas in the cores of these groups is not in hydrostatic equilibrium, perhaps because of gas motions or because the gas is over-pressured due to recent heating, as has been suggested for some elliptical galaxies [11]. AWM 4 unfortunately cannot be compared to these systems, as the resolution of the *XMM* temperature profile is too coarse to allow it to be traced into the core. However two of the apparently undisturbed groups also have very low central M/L. If AGN activity is responsible, this suggests that even apparently relaxed systems may not produce reliable mass profiles.

Acknowledgement. EO'S acknowledges support from NASA grant AR4-5012X.

References

1. J. H. Croston, M. J. Hardcastle, M. Birkinshaw, 2005, MNRAS, 357, 279
2. E. O'Sullivan, J. M. Vrtilek, J. C. Kempner et al, 2005, MNRAS, 357, 1134
3. U. Morita, Y. Ishisaki, N. Y. Yamasaki et al, 2006, PASJ, 58, 719
4. D. M. Koranyi, M. J. Geller, 2002, AJ, 123, 100
5. C. Jones, W. R. Forman, A. Vikhlinin et al, 2002, ApJ, 567, L155
6. S. W. Allen, R. J. H. Dunn, A. C. Fabian et al, 2006, MNRAS, in press (astro-ph/0602549)
7. E. O'Sullivan, J. M. Vrtilek, J. C. Kempner, 2005, ApJ, 624, L77
8. R. P. Kraft, W. R. Forman, E. Churazov et al, 2002, ApJ 601, 221
9. T. J. Ponman, A. J. R. Sanderson, A. Finoguenov, 2003, MNRAS, 343, 331
10. P. Tozzi, C. Norman, 2001, ApJ, 546, 63
11. S. Diehl, T. S. Statler, ApJ, in press (astro-ph/0606215)

Part VI

Chemical Abundances in Cool Cores

Helium and Iron in X-ray Galaxy Clusters

S. Ettori

INAF, Osservatorio astronomico, via Ranzani 1, 40127 Bologna (Italy)
stefano.ettori@oabo.inaf.it

1 The metals in X-ray galaxy clusters

The X-ray emitting hot plasma in galaxy clusters represents the 80 per cent of the total amount of cluster baryons. It is composed by a primordial component polluted from the outputs of the star formation activity taking place in galaxies, that are the cold phase accounting for about 10 per cent of the cluster barionic budget (see Fig. 1).

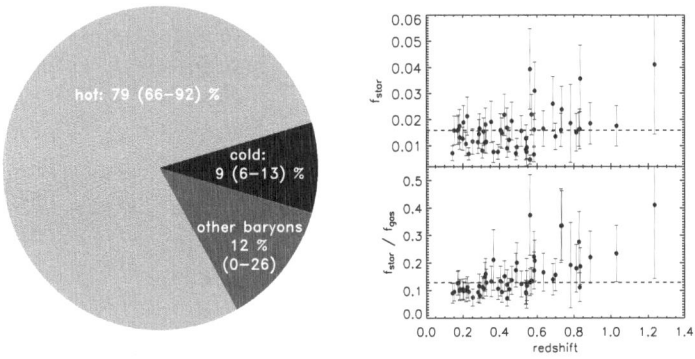

Fig. 1. (Left) Updated version of the cluster baryonic pie presented in [11] and obtained by using $f_{bar,\mathrm{WMAP}} = 0.176 \pm 0.015$ from the best-fit results of the WMAP 3-yrs data ([21]), a depletion parameter $Y = 0.920 \pm 0.023$ and the gas and stellar mass fractions shown on the right which refer to a sample of 58 clusters with $T_{gas} > 4$ keV, redshift in the range $0.14 - 1.26$ and discussed in Ettori et al. (2006, in prep.). The dashed lines indicates the median value of $f_{star} = 0.016$ and $f_{star}/f_{gas} = 0.129$.

As reference, a number density of 9.77×10^{-2} ions of helium and 4.68×10^{-5} ions of iron is expected for each atom of hydrogen in a hot plasma with solar abundance (as in [1]; for the most recent estimates in [10] and [3], the number density relative to H is 8.51×10^{-2} for He, 3.16×10^{-5} and 2.82×10^{-5} for Fe, respectively; see Table 1). These values imply a mass fraction of $(0.707, 0.738)$ for H, $(0.274, 0.250)$ for He and $(0.019, 0.012)$ for heavier elements, accordingly to ([1], [3]), respectively.

I discuss here some speculations, and relevant implications, of the sedimentation of helium in cluster cores (Sect. 2; details are presented in [14]) and a history of the metal accumulation in the ICM, with new calculations (with respect to the original work in [13]) following the recent evidence of a bi-modal distribution of the delay time in SNe Ia (see Sect. 3).

2 Helium in X-ray galaxy clusters

Diffusion of helium and other metals can occur in the central regions of the intracluster plasma under the attractive action of the gravitational potential, enhancing their abundances on time scales comparable to the cluster age. For a Boltzmann distribution of particles labeled 1 with density n_1 and thermal velocity $v_{\rm th} = (2kT/A_1 m_{\rm p})^{1/2}$ in a plasma with temperature kT in hydrostatic equilibrium with a NFW potential $g(r)$, the drift velocity of the heavier ions 2 with respect to 1 is given by ([22])

$$v_{\rm sed}(r) = \frac{3m_{\rm p}^2\, A_1 A_2\, v_{\rm th}^3\, g(r)}{16\pi^{1/2}e^4\, Z_1^2 Z_2^2\, n_1\, \ln\Lambda}, \tag{1}$$

where $\log\Lambda$ is the Coulomb logarithm. In [14], we have studied the effects of the sedimentation of helium nuclei on the X-ray properties of galaxy clusters.

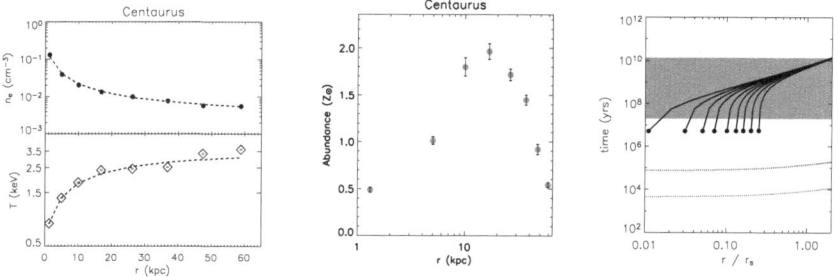

Fig. 2. (Left) Best joint-fit of the gas density and temperature profile of the de-projected data of Centaurus (from [19]) with a modified NFW gas profile. (Middle) Metal abundance profile with the characteristic central drop observed in spatially well resolved cool-core clusters. (Right) H-He (thick dotted line) and H-Fe (thin dotted line) equipartition time in the Centaurus cluster. The dots mark the radius $r_{\rm in}$ at which He is accumulated from the regions beyond with a sedimentation time represented by the solid line: $t_{\rm sed} = \int_{r_{\rm in}}^{r_{\rm out}} dr/v_{\rm sed}(r)$. The shaded region ranges between a $t_{\rm cool} = 2 \times 10^7$ yrs and the age of the Universe (details in [14]).

We have estimated the gravitational acceleration by fitting to the observed deprojected gas density and temperature profiles some functional forms that well reproduce the central steepening and are obtained under the assumption that the plasma is in the hydrostatic equilibrium with a NFW potential (left panel in Fig. 2). The observed gas density and temperature values do not allow metals to settle down in cluster cores over timescales shorter than few 10^9 years. On the other hand, by assuming that in the same potential the gas that is now describing a cool core was initially isothermal, the sedimentation times are reduced by 1–2 order of magnitude within $0.2r_{200}$. The sedimentation of helium can then take place in cluster cores. Even modest enhancement in the helium abundance affects (i) the relative number of electron and ions $\frac{n_e}{n_p} = c_M = \sum M_i Z_i N_i = 1 + 2M_{\mathrm{He}}N_{\mathrm{He}} + \sum_{i \neq \mathrm{H,He}} M_i Z_i N_i$, (ii) the atomic mean molecular weight $\mu = \left(\sum_i \frac{X_i(1+Z_i)}{A_i} \right)^{-1}$ and all the X-ray quantities that depend on these values, such as the emissivity ϵ, the gas mass density $\rho_{\mathrm{g}} \propto \mu(1+c_M)$, the total gravitating mass, $M_{\mathrm{tot}} \propto \mu^{-1}$, the gas mass fraction, $f_{\mathrm{gas}} \propto \mu^2(1 + c_M)$. Moreover, we show in [14] that if we model a super-solar abundance of helium with the solar value, we underestimate the metal (iron) abundance and overestimate the model normalization or emission measure. For example, with $M_{\mathrm{He}} = 3$, the measured iron abundance and emission measure are ~ 0.65 and 1.5 times the input values, respectively. Thus, an underestimated excess of the He abundance might explain the drops in metallicity observed in the inner regions of cool-core clusters (see, e.g., middle panel in Fig. 2). It is worth noticing that the diffusion of helium can be suppressed, however, by the action of confinement due to reasonable magnetic fields (see [7]), or limited to the very central region ($r < 20$ kpc) by the turbulent motion of the plasma.

3 Iron in X-ray galaxy clusters

The iron abundance is nowadays routinely determined in nearby systems thanks mostly to the prominence of the K-shell iron line emission at restframe energies of 6.6-7.0 keV (Fe XXV and Fe XXVI). Furthermore, the few X-ray galaxy clusters known at $z > 1$ have shown well detected Fe lines, given sufficiently long (> 200 ksec) *Chandra* and *XMM-Newton* exposures ([18], [15], [25]). It is more difficult to assess the abundance of other prominent metals that should appear in an X-ray spectrum at energies (observer rest frame) between ~ 0.5 and 10 keV, such as oxygen (O VIII) at (cluster rest frame) 0.65 keV, silicon (Si XIV) at 2.0 keV, sulfur (S XVI) at 2.6 keV, and nickel (Ni) at 7.8 keV. In [13], we infer from observed and modeled SN rates the total and relative amount of metals that should be present in the ICM both locally and at high redshifts. Through our phenomenological approach, we adopt the models of SN rates as a function of redshift that reproduce well both the very recent observational determinations of SN rates at $z > 0.3$ ([8], [6]) and the measurements of the star formation rate derived from UV-luminosity densities and IR data sets. We then compare the products of the enrichment process to the constraints obtained through X-ray observations of galaxy clusters up to $z \sim 1.2$.

Table 1. Adopted values for the average atomic weight (W), solar abundance by number with respect to H (A, from [1] and [3]; $R = A_{A05}/A_{AG89}$), total synthesized isotopic mass per SN event (m_{Ia} from deflagration model W7 and m_{CC} integrated over the mass range $10 - 50 M_\odot$ with a Salpeter IMF; see [17]) and corresponding abundance ratios by numbers with respect to Fe, $Y_i = m_i/(W_i A_i) \times (W_{Fe} A_{Fe})/m_{Fe}$.

metal	W	A_{AG89}	A_{A05}	R	m_{Ia} $M_\odot$	Y_{Ia}	m_{CC} $M_\odot$	Y_{CC}
He	4.002	9.77e-2	8.51e-2	0.87	–	–	–	–
Fe	55.845	4.68e-5	2.82e-5	0.60	0.743	–	0.091	–
O	15.999	8.51e-4	4.57e-4	0.54	0.143	0.037	1.805	3.818
Si	28.086	3.55e-5	3.24e-5	0.91	0.153	0.538	0.122	3.526
S	32.065	1.62e-5	1.38e-5	0.85	0.086	0.585	0.041	2.284
Ni	58.693	1.78e-6	1.70e-6	0.95	0.141	4.758	0.006	1.647

The observed iron mass is obtained from [9] and [25] (see also [12]) as

$$M_{Fe,obs} = 4\pi A_{Fe} W_{Fe} \int_0^R Z_{Fe}(r)\rho_H(r) r^2 dr, \qquad (2)$$

where $Z_{Fe}(r)$ is the radial iron abundance relative to the solar value A_{Fe}, W_{Fe} is the iron atomic weight and $\rho_H(r)$ is the hydrogen mass density. To recover the history of the metal accumulation in the ICM, we use the models of the cosmological rates (in unit of SN number per comoving volume and rest-frame year) of Type Ia, r_{Ia}, and core-collapse supernovae, r_{CC}, as presented in [8] and [23]. We estimate then the iron mass through the equation

$$M_{Fe,SN} = M_{Fe,Ia} + M_{Fe,CC} = \sum_{dt,dV} (m_{Fe,Ia} r_{Ia}(dt) + m_{Fe,CC} r_{CC}(dt)) \, dt dV \quad (3)$$

where $m_{Fe,Ia}$ and $m_{Fe,CC}$ are quoted in Table 1, dt is the cosmic time elapsed in a given redshift range, dV is the cluster volume defined as the volume corresponding to the spherical region that encompasses the cosmic background density, $\rho_b = 3H_0^2/(8\pi G) \times \Omega_m$ ($\approx 4 \times 10^{10} M_\odot$ Mpc^{-3} for the assumed cosmology), with a cumulative mass of $M_{vir} \approx 6.8 (kT/5keV)^{3/2} \times 10^{14} M_\odot$ ([2]). We obtain that these SN rates provide on average a total amount of iron that is marginally consistent with the value measured in galaxy clusters in the redshift range $0 - 1$, and a relative evolution with redshift that is in agreement with the observational constraints up to $z \approx 1.2$. We predict metal-to-iron ratios well in agreement with the X-ray measurements obtained in nearby clusters implying that (1) about half of the iron mass and > 75 per cent of the nickel mass observed locally are produced by SN Ia ejecta, (2) the SN Ia contribution to the metal budget decreases steeply with redshift and by $z \approx 1$ is already less than half of the local amount and (3) a transition in the abundance ratios relative to the iron is present at z between ~ 0.5 and 1.4, with SN CC products becoming dominant at higher redshifts.

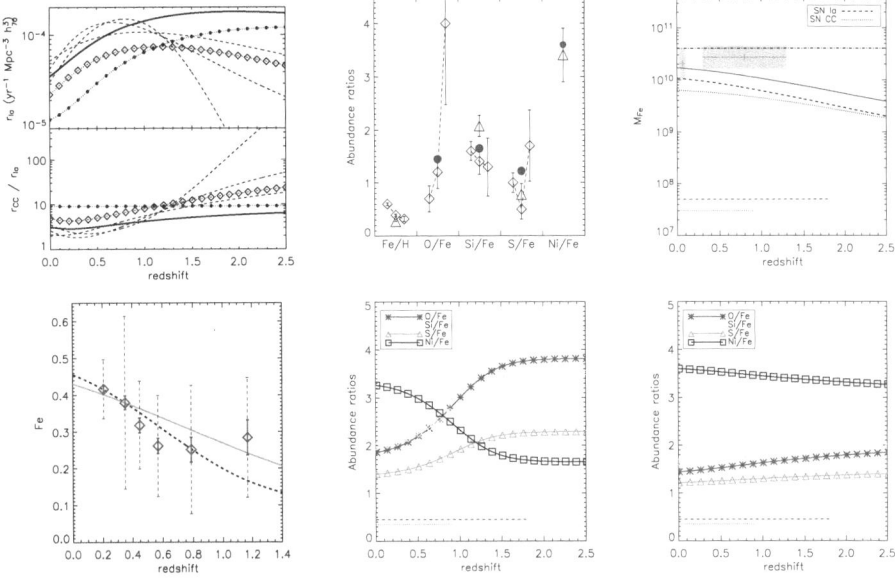

Fig. 3. (From left to right, from top to bottom): (A) SN Ia rates and number ratio of SNe CC to Ia as a function of z for the delay time distribution functions $\phi(t_d)$ considered in [8] (dashed lines) and [16] (the solid line is the sum of the contribution from "tardy"/squares and "prompt"/dots population). (B) Predictions (*big dots*; using $\phi(t_d)$ from [16]) of the local metal ratios compared with the X-ray measurements in [24] and [5]. (C) Accumulation history of the iron mass, M_{Fe}, as obtained from eq. (3) with rates from [16]. The solid line shows the sum of SN Ia (*dashed line*) and SN CC (*dotted line*) contribution. The accumulation history is here compared to the total iron mass measured in samples of local ([9]) and high−z ([25]) clusters for a typical object at 5 keV. (D) Observed ([4]) and predicted (as in panel A and normalized to the observed value at $z \approx 0.4$) iron abundance as function of redshift. (E-F) The last two plots show the metal abundance ratios as function of redshift as predicted from a $\phi(t_d)$ described by a "narrow" Gaussian with $\tau = 4$ Gyr and $\sigma_{t_d} = 0.2\tau$ as in [8] (E; see the corresponding rates in panel A) and from a bi-modal SN Ia distribution as in [16] (F). Horizontal lines indicate the redshift region where the SN rates are actually measured. These calculations refer to the solar abundance values in [1] for a direct comparison with the data. By adopting the compilation in [3], the trends are confirmed with the most relevant change being the predicted Ni/Fe≈ 2 at $z = 0$.

Recently, Mannucci et al. (2006; see also [20]) argue that the data on (i) the evolution of SN Ia rate with redshift, (ii) the enhancement of the SN Ia rate in radio-loud early type galaxies and (iii) the dependence of the SN Ia rate on the colours of the parent galaxies suggest the existence of two populations of progenitors for SN Ia, one half (dubbed "prompt") that is expected to occur soon after the stellar birth on time scale of 10^8 years and is able to pollute significantly their environment with iron even at high redshift, the other half ("tardy") that has a delay time function consistent with an exponential function with a decay time of ~ 3 Gyrs. Due to the larger production of SN Ia at higher redshift (see panel A in Fig. 3), the rates suggested by Mannucci et al. predict a more flat distribution of the Fe abundance as function of z than the rates tabulated in [8], but still show a negative evolution partially consistent with the measurements up to $z \approx 1.2$ (panel D in Fig. 3). Furthermore, these rates provide a better agreement than the ones in [8] both with local abundance ratios (panel B in Fig. 3) and with the overall production of Fe (panel C), that at $z = 0$ has been released from SNe Ia by more than 60 per cent. A potential discriminant is the behaviour of the abundance ratios relative to Fe in the ICM as a function of redshift, with the rates in [16] providing values of O/Fe and Ni/Fe that are a factor of 2 different from the predictions of the single-population delay time distribution function (see panels E and F).

References

1. Anders E., Grevesse N.: Geochimica et Cosmochimica Acta, 53, 197 (1989)
2. Arnaud M., Pointecouteau E., Pratt G.W.: A&A, 441, 893 (2005)
3. Asplund M., Grevesse N., Sauval A.J.: "Cosmic abundances as records of stellar evolution and nucleosynthesis", eds. Bash F.N. and Barnes T.G., ASP Conf. Series, 336, 25 (2005)
4. Balestra I. et al.: A&A, in press (2006; astro-ph/0609664)
5. Baumgartner W.H. et al.: ApJ, 620, 680 (2005)
6. Cappellaro E. et al.: A&A, 430, 83 (2005)
7. Chuzhoy L., Loeb A.: MNRAS, 349, L13 (2004)
8. Dahlen T. et al.: ApJ, 613, 189 (2004)
9. De Grandi S., Ettori S., Longhetti M., Molendi S.: A&A, 419, 7 (2004)
10. Grevesse N., Sauval A.J.: Space Science Rev., 85, 161 (1998)
11. Ettori S.: MNRAS, 344, L13 (2003)
12. Ettori S., Tozzi P., Borgani S., Rosati P.: A&A, 417, 13 (2004)
13. Ettori S.: MNRAS, 362, 110 (2005)
14. Ettori S., Fabian A.C.: MNRAS, 369, L42 (2006)
15. Hashimoto Y. et al.: A&A, 417, 819 (2004)
16. Mannucci F., Della Valle M., Panagia N.: MNRAS, 370, 773 (2006)
17. Nomoto K. et al.: Nucl. Phys. A, 621, 467 (1997)
18. Rosati P. et al.: AJ, 127, 230 (2004)
19. Sanders J.S., Fabian A.C.: MNRAS, 331, 273 (2002)
20. Scannapieco E., Bildsten L.: ApJ, 629L, 85 (2005)
21. Spergel D.N. et al.: ApJ, submitted (2006; astro-ph/0603449)
22. Spitzer L.: *Physics of Fully Ionized Gases* (Interscience Publishers, New York 1956)
23. Strolger L.G. et al.: ApJ, 613, 200 (2004)
24. Tamura T. et al.: A&A, 420, 135 (2004)
25. Tozzi P. et al.: ApJ, 593, 705 (2003)

Tracing the Evolution in the Iron Content of the ICM

I. Balestra[1], P. Tozzi[2,3], S. Ettori[4], P. Rosati[5], S. Borgani[3,6], V. Mainieri[1,5] and C. Norman[7]

[1] Max-Planck-Institut für Extraterrestrische Physik, Postfach 1312, 85741 Garching, Germany balestra@mpe.mpg.de
[2] INAF, Osservatorio Astronomico di Trieste, via G.B. Tiepolo 11, I–34131, Trieste, Italy
[3] INFN, National Institute for Nuclear Physics, Trieste, Italy
[4] INAF, Osservatorio Astronomico di Bologna, via Ranzani 1, I–40127, Bologna, Italy
[5] European Southern Observatory, Karl-Schwarzschild-Strasse 2, D-85748 Garching, Germany
[6] Dipartimento di Astronomia dell'Università di Trieste, via G.B. Tiepolo 11, I–34131, Trieste, Italy
[7] Department of Physics and Astronomy, Johns Hopkins University, Baltimore, MD 21218

Summary. We present a Chandra analysis of the X-ray spectra of 56 clusters of galaxies at $z > 0.3$, which cover a temperature range of $3 > kT > 15$ keV. Our analysis is aimed at measuring the iron abundance in the ICM out to the highest redshift probed to date. We find that the emission-weighted iron abundance measured within $(0.15 - 0.3)\,R_{vir}$ in clusters below 5 keV is, on average, a factor of ~ 2 higher than in hotter clusters, following $Z(T) \simeq 0.88\,T^{-0.47}\,Z_\odot$, which confirms the trend seen in local samples. We made use of combined spectral analysis performed over five redshift bins at $0.3 > z > 1.3$ to estimate the average emission weighted iron abundance. We find a constant average iron abundance $Z_{Fe} \simeq 0.25\,Z_\odot$ as a function of redshift, but only for clusters at $z > 0.5$. The emission-weighted iron abundance is significantly higher ($Z_{Fe} \simeq 0.4\,Z_\odot$) in the redshift range $z \simeq 0.3 - 0.5$, approaching the value measured locally in the inner $0.15\,R_{vir}$ radii for a mix of cool-core and non cool-core clusters in the redshift range $0.1 < z < 0.3$. The decrease in Z_{Fe} with z can be parametrized by a power law of the form $\sim (1 + z)^{-1.25}$. The observed evolution implies that the average iron content of the ICM at the present epoch is a factor of ~ 2 larger than at $z \simeq 1.2$. We confirm that the ICM is already significantly enriched ($Z_{Fe} \simeq 0.25\,Z_\odot$) at a look-back time of 9 Gyr. Our data provide significant constraints on the time scales and physical processes that drive the chemical enrichment of the ICM.

1 Properties of the sample and spectral analysis

The selected sample consists of all the public *Chandra* archived observations of clusters with $z \geq 0.4$ as of June 2004, including 9 clusters with $0.3 < z < 0.4$. We used the XMM-*Newton* data to boost the S/N only for the most distant clusters in our current sample, namely the clusters at $z > 1$.

We performed a spectral analysis extracting the spectrum of each source from a region defined in order to maximize the S/N ratio. As shown in Fig. 1, in most cases the extraction radius R_{ext} is between 0.15 and 0.3 virial radius R_{vir}. The spectra were analyzed with XSPEC v11.3.1 [2] and fitted with a single-temperature mekal model [7, 8] in which the ratio between the elements was fixed to the solar value as in [1].

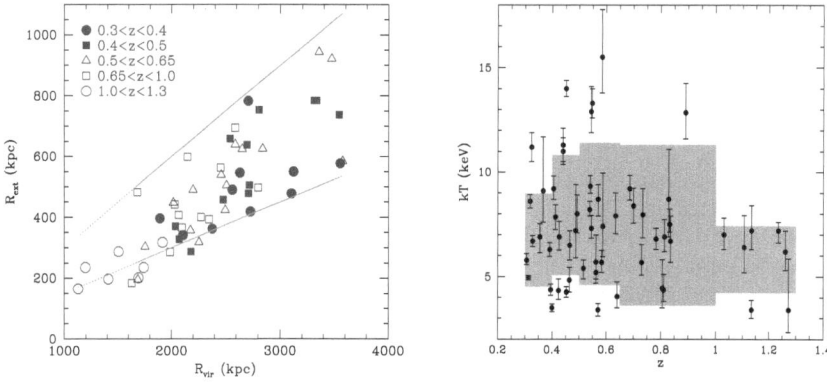

Fig. 1. *Left:* extraction radius R_{ext} versus R_{vir}. Lower and upper lines show $R_{ext} = 0.15\,R_{vir}$ and $R_{ext} = 0.3\,R_{vir}$, respectively. *Right:* temperature vs redshift. Shaded areas show the *rms* dispersion around the weighted mean in different redshift bins.

We show in Fig. 1 the distribution of temperatures in our sample as a function of redshifts (error bars are at the 1σ c.l.). The Spearman test shows no correlation between temperature and redshift ($r_s = -0.095$ for 54 d.o.f., probability of null correlation $p = 0.48$). Fig. 1 shows that the range of temperatures in each redshift bin is about $6 - 7$ keV. Therefore, we are sampling a population of medium-hot clusters uniformly with z, with the hottest clusters preferentially in the range $0.4 < z < 0.6$.

Our analysis suggests higher iron abundances at lower temperatures in all the redshift bins. This trend is somewhat blurred by the large scatter. We find a more than 2σ negative correlation for the whole sample, with $r_s = -0.31$ for 54 d.o.f. ($p = 0.018$). The correlation is more evident when we compute the weighted average of Z_{Fe} in six temperature intervals, as shown by the shaded areas in Fig. 2.

2 The evolution of the iron abundance with redshift

The single-cluster best-fit values of Z_{Fe} decrease with redshift. We find a $\sim 3\sigma$ negative correlation between Z_{Fe} and z, with $r_s = -0.40$ for 54 d.o.f.

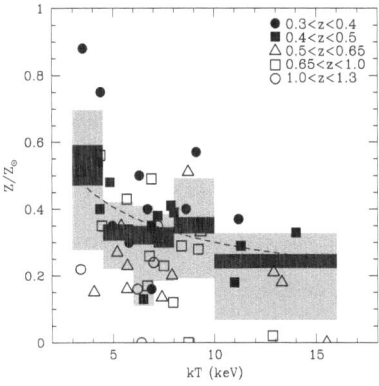 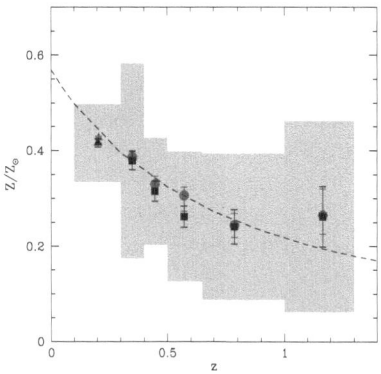

Fig. 2. *Left:* scatter plot of best-fit Z_{Fe} values versus kT. The dashed line represents the best-fit $Z - T$ relation ($Z/Z_\odot \simeq 0.88\,T^{-0.47}$). Shaded areas show the weighted mean (blue) and average Z_{Fe} with *rms* dispersion (cyan) in 6 temperature bins. *Right:* mean Z_{Fe} from combined fits (red circles) and weighted average of single-source measurements (black squares) within 6 redshift bins. The triangles at $z \simeq 0.2$ are based on the low-z sample described in Sect. 2. Error bars refer to the 1σ c.l.. Shaded areas show the *rms* dispersion. The dashed line indicates the best fit over the 6 redshift bins for a simple power law of the form $\langle Z_{Fe} \rangle = Z_{Fe}(0)\,(1+z)^{-1.25}$.

($p = 0.0023$). The decrease in Z_{Fe} with z becomes more evident by computing the average iron abundance as determined by a *combined spectral fit* in a given redshift bin. This technique is similar to the stacking analysis often performed in optical spectroscopy, where spectra from a homogeneous class of sources are averaged together to boost the S/N, thus allowing the study of otherwise undetected features. In our case, different X-ray spectra cannot be stacked due to their different shape (different temperatures). Therefore, we performed a simultaneous spectral fit leaving temperatures and normalizations free to vary, but using a unique metallicity for the clusters in a narrow z range.

The Z_{Fe} measured from the *combined fits* in 6 redshift bins is shown in Fig. 2. We also computed the weighted average from the single cluster fits in the same redshift bins. The best-fit values resulting from the *combined fits* are always consistent with the weighted means within 1σ (see Fig. 2). This allows us to measure the evolution of the average Z_{Fe} as a function of redshift, which can be modelled with a power law of the form $\sim (1+z)^{-1.25}$.

Since the extrapolation of the average Z_{Fe} at low-z points towards $Z_{Fe}(0) \simeq 0.5\,Z_\odot$, we need to explain the apparent discrepancy with the oft-quoted canonical value $\langle Z_{Fe} \rangle \simeq 0.3\,Z_\odot$. The discrepancy is due to the fact that our average values are computed within $r \simeq 0.15\,R_{vir}$, where the iron abundance is boosted by the presence of metallicity peaks often associated to cool cores. The regions chosen for our spectral analysis, are larger than the typical size of the cool cores, but smaller than the typical regions adopted in studies of local samples. In order to take into account aperture effects, we selected a small subsample of 9 clusters at redshift $0.1 < z < 0.3$, including 7 cool-core and 2 non cool-core clusters, a mix that is representative of the low-z population. These clusters are presently being analyzed for a separate project aimed at obtaining spatially-resolved spectroscopy (Baldi et al., in preparation). Here we analyze a region within $r = 0.15\,R_{vir}$ in order to probe the same regions probed at high redshift. We used this small control sample to add a low-z point in our Fig. 2, which extends the Z_{Fe} evolutionary trend.

3 Discussion

We investigate whether the evolution of Z_{Fe} could be due to an evolving fraction of clusters with cool cores, which are known to be associated with iron-rich cores [5] and which amount to more than 2/3 of the local clusters [4]. In order to use a simple characterization of cool-core clusters in our high-z sample, we computed the ratio of the fluxes emitted within 50 and 500 kpc ($C = f(r < 50\,kpc)/f(r < 500\,kpc)$ computed as the integral of the surface brightness in the $0.5 - 5$ keV band (observer frame). This quantity ranges between 0 and 1 and it represents the relative weight of the central surface brightness. Higher values of C are expected if a cool core is present. If the decrease in Z_{Fe} with redshift is associated to a decrease in the number of cool-core clusters for higher z, we would expect to observe a positive correlation between Z_{Fe} and C and a negative correlation between C and z. In Fig. 3 we plot Z_{Fe} as a function of C for our sample. We find that in our sample there is no correlation between metallicity and C with a Spearman's coefficient of $r_s = 0.02$ (significance of $\sim 0.2\sigma$) nor one between C and z ($r_s = -0.11$, a level of confidence of 0.8σ).

The absence of strong correlations between C and Z_{Fe} or between C and z suggests that the mix of cool cores and non cool cores over the redshift range studied in the present work cannot justify the observed evolution in the iron abundance. We caution, however, that a possible evolution of the occurrence of cool-core clusters at high redshift may still partially contribute to the observed evolution of Z_{Fe}. In other words, whether the observed evolution of Z_{Fe} is contributed entirely by the evolution of the mass of iron or is partially due to a redistribution of iron in the central regions of clusters is an open issue to be addressed with a proper and careful investigation of the surface brightness of the high-z sample.

A final check is provided by the scatter plot of Z_{Fe} versus R_{ext}/R_{vir}, shown in Fig. 3. We do not detect any dependence of Z_{Fe} on the extraction radius adopted for the spectral analysis. In particular, we find that clusters with smaller extraction radii do not show higher Z_{Fe} values.

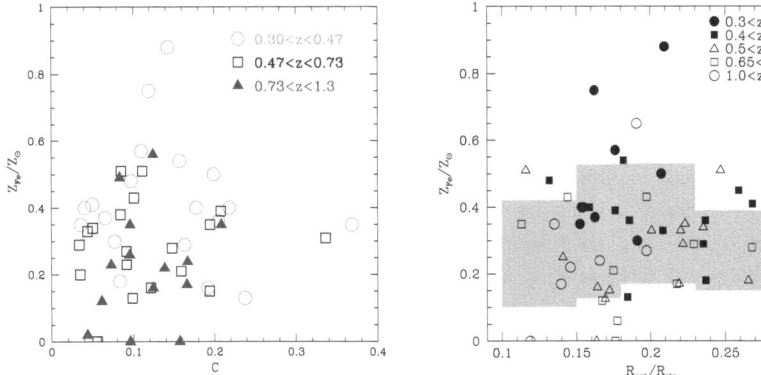

Fig. 3. *Left:* iron abundance plotted versus $C = f(r < 50\,kpc)/f(r < 500\,kpc)$. Clusters within different redshift bins are coded with different symbols. *Right:* iron abundance plotted versus the ratio R_{ext}/R_{vir}. Shaded areas show the *rms* dispersion around the average iron abundance in four bins.

4 Conclusions

We have presented the spectral analysis of 56 clusters of galaxies at intermediate-to-high redshifts observed by *Chandra* and XMM-*Newton* [3]. This work improves our first analysis aimed at tracing the evolution of the iron content of the ICM out to $z > 1$ [9], by substantially extending the sample. The main results of our work can be summarized as follows:

- We determine the average ICM iron abundance with a $\sim 20\%$ uncertainty at $z > 1$ ($Z_{Fe} = 0.27 \pm 0.05\,Z_\odot$), thus confirming the presence of a significant amount of iron in high-z clusters. Z_{Fe} is constant above $z \simeq 0.5$, the largest variations being measured at lower redshifts.
- We find a significantly higher average iron abundance in clusters with $kT < 5$ keV, in agreement with trends measured in local samples. For $kT > 3$ keV, Z_{Fe} scales with temperature as $Z_{Fe}(T) \simeq 0.88\,T^{-0.47}$.
- We find significant evidence of a decrease in Z_{Fe} as a function of redshift, which can be parametrized by a power law $\langle Z_{Fe} \rangle \simeq Z_{Fe}(0)\,(1+z)^{-\alpha_z}$, with $Z_{Fe}(0) \simeq 0.54 \pm 0.04$ and $\alpha_z \simeq 1.25 \pm 0.15$. This implies an evolution of more than a factor of 2 from $z = 0.4$ to $z = 1.3$.

We carefully checked that the extrapolation towards $z \simeq 0.2$ of the measured trend, pointing to $Z_{Fe} \simeq 0.5\,Z_\odot$, is consistent with the values measured within a radius $r = 0.15\,R_{vir}$ in local samples including a mix of cool-core and non cool-core clusters. We also investigated whether the observed evolution is driven by a negative evolution in the occurrence of cool-core clusters with strong metallicity gradients towards the center, but we do not find any clear evidence of this effect. We note, however, that a proper investigation of the thermal and chemical properties of the central regions of high-z clusters is necessary to confirm whether the observed evolution by a factor of ~ 2 between $z = 0.4$ and $z = 1.3$ is due entirely to physical processes associated with the production and release of iron into the ICM, or partially associated with a redistribution of metals connected to the evolution of cool cores.

Precise measurements of the metal content of clusters over large look-back times provide a useful fossil record for the past star formation history of cluster baryons. A significant iron abundance in the ICM up to $z \simeq 1.2$ is consistent with a peak in star formation for proto-cluster regions occurring at redshift $z \simeq 4 - 5$. On the other hand, a positive evolution of Z_{Fe} with cosmic time in the last 5 Gyrs is expected on the basis of the observed cosmic star formation rate for a set of chemical enrichment models. Present constraints on the rates of SNe type Ia and core-collapse provide a total metal produciton in a typical X-ray galaxy cluster that well reproduce (i) the overall iron mass, (ii) the observed local abundance ratios, and (iii) the measured negative evolution in Z_{Fe} up to $z \simeq 1.2$ [6].

References

1. Anders, E. & Grevesse, N. 1989, Geochim. Cosmochim. Acta, 53, 197
2. Arnaud, K. A. 1996, in ASP Conf. Ser. 101: Astronomical Data Analysis Software and Systems V, ed. G. H. Jacoby & J. Barnes, 17
3. Balestra, I., Tozzi, P., Ettori, S., et al. 2006, A&A, in press (astro-ph/0609664)
4. Bauer, F. E., Fabian, A. C., Sanders, J. S., et al. 2005, MNRAS, 359, 1481
5. De Grandi, S., Ettori, S., Longhetti, M., & Molendi, S. 2004, A&A, 419, 7
6. Ettori, S. 2006, in these proceedings (astro-ph/0610466)
7. Kaastra, J. S. 1992, (Internal SRON–Leiden Report, updated version 2.0)
8. Liedahl, D. A., Osterheld, A. L., & Goldstein, W. H. 1995, ApJl, 438, L115
9. Tozzi, P., Rosati, P., Ettori, S., et al. 2003, ApJ, 593, 705

Chemical Gradients in Galaxy Clusters and the Multiple Ways of Making a Cold Front

R. A. Dupke

Dept. Astronomy, University of Michigan, rdupke@umich.edu

1 Introduction

Cold fronts were originally interpreted as being the result of subsonic/transonic motions of head-on merging substructures with suppressed thermal conduction (Markevitch et al. 2000, 2001; Vikhlinin et al. 2001). This merger core remnant model is theoretically justified (e.g. Bialek & Evrard 2002; Nagai & Kravtsov 2003) and holds relatively well for clusters that have clear signs of merging, such as 1E0657-56 (Markevitch et al. 2002), but they do not work well for the increasing number of cold fronts found in clusters that do not show clear merging signs such as, among others, A496 (Dupke & White 2003, hereafter DW03), A1795 (Markevitch et al. 2001), RXJ1720.1+2638 (Mazzotta et al. 2001). This prompted the re-evaluation of other models for cold front generation, such as oscillation of the cD+central gas (Fabian et al. 2001; DW03), hydrodynamic gas sloshing due to scattering of a smaller halos (Tittley & Henriksen 2005; Ascasibar & Markevitch 2006, hereafter AM06).

As pointed out by DW03, different models for cold front formation can be discriminated through the analysis of SN Ia/II contamination across the front. If the cold front is generated by a merger core remnant we should expect the front to be accompanied by a characteristic discontinuity of metal enrichment histories, which can be determined through measurements of elemental abundance ratios. DW03 performed a chemical analysis of the cold front in A496. With an effective exposure of $\sim$9 ksec, they found no clear chemical discontinuities uniquely related to the cold front itself. Here we report the results of a deeper (effectively 55 ksec) observation of that cluster (Fig. 1a) that allowed us to produce high quality maps of the gas parameters and to compare more closely the observations with the predictions given by different models for cold front formation. We found for the first time a "cold arm" characteristic of a flyby of a massive DM halo near the core of the cluster. The cold arm is accompanied by an enhanced SN II Fe mass fraction, inconsistent with the merger core remnant scenario. We assume $H_0 = 70$ km s^{-1}Mpc^{-1}, $\Omega_0 = 1$ and $1'' \approx 0.66$ kpc.

The temperature and abundance ratio maps are shown in Figs 1b,c, with X-ray surface brightness contours overlaid. One striking feature that can be seen in the temperature map is a "cold spiral arm" that departs from the core to the N-NW up to the cold front position and runs along the cold front to the E-NE. Then it becomes more diffuse towards the S. The temperature of the arm is ∼3.0±0.2 keV. The temperatures of the surrounding regions of the cold arm are 3.8 keV and 4.3 keV towards the inner and outer cluster regions, respectively. This arm is definitely associated to the N (main) cold front and to a lesser extend to the W cold front. The arm departs from a boxy low temperature region, the edges of which appear coincides the near-core SE and S cold fronts. The overall temperature error along the "cold arm" is 10%. The higher southern temperatures near the CCD border have also high corresponding errors ($\gtrsim$ 1 keV). There is a cold tail starting 2.3′ SW of the cluster's center that may extend S for more 5′ (Tanaka et al. 2006).

In most simulations analyzed, AM06 found long lasting cold spiral arms coinciding with the cold fronts, sustained for many Gyrs after the sub-halo fly-by. In particular, their case for a DM perturber produces properties very similar to those observed in A496. In AM06 a pure dark matter halo with $\frac{1}{5}$ of the mass of the main cluster flies by with an impact parameter of 500 kpc and with closest approach at t∼1.37 Gyr (Fig 1d). Their simulations also seem to indicate the presence of milder cold fronts in the opposite side closer to cluster's core. These are a clear predictions that are corroborated well in A496 and suggest strongly that a flyby dark matter halo created the cold fronts in this cluster. A prediction of this model is the presence of a DM halo in the outskirts of the cluster without significant X-ray emitting gas. From the simulations, the position of that clump at epoch (t=2 Gyr, i.e., now) would be towards its apocenter at North, the same general direction of the main cold front (Dupke et al. 2006).

Acknowledgement. We acknowledge support from NASA through *Chandra* award number GO 4-5145X, NNG04GH85G and GO5-6139X and NASA grant NAG 5-3247.

References

1. Ascasibar, Y. & Markevitch, M., 2006, ApJ, in Press, astro-ph 0603246
2. Bialek, J. J., Evrard, A.E. & Mohr, J.J., 2002, ApJ, **578**, 9
3. Dupke, R. A. & White, R. E. III, 2003, ApJL, **583**, 13
4. Dupke, R.,, White, R. III, & Bregman J. 2006, ApJ, to be Submitted
5. Fabian, A. C., Sanders, J. S., Ettori, S., et al. 2001, MNRAS **321**, 33
6. Markevitch M. et al. 2000, ApJ, **541**, 542
7. Markevitch, M., Vikhlinin, A., & Mazzotta, P. 2001, ApJL, **562**, L153
8. Markevitch, M. et al. 2002, ApJL, 567, 27
9. Mazzotta, P., Markevitch, M., Vikhlinin, A., et al. 2001, ApJ, **555**, 205
10. Nagai, D. & Kravtsov, A. V., 2003, ApJ, **587**, 514
11. Tanaka, T., Kunieda, H., Hudaverdi, M., et al. 2006, PASJ, **58**, 703
12. Tittley, E. R., & Henriksen, M. 2003, ApJ, **563**, 673
13. Vikhlinin, A., Markevitch, M., & Murray, S. 2001, ApJ, **551**, 160

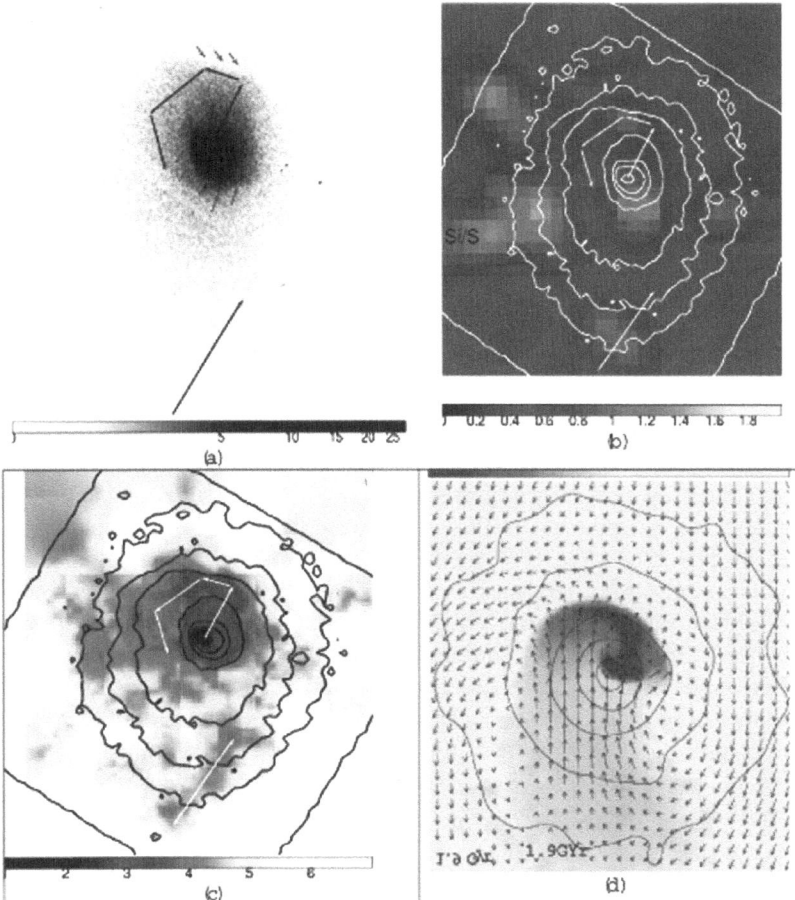

Fig. 1. (a) *Chandra* image of Abell 496. North is up. Blue arrows show the position of the northern (main), southern and eastern cold fronts. Black arrows near the center show the position of the low temperature and high SN II Fe mass fraction arm and the long arrow to the south shows the position of the cold "tail" in Figs a,b,c. (b) silicon to sulfur map. We also overlay the X-ray contours. The CCD border is also shown as the most external contour. Values shown for regions outside the CCD borders are an artifact of this type of code and should be neglected. The units are pixels and 1 pixel=0.5 arcsec. (c) Same as (b) but for gas temperatures (d) Zoom in of the core of a simulated cluster 0.53 Gyr after the flyby of the dark matter halo, from AM06. Yellow is ∼7–9 keV and blue 2 keV. DM density contours are overlaid and arrows indicate gas velocity (the longest corresponding to 500 km s^{-1}). The size of the panel is 250 kpc, similar to the size of Acis-S3 CCD borders at the redshift of the cluster (∼320 kpc). The figure has been flipped vertically to match the configuration of the cold front in A496.

Stochastic Gas motions in Galaxy Clusters

P. Rebusco[1], E. Churazov[1,2], H. Böhringer[3] and W. R. Forman[4]

[1] Max-Planck-Institut für Astrophysik, Karl-Schwarzschild-Strasse 1, 85748 Garching, Germany pao@mpa-garching.mpg.de
[2] Space Research Institute (IKI), Profsoyuznaya 84/32, Moscow 117810, Russia
[3] MPI für extraterrestrische Physik, Giessenbachstr. 1, 85748 Garching, Germany
[4] Harvard-Smithsonian Center for Astrophysics, 60 Garden st., Cambridge, MA 02138, USA

1 Introduction

Peaked abundance profiles are often observed in clusters with cool cores. These peaks are probably produced by the stars of the brightest cluster galaxy (BCG) after the cluster/group was assembled. The comparison of the optical light and the abundance profiles shows that the former are much steeper than the latter (see Fig.1), suggesting that there must be a mixing of the injected metals. Another feature of cool core clusters is a short gas cooling time (of the order of 10^8 yr), which could lead swiftly to temperatures lower than 1 keV. However such temperatures are not observed. The gas radiative cooling losses can be compensated by the dissipation of the gas motions. We estimate the parameters of stochastic gas motions in the core of cool core clusters/groups, under the assumption that metals are spread through the gas by the same motions. We also evaluate an impact of such motions on the width of the X-ray emission lines.

2 Iron Enrichment and Stochastic Diffusion

Consider the evolution of the iron abundance profile in the gas involved in stochastic motions with a characteristic spatial scale l and velocity scale v. Moreover assume that the gas density and temperature are constant with time. In the diffusion approximation the iron radial profile obeys:

$$\frac{\partial na}{\partial t} = \nabla \cdot (Dn\nabla a) + S, \tag{1}$$

where $n(r)$ is the gas density, $a(r,t)$ is the iron abundance and $S(r,t)$ is the iron injection from the BCG alone. $D = const$ is the diffusion coefficient, of the order of $\frac{vl}{3}$, with v being the characteristic velocity and l the characteristic length scale. The total iron injection rate within a given radius r can be written as $s(< r,t) = 4\pi \int_0^r \rho x^2 S dx \propto \left(\frac{L_B(<r)}{L_\odot^B} \right)$ ([1]). The diffusion coefficients listed in Table 1 are obtained by integrating equation 1 starting from zero abundance at all radii and by comparing the observed and predicted abundance profiles (e.g. see Fig.1, left).

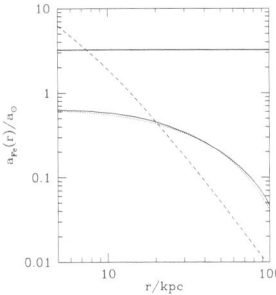
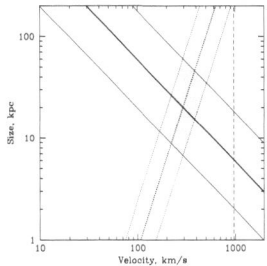

Fig. 1. Left: M87 - the solid line shows the abundance profile adopted, from which a constant value of $\sim 0.2\,a_{\odot}$ is subtracted. For comparison we show the expected iron abundance profile (long dashed line) due to the ejection of metals from the BCG, calculated assuming that the ejected metal distribution follows the optical light: it is much more peaked than the observed one. The horizontal line indicates the maximum value that the abundance can take, over which our approximation is not valid anymore. The short dashed line shows the profile derived with the additional effect of diffusion ($D \sim 8 \times 10^{28}\ cm^2/s$). Right: Perseus cluster - range of v and l which provide the required diffusion and dissipation rates. Along the thick solid line the diffusion coefficient is equal to $2 \times 10^{29}\ cm^2\ s^{-1}$. Along the thick dotted line Γ_{diss} is equal to the gas cooling rate. The thin solid and the dotted lines correspond to a variation of the coefficients by factors of 0.3 and 3. The vertical dashed line shows the sound velocity in the gas.

3 Cooling and Heating

Let us assume that the dissipation of turbulent motions is the dominant source of heat and that the heating rate Γ_{diss} is equal to the gas cooling rate Γ_{cool}. For a given v and l the diffusion coefficient and the dissipation rate can be written as $D \sim 0.11vl$ and $\Gamma_{diss} \sim 0.4\rho v^3/l$ [2]. Using the value of D, determined in section 1 and setting $\Gamma_{diss} \approx \Gamma_{cool}$ at $r_{cool}/2$ we can get constraints on v and l (e.g. see Fig.1, right). The intersection of the two bands gives the locus of the combinations of l and v such that on one hand D is approximately equal to the required value and on the other hand the dissipation rate is approximately equal to the cooling rate. In Table 1 the estimated values of D, v and l are listed. l are of the order of 10 kpc and v of the order of few hundred km/s: there is no obvious trend with the temperature of the source.

4 Linewidth profile

Turbulent gas motions in the ICM can significantly affect the X-ray line profiles. For lines of heavy ions the turbulent broadening dominates over the thermal broadening, hence iron ions are indicative tracers of turbulent

Table 1. Estimate of the turbulence parameters for each source in our sample: name (1), diffusion coefficient (2), length-scale (3) and velocity (4) of the gas motion.

Name	$D\ [cm^2/s]$	$l\ [kpc]$	$v\ [km/s]$
NGC 5044	9.0×10^{28}	11	245
NGC 1550	1.2×10^{29}	16	225
M87	8.0×10^{28}	10	260
AWM4	4.5×10^{29}	36	376
Centaurus	1.5×10^{28}	2	260
AWM7	6.0×10^{27}	1	226
A1795	1.0×10^{29}	12	300
Perseus	2.0×10^{29}	21	300

velocity fields in cluster cores. Figure 2 shows the expected width of the 6.7 kev iron line in the Perseus cluster for the isotropic velocity (of the order of few hundred km/s) required to offset the cooling with the dissipation of the gas motions at each radius (i) for constant spatial scale l ($l \sim 20$ kpc, solid line) and (ii) for the parametrisation (first proposed by Dennis & Chandran 2005) $l = \alpha r$ kpc ($\alpha = 0.3$, dotted line). XEUS and Constellation X missions

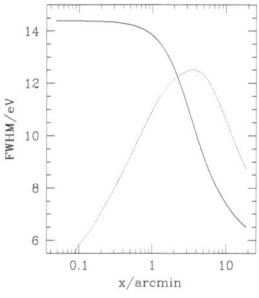

Fig. 2. Perseus cluster: linewidth as a function of the projected radius. See section 4 for an explanation.

are expected to attain an energy resolution of 1-2 eV for a spatial resolution of few arc-seconds, opening the possibility test the different hypothesis on the origin of core gas motion.

References

1. H.Böhringer, K. Matsushita, E. Churazov, A. Finoguenov, Y. Ikebe: A&A, **416**, 21 (2004)
2. T.Dennis, B. Chandran: ApJ, **622**, 205 (2005)
3. P. Rebusco, E. Churazov, H. Böhringer, W. Forman: MNRAS, **359**, 1041 (2005)
4. P. Rebusco, E. Churazov, H. Böhringer, W. Forman: MNRAS, **372**, 1840 (2006)

Carbon and Nitrogen in the X-ray Emitting Hot Gas of M 87

N. Werner[1], H. Böhringer[2], J.S. Kaastra[1], J. de Plaa[1], A. Simionescu[2] and J. Vink[3]

[1] SRON Netherlands Institute for Space Research, Utrecht, the Netherlands
[2] Max-Planck-Institut für extraterrestrische Physik, Garching, Germany
[3] Astronomical Institute, Utrecht University, Utrecht, the Netherlands

Summary. We compare the abundance ratios of carbon, nitrogen, oxygen, and iron in the inter-stellar medium (ISM) of M 87 with those in the stellar population of our Galaxy. The relative contribution of core-collapse supernovae to the enrichment of the ISM in M 87 is significantly lower than in the Milky Way. The abundance ratios indicate that the enrichment of the ISM by iron through Type Ia supernovae and by carbon and nitrogen is occurring in parallel. This suggests that the main sources of carbon and nitrogen in M 87 are the low- and intermediate-mass asymptotic giant branch stars.

1 Introduction

Although it is well-known that elements from oxygen up to the iron group are produced primarily in supernovae, the main sites of the carbon and nitrogen production are still debated. It is believed that both elements are produced by a wide range of sources. The main question is whether their production is dominated by the winds of short-lived massive stars or by the longer-lived progenitors of the asymptotic giant branch (AGB) stars.

We present a study of chemical abundances in the giant elliptical galaxy M 87 using high-resolution spectra obtained with the Reflection Grating Spectrometers (RGS) during two deep XMM-Newton observations. The high statistics allow us, for the first time, to determine relatively accurate abundance values for carbon and nitrogen, and compare them with those measured in the stellar population of our Galaxy.

2 The total spectrum

M 87 was observed with XMM-Newton on June 19 2000 for 60 ks and re-observed in January 2005 with an exposure time of 109 ks. The extraction regions for the RGS spectra are rotated with respect to each other by only 7.3°, which allows us to combine the spectra obtained during the two observations. The total useful exposure time of the combined dataset is 124 ks. We fit these spectra simultaneously. We model the plasma with two thermal

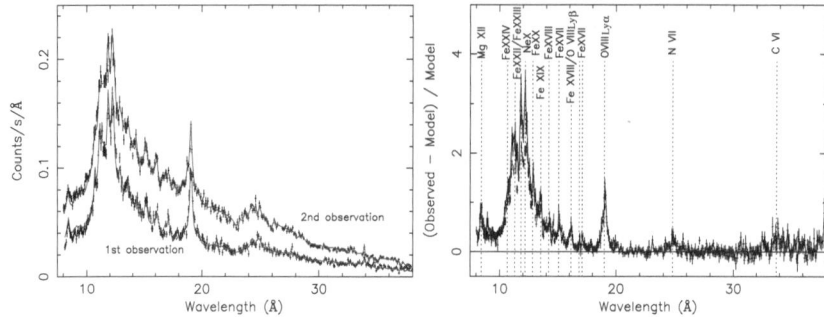

Fig. 1. *Left panel:* 1st order RGS spectra of M 87 obtained during two observations. The AGN in the core of M 87 was much brighter during the second observation. The continuous line shows the best fit model. *Right panel:* Residuals of the fit to the combined RGS spectrum with line emission set to zero in the model. We indicate the positions of all the detected spectral lines.

(MEKAL) components and the AGN emission with a separate power-law for both observations. The fit to the total spectrum is shown on the left panel of Fig. 1. All detected spectral lines are shown on the right panel of Fig. 1. For more details on the XMM-Newton RGS study of M 87 see [5].

3 Carbon and nitrogen abundances and enrichment by AGB stars

In Fig. 2 we compare the C/Fe and O/Fe enrichment in M 87 with the Galactic thin and thick disk stars of [1] and the N/O and N/Fe enrichment with the sample of Galactic stars of [2] and [4].

The decrease of the O/Fe ratio in the Galactic stars of about a factor of 5 over the chemical evolution history represented by [Fe/H] is in general interpreted by an early enrichment by oxygen from core-collapse supernovae (SN_{CC}) and a subsequent production of iron by type Ia supernovae (SN Ia), which go off on much longer time scales than the SN_{CC}. In contrast, the flat

Table 1. Fit results for the combined RGS spectra of M 87. The iron abundance is given with respect to proto-solar values of [3] and the other abundances with respect to iron. Emission measures ($Y = \int n_e n_H dV$) are given in 10^{64} cm^{-3}.

Y_1	Y_2	kT_1 (keV)	kT_2 (keV)	C/Fe
0.86 ± 0.03	9.97 ± 0.13	0.80 ± 0.01	1.67 ± 0.02	0.73 ± 0.11
N/Fe	O/Fe	Ne/Fe	Mg/Fe	Fe
1.62 ± 0.19	0.60 ± 0.03	1.40 ± 0.11	0.70 ± 0.05	0.98 ± 0.02

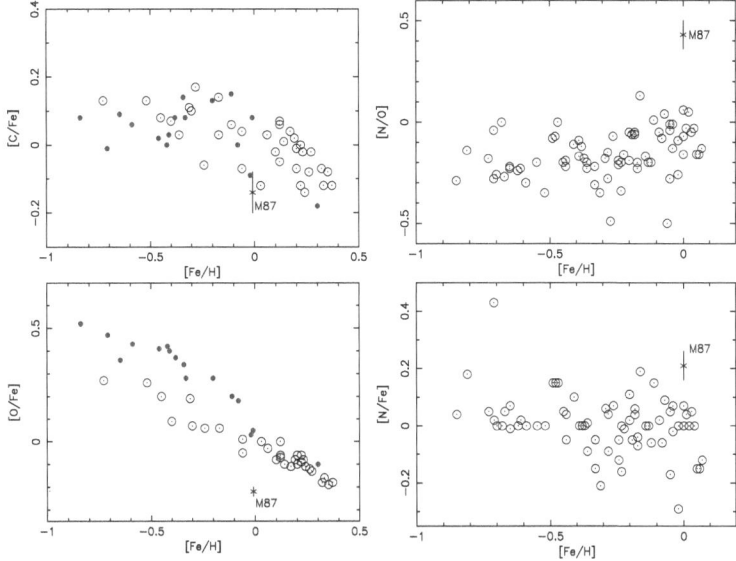

Fig. 2. Comparison of the [C/Fe] and [O/Fe] enrichment in M 87 with the enrichment of the Galactic thin and thick disk stars of [1]. Thin and thick disk stars are marked by open and filled circles, respectively. The [N/Fe] and [N/O] enrichment in M 87 is compared with the Galactic stellar population of [2] and [4]. The value for M 87 is indicated by the asterisk.

C/Fe ratio then implies that the enrichment by Fe through SN Ia and the C pollution has occurred very much in parallel [1].

In the ISM of M 87 the O/Fe ratio is lower and N/O ratio is higher than in the stellar population of our Galaxy, which shows that the relative contribution of SN_{CC} in the old stellar population of M 87 was significantly lower. This suggests that the star formation in M 87 has essentially stopped. The C/Fe and N/Fe ratios in the ISM of M 87 are consistent with those found in the stellar population of our Galaxy. This suggests, that while oxygen (and probably also some of the nitrogen) was supplied by core collapse supernovae early in the enrichment history, nitrogen and carbon are continuing to be supplied by intermediate- and low-mass AGB stars on a time scale similar to the enrichment by iron.

References

1. Bensby, T. & Feltzing, S. 2006, MNRAS, 367, 1181
2. Chen, Y. Q., Nissen, P. E., Zhao, G., Zhang, H. W., & Benoni, T. 2000, A&AS, 141, 491
3. Lodders, K. 2003, ApJ, 591, 1220
4. Shi, J. R., Zhao, G., & Chen, Y. Q. 2002, A&A, 381, 982
5. Werner, N., Boehringer, H., Kaastra, J. S., et al. 2006, A&A, 459

Part VII

Optical and Sub-mm Observations of Cool Cores

IFU Observations of Hα in Brightest Central Galaxies of Cooling Flow Clusters

L.O.V. Edwards and C. Robert

Département de Physique, Génie Physique, et Optique, and Observatoire du mont Mégantic, Université Laval, Québec, QC G1K 7P4, Canada
louise.edwards.1@ulaval.ca

Summary. We are conducting a mini-survey of ten brightest cluster galaxies (BCGs) using Gemini Integral Field Unit observations in the optical around Hα at 656.3 nm. Most of our targets are in cooling flow (CF) clusters, however, we include some control BCGs which are in clusters without CFs. In the Hα emitting galaxies, we find that there is much diversity in terms of the structure of the Hα emission and other line ratios. Here we present the spectra and Hα images for Abell 1060 and Abell 1204.

1 Introduction

Many cooling flow (CF) clusters show Hα emission in their brightest central galaxy (BCG). The morphology is diverse: tails are seen [6], as are cases where the emission is highly concentrated [4], and others where it is more filamentary [2]. Very often these BCGs harbour powerful AGN, whose jets are thought to play an important role in heating the cluster and ionizing gas at the center of the BCG. Diffuse emission is a characteristic sign of a young stellar population, which in turn may be related to the cluster CF. In order to decode the relative importance of the many concurrent processes, we require detailed observations.

Here we describe the Hα observations of two CF cluster BCGs with the most morphologically interesting emission: Abell 1060 and Abell 1204. We are publishing a full analysis of all ten clusters in the near future.

2 Gemini IFU Observations and Data Reduction

The GMOS integral field spectrographs (IFUs) on the Gemini Telescopes harbor a lenslet array with 1000 elements. These spectra reveal a detailed picture of the star formation history and allow for a clear spatial distinction between the different stellar populations within the 5" × 7" field of view.

Table 1 describes our target clusters. This table states whether or not the cluster is a CF [8, 9], if Hα emission has been detected previously [3], if there is molecular gas [5], or dust [7], and whether there are detections of radio emission [1]. We count only clusters with mass deposition rates (from X-ray measurements) higher than 10 solar masses per year as CF clusters.

Table 1. Clusters Observed with GMOS IFU

Cluster	Type	Emission	Molecular Gas	Dust	Radio
A1060	CF	yes		yes	no
A754	nCF	no			yes
A1651	CF	no			
Hydra-A	CF	yes	yes		yes
Ophiucus	CF	yes			yes
A1795	CF	yes	yes		yes
A1204	CF	yes			
MKW3S	CF	yes			yes
A795	nCF	yes			
A1668	nCF	yes			

Each cluster was observed for 60 minutes. We use the R400 grating in combination with the r filter for Abell 1060, and with the i filter Abell 1204. The Gemini IRAF package *gemtools* was used to do the bias subtraction, cosmic ray rejection, and flat fielding.

3 Results and Discussion

We present the spectra and Hα images of two of our CF cluster BCGs in Figures 1 - 4. At the distance of the clusters, 1" corresponds to 0.25 kpc and 2.80 kpc for Abell 1060 and Abell 1204, respectively (for h_o=0.7).

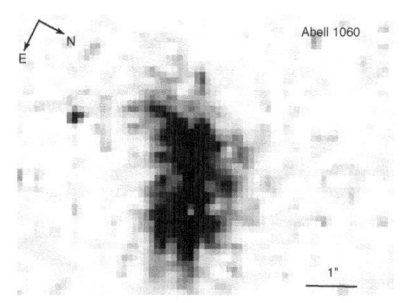

Fig. 1. Hα image of the BCG of Abell 1060

Fig. 2. Hα image of the BCG of Abell 1204

Abell 1060 The BCG shows Hα emission with nebular line ratios characteristic of star formation. The strongest Hα emission is spatially correlated with a large dust lane seen we see in our r-band image.

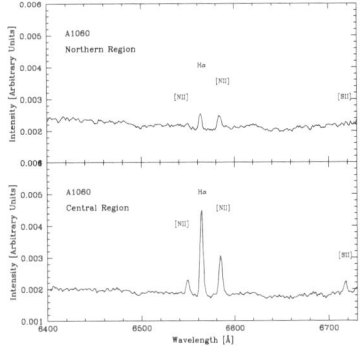

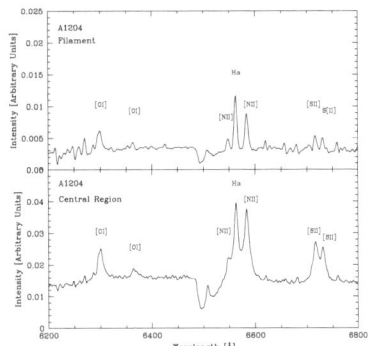

Fig. 3. Integrated spectra of the BCG of Abell 1060. Twenty spectra centered on the Hα emission of Figure 1 are combined to give the lower spectrum. Twenty spectra away from the Hα emission combine to give the top spectrum.

Fig. 4. Integrated spectra of the BCG of Abell 1204. Seventy spectra from the central emitting regions of Figure 2 have been added to give the lower spectrum, and seventy spectra from the North-Eastern wisp give the top spectrum.

Abell 1204 A wisp of Hα emission is seen to the NE of Figure 2. The BCG is near to several smaller galaxies in this cluster. Wilman et al [10] find that the systems with the most disturbed Hα emission morphology harbor a secondary galaxy close to the BCG. A1204 may be one such system. The emission lines in the central regions of the BCG are broad, and our ongoing kinematic study will help to separate multiple velocity components.

References

1. Bîrzan, L., Rafferty, D. A.,McNamara, B. R. et al: ApJ, **607**, 800 (2004)
2. Conselice, C. J., Gallagher, III, J. S., & Wyse, R. F. G.:AJ, **123**, 2246 (2002)
3. Crawford, C. S., Allen, S. W., Ebeling, H., et al: MNRAS **306**, 857 (1999)
4. Donahue, M., Mack, J., Voit, G. M. et al: ApJ, **545**, 670 (2000)
5. Edge, A. C.,Wilman, R. J., Johnstone, R. M., et al: MNRAS, **337**, 49 (2002)
6. Fabian, A.C., Sanders, J.S., Ettori, S., et al: MNRAS, **321**, 33 (2001)
7. Laine, S., van der Marel, R. P., Lauer, T. R. et al: AJ, **125**, 478 (2003)
8. Peres, C. B., Fabian, A. C., Edge, A. C., et al: MNRAS, **298**, 416 (1998)
9. White, D.A.: MNRAS, **312**, 663 (2000)
10. Wilman, R.J., Edge, A.C., & Swinbank, A.M.:MNRAS, **371**, 93 (2006)

Ionized Gas in Cluster Cores

N. A. Hatch, C. S. Crawford and A. C. Fabian

Institute of Astronomy, University of Cambridge, Madingley Road, Cambridge, CB3 0HA, hatch@strw.leidenuniv.nl

1 Introduction

Luminous optical and UV line-emitting nebulae are commonly found surrounding brightest cluster galaxies (BCGs) that reside at the centre of cooling flow clusters. These massive galaxies act as laboratories where one can study gas cooling from the intracluster medium (ICM) and accreting onto the galaxy, a situation which is assumed to be much more common in the high redshift universe. They can be used to test models of galaxy formation and feedback. Many ionized nebulae are filamentary and extend over tens of kpc from the host galaxy. Some filaments are co-spatial with soft X-ray features implying that the nebular gas interacts with the surrounding ICM [1, 2]. The BCGs which are surrounded by a line-emitting nebula also contain large reservoirs of cold molecular hydrogen [3]. The origin and source of ionization of the nebular gas is unclear, but are thought to be linked to the cluster environment as there is a strong correspondence between the presence of a nebula and the brevity of the radiative cooling time of the cluster core [4].

These nebulae can be used as tools to investigate the complex relationship between the BCG and the cluster. Since the filling fraction of the nebular gas is low, the filaments are coupled to the large-scale motion of the ICM. Therefore the filaments can act as streamlines, tracing the flow pattern in the cluster core and the interaction between the buoyantly rising radio bubbles and the ICM [5]. Additionally, if star formation in the BCG ionizes the nebula, the emission-line intensity can measure the amount of gas that is able to cool and form stars [6]. Thus establishing the excitation and origin of these nebulae will help us understand heating and cooling in cluster cores, and may unveil details of the feedback process that regulates black hole and massive galaxy growth.

In these proceedings we introduce an integral field spectroscopic study of the emission-line nebulae that surround BCGs in cooling flow clusters. We present maps of the morphology, kinematics and ionization state of three nebulae to investigate their formation and source of heating.

2 Sample properties and data reduction

The target BCGs were selected from the *ROSAT* Brightest Cluster Sample optical follow-up of the brightest cluster galaxies [4]. All the target galaxies lie within the centre of clusters which exhibit bright centrally peaked X-ray emission and have cool cores. We refer to the central galaxy by the name of the cluster in which it resides.

The data were obtained with the Oasis integral field unit (IFU) instrument on the William Herschel Telescope on the nights of 2005 Sept 22, 27 and 2005 Dec 6. The seeing conditions were 0.8-1.5 arcsec but significant improvement was achieved using the adaptive optics system NAOMI. The IFU field-of-view is 10.3×7.4 arcsec, which is divided into ~1100 lenslets. Table 1 lists the observations.

Cluster name	RA (J2000)	Dec (J2000)	Redshift	Exposure time (sec)	Grating	Resolution
A262	01 52 46.5	36 09 08	0.0163	4×900	MR_661	1650
A496	04 33 37.7	-13 15 39	0.0329	1000+900	MR_661	1650
2A0335-096	03 38 40.5	09 58 12	0.0349	4×900	MR_661	1650
A1068	10 40 44.4	39 57 12	0.1375	6×900	MR_735	1840
A2390	21 53 36.7	17 41 45	0.228	4×900	MR_807	2020

Table 1. Summary of the IFU observations.

The data were processed using the OASIS dedicated reduction package XOASIS (version 6.2). After preprocessing, the spectra of each frame were extracted to form an IFU datacube. Each datacube underwent data-reduction including flat-fielding, cosmic ray removal, sky-subtraction and was then re-sampled to a spatial scale of 0.2 arcsec per lenslet. The spectral data from each lenslet were fitted with a sextuplet Gaussian model for the lines of Hα, [N$_{II}$]$\lambda\lambda$6548, 6584, [O$_I$]$\lambda\lambda$6300, 6363, and [S$_{II}$]$\lambda\lambda$6717, 6731, in which all the emission-lines were forced to have the same redshift and line-width, and the relative normalisation of [N$_{II}$]λ6548 was fixed at a third of the [N$_{II}$]λ6584 emission-line. The line-of-sight velocities of the emission-line nebulae were obtained from the Doppler shifts of the strong Hα and [N$_{II}$] emission-lines. The zero-point of the line-of-sight velocity is defined as the redshift of the central IFU lenslet. All line-widths discussed are FWHM (full width at half maximum) and have been corrected for instrumental broadening ($\sim$215-260km s^{-1}).

3 Results

We present results from three of the five observed nebulae: A262, 2A 0335+096 and A2390.

3.1 A262

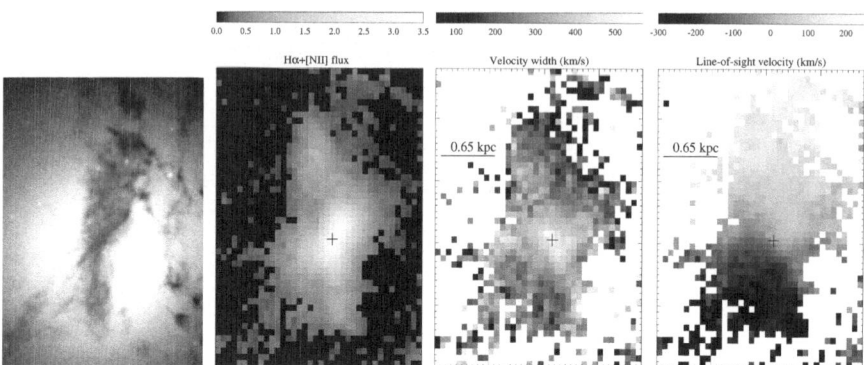

Fig. 1. 7.4×10.3 arcsec images of the BCG of A262. From left to right: HST ASC image using filter F435w, Hα +[Nɪɪ] flux (with colour bar scaling in units of $10^{-16}\,\mathrm{erg\,cm^{-2}s^{-1}arcsec^{-2}}$, line-width [km s^{-1}] and line-of-sight velocity [km s^{-1}]. The cross marks the emission-line flux peak which indicates the galaxy centre. North is up, East is left.

The Hα+[Nɪɪ] flux, line-width and line-of-sight velocity structure of the nebula surrounding the BCG of A262 are presented in Fig. 1, together with a Hubble Space Telescope (HST) snapshot of the galaxy. The line-emission traces the dust lanes that bisect the galaxy. At the centre of the nebula there is a bright Hα bar, approximately 0.6 kpc long, at a position angle of −58 degrees. The line-width peak is co-spatial with the Hα peak, but extends orthogonally to the Hα bar, in the same direction as the radio source axis, which has a position angle of ∼116 degrees. An interaction between the radio source and the ionized gas is likely to have caused the increase in line-width in this region. The ionized nebula is rotating with a peak-to-peak velocity of 550 km s^{-1}, and there is some gradual twisting in the kinematics at the core. This BCG contains an underlying reservoir of $4 \times 10^8\,\mathrm{M_\odot}$ of cold molecular hydrogen [7] which was detected through observations of CO(1-0) and CO(2-1) emission-lines that have 'double-horn' profiles with a FWHM of 550km s^{-1}. This indicates that the molecular gas also lies in a rotating disk. As the amplitude of the rotation velocity of the molecular gas is similar to that of the ionized nebula, it is possible the molecular and ionized gas are different temperature components of the same gas reservoir. Although the ionized gas is more readily observed, the low Hα luminosity means that it comprises only a tiny fraction (0.03%) of the visible gas mass.

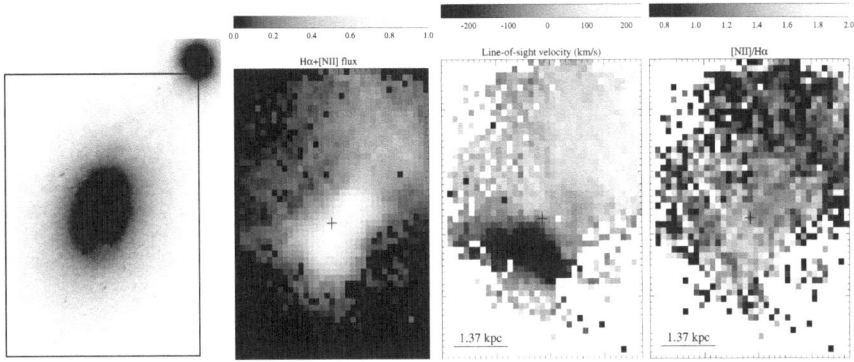

Fig. 2. From left to right: HST F606w image of the core of the cluster 2A 0335+096. The black box is centred on the brightest cluster galaxy and is 7.4×10.3 arcsec² matching the size of the IFU field-of-view. A secondary galaxy is visible to the Northwest just beyond the IFU field-of-view. Hα+[NII] flux (with colour bar scaling in units of 10^{-16} erg cm^{-2}s^{-1}arcsec^{-2}) of the nebula surrounding the central galaxy. Line-of-sight velocity and map of the [NII]λ6584/Hα ratio. The cross marks the galaxy centre defined by the continuum peak. North is up, East is left.

3.2 2A 0335+096

Fig. 2 displays a HST snapshot, the emission-line map, line-of-sight velocity and [NII]/Hα map of 2A 0335+096. Two separate Hα+[NII] central knots are visible forming a bar-like morphology that was first noted by [8]. Bright diffuse line-emission extends Northwest toward the secondary galaxy seen just beyond the IFU field-of-view in the HST image. The nebular gas that extends Northwest towards the secondary galaxy has a bulk velocity that is radially increasing in redshift up to 220 km s^{-1}. The secondary galaxy is redshifted by approximately 212 km s^{-1} relative to the central galaxy nucleus [9], so the Northwest emission not only extends towards but also matches the velocity of the secondary galaxy. Gelderman et al. [9] also note abrupt changes in the line ratios and line-widths at the position of the secondary galaxy, providing further evidence that the secondary galaxy (which lies 4.5 kpc away in projection from the BCG nucleus) has previously disturbed the molecular gas reservoir of the BCG, and formed the Northwest extension of the nebula.

The two Hα nuclei have different line-of-sight velocities: the Southeastern nucleus is blueshifted by -250 km s^{-1} compared to the Northwestern nucleus. They also have differing line ratios, the Southeast nucleus has larger forbidden-to-Hα line ratios than the Northwest nucleus. Neither of the Hα nuclei are co-spatial with the continuum and line-width peak, which is marked in Fig. 2 by a cross. The ionization state of the gas is not uniform, generally the ratio of [NII]/Hα is largest in the BCG centre and decreases in the extended regions.

3.3 A2390

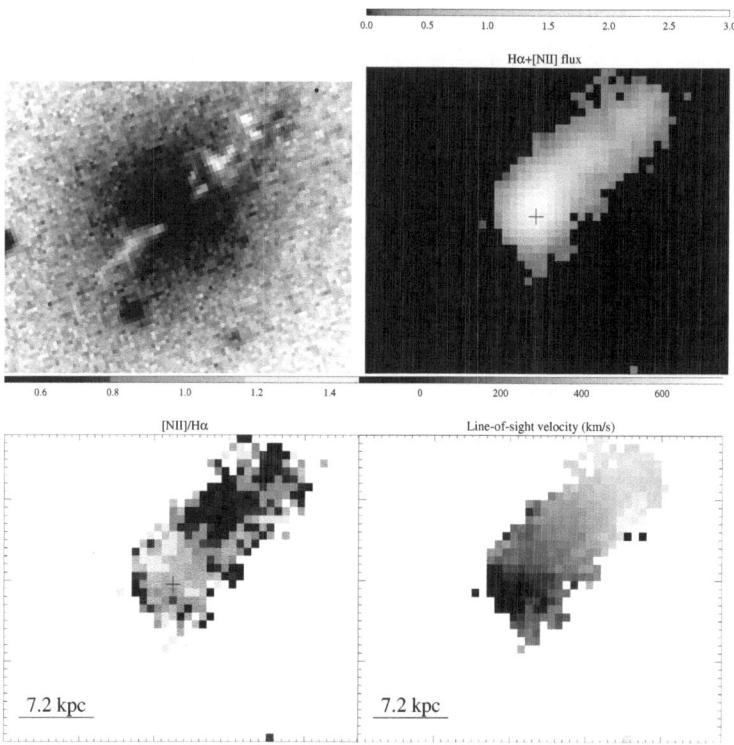

Fig. 3. Images of A2390, from top left clockwise: HST F555w/F814w image centred on the brightest cluster galaxy and matching the size of the IFU field-of-view, Hα+[NII] flux (with colour bar scaling in units of $10^{-16}\,\mathrm{erg\,cm^{-2}s^{-1}arcsec^{-2}}$), line-of-sight velocity [$\mathrm{km\,s^{-1}}$] and [NII]λ6584/Hα line ratio. The cross marks the galaxy centre defined by the continuum peak. North is up, East is left.

Fig. 3 presents a F555w/F814w HST snapshot of A2390, together with a emission-line map, the line-of-sight kinematics and a [NII]/Hα line ratio map. Blue knots and dust features extend from the nucleus at a position angle of approximately −45 degrees, both Northwest and Southeast, whilst the emission-line nebula extends only to the Northwest. The kinematics of A2390 are ordered: the line-of-sight velocity gradually increases from the galaxy centre towards the Northwest to velocities of $700\,\mathrm{km\,s^{-1}}$. Both AGN and starbursts are energetically plausible methods to cause such an outflow, as this galaxy has a large star formation rate [10], and is host to a relatively strong radio source. The range of line ratios in A2390 is narrow, similar to high luminosity BCGs studied by [6], but the ionization state is not uniform. The dusty bar that bisects the galaxy at a position angle of approximately

-53 degrees and extends to the Northwest has the largest value of [NII]/Hα and also the largest emission-line flux, whilst the brightest blue knot approximately 2 arcsec Northwest from the galaxy centre has the lowest [NII]/Hα ratio. The blue knots are regions of star formation and stellar UV may ionize the nearby nebular gas. This produces low forbidden-to-Hα line ratios, but it does not greatly increase the Hα+[NII] flux. Therefore stellar UV from hot young stars is not the only source of ionization in this galaxy, but a contributing factor that changes the line ratios.

4 Discussion

We have presented IFU observations that map the morphology, kinematics and ionization state of three nebulae that surround BCGs at the centre of cooling flow clusters. Smooth velocity shears and bulk flows of a few hundred km s^{-1} occur in all nebulae, but there is no general trend in the kinematics. Instead we find that multiple phenomena can influence the nebulae, including interactions with nearby galaxies, as seen in 2A 0335+096, and AGN or starburst driven outflows as postulated for A2390. The nebula surrounding A262 shows no signs of disturbance as it has the same rotation velocity as the underlying molecular reservoir. Therefore the formation of a nebula may not rely on any disturbance of the molecular reservoir, but merely the excitation of the gas. All the nebulae have low-ionization spectra, with large forbidden-to-Hα line ratios, but the nebular ionization state is not uniform: regions with excess blue or UV light have lower forbidden-to-Hα line ratios. The presence of stellar UV photons influences the observed line ratios, but the lack of any correlation between the line ratios and the emission-line flux implies it cannot be the dominant ionization source; there must exist another source of heating that produces the low-ionization spectral features.

References

1. A.C. Fabian, J.S. Sanders, C.S. Crawford, et al.: MNRAS **344**, L48 (2003)
2. C.S. Crawford, N.A. Hatch, et al.: MNRAS **363**, 216 (2005)
3. A.C. Edge: MNRAS **328**, 782 (2001)
4. C.S. Crawford, S.W. Allen, H. Ebeling, et al.: MNRAS **306**, 857 (1999)
5. N.A. Hatch, C.S. Crawford, R.M. Johnstone, et al.: MNRAS **367**, 433 (2006)
6. R.J. Wilman, A.C. Edge, A.M. Swinbank: MNRAS **371**, 93 (2006)
7. I. Prandoni, R.A. Laing, et al.: *ArXiv Astro-ph eprints* 0608577 (2006)
8. W. Romanishin, P. Hintzen: ApJ **324**, L17 (1988)
9. R. Gelderman: Tracking the Mass in Cluster Cooling Flows: Kinematics and Ionization in the cD cluster 2A 0335+096. In *ASP Conf. Ser. 88: Clusters, Lensing, and the Future of the Universe*, ed by V. Trimble and A. Reisenegger, pp168
10. E. Egami et al.: ApJ **647**, 922 (2006)

Observing the Cold Molecular Gas in Cooling Flow Clusters

P. Salomé[1], F. Combes[2] and Y. Revaz[2]

[1] IRAM, 300 rue de la piscine F-38400 St Martin d'Hères FRANCE
salome@iram.fr
[2] LERMA, Observatoire de Paris F-75014 Paris FRANCE

1 Single dish detections of very cold gas tracers around cD galaxies

1.1 Long search for a cold residual in cooling flows

During the last decades, many attempts have been carried out to detect a cold gas reservoir in galaxy cluster centers [9, 12, 13, 2]. But the cold residual of the hot cooling Intra-cluster medium, expected at larger wavelength, was not found apart from the single case of the Perseus cluster core [11].

More recently, thanks to the improved performance of heterodyne receivers, this very cold gas has been detected through CO(1-0), CO(2-1) and CO(3-2) emission lines, in a total of 23 cluster cores [8, 15] with the IRAM 30m telescope, the JCMT and the CSO.

These results have shown the existence of large amount of cold gas inside a $\sim$50 kpc region around several central cluster cD galaxies. The molecular gas mass deduced with standard conversion factor is $M_{gas} = 10^{8-11.5}$ M$\odot$. The CO line ratios are typical of an optically thick cold and dense molecular gas.

1.2 Limits of single dish observations

Single dish observations cannot solve the issue of the origin of this very cold gas, which may be coming from tidal stripping of a companion galaxy passing by the central cDs, or could be the cold residual gas expected in a cooling flow scenario. Most of the CO line detections were achieved at the limit of the actual performance of the instruments (3-5 σ detections and upper limits). It is now important to increase the catalogue of sources observed in the millimeter. To do so, we recall that there is a strong correlation of CO with the Hα luminosity. Optical filaments clearly appear as the best tracers of cold gas in cooling flow cluster cores. The other limitation of single dish observations is the limited spatial resolution (22$''$ and 11$''$ at 3mm and 1mm respectively) which gives only a small field of view for the nearest objects.

We have tried to constrain the origin of the cold gas detected by computing a gas to dust ratio in order to look for a dust depletion expected if the gas is cooling from the ICM, but continuum observations performed at 1.2mm observations (IRAM bolometer MAMBO) were not conclusive because of the very good weather required. However, from preliminary results it is clear that it is difficult to disentangle the dust emission from the AGN synchrotron continuum and a deeper study should be carried out (on a reduced sample of sources) in the submillimeter window. We also observed some other molecule tracers and some higher CO transition lines (Edge et al., in prep) reaching very quickly the limit in sensitivity of actual receivers.

So the best way to study cold molecular gas in cooling flow was to focus on the mapping of some of the detected clusters, in order to better understand the morphology and dynamics of the cold gas (see also [7]).

2 Mapping cooling flow cluster cores

2.1 Abell 1795

Abell 1795 is the first cooling flow cluster of galaxies that we have imaged through CO(1-0) and CO(2-1) emission line with the Plateau de Bure interferometer, giving a first idea of the peculiar morphology and dynamics of the very cold gas around a cD galaxy. The gas is found to be located in 2 regions : one along the northern radio lobe (emitted by the central radio source) and a second at the position of the central galaxy [16]. We have shown an indication of a velocity gradient from the northern position (at the velocity of the cluster), towards the position and velocity of the central object.

The PdB map of Abell 1795 shows the morphological similarities of the CO(2-1) and the optical emission. These results suggest that the hot intergalactic gas could have been pushed and compressed by the radio lobes in expansion. Along the lobes, where the gas is denser, the cooling is more efficient and the gas can cool down very quickly to very low temperatures. This compression of the hot ICM may also be due to the oscillation of the central galaxy in the cluster potential well. At this stage, no clear conclusion can be derived from these observations alone, and modelling the dynamics of the cooling gas is of current interest.

The same association of gas at different wavelengths, avoiding radio lobe cavities, also exist in NGC 1275. More clusters have to be mapped in order to better understand the origin and the fate of the molecular gas observed, as it is a clue to constrain the intermittent cooling flow scenarios.

New observations of Abell 1795 have been performed with the PdB interferometer (Salome et al, 2007, in prep.). We have observed the long North-South filament (about 40 kpc long). We have found signatures of molecular emission in the brightest part of that filament, where the H_α and the X-ray emission is the strongest, and where star formation activity has been found. This supports the cooling wake scenario where, as suggested by [6], the hot ICM is cooling down along the cD trajectory in the cluster center.

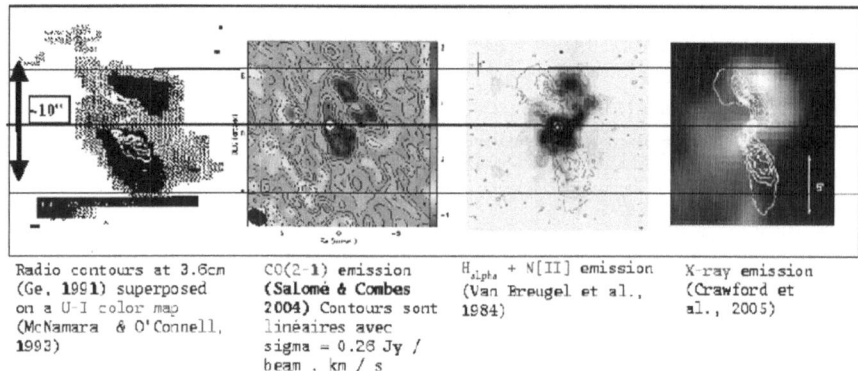

Radio contours at 3.6cm (Ge, 1991) superposed on a U-I color map (McNamara & O'Connell, 1993)

CO(2-1) emission (Salomé & Combes 2004) Contours sont linéaires avec sigma = 0.26 Jy / beam . km / s

H$_{alpha}$ + N[II] emission (Van Breugel et al., 1984)

X-ray emission (Crawford et al., 2005)

Fig. 1. Abell 1795 cluster center. Comparison of CO(2-1) integrated emission map with other wavelength images. A very similar two component structure is seen in X-ray, H$_\alpha$, CO and in the optical.

2.2 RXJ 0821+07

The second cD galaxy we observed is at the center of the RXJ0821+07 cluster, which was discovered in the extended Brightest Cluster Sample [5]. The galaxy redshift is 0.11 [4]. The cD harbors a very weak radio source (2.26 mJy at 1.4 Ghz). Chandra images showed that the gas is cooling within a radius of 20 kpc at the rate of a few 10 M$_\odot$/yr [1]. The galaxy is embedded in an extended H$_\alpha$ nebula which is spatially coincident with the X-ray emitting region and also with clumps of blue continuum away from the cD galaxy towards the NW. An elliptical galaxy is also observed, in the optical, at 4" SE with a redshift of +77km/s in the cD rest frame.

The CO emitting region is not coincident with the cD galaxy. The peak in CO is   2" NW away from the cD center (6 kpc), in the same direction as the Halpha and the X-ray asymmetries. The higher spatial resolution at 1.3mm shows there are two resolved clumps of molecular gas, one of them coinciding with an optical blob. The molecular gas emission is detected around +270 km/s by comparison to the cD galaxy. This is consistent with the single dish spectra obtained at the IRAM 30m telescope.

So, the cold gas is not associated with the central galaxy. The mass of cold gas derived from the single dish observations is M$_{gas}$ = 1.3 × 10^{10} M$_\odot$. The flux retrieved wth the interferometer is 7.4 Jy.km/s in CO(1-0) and 10.1 Jy.km/s in CO(2-1), which is very close to the values found with the 30m telescope. The molecular gas component is not extended, contrary to that of Abell 1795. The possibility of an interaction with the second galaxy is being studied with dynamical models of tidal stripping processes. The observed filament may also be a cooling wake, as suggested for Abell 1795. The modelling in progress will enable comparison of these two scenarios.

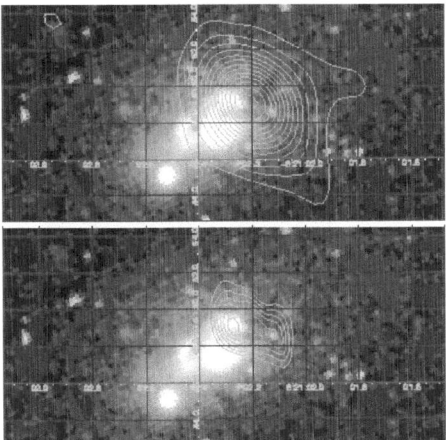

Fig. 2. RXJ 0821+07 : 20.5 × 16 arcsec HST image of the cluster core, the galaxy is at the center of the field [1]. Contours are the molecular gas integrated emissions. On the top : CO(1-0), rms = 0.28 Jy/beam.km/s. On the bottom, CO(2-1), rms = 0.56 Jy/beam.km/s. Levels are by 1σ.

3 Filaments and cavities in the cold gas emission

3.1 Deeper study of Perseus filaments

The giant galaxy NGC 1275 lies in the center of the X-ray brightest cluster of galaxies, at a redshift of 0.0183. This cluster harbors a strong cooling flow and has remained the only cooling flow cluster core revealed in CO for about 10 yrs.

We have observed the Perseus cluster center with an heterodyne camera array (HERA) of 3 x 3 receivers in two polarisations (18 receivers in total) installed on the IRAM 30m telescope. This array works at 1.3mm, giving access to the CO(2-1) emission line at the cluster redshift. A large map was built, covering the entire H_α filamentary structure pointed out by [3]. For the first time, clear evidence of molecular filaments far away from the galaxy center is shown.

The CO(2-1) emission avoids radio lobes and X-ray cavities, and coincides with the large H_α filaments around NGC 1275 towards the East and West of the galaxy. The molecular gas is not in rotation in the galaxy potential well. Velocities agree with H_α and excited H2 velocities computed by [10].

Region	Off1	Off2	Pos2	Pos11	East	Off3
Mgas $(10^8 M_\odot)$	1.1	1.1	0.7	0.9	8.3	4.9

Table 1. Molecular gas masses in the filaments, using a standard Ico/N(H2) conversion ratio. The total mass found with HERA is $4. \times 10^{10} M_\odot$.

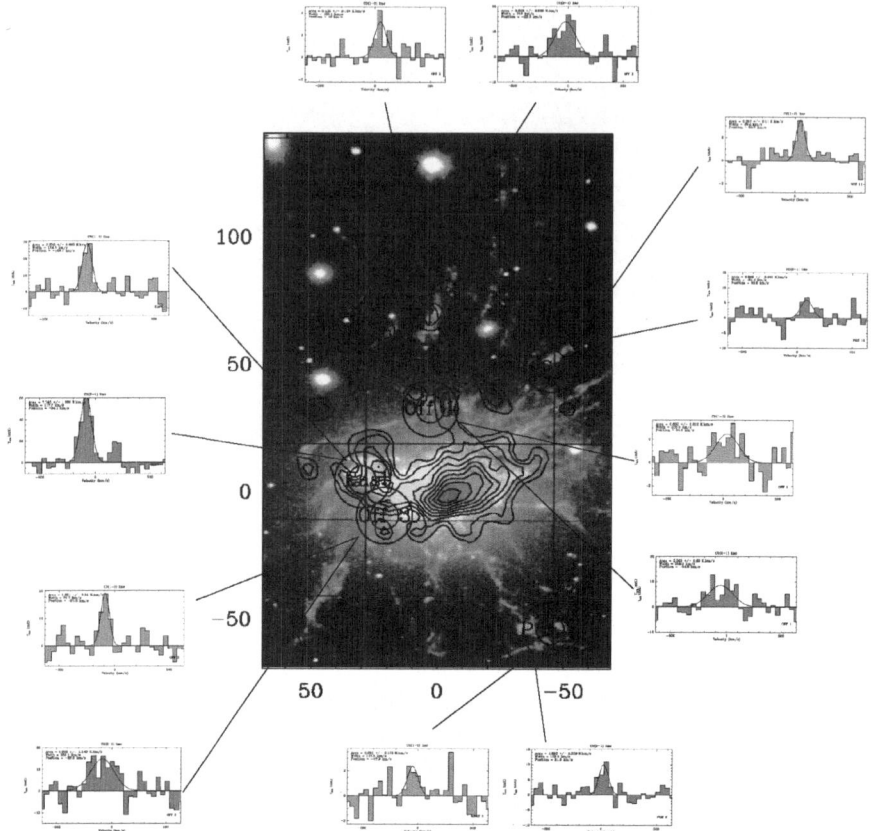

Fig. 3. H$_\alpha$ filamentary structure observed around NGC 1275 by [3]. Overlaid in contours is the CO(2-1) emission line map obtained by [17] with HERA (HEterodyne Receiver Array installed on the IRAM 30m telescope. We have re-observed selected regions (identified on the image) through CO(1-0) and CO(2-1) emission lines with standard receivers on the 30m telescope. The spectra obtained for those regions are shown here, with the results of a Gaussian fit at each frequency.

More observations were dedicated last summer to confirm the presence of molecular gas in the filaments. CO(1-0) and CO(2-1) emission lines have been detected (Salome et al 2007, in prep.) around the central galaxy, in six different regions, as shown in Fig. 3. Different wavelength studies suggest that we see cooling filaments around the central cluster galaxy.

Are these filaments due to uplifted cold gas, dragged out of the center by the radio lobes expansion in the intra-cluster medium ? Are these filaments due to cooling ICM gas flowing on the central cD ? There is no clear answer now and they are probably due to a mixture of these two phenomena. Detailed modelling of filament formation is in progress now [14], showing that the

AGN, invoked to re-heat the ICM, might also cause the formation of cooling filaments, eventually fueling the central galaxy.

4 Conclusions and perspectives

The higher spatial resolution of the New generation Plateau de Bure interferometer, and the increased sensitivity of the new receivers will enable deeper mapping of X-ray identified cooling cores before ALMA first science.

ALMA will be the ideal instrument to statistically study the cooling flow problem through the millimeter window. It will be possible to map simultaneously several molecular clouds for each cluster, and to study the physics of molecular clouds in such complex environments. The large number of clusters visible will permit us to probe the evolutionary processes of cold gas accretion on galaxy (central cD, elliptical galaxies) and help to better constrain the galaxy formation models with AGN feedback heating.

References

1. C.M. Bayer-Kim, C.S. Crawford, S.W. Allen, A.C. Edge, A.C. Fabian : MNRAS **337**, 938 (2002)
2. J. Braine, C. Dupraz : A&A **283**, 407 (1994)
3. C. Conselice, J. Gallagher, III. Wyse, R. Rosemary : AJ **122**, 2281(2001)
4. C. Crawford et al. : MNRAS **274**, 75 (1995)
5. H. Ebeling : MNRAS **318**, 333 (2000)
6. A.C.Fabian, J.S. Sanders, S. Ettori et al. : MNRAS, **321**, L33 (2001)
7. A.C. Edge, D.T Frayer : ApJL **594**, L13 (2003)
8. A.C. Edge : MNRAS **328**, 662 (2001)
9. D.A. Grabelsky, M.P. Ulmer : ApJ **355**, 401 (1990)
10. N.A Hatch, C.S. Crawford, R.M. Johnstone, A.C. Fabian : MNRAS **367**, 433 (2006)
11. B. Lazareff, A. Castets, D.W. Kim., M. Jura : ApJL **336**, L13 (1989)
12. B.R. McNamara, W. Jaffe : A&A **281**, 673 (1994)
13. C.P. O'Dea, S.A. Baum, P.R. Maloney, L.J. Tacconi, W.B. Sparks : ApJ **422**, 467 (1994)
14. Y. Revaz, et al. these proceedings
15. P. Salome, F. Combes : A&A **412**, 657 (2003)
16. P. Salome, F. Combes : A&A **415L**, 1 (2004)
17. P. Salome, F. Combes, A. Edge et al. : A&A **454**, 437 (2006)

Formation of Cold Molecular Filaments in Cooling Flow Flusters

Y. Revaz[1], F. Combes[1] and P. Salomé[2]

[1] LERMA, Observatoire de Paris, 61 av. de l'Observatoire, 75014 Paris, France
[2] IRAM, 300 rue de la piscine, 38400 St-Martin d'Hères, France

1 Introduction

New CO observations at the center of cooling flow in Perseus cluster [4] show a clear correlation of the molecular gas with the previously detected H-α filaments [1]. In this contribution, we present high resolution multi-phase simulations of the Perseus Cluster, taking into account the AGN feedback in form of hot buoyant bubbles. These simulations show that a significant amount of gas can cool far from the center. The AGN feedback provides some heating, but also triggers compression of the hot gas, favouring cooling (positive feedback), even at high radius ($R > 30\,\mathrm{kpc}$). The cooled gas flows into the cluster core forming the observed filaments.

2 Initial Conditions and gas physics

Our cluster model is designed to fit the Perseus X-ray data [5]. The total mass distribution profile follows a pseudo-isothermal sphere with a central electronic density of 5×10^{-2} and a core radius of 40 kpc. The mass distribution is truncated at 2 Mpc and the total mass is $5.5 \times 10^{14}\,\mathrm{M_\odot}$. The gas corresponding to 15% of the total mass has an initial temperature of $2.8 \times 10^7\,\mathrm{K}$.

The bulk of the intra-cluster gas is at high temperature ($10^7\,\mathrm{K}$), and is well modeled by an ideal gas with adiabatic index of 5/3. This hot phase has been computed using an SPH technique. For lower temperatures, SPH is not suited to render the clumpiness of the gas. Below $10^4\,\mathrm{K}$, the gas is treated as semi-collisional, using a sticky particle technique [2]. Star formation may occur in this cold phase, following a Schmidt law. The cooling of the hot gas is computed using the standard cooling function [8], assuming an abundance of one third of solar.

Following [6], the AGN feedback is modeled by injecting energy in hot intra-cluster gas bubbles which are then driven buoyantly at higher radius. Bubbles are defined by their position, diameter, temperature, over-pressure and angle. They are generated in symmetric pairs, on average each 200 Myr.

The simulations have been run using a modified version of the Gadget2 Tree-SPH code (Springel, 2005), including cooling, star formation and multiphase SPH/Sticky gas. In each simulations the gas is represented by 4,194,304 SPH particles. The dark matter is modeled by a fixed outer potential.

3 Results

Fig. 1. shows the evolution of an isolated bubble. The bubble has an initial temperature of 10^8 K. It reaches 30 kpc and takes the characteristic mushroom shape as observed in H-α and X-ray [3]. A substantial amount of cooler gas (about 1/2 of the bubble mass) is dragged by the rising bubble and forms the trunk of the mushroom. Between $t = 300 - 500$ Myr, this gas falls back into the cluster center, while the head of the mushroom stops and dissipates into the intra-cluster medium.

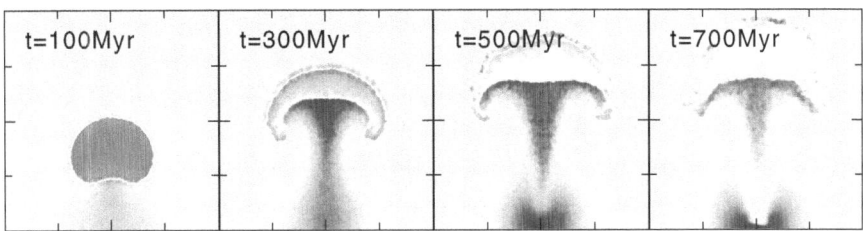

Fig. 1. Temperature map of an isolated bubble, from 10^4 K to 2×10^8 K. The box dimension is 200×200 kpc.

3.1 Global evolution and cold gas formation

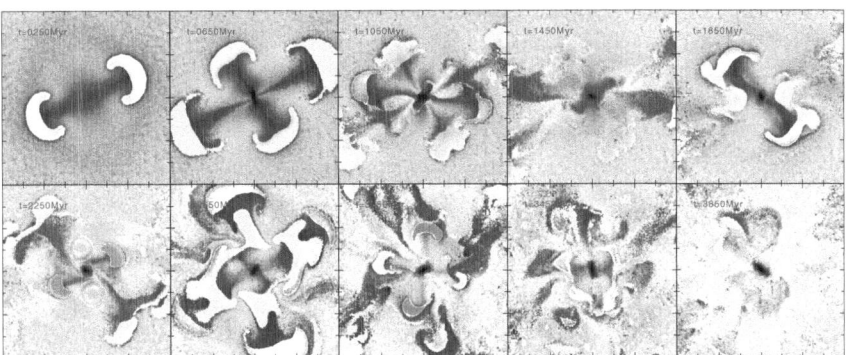

Fig. 2. Global evolution of the cluster temperature during 4 Gyr. The box dimension is 400×400 kpc.

In addition to slowing down the cooling flow at the center of the cluster, our simulations also show how AGN feedback may trigger cold gas formation at high radius ($R > 30$ kpc). Physical conditions for efficient cooling of intra-cluster gas occur either when the gas is less than 10^7 K or when its density is sufficiently high. Buoyant bubbles are responsible for strong inhomogeneities in temperature (see Fig.) as well as in density of the intra-cluster medium. In Fig. 3 left, we show how cooler falling gas (trunk of a bubble, see Fig.) is trapped between an old and a new rising bubble and is compressed to reach a state where its cooling time is a fraction of Gyr. In Fig. 3 right, the cluster is seen 300 Myr later. The short cooling time gas of Fig. 3 left, has now cooled down below 10^4 K. As it is not supported by the hot gas pressure, it falls radially to the center, forming a filament like structure (50 to 100 kpc) of mass $1.5 \times 10^8 M_\odot$. In the filament, the gas density is not high enough to form stars.

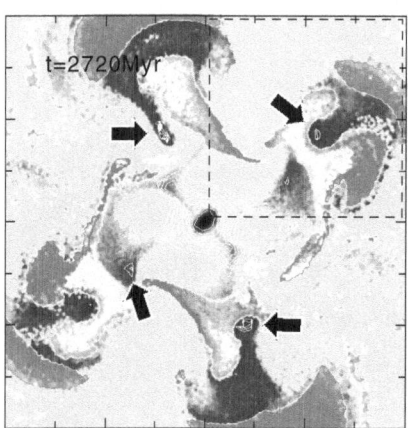 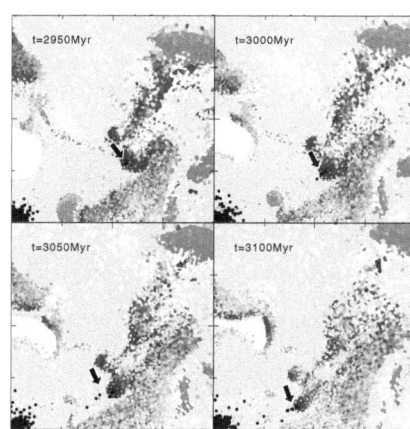

Fig. 3. *Left* : Gas temperature at $t = 2720$ Myr. The white contours pointed by black arrows indicate the position of the gas having a short cooling time $\tau_c < 1.1$ Gyr that will be transformed into cold gas. The upper left box corresponds to the region zoomed in the right part of the figure. *Right* : Gas temperature between $t = 2950$ and $t = 3100$ Myr, from 10^4 to 2×10^8 K . The small black dots pointed by the black arrow represents cold gas ($T < 10^4$ K) falling into the cluster center.

References

1. Conselice C.J., Gallagher J.S., Wyse R.F.G., 2001, ApJ, 122, 2281
2. Combes F., Gerin M., 1985, A&A, 150, 327
3. Fabian A.C., Sanders J.S., Crawford C.S., Conselice C.J., Gallagher J.S., Wyse R.F.G., 2003, MNRAS, 344L, 48
4. Salomé P., Combes F., Edge A.C., Crawford C., Erlund M., Fabian A.C., Hatch N.A., Johnstone R.M.,Sanders J.S., Wilman R., 2006, in press
5. Sanders J.S., Fabian A.C., Allen S.W., Schmidt R.W., 2004, MNRAS, 349, 952
6. Sijacki D., Springel V., 2006, MNRAS, 366, 397
7. Springel V., 2005, MNRAS, 364, 1105
8. Sutherland R.S., Dopita M.A., 1993, ApJS, 88, 253

Hyperfine Structure Radio Lines from Hot Gas in Elliptical Galaxies and Groups of Galaxies

D. Docenko[1] and R.A. Sunyaev[1,2]

[1] Max-Planck-Institut für Astrophysik, Karl-Schwarzschild-Str. 1, Postfach 1317, 85741 Garching, Germany; e-mail: dima@mpa-garching.mpg.de
[2] Space Research Institute, Russian Academy of Sciences, Profsoyuznaya ul. 84/32, Moscow 117810, Russia

Summary. Observations of intracluster gas bulk motions and turbulence in clusters and groups of galaxies should provide important information about physical conditions in the intergalactic gas. Thanks to the high spectral resolution of radio detectors, hyperfine structure (HFS) lines of highly-charged ions might help in this respect. Though not yet observed, intensities of emission and absorption HFS lines from the hot interstellar medium of elliptical galaxies and the intracluster medium of groups of galaxies are estimated to be on the limit of the abilities of existing instruments. The most interesting lines for this purpose are discussed.

1 Introduction

Three major spectral bands (X-ray, radio and microwave) are used nowadays for astrophysical observations to derive information on dynamics and state of the intracluster medium (ICM). Till now none of these methods allows us to observe ICM turbulence and bulk motions directly.

Following [6], we propose observations of hot gas using radio HFS spectral lines of high-Z ions. They might probe not only gas temperature and metallicity, but also its motions thanks to the high spectral resolution of radio telescopes. We have partly discussed this topic in [7], where the same method was applied mostly to studies of warm-hot intergalactic medium and supernova remnants. The best prospect has the HFS line of ^{14}N VII ion. Unfortunately, it is rather strongly absorbed in the Earth atmosphere, if observed from nearby objects. With increasing redshift, the HFS line is moving out of the absorption band and at $z > 0.15$ the effect of the atmosphere finally becomes inessential [6].

2 Hyperfine structure transitions

In our analysis we include ions being most abundant in the temperature range $10^6 - 10^7$ K and having HFS lines in the millimeter band, i.e., ^{14}N VII, ^{25}Mg X and ^{29}Si XII. The wavelengths λ of HFS transitions are taken from [5]. Absorption cross-sections σ and transition rates A, computed from these

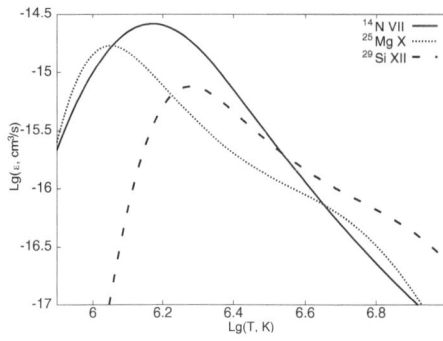

Fig. 1. Emissivities $\varepsilon(T)$, cm^3/s, of HFS transitions as functions of temperature.

data, are given in Table 1. HFS level excitation rates were obtained using data from [8, 2].

Due to collisions and interaction with the cosmic microwave background (CMB) radiation the HFS line optical depth and emissivity diminish. The emissivity, on the other hand, is enhanced by the photo-excitation of the same ion resonant transitions [1, 6] having large scattering cross-sections. Line emissivities are shown in Figure 2, accounting for the CMB effect (i.e., multiplied by $D(T_R)$ factor, see Table 1) and assuming collisional ionization equilibrium.

Table 1. Parameters of transitions between the ground state hyperfine sublevels

Ion	X_{iso}	λ, mm	A, s^{-1}	σ^*, cm^2	$D(2.7$ K$)$	n_{cr}, cm^{-3}
^{14}N VII	8.3×10^{-5}	5.652	1.992×10^{-10}	5.39×10^{-19}	0.340	3
^{25}Mg X	3.8×10^{-6}	6.680	2.111×10^{-10}	3.37×10^{-19}	0.413	0.05
^{29}Si XII	1.7×10^{-6}	3.725	1.566×10^{-9}	2.01×10^{-19}	0.701	0.5

* for a velocity dispersion of 30 km/s

3 Emission and absorption lines

To have a high brightness temperature T_b in HFS emission lines, the plasma should have a high linear emission measure $\int n_e^2 dl$ and appropriate temperature. Besides, less dense objects will be brighter due to diminishing of the emissivity with density (characteristic "critical density" n_{cr} is given in Table 1). In Table 2 we give T_b estimates from the nearby cool and X-ray bright group of galaxies NGC 3557 ($kT = 0.24$ keV, $\lg(L_X$, erg/s$) = 41.11$, $z = 0.01$) observed by ROSAT [3], assuming thermal velocities and abundances of 0.3 solar.

The detection of emission lines arising due to resonant scattering of the quasar radio emission in the hot gas surrounding it [4] might also be very interesting. As an example, we estimate that such additional flux in the ^{14}N VII

Table 2. Expected HFS emission line brightness temperatures T_b and absorption line central optical depths τ from the group of galaxies NGC 3557

Ion	^{14}N VII	^{25}Mg X	^{29}Si XII
T_b, μK	400	60	80
τ	1×10^{-3}	5×10^{-6}	3×10^{-5}

line center from the 3C 273 quasar ($z = 0.158$, $S \approx 30$ Jy) will constitute about 0.5 mJy (assuming gas parameters similar to ones observed in elliptical galaxies).

Despite low cross-sections, radio source absorption line optical depths τ, arising in ISM and ICM, may reach observable levels. Typical τ in the ^{14}N VII line is about 10^{-4}; for other lines it is lower due to smaller isotopic abundances. In Table 2 we give also τ estimates from central source in the NGC 3557 group.

4 Estimates of detectability and conclusions

Neglecting systematic effects, the *Green Bank Telescope* in a reasonable integration time of 10 hours and with a velocity resolution of 10 km/s will provide a 3-σ detection limit for the redshifted ^{14}N VII line flux of about 0.3 mJy. This corresponds to $T_b \approx 0.3$ mK and to an optical depth of $\tau \approx 2 \times 10^{-5}$, if observed from a 15 Jy source. So, the telescope would be able to detect ^{14}N VII line if a group of galaxies similar to NGC 3557 would be known at $z > 0.15$.

We conclude that, thanks to the high abundance of ^{14}N, the HFS line of ^{14}N VII is the best candidate to be observed in absorption in spectra of the brightest radio sources at $z > 0.15$, so that it is unaffected by atmospheric absorption. Its typical optical depth is about 10^{-4} being within the reach of existing instruments. The brightness temperature in emission is at the limit of modern observing possibilities.

Observations of hyperfine structure lines might allow to measure motions of the hot gas directly, thus helping to understand details of its heating.

References

1. G.B. Field, 1958, in Proceedings of the Institute of Radio Engineers, 46, 240 (1958)
2. V.I. Fisher, Y.V. Ralchenko, V.A. Bernshtam et al, 1997, Phys. Rev. A, 56, 3726
3. J.P.F. Osmond, T.J. Ponman, 2004, MNRAS, 350, 1511
4. S.Y. Sazonov, J.P. Ostriker, R.A. Sunyaev, 2004, MNRAS, 347, 144
5. V.M. Shabaev, M.B. Shabaeva, I.I. Tupitsyn, 1995, Phys. Rev. A, 52, 3686

6. R.A. Sunyaev, E.M. Churazov, 1984, Sov. Ast. Lett., 10, 201
7. R.A. Sunyaev, D. Docenko, 2006, astro-ph/0608256
8. H.L. Zhang, D.H. Sampson, 1997, MNRAS, 292, 133

Part VIII

Numerical Simulations and Cosmological Applications

Heating, Cooling and Enrichment in Clusters with Hydrodynamical Himulations

S. Borgani

Department of Astronomy, University of Trieste, via Tiepolo 11, I-34131 Trieste
borgani@oats.inaf.it

1 Introduction

Clusters of galaxies form from the collapse of exceptionally high density perturbations having typical size of ~ 10 comoving Mpc. As such, they mark the transition between two distinct regimes in the study of the formation of cosmic structures. The evolution of structures involving larger scales is mainly determined only by the effect of gravitational instability driven by the dark matter component and, therefore, it retains memory of the initial conditions. On the other hand, galaxy–sized structures, which form from initial fluctuations on scales of ~ 1 Mpc, form under the combined action of gravity and of complex gas–dynamical and astrophysical processes. On such scales, gas cooling, star formation and the subsequent release of energy and metal feedback from supernovae (SN) and AGN have a deep impact on the observational properties of the galaxy population.

In this sense, clusters of galaxies can be used as invaluable cosmological tools and astrophysical laboratories (see refs. [1, 2] for two reviews). These two aspects are clearly interconnected. On the one hand, the number density of galaxy clusters and their overall baryonic content provide in principle powerful constraints on cosmological parameters. On the other hand, for such constraints to be robust, one has to understand in detail a number of astrophysical processes. In this framework, hydrodynamical simulations, which include gas–dynamical effects, represent a powerful means to study in detail the existing interplay between cosmological evolution and astrophysical processes, which determine the observational properties of galaxy clusters. In the last years, code efficiency and supercomputing capabilities have opened the possibility to perform simulations over fairly large dynamic ranges, thus allowing resolution on scales of a few kiloparsecs, which are relevant for the formation of single galaxies, while capturing the global cosmological environment on scales of tens or hundreds of Megaparsecs, which are relevant for the evolution of galaxy clusters.

In this contribution, I will review the recent advancement performed in this field of computational cosmology. In particular I will focus my discussion on the comparison between observed and simulated properties of galaxy clusters, both in the X–ray and in the optical/near–IR bands. As already mentioned, the formation and evolution of galaxies are determined by process of gas cooling, star formation and feedback. In turn, these processes also alter the thermodynamical status of the hot intra–cluster medium (ICM), which is studied through X–ray observations. In this sense, a comprehensive description of the evolution of cosmic baryons in their diffuse and condensed phases necessarily requires a multi–wavelength approach.

2 Temperature profiles

Spatially resolved spectroscopic observations from X–ray satellites have opened the possibility of determining in detail the temperature structure of the ICM. ASCA observations [3] have established that temperature profiles are characterized by negative gradients. Beppo–SAX observations [4] have shown that such negative gradients do not extend down to the innermost cluster central regions, where instead an isothermal regime is observed, possibly followed by a decline of the temperature towards the center, at least for relaxed clusters. The much improved sensitivity of the Chandra satellite provides now a much more detailed picture of the central temperature profiles [5, 6]). Relaxed clusters are generally shown to have a smoothly declining profile toward the centre, reaching values which are about half of the overall virial cluster temperature in the innermost sampled regions, with non relaxed clusters having, instead, a greater diversity of temperature profiles. At the same time, observations with the XMM–Newton satellite demonstrated that the central regions of relaxed clusters are characterized by a spectrum with no significant or, at most, weak emission lines associated with soft atomic energy transitions [7, 8]). The consequence of this result is that a small amount of cooled gas should be present in the cluster cooling regions, thus implying a small value for the mass deposition rate. While this is in contradiction with the standard model of cooling flows [9], it is in much better agreement with optical observations on the brightest cluster galaxies, which imply a fairly low rate of star formation [10]. Therefore, the emerging picture suggests that gas cooling is responsible for the decline of the temperature in the central regions of clusters, while some mechanism of energy feedback should be responsible for preventing gas overcooling, thereby suppressing the mass deposition and the resulting star formation.

As for hydrodynamical simulations, they have shown to be generally rather successful in reproducing the negative temperature gradients outside the "cool core" regions [3, 12, 13], where gas cooling is unimportant. On the other hand, including gas cooling in hydrodynamical simulations has been shown to produce temperature profiles which, in the core regions, are generally steeper than those in non–radiative simulations. This is illustrated in the left panel of Figure 1 where we show the effect of introducing radiative cooling and different forms of feedback in simulations of a galaxy cluster [11]. While the temperature profiles are left essentially unchanged for $R \gtrsim 0.2R_{vir}$, the effect of cooling is that of steepening the profiles in the innermost regions. The

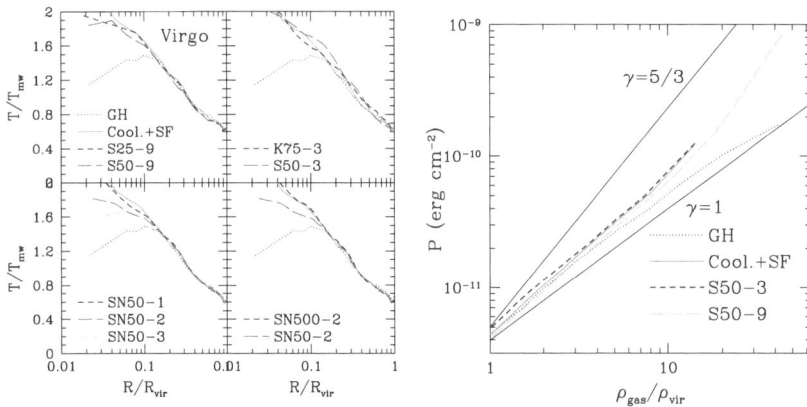

Fig. 1. Left panel: temperature profiles from hydrodynamical simulations of galaxy clusters. In all panels the dotted and the solid curves correspond to a non radiative run and to a run including cooling and star formation, respectively. The other curves represent results for different recipes of gas heating. Right panel: the relationship between gas density and pressure for a subset of the simulations shown in the right panel. The two straight lines are for polytropic models with $\gamma = 1$ and $\gamma = 5/3$ (from [11]).

counter–intuitive conclusion that cooling gives rise to a temperature increase in cluster cores has actually a fairly simple explanation. If not counteracted by some sort of feedback, cooling is so efficient at removing gas from the hot phase, it leads to a suppression of pressure support at the cluster centre. As a consequence, more external gas falls toward the centre, thus undergoing heating by adiabatic compression. This effect is illustrated in the right panel of Fig.1, where we show the relation between gas density and pressure for a subset of the simulations whose temperature profiles are reported in the left panel. The two solid lines in this plot show the relation for an adiabatic effective equation of state, $p \propto \rho_{gas}^{5/3}$, and for an isothermal one, $p \propto \rho_{gas}$. Quite apparently, the ICM in the non–radiative run behaves in a nearly isothermal way in the central regions, with a slope approaching $\gamma \simeq 1$. On the other hand, the effect of cooling is clearly that of steepening this relation at the centre, such that it approaches the adiabatic slope. This is exactly what is expected for gas undergoing adiabatic compression.

In Figure 2 I show a comparison between the observed temperature profiles and those predicted by simulations, which include gas cooling, star formation and energy feedback from galactic winds powered by SN [13]. Although this sort of feedback has been shown to produce a realistic cosmic star formation history [14], it is not efficient enough to regulate gas cooling and, therefore, to produce realistic temperature profiles. They are always

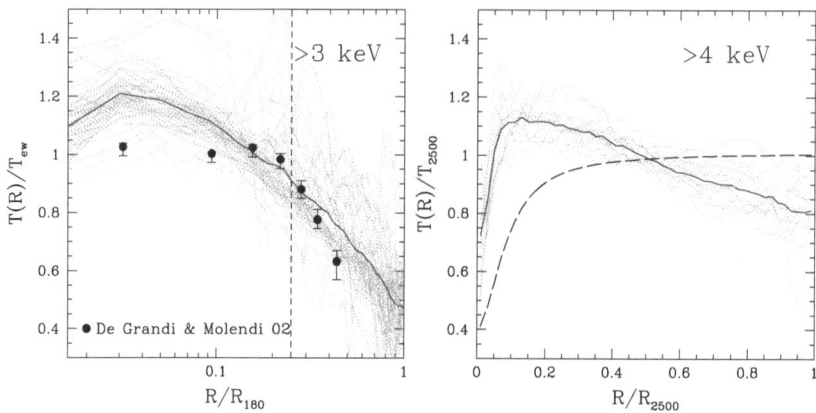

Fig. 2. Left panel: comparison between temperature profiles from Beppo–SAX observations [4] and from a cosmological simulation including radiative cooling, star formation and SN feedback. Right panel: comparison between the same set of simulations and the best–fit curve to the observed inner Chandra temperature profiles [5]. Figure taken from [13].

steeper, and have an inversion from negative to positive gradients at too small cluster-centric distances, with respect to observations.

The fact that overcooling and incorrect temperature profiles are two aspects of the same problem is also shown by the right panel of Figure 3. Here we compare the fraction of baryons converted into stars for the same set of simulated clusters and for observed clusters. Quite apparently, simulations still show an excess of cooling inside clusters, despite the fact that the cosmic values of the stellar fraction is consistent with observational estimates [14]. Since cooling is a runaway process, its efficiency increases with resolution. Therefore, one may wonder whether this estimate of the star fraction from simulations is reliable or represents a lower limit, due to the finite force and mass resolutions. In fact, the left panel of Figure 3 suggests that this not the case. This plot shows the dependence of the star fraction on the mass resolution for simulations of four clusters, which include feedback from galactic winds. Only for one of them, results are also shown in the case that galactic winds are turned off. Remarkably, the presence of efficient feedback makes cooling efficiency independent of resolution, while excluding it makes the runaway cooling reappear. These results demonstrate that, even if feedback is successful in terms of preventing the cooling runaway, it is not guaranteed to provide the correct thermal structure in the central regions of galaxy clusters.

Vice versa, the fact that a feedback mechanism is able to produce the correct temperature profiles does not guarantee that it also prevents overcooling. For instance, the simulations presented in [15] used a selective heating model, in which the energy release from SN is used to target ad hoc those gas particles which are just about to undergo cooling, increasing their entropy to

a critical value. With this approach, they were able to produce temperature profiles which are in better agreement with observations. However, the
resulting stellar fraction was still found to be too high, $\gtrsim 25$ per cent.

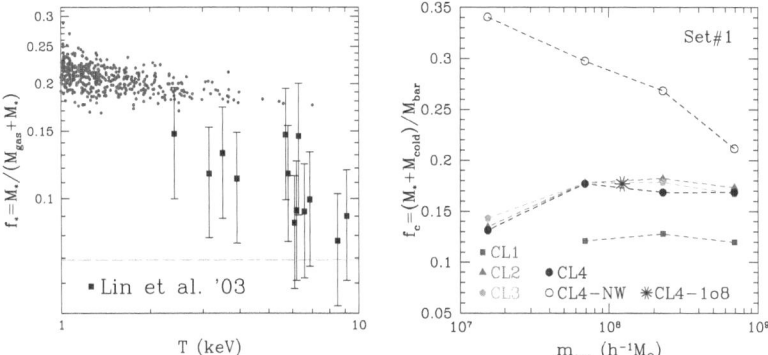

Fig. 3. Left panel: a comparison between observed (points with errorbars; from
[16]) and the simulated fraction of cluster baryons contained in stars (from [13]).
Right panel: the dependence of the stellar fraction on resolution for simulations of
four clusters, both including galactic winds (filled symbols) and without galactic
winds (open circles; from [17]).

3 The metal content of the ICM

Deciding whether SN feedback is the main process responsible for the heating
of the ICM relies on our knowledge of the past star formation history in the
cluster and proto–cluster environment. In this respect, the metal content of
the ICM provides an invaluable tool to trace the number of SN-Ia and SN-II
which, with different yields, contributed to the chemical enrichment [18].

ASCA and Beppo–SAX observations have provided information on the
global metal content of the ICM for a fairly large number of clusters [19],
while allowing to trace the spatial distribution of Iron for a few tens of objects
[20, 21]. These results have further confirmed the direct relationship between
star formation, dynamical status of the cluster and metallicity structure.
More recently, the much improved sensitivity of Chandra and XMM–Newton
have provided a quantum leap in the level of details in the description of the
ICM enrichment pattern [22, 6]. At the same time, the new generation of X–
ray satellites opened the possibility of tracing the evolution of the ICM Iron
content out to the largest redshifts, $z \simeq 1.3$, at which galaxy clusters have
been detected so far [23]. For nearby clusters, these observations have shown
that significant metallicity gradients exist in dynamically relaxed, cool–core
clusters. They generally show a central enhancement of the Iron distribution,

which is associated with the star formation occurring in the BCG. These gradients are steeper for relatively cooler systems, thus supporting the enhanced role that ram–pressure stripping should play in redistributing metals in the hotter atmospheres of richer clusters. For distant clusters, evidence is now emerging that a significant increase of the Iron content took place since $z \simeq 0.5$, with less significant evolution occurring at earlier epochs. Establishing whether this evolution is mainly driven by newly produced metals or by a redistribution of metals associated with the creation of cool cores, is at the moment a matter of debate.

Modeling the process of chemical enrichment with cosmological hydrodynamical simulations is generally not an easy task. On the one hand, it requires that simulation codes incorporate a physically motivated sub–resolution description of the star formation process, which provides a realistic description of the star formation history. On the other hand, these simulations must also include advanced models of stellar evolution, so as to account for the metals produced by different stellar populations and for the delay times with which metals and energy are released into the ICM since the star formation episode [24, 25, 26, 27]. In general, such code refinements imply an increase in the number of parameters, which describe the sub–resolution physics, to be determined through detailed comparisons with observational data. In this respect, hybrid approaches, based on combining N–body or non–radiative simulations with semi–analytical models of galaxy formation, offer in principle more flexible tools to explore this parameter space [28, 29]. However, by their nature, these approaches necessarily provide an approximate description of the chemo–dynamical properties of the ICM.

In Figure 4 we show a comparison between the chemical properties of the ICM for simulated and observed clusters. These simulations have been carried out using the Tree+SPH GADGET-2 code [30], with the implementation of stellar evolution and chemical enrichment provided by [26]. The left panel shows the relation between the ICM temperature and Iron and Silicon abundances. The simulations shown here have been carried out assuming a Salpeter IMF [31]. Observations refer to the ensemble of clusters observed with ASCA [19]. Error bars on the observational points represent the uncertainties after a stacking of the X–ray spectra within different temperature intervals. As such, these errors are purely statistical and do not include the effect of any intrinsic scatter in the metallicity–temperature relations. For the Iron abundance, the agreement between observations and simulations is quite encouraging, although there is a tendency towards lower abundance for cooler systems, which also have a larger scatter. In general the decreasing temperature dependence for $T_X \gtrsim 2$ keV is stronger in data than in simulations. For the Silicon abundance, simulation results agree, within the scatter, with observations of cool systems. The agreement is not as good for rich systems, although larger uncertainties arise in the observational analysis of hotter systems.

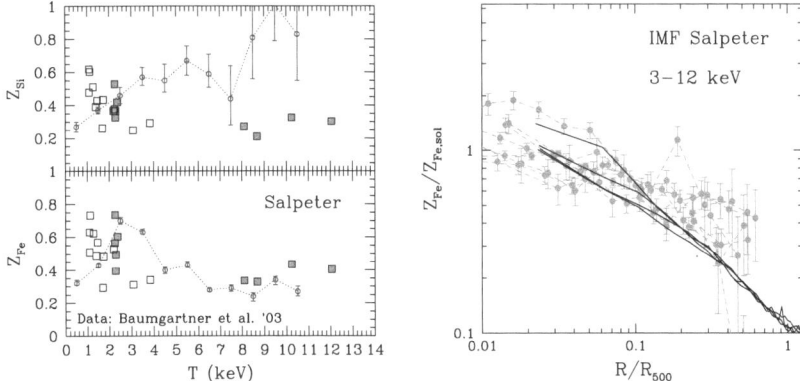

Fig. 4. Left panel: A comparison between the observed and the simulated relation between cluster metallicity and temperature. Upper and lower panels are for Silicon and Iron, respectively. Open circles with errorbars refer to the ASCA data [19], while squares are for simulations. Right panel: a comparison between observed and simulated profiles of the Iron abundance. Points with errorbars refer to Chandra observations for clusters with $T_X > 3$ keV [6], while the solid curves are for simulated clusters.

A point worth mentioning here is that the metallicity in simulations is computed by weighting each gas particle according to its X–ray emissivity. It is not clear whether this prescription is directly comparable with that used in observations, where the metallicity are computed from a spectral fitting procedure. It is clear that a proper comparison would require producing spectra from simulated clusters, which include a realistic observational noise and response function of the instrument, thus similar in spirit to that done to test the reliability of the spectroscopic temperature in hydrodynamical simulations [32, 33].

The right panel of Fig.4 provides the comparison between the profiles of Iron abundance for the four simulated clusters having high temperature and for the subset of hot clusters observed with Chandra [6]. This comparison shows a reasonable agreement between simulations and observations, although the profiles in simulated clusters tend to be somewhat steeper, and fall below the observed ones at $R/R_{500} \gtrsim 0.1$. Reproducing the details of the distribution of metals in the ICM must be considered in general as a challenging task for hydrodynamical simulations. In fact, the way in which metals are transported and diffused from the star forming regions depends on a number of complex physical processes, such as ram–pressure stripping [34], galactic winds (included in our simulations) and transport by buoyant bubbles [35], diffusion by turbulence [36] and by thermal motions, as well as gas viscosity [37]. While stochastic gas motions and ram–pressure stripping are included

346   S. Borgani

at some level in simulations, like those presented here, the level of accuracy
with which they are described depends highly on the details concerning the
numerics and the resolution achievable.

4 Properties of the galaxy population

The physical processes determining the thermodynamical and chemical properties of the ICM are inextricably linked to those determining the pattern
of star formation in cluster galaxies. In this sense, a successful description
of galaxy clusters requires a multiwavelength approach, which is able at the
same time to account for both the X–ray properties of the diffuse hot gas and
of the optical/near–IR properties of the galaxy populations. For instance, it is
clear that a numerical modeling of the ICM chemical enrichment can be considered as reliable as long as it also reproduces galaxy luminosity functions
and colors. Only with the recent advances in supercomputing capabilities and
code efficiency it has been possible to start studying global properties of the
galaxy population in cosmological simulations of galaxy clusters [38, 39].

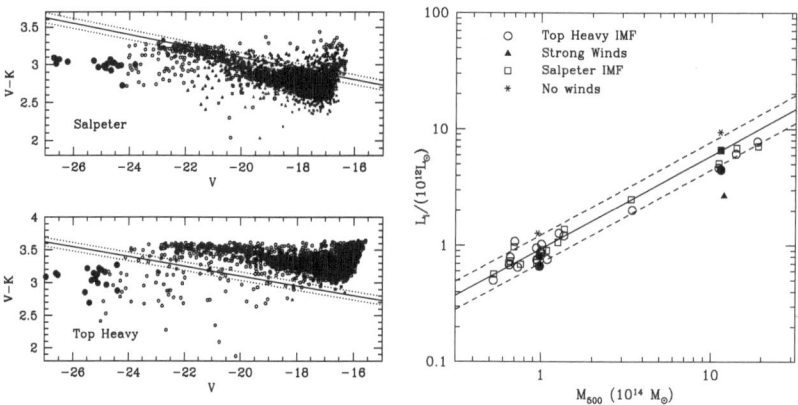

Fig. 5. Left panel: The color–magnitude diagram for simulated cluster galaxies
(symbols), compared with observations [40]. The large dots represent the BCGs of
the simulated clusters. Upper and lower panels refer to simulations performed with
a Salpeter IMF and with a top–heavier IMF, respectively. Right panel: the relation
between i–band luminosity and total cluster mass within R_{500}. Squares and circles
represent simulations based on assuming a Salpeter [31] and a top–heavy IMF [41],
respectively. The straight line is the best fit to the observed relation [42], with
dashed lines indicating the corresponding 1σ scatter (from [39]).

To tackle with this problem we have used the simulations discussed in
the previous Section to identify galaxies as clumps of star particles. In these

simulations, each star particle is treated as simple stellar population (SSP), being characterized by a formation time and a chemical composition. Using the GALAXEV spectro-photometric code [43], we compute broad band luminosities for each of them. In this way, we finally obtain luminosities and colors for the simulated galaxy population, to be then compared with observational data [39].

Figure 5 shows two examples of this comparison between observed and simulated properties of the cluster galaxy population. In the left panel we show the color–magnitude relation (CMR) of simulated galaxies, assuming both a Salpeter IMF [31] and a top–heavy IMF [41]. Quite apparently, the former performs better in reproducing galaxy colors, while the latter predicts too red colors as a consequence of the larger metal production from SN-II. These results also show that, brighter redder galaxies are on average more metal rich, therefore demonstrating that the CMR corresponds in fact to a metallicity sequence. However, a clear failure of these simulations lies in the fact that the BCGs are always much bluer than observed. This is due to an excess of recent star formation, which, in BCGs in $\sim 10^{15} M_\odot$ simulated clusters, can be as high as 500–1000 $M_\odot$/yr. This excess of star formation is clearly the consequence of overcooling which is not prevented by the action of galactic winds at the centre of clusters. In this sense, incorrect BCG colors and temperature profiles in cluster core regions are two sides of the same problem, i.e. the difficulty of balancing the cooling runaway within the largest halos.

On the other hand, our results show that the BCG is an exception, since the observed properties of the bulk of the galaxy population are well reproduced by simulations. This is also confirmed by the results on the mass–luminosity relation shown in the right panel of Figure 5. Simulations and observations provide quite consistent results, thus lending further support to the reliability of numerical simulations in describing the general properties of the population of cluster galaxies.

Acknowledgement. It is a great pleasure to thank the organizers of the "Cooling vs. Heating" Conference, and especially Hans Böhringer, for having given me the opportunity of presenting this review.

References

1. Rosati, P., Borgani, S., & Norman, C. 2002, ARAA, 40, 539
2. Voit, G. M. 2005, Reviews of Modern Physics, 77, 207
3. Markevitch, M., Forman, W. R., Sarazin, C. L., & Vikhlinin, A. 1998, ApJ, 503, 77
4. De Grandi S., Molendi S., 2002, ApJ, 567, 163
5. Allen S.W., Schmidt R.W., Fabian A.C., 2001, MNRAS, 328, L37
6. Vikhlinin, A., Markevitch, M., Murray, S. S., Jones, C., Forman, W., & Van Speybroeck, L. 2005, ApJ, 628, 655

7. Peterson, J. R., et al. 2001, A&A, 365, L104
8. Böhringer, H., Matsushita, K., Churazov, E., Ikebe, Y., & Chen, Y. 2002, A&A, 382, 804
9. Fabian, A. C. 1994, ARAA, 32, 277
10. Rafferty, D. A., McNamara, B. R., Nulsen, P. E. J., & Wise, M. W. 2006, ArXiv Astrophysics e-prints, arXiv:astro-ph/0605323
11. Tornatore L., Borgani S., Springel V., Matteucci F., Menci N., Murante G., 2003, MNRAS, 342, 1025
12. Loken, C., Norman, M. L., Nelson, E., Burns, J., Bryan, G. L., & Motl, P. 2002, ApJ, 579, 571
13. Borgani S., Murante G., Springel V., et al., 2004, MNRAS, 348, 1078
14. Springel, V., & Hernquist, L. 2003, MNRAS, 339, 312
15. Kay, S. T., Thomas, P. A., Jenkins, A., & Pearce, F. R. 2004, MNRAS, 355, 1091
16. Lin Y.-T., Mohr, J.J., Stanford S.A., 2003, ApJ, 591, 749
17. Borgani, S., et al. 2006, MNRAS, 367, 1641
18. Renzini, A. 1997, ApJ, 488, 35
19. Baumgartner, W. H., Loewenstein, M., Horner, D. J., & Mushotzky, R. F. 2005, ApJ, 620, 680
20. Finoguenov, A., Arnaud, M., & David, L. P. 2001, ApJ, 555, 191
21. De Grandi, S., Ettori, S., Longhetti, M., & Molendi, S. 2004, A&A, 419, 7
22. Tamura, T., Kaastra, J. S., den Herder, J. W. A., Bleeker, J. A. M., & Peterson, J. R. 2004, A&A, 420, 135
23. Tozzi, P., Rosati, P., Ettori, S., Borgani, S., Mainieri, V., & Norman, C. 2003, ApJ, 593, 705
24. Lia, C., Portinari, L., & Carraro, G. 2002, MNRAS, 330, 821
25. Valdarnini, R. 2003, MNRAS, 339, 1117
26. Tornatore, L., Borgani, S., Matteucci, F., Recchi, S., & Tozzi, P. 2004, MNRAS, 349, L19
27. Romeo, A. D., Sommer-Larsen, J., Portinari, L., & Antonuccio-Delogu, V. 2006, MNRAS, 371, 548
28. Nagashima, M., Lacey, C. G., Okamoto, T., Baugh, C. M., Frenk, C. S., & Cole, S. 2005, MNRAS, 363, L31
29. Cora, S. A. 2006, MNRAS, 368, 1540
30. Springel, V. 2005, MNRAS, 364, 1105
31. Salpeter, E. E. 1955, ApJ, 121, 161
32. Mazzotta, P., Rasia, E., Moscardini, L., & Tormen, G. 2004, MNRAS, 354, 10
33. Vikhlinin, A. 2006, ApJ, 640, 710
34. Domainko, W., et al. 2006, A&A, 452, 795
35. Sijacki, D., & Springel, V. 2006, MNRAS, 366, 397
36. Rebusco, P., Churazov, E., Böhringer, H., & Forman, W. 2005, MNRAS, 359, 1041
37. Sijacki, D., & Springel, V. 2006, MNRAS, 371, 1025
38. Romeo, A. D., Portinari, L., & Sommer-Larsen, J. 2005, MNRAS, 361, 983
39. Saro, A., Borgani, S., Tornatore, L., Dolag, K., Murante, G., Biviano, A., Calura, F., & Charlot, S. 2006, ArXiv Astrophysics e-prints, arXiv:astro-ph/0609191
40. Bower, R. G., Lucey, J. R., & Ellis, R. S. 1992, MNRAS, 254, 601
41. Arimoto, N., & Yoshii, Y. 1987, A&A, 173, 23
42. Popesso, P., Biviano, A., Böhringer, H., & Romaniello, M. 2006, A&A, 445, 29
43. Bruzual, G., & Charlot, S. 2003, MNRAS, 344, 1000

Beyond the Cool Core: The Formation of Cool Core Galaxy Clusters

J.O. Burns[1], E.J. Hallman[1], B. Gantner[1], P.M. Motl[2] and M. L. Norman[3]

[1] Center for Astrophysics and Space Astronomy, Department of Astrophysical & Planetary Science, University of Colorado, Boulder, CO 80309

[2] Department of Physics and Astronomy, Louisiana State University, Baton Rouge, LA 70803

[3] Center for Astrophysics and Space Sciences, University of California-San Diego, 9500 Gilman Drive, La Jolla, CA 92093

Summary. Why do some clusters have cool cores while others do not? In this paper, cosmological simulations, including radiative cooling and heating, are used to examine the formation and evolution of cool core (CC) and non-cool core (NCC) clusters. Numerical CC clusters at $z = 0$ accreted mass more slowly over time and grew enhanced cool cores via hierarchical mergers; when late major mergers occurred, the CCs survived the collisions. By contrast, NCC clusters of similar mass experienced major mergers early in their evolution that destroyed embryonic cool cores and produced conditions that prevent CC re-formation. We discuss observational consequences.

1 Introduction

Recently, 49% of clusters were identified as having cool cores in a flux-limited sample, HIFLUGCS, based upon *ROSAT* and *ASCA* observations [1]. The earliest and simplest model assumed these clusters to be spherical, isolated systems where "cooling flows" formed; as radiating gas loses pressure support, cooling gas flows inwards to higher density values which further accelerates the cooling rate (e.g., [2]). However, the predicted end-products of this mass infall (e.g., star formation, HI, CO) have not been observed and the central temperatures indicate that the gas at the cores has cooled only moderately (generally only 30-40% of the virial temperature [3]). Furthermore, this model failed to consider the important effects of mergers and on-going mass accretion from the surrounding supercluster environment.

We have performed numerical simulations of the formation and evolution of clusters in a cosmological context using the adaptive mesh refinement N-body/hydro code *Enzo*, aimed at further understanding cool cores [4]. We find that most mergers are oblique with halos spiraling into the cluster center and gently bequeathing cool gas; this allows the cool core to strengthen over time. Thus, in this model, cool cores themselves grow hierarchically via the merger/accretion process. Also, a realistic model of cool cores must involve heating, as well as cooling, that "softens" the cores (i.e., reduces density and increases temperature) potentially by star formation or by AGNs. This softening would increase the susceptibility of disruption of the cool core by subsequent mergers.

In this paper, we present new enhanced-physics *Enzo* cosmology simulations that include radiative cooling, star formation (i.e., a mass sink for cold gas), and heating (modeled via kinetic energy input from Type II supernovae). In addition to the more realistic baryonic physics, these simulations are superior to previous numerical models (e.g., [5, 6]) due to their bigger volumes and larger samples of clusters. With hundreds of clusters of mass $> 10^{14} M_\odot$, we have the capability for the first time to examine evolutionary effects in these numerical clusters with adequate statistics.

2 The Formation of CC and NCC Clusters

Enzo is an adaptive mesh refinement cosmological simulation code that couples an N-body particle-mesh (PM) solver with an Eulerian block-structured AMR method for ideal gas dynamics. Enzo has physics modules which model radiative cooling of both primordial and metal-enriched gases, and prescriptions for star formation and feedback [7].

Our AMR simulations have produced a sample of 1494 numerical clusters with $0 < z < 1$ and $M_{virial} > 10^{14} M_\odot$. Only $\approx 10\%$ of this sample of numerical clusters have cool cores. This fraction is smaller than observed. We believe that our specific recipe for star formation and heating has over-softened the cores making them too vulnerable to disruption. This will be addressed in subsequent simulations.

In Figure 1, we show examples of the evolution of a CC and an NCC cluster which represent well the general scenarios for how such clusters form. This figure illustrates the different histories of two clusters which have the same final mass at $z = 0$, $M_{virial} = 5 \times 10^{14} M_\odot$.

Many clusters start off with cool cores early in their history when initial conditions produce central densities and temperatures that allow the gas to radiatively cool. Depending upon the sequence and magnitude of cluster collisions, cool cores will be destroyed in some clusters and preserved in others. NCC clusters often undergo major mergers early in their history. From z=0.65 to z=0.5, the mass of the NCC cluster in Fig. 1 increased by 110%. Such mergers destroy the cool cores, leaving behind hotter, thermalized, moderately dense cores. Subsequently, cool halos infalling into these NCC clusters do not survive passage through the central parts of the clusters and the central conditions do not allow cool cores to re-establish. NCC clusters continue to experience minor mergers as they continue to slowly evolve (typically mass increases 10-20% over Gyr time frames). We suggest that such an early major merger produced the complex characteristics observed today for the Coma cluster.

On the other hand, Figure 1 suggests that CC clusters evolve differently. *Many CC clusters do not experience major mergers early in their history but*

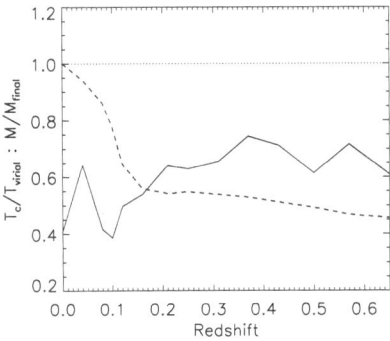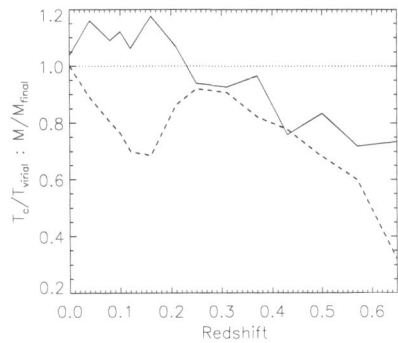

Fig. 1. The time history of the evolution of a CC (left) and an NCC (right) numerical cluster. Dashed lines indicate the total mass in units of the virial mass normalized to 1 at z=0. Solid line is the central temperature normalized by the virial temperature. Note that the dip in mass at $z \approx 0.15$ in the NCC cluster is an artifact of how we measure the cluster mass within the virial radius; it is caused by the passage of an infalling subcluster through the core and back out beyond r_{virial} with a final acquisition into the cluster by z=0.

rather grow slowly such that the cool cores increase in mass and stability. The CC cluster at z=0.65 began with a higher mass than the NCC at the same redshift. From z=0.65 to z=0.1, mergers increased the mass of the CC cluster by only $\approx$20%. At $z \approx 0.1$, substantial multiple mergers occurred increasing the mass by $\approx$50% by z=0. But by this epoch, the cool core is so well established that it survives ever larger impacts. In fact, the cool gas from the infalling halos near the present epoch merges with the cool core such that it continues to grow.

3 Consequences of Evolutionary Differences in CC and NCC Clusters

The large-scale X-ray surface brightness profiles (S_X) of our simulated NCC clusters have S_X distributions which fit very well to β-models ($S_X \propto [1 + (r/r_c)^2]^{1/2-3\beta}$) out to the cluster virial radius as we recently described [8]. This is not true for CC clusters in our samples. As detailed in [8], the β-model does a poor job of extrapolating the surface brightness from r_{500} to r_{200} (subscripts refer to overdensities relative to critical density; $r_{200} \approx r_{virial}$) for CC clusters. The CC cluster profile has a typically $\approx$35% steeper slope than the β-model in the outer part of the cluster. We view this as a further signature of recent merger activity. The result is a bias or overestimate of the density profile, leading to a significant overestimate in mass.

These results are typical for CC clusters in our sample. We have fit β-models to the synthetic X-ray surface brightness profiles for all the numerical clusters with $0 < z < 0.5$ and $M > 5 \times 10^{14} M_\odot$ ($\approx$800 clusters). The fits

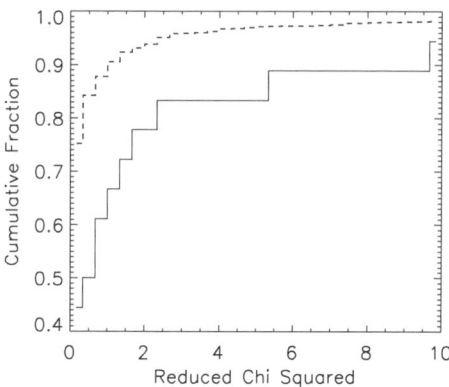

Fig. 2. Histograms of the quality of fit, as measured by the reduced χ^2, for β-model fits to the X-ray surface brightness profiles for $\approx$800 of our numerical clusters with $M > 5 \times 10^{14} M_\odot$. Dotted line is NCC clusters and solid line is CC clusters.

excluded the core regions with $r < 100\ h^{-1}$ kpc and extended out to r_{500} to mimic what is typically available for X-ray observations. We then extrapolated the fit out to r_{200} and calculated the reduced χ^2 between the β-model and the actual profile for the numerical clusters. A histogram of the cumlative fraction of the χ^2 distribution for NCC and CC clusters is presented in Figure 2. There is a significant difference between the quality of the β-model fits between these two types of clusters. Nearly 90% of the NCC clusters have reduced $\chi^2 < 1$ whereas only 60% of CC clusters have values < 1. This results in a mass overestimate or bias of typically a factor of 3-5 for CC clusters. Thus, there is a breakdown in the assumptions of a single β-model (i.e., dynamical equilibrium as assumed in simple cooling flow model) even outside the core for CC clusters. Similar results for a sample of cool core clusters observed by *Chandra* have been found [9].

Our numerical CC clusters also generally show a higher incidence of cold substructure outside the core than do their NCC counterparts when measured relative to the virial temperatures. The lower temperature gas is composed of both compact infalling halos as well as diffuse cool gas. We predict that this signature will be apparent in hardness ratio maps. As shown in Fig. 3 for two typical cases drawn from our simulations, the hard-to-soft band ratios (2-8 keV/0.7-2 keV) illustrate the abundance of cooler gas beyond the cores in CC clusters. We find that this prediction agrees well with the hardness ratio distribution of several Abell clusters observed by *Chandra*.

4 Conclusions

Our cosmological numerical simulations suggest that

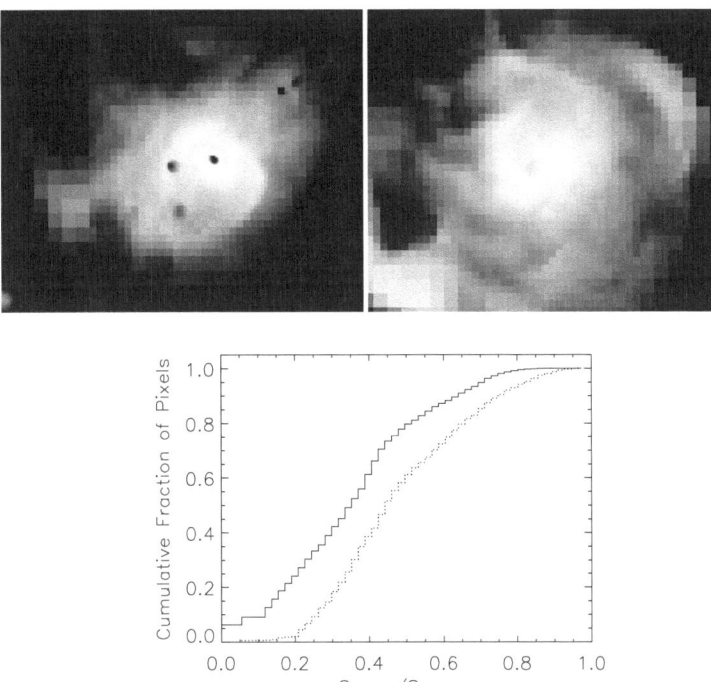

Fig. 3. Maps (3 h^{-1} Mpc on a side) of the X-ray hardness ratio $(2-8\,\text{keV})/(0.7-2\,\text{keV})$ for numerical clusters with a cool core (left) and without a cool core (right). Dark are smaller (0.2) HR values (corresponding to cool temperatures) and white are large (> 0.8) HR values (hot temperatures). The clusters have similar masses of $\approx 5 \times 10^{14} M_\odot$. At the bottom, the histogram is the distribution of hardness ratios for the two clusters above; dotted is for the NCC cluster and solid is for the CC cluster. The central 100 kpc of each cluster (cool core region) was excluded in constructing this histogram.

- Non-cool core (NCC) clusters suffer early major mergers when embryonic cool cores are destroyed. Cool core (CC) clusters, on the other hand, grow more slowly without early major mergers.
- The X-ray surface brightness profiles for NCC clusters are well fit by single β-models whereas the outer emission for CC clusters is biased low compared to β-models. The resulting densities and masses of CC clusters estimated from single β-model extrapolations are biased high by factors of 3-5.
- CC clusters have more cool gas beyond the cores than do NCC clusters, reflected in hardness ratio maps.

Acknowledgement. This work was supported in part by grants from the U.S. National Science Foundation (AST-0407368) and NASA (TM3-4008A).

References

1. Y. Chen & et al., In H. Böhringer, G.W. Pratt, A. Finoguenov, P. Schuecker, eds., *These proceedings*, 2006.
2. A. C. Fabian, 2002, in M. Gilfanov, R. Sunyaev, E. Churazov, eds., *Lighthouses of the Universe: The Most Luminous Celestial Objects and Their Use for Cosmology*, pp24
3. M. Donahue & G. M. Voit, 2004, in J. S. Mulchaey, A. Dressler, & A. Oemler, eds., *Clusters of Galaxies: Probes of Cosmological Structure and Galaxy Evolution*, pp143
4. P. M. Motl, J. O. Burns, C. Loken, M. L. Norman, & G. Bryan, 2004, ApJ, 606, 635
5. P. M. Motl, E. J. Hallman, J. O. Burns, & M. L. Norman, 2005, ApJ, 623, L63
6. A. V. Kravtsov, D. Nagai, & A. A. Vikhlinin, 2005, ApJ, 625, 588
7. B. W. O'Shea, G. Bryan, J. Bordner, M. L. Norman, T. Abel, R. Harkness, & A. Kritsuk, 2005, in Adaptive Mesh Refinement: Theory and Applications, (Springer, Berlin)
8. E. J. Hallman, P. M. Motl, J. O. Burns, & M. L. Norman, 2006, ApJ, 648, 852
9. A. Vikhlinin, A. Kravtsov, W. Forman, C. Jones, M. Markevitch, S. S. Murray, & L. Van Speybroeck, 2006, ApJ, 640, 691

AMR Simulations of the Cosmological Light Cone: SZE Surveys of the Synthetic Universe

E. J. Hallman[1], B. W. O'Shea[2], M. L. Norman[3], R. Wagner[3] and J. O. Burns[1]

[1] Center for Astrophysics and Space Astronomy, Department of Astrophysical & Planetary Science, University of Colorado, Boulder, CO 80309
[2] Los Alamos National Laboratory, Los Alamos, NM 87501
[3] Center for Astrophysics and Space Sciences, University of California-San Diego, 9500 Gilman Drive, La Jolla, CA 92093

Summary. We present preliminary results from simulated large sky coverage (~100 square degrees) Sunyaev-Zeldovich effect (SZE) cluster surveys using the cosmological adaptive mesh refinement N-body/hydro code Enzo. We have generated simulated light cones to match the resolution and sensitivity of current and future SZE instruments. These simulations are the most advanced calculations of their kind. The simulated sky surveys allow a direct comparison of large N-body/hydro cosmological simulations to current and pending sky surveys. Our synthetic surveys provide an indispensable guide for observers in the interpretation of large area sky surveys, and will develop the tools necessary to discriminate between models for cluster baryonic physics, and to accurately determine cosmological parameters.

1 Background and Method

Clusters of galaxies form from the highest peaks in the primordial spectrum of density perturbations generated by inflation in the early universe. They are the most massive virialized structures in the universe, and as such are rare objects. The number density of galaxy clusters as a function of mass and redshift is strongly dependent on a number of cosmological parameters [1, 2]. Cluster survey yields depend on the value of the minimum flux probed as a function of redshift, the growth function of structure, and the redshift evolution of the comoving volume element [3].

The simulation used to generate the light cones described in this contribution is of a 512 Mpc/h comoving volume of the universe, with the following cosmological parameters: $(\Omega_b, \Omega_{CDM}, \Omega_\Lambda, \text{h}, \text{n}, \sigma_8) = (0.04, 0.26, 0.7, 0.7, 1.0, 0.9)$. The simulation was initialized on a 512^3 root grid with 512^3 dark matter particles, corresponding to a dark matter (baryon) mass resolution of 7.2×10^{10} (1.1×10^{10}) $M_\odot/\text{h}$ and an initial comoving spatial resolution of 1 Mpc/h. The simulation was then evolved to z=0 using a maximum of 4 levels of adaptive mesh refinement. This simulation results in a higher dynamic range than achieved by any previous AMR cosmological simulation representing such a large physical volume.

These light cone simulations are run with both dark matter and baryons, unlike most previous similar studies. It has been shown in recent studies using both simulations e.g., [4] and high resolution X-ray observations of galaxy clusters e.g., [5, 6] that many clusters show strong departures from both equilibrium and isothermality. These deviations can have a strong impact on both the observable and derived properties of clusters. We have previously shown that deviations of factors of 10 or more are common in SZE and X-ray observables during even low mass ratio mergers. Thus, in order to properly simulate sky surveys, it is critically important to self-consistently include baryons in numerical simulations.

The light cone discussed here was generated from the above simulation using a stacking method similar to that used by [7], but with 100 times the angular coverage of the best previous N-body+hydro light cone of [8]. We simulate an SZE observation of a 100 square degree region of the sky, looking at the corresponding comoving volume of the universe from z=0.1 to z=2.75. We use 26 discrete volumes of $\delta_z \simeq 0.1$. We have generated SZE Compton y parameter images that are 2048 pixels on a side and cover 10 degrees x 10 degrees at a resolution of approximately 17.5 arcseconds/pixel. The images are then degraded using Gaussian smoothing to the resolution of several experiments which will be providing SZE data in the very near future, including APEX-SZ, the South Pole Telescope, the Planck Surveyor and the Atacama Cosmology Telescope.

Acknowledgement. This work was supported in part by a grant from the U.S. National Science Foundation (AST-0407368).

References

1. L. Wang and P. J. Steinhardt, 1998, ApJ, 508, 483
2. Z. Haiman, J. J. Mohr, and G. P. Holder, 2001, ApJ, 553, 545
3. P. Rosati, S. Borgani, and C. Norman, 2002, ARAA, 40, 539
4. E. Rasia, S. Ettori, L. Moscardini, P. Mazzotta, S. Borgani, K. Dolag, G. Tormen, L. M. Cheng, and A. Diaferio, 2006, MNRAS, 369, 2013
5. A. Vikhlinin, M. Markevitch, S. S. Murray, C. Jones, W. Forman, and L. Van Speybroeck, 2005, ApJ, 628, 655
6. M. Markevitch, A. H. Gonzalez, L. David, A. Vikhlinin, S. Murray, W. Forman, C. Jones, and W. Tucker, 2002, ApJ, 567, L27
7. A. C. da Silva, D. Barbosa, A. R. Liddle, and P. A. Thomas, 2000, MNRAS, 317, 37
8. V. Springel, M. White, and L. Hernquist, 2001, ApJ, 549, 681

Fig. 1. 10 × 10 degree synthetic SZE Compton y survey image from Enzo AMR simulation of comoving 512 Mpc/h volume. The simulation contains 512^3 root grid zones and 512^3 dark matter particles with up to 4 levels of dynamic refinement. The image contains 2048^2 pixels for an angular resolution of ∼0.3 arcmin/pixel. Image contains structures from z=2.75 to z=0.1.

Modeling Chandra X-ray Observations of Galaxy Clusters using Cosmological Simulations

D. Nagai[1], A. V. Kravtsov[2] and A. Vikhlinin[3,4]

[1] Theoretical Astrophysics, California Institute of Technology, Mail Code 130-33, Pasadena, CA 91125 daisuke@caltech.edu
[2] Department of Astronomy and Astrophysics, KICP, & EFI, The University of Chicago, 5640 South Ellis Ave., Chicago, IL 60637
[3] Harvard-Smithsonian Center for Astrophysics, 60 Garden Street, Cambridge, MA 02138
[4] Space Research Institute, 8432 Profsojuznaya St., GSP-7, Moscow 117997, Russia

Summary. X-ray observations of galaxy clusters potentially provide powerful cosmological probes if systematics due to our incomplete knowledge of the intracluster medium (ICM) physics are understood and controlled. In this paper, we study the effects of galaxy formation on the properties of the ICM and X-ray observable-mass relations using high-resolution self-consistent cosmological simulations of galaxy clusters and comparing their results with recent *Chandra* X-ray observations. We show that despite complexities of their formation and uncertainties in their modeling, clusters of galaxies both in observations and numerical simulations are remarkably regular outside of their cores, which holds great promise for their use as cosmological probes.

1 Testing X-Ray Measurements of Galaxy Clusters

X-ray observations with *Chandra* and *XMM-Newton* enable us to study properties of the ICM with unprecedented detail and accuracy and provide important handles on ICM modeling and associated systematics. Their superb spatial resolution and sensitivity enable accurate X-ray brightness and temperature measurements at a large fraction of the cluster virial radius and also make it simple to detect most of the small-scale X-ray clumps. Despite this recent observational progress, the biases in the determination of the key cluster properties remain relatively uncertain.

We therefore assess the accuracy of the X-ray measurements of galaxy cluster properties using mock *Chandra* analyses of cosmological cluster simulations and analyzing them using a model and procedure essentially identical to that used in real data analysis [1, 2]. The comparison of the true and derived cluster properties provides an assessment of biases introduced by the X-ray analysis. We examine the bias in mass measurements separately for dynamically relaxed and non-relaxed clusters, identified based on the overall structural morphology of their *Chandra* images, which mimics the procedure

Fig. 1. Mock *Chandra* images of one of the relaxed (*top*) and unrelaxed (*bottom*) simulated clusters at z=0. The detectable extended X-ray sources, indicated by ellipses, are detected and masked out from the analysis. The physical size of the images is $5\,h^{-1}$ Mpc.

used by observers. The typical examples of systems classified as relaxed or unrelaxed are shown in Figure 1. To check for any redshift dependence in such biases, we also analyze the simulation outputs at $z = 0$ and 0.6. The simulations and analysis procedures are fully described in [3].

Figure 2 illustrates that the X-ray analysis provides accurate reconstruction of the 3D properties of the ICM for the nearby, relaxed clusters. The strongest biases we find are those in the hydrostatic mass estimates, which is biased at a level of about 13% at r_{500c} even in the relaxed clusters. We find that the biases are primarily due to additional pressure support provided by subsonic bulk motions in the ICM, ubiquitous in our simulations even in relaxed systems [4, 5, 6]. These biases are related to physics explicitly missing

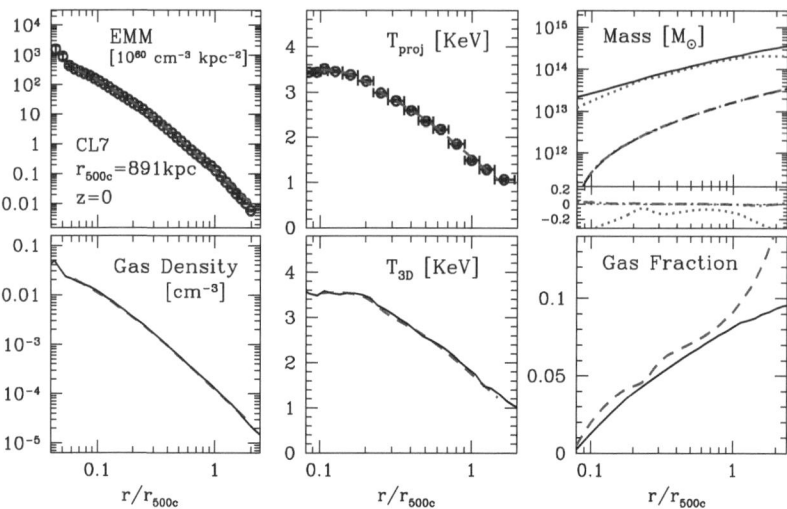

Fig. 2. The mock *Chandra* analyses of one of the relaxed clusters at $z = 0$ with $M_{500c} = 1.41 \times 10^{14} h^{-1} M_\odot$. In the *left* and *middle* panels, the best-fit model (*dashed* lines) recovers well both the projected profiles and the actual 3D gas profiles. In the *upper-right* panel, the reconstructed M_{gas} profile (*dot-dashed* line) is accurate to a few percent in the entire radial range shown. The hydrostatic M_{tot} estimate (*dotted* line), on the other hand, is biased low by about 5%–10% in the radial range, $[0.2, 1.0]r_{500c}$. The *lower-right* shows that measured cumulative f_{gas} is biased high by $\approx 10\%$ in the radial range of $[0.2, 1.0]r_{500c}$ for this cluster, and it is primarily due to the bias in the hydrostatic mass estimate.

from the hydrostatic method (e.g., turbulence), and not to deficiencies of the X-ray analysis. Gas fraction determinations are therefore biased high. The bias increases toward cluster outskirts and depends sensitively on its dynamical state, but we do not observe significant trends of the bias with cluster mass or redshift. We also compute a X-ray spectral temperature (T_X), a value derived from a single-temperature fit to the integrated cluster spectrum excluding the core ($< 0.15r_{500c}$) and detectable small-scale clumps [7, 8].

2 Effects of Galaxy Formation on the ICM Profiles

Next, we investigate the effects of galaxy formation on the ICM properties and compare the results of simulations with recent *Chandra* X-ray observations of nearby relaxed clusters. The impact of galaxy formation on the properties of ICM are investigated by comparing simulations performed with and without the processes of galaxy formation (e.g., gas cooling, star formation, stellar feedback and metal enrichment), which we refer to as the cooling+SF (CSF)

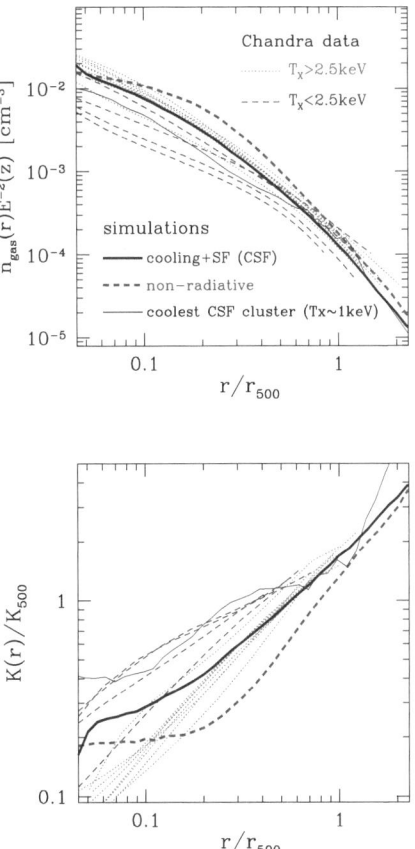

Fig. 3. The ICM gas density and entropy profiles of simulated clusters and Chandra X-ray observations of nearby ($z \approx 0$) relaxed clusters.

and non-radiative runs, respectively. Fig. 3 shows that gas cooling and star formation modify both the normalization and the shapes of the ICM profiles. What happens is that the removal of low-entropy gas in the inner region raises the level of entropy and lowers the gas density [9]. The effects are strongly radial dependent and increase toward the inner regions down to about $\sim$ $0.1r_{500c}$, inside which the observed properties are not well reproduced in the simulations. On the other hand, the ICM properties outside the cores in the cooling CSF simulations and observations agree quite well. At r_{500c}, both the ICM density and entropy profiles of different mass systems converge, indicating that the clusters are self-similar in the outskirts. Note that the non-radiative simulations predict an overall shape of the density and entropy profiles which is inconsistent with observations.

3 X-ray observable-mass relations

For cosmological application, it is important to understand the relations between X-ray observables and cluster mass. In Fig. 4, we present recent comparisons of two X-ray proxies for the cluster mass — the spectral temperature, T_X, and the new proxy, Y_X, defined as a simple product of T_X and M_g [10]. Analogously to the integrated Sunyaev-Zel'dovich flux, Y_X is related to the total thermal energy of the ICM.

The $M_{500} - T_X$ relation has a $\sim 20\%$ scatter in M_{500} around the mean relation, most of which is due to unrelaxed clusters. The unrelaxed clusters also have temperatures biased low for a given mass. This is likely because during mergers, the mass of the system has already increased but only a fraction of the kinetic energy of merging systems is converted into the thermal energy of gas, due to incomplete relaxation [11]. The slope and evolution of the $M_{500} - T_X$ relation are also quite close to the self-similar model.

The $M_{500} - Y_X$ relation shows the scatter of only $\approx 7\%$. Note that this value of scatter includes clusters at both low and high-redshifts and both relaxed and unrelaxed systems. In fact, the scatter in $M_{500} - Y_X$ for relaxed and unrelaxed systems is indistinguishable within the errors. Y_X is therefore a robust mass indicator with remarkably low scatter in M_{500} for fixed Y_X, regardless of whether the clusters are relaxed or not. The redshift evolution of the $Y_X - M_{500}$ relation is also close to the simple self-similar prediction, which makes this indicator a very attractive observable for studies of cluster mass function with X-ray selected samples, because it indicates that the redshift evolution can be parameterized using a simple, well-motivated function.

Finally, the results of the simulations are compared to the observational results. In both relations, the observed clusters show a tight correlation with a slope close to the self-similar value. The normalization for our simulated sample agrees with the *Chandra* measurements to $\approx 10 - 15\%$. This is a considerable improvement given that significant disagreement existed just several years ago [12, 13]. The residual systematic difference in the normalization is likely caused by non-thermal pressure support from bulk gas motions, which is unaccounted for in X-ray hydrostatic mass estimates. The much improved agreement of simulations and observations in these relations gives us confidence that the clusters formed in modern simulations are sufficiently realistic and thus can be meaningfully used for interpretation of observations. The existence of tight relations of X-ray observables, such as Y_X, and total cluster mass and the simple redshift evolution of these relations hold promise for the use of clusters as cosmological probes.

References

1. Vikhlinin, A., Markevitch, M., Murray, S. S., Jones, C., Forman, W., & Van Speybroeck, L. 2005, ApJ, 628, 655

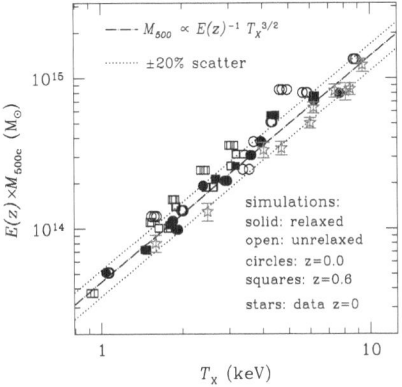

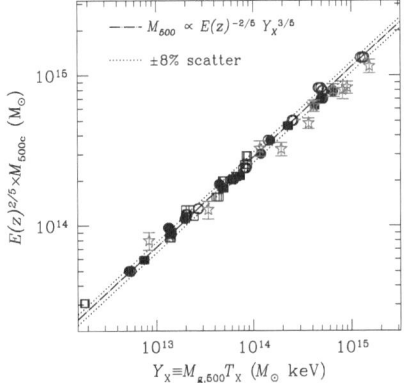

Fig. 4. Correlation between the total mass, M_{500c} and X-ray spectral temperature, T_X (*left panel*) and the integrated X-ray pressure, Y_X (*right panel*). Separate symbols indicate relaxed and unrelaxed simulated clusters, and also z=0 and 0.6 samples. The figures include points corresponding to three projections of each cluster. The *dot-dashed* lines are the power law relation with the self-similar slope fit for the sample of relaxed clusters. The *dotted* lines indicate 20% and 8% scatter, respectively. The data points with errorbars are Chandra measurements of nearby relaxed clusters.

2. Vikhlinin, A., Kravtsov, A., Forman, W., Jones, C., Markevitch, M., Murray, S., & Van Speybroeck, L. 2006, ApJ, 640, 691
3. Nagai, D., Kravtsov, A.V., & Vikhlinin, A., 2006, ApJ, in press (astro-ph/0609247)
4. Faltenbacher, A., Kravtsov, A. V., Nagai, D., & Gottlöber, S. 2005, MNRAS, 358, 139

5. Rasia, E., Ettori, S., Moscardini, L., Mazzotta, P., Borgani, S., Dolag, K., Tormen, G., Cheng, L. M., & Diaferio, A. 2006, MNRAS, 369, 2013

6. Lau, E., Kravtsov, A. V., & Nagai, D. 2006, in preparation

7. Mazzotta, P., Rasia, E., Moscardini, L., & Tormen, G. 2004, MNRAS, 354, 10

8. Vikhlinin, A. 2006, ApJ, 640, 710

9. Voit, G. M., Bryan, G. L. 2001, Nature, 2001, 414, 425

10. Kravtsov, A. V., Vikhlinin, A. A., & Nagai, D. 2006, ApJ, 650, 128

11. Mathiesen, B. F. & Evrard, A. E. 2001, ApJ, 546, 100

12. Finoguenov, A., Reiprich, T. H., & Böhringer, H. 2001, A&A, 368, 749

13. Pierpaoli, E., Borgani, S., Scott, D., & White, M. 2003, MNRAS, 342, 163

Observing Metallicity in Simulated Clusters with X-MAS2

E. Rasia[1,2], P. Mazzotta[3], H. Bourdin[3], S. Ettori[4], S. Borgani[5], K. Dolag[6], L. Moscardini[2], J.L. Sauvageot[7] and L. Tornatore[8]

[1] Department of Physics, 450 Church St., University of Michigan, Ann Arbor, MI 48109-1120 USA; rasia@umich.edu
[2] Dipartimento di Astronomia, Università di Bologna, via Ranzani 1, I-40127 Bologna, Italy; lauro.moscardini@unibo.it
[3] Dipartimento di Fisica, Università di Roma Tor Vergata, via della Ricerca Scientifica 1, I-00133 Roma, Italy; pasquale.mazzotta@roma2.infn.it, herve.bourdin@roma2.infn.it
[4] INAF, Osservatorio Astronomico di Bologna, via Ranzani 1, I-40127 Bologna, Italy; stefano.ettori@bo.astro.it
[5] Dipartimento di Astronomia, Università di Trieste, via Tiepolo 11, I-34131 Trieste, Italy; borgani@ts.astro.it
[6] Max-Planck-Institut für Astrophysik, Karl-Schwarzschild Strasse 1, D-85748 Garching bei München, Germany; kdolag@mpa-garching.mpg.de
[7] CEA, DSM, DAPNIA, Service d'Astrophysique, CE Saclay, 91191, Gif-sur-Yvette Cedex, France; jsauvageot@cea.fr
[8] SISSA, via Beirut 4, 34014, Trieste, Italy; torna@sissa.it

1 X-MAS2

X-MAS2 is the new version of the X-ray Map Simulator (X-MAS, [1, 2]). The tool simulates X-ray observations of hydro simulated galaxy clusters and consists of two main parts: firstly it generates differential flux maps, and after simulates a particular X-ray detector observation. Compared with the previous version: in the **first unit** we basically i) modify the distribution technique, ii) improve the algorithm increasing the computational velocity, iii) include the treatment of metal lines, and iv) the possibility to use APEC/VAPEC model in addition to MEKAL/VMEKAL model to calculate the emissivity of each particle. The **second part** has been more extensively changed in order to simulate Chandra observations in ACIS-I mode and XMM-Newton observations with all the EPIC cameras (MOS1, MOS2 and PN).

We present here the first results based on 3 simulated clusters, a more extensive work will appear in a forthcoming paper.

2 Simulated cluster sample

The simulations have been carried out with GADGET-2 [3] and include radiative cooling, heating by an ultraviolet background, and a treatment of star formation, feedback processes and metal enrichment from Type Ia and II Supernovae. The analytical fitting formulas for stellar yields of the two types of Supernovae and Planetary Nebulae are provided by [4]. The formulation for the supernovae Ia rate has been calculated as in [5]. Besides H and He the current version of the code follows the production of Fe, C, N, O, Si, Mg, and S [6]. The three clusters can be classified as Hot-Relaxed, Hot-Perturbed and Cold-Perturbed objects. The results are very similar from cluster to cluster. The plots refer to the first one. In the left panel of Fig. 1 we present the metal map as obtained directly from the simulation.

3 Spectral Analysis

Thanks to X-MAS2 we produce EPIC MOS1, MOS2 and PN event files for each cluster (right panel of Fig.1 for an example of MOS1 and PN images). In all the cases we mask the inner region in order to avoid the influence of the central cooling region, moreover we select and exclude all the subclumps clearly detectable from the X-ray images. We extract the spectra from circular annuli, rebin and analyze them in the [0.5 8] keV energy band. A thermal (VMEKAL) model absorbed by the Galactic column density is fitted to the data by using the C-STAT statistic in the XSPEC package; the free parameters are: temperature, normalization, oxygen, magnesium, silicium and iron.

4 Results

We compare in Fig. 2 the values of temperature, iron, silicium, magnesium and oxygen obtained from the spectra with the input of our simulation.

Temperature and Iron. The spectroscopic temperature and the spectroscopic iron agree well with the input values (spectroscopic-like weighted temperature, [7], and emission-weighted iron). Some points are more than 3 σ discrepant due to the high number of counts per bin (10^5 for MOS1-MOS2 and 2 10^5 for PN). The relative difference is always smaller than 5% for the temperature and 20% for the iron.

Silicium and Magnesium are more difficult to detect, thus their error bars are bigger. However there is a clear indication that they follow the input profile.

Oxygen The spectroscopic values are often greater than the input, particularly in the external regions. This could be caused by the presence of small cold blobs along the line of sight. In fact if only 1/100 of the emission of a T_1=10 keV plasma is due to a second component of T_2=0.2 keV, the spectroscopic detection of oxygen is twice the input.

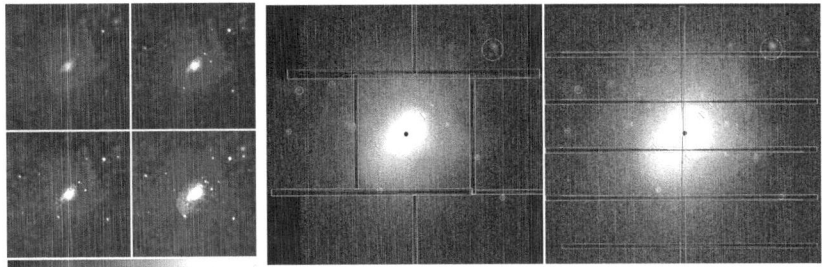

Fig. 1. Left: Fe, Si, Mg, O maps (clockwise order, starting from top-left); Right: MOS1 and PN images

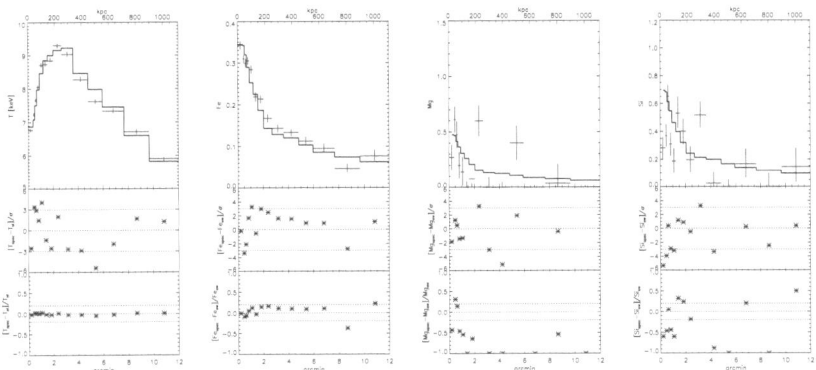

Fig. 2. Comparison between simulated and spectroscopic values of temperature, Fe, Mg and Si.

References

1. A. Gardini, E. Rasia, P. Mazzotta, G. Tormen, S. De Grandi, L. Moscardini, 2004, *MNRAS*, **351**, 505
2. E. Rasia, et al., 2006, *MNRAS*, **369**, 2013
3. V. Springel, 2005, *MNRAS*, **364**, 1105
4. S. Recchi, F. Matteucci, A. D'Ercole, 2001, *MNRAS*,**322**,800
5. F. Matteucci, 2001, *Nature*, **414**, 253
6. L. Tornatore, S. Borgani, F. Matteucci, S. Recchi, P, Tozzi, 2004, *MNRAS*, **349**, 19
7. P. Mazzotta, E. Rasia, L. Moscardini, G. Tormen, 2004, *MNRAS*, **354**, 10

N-body + Hydrodynamical Simulations of Merging Clusters of Galaxies: Comparison with 1E 0657-56

M. Takizawa

Department of Physics, Yamagata University, Kojirakawa-machi 1-4-12,
Yamagata 990-8560, Japan takizawa@sci.kj.yamagata-u.ac.jp

Summary. We investigate the X-ray and mass distribution in the merging galaxy cluster 1E 0657-56 . A clear off-set of an X-ray peak from a mass peak [1] is first reproduced in the N-body + hydrodynamical simulations. We estimate the ram pressure-stripping conditions of the substructure in mergers of two NFW dark halos using a simple analytical model. The characteristic X-ray and mass structures found in 1E 0657-56 suggest that neither the ram pressure nor the gravitational bound force overwhelms the other.

1 Introduction

Merging clusters are the sites of structure formation in the universe that can be investigated in detail via different types of observations. 1E 0657-56 is one of the most well-known examples of a merging cluster. There are two peaks in both the X-ray surface brightness distribution [2] and galaxy distribution [3], but their positions do not agree with each other. Recently, the mass distribution in 1E 0657-56 was investigated through weak gravitational lensing [1]. They show clear offsets of the mass density peaks from the X-ray peaks, and that the mass distribution is quite similar to the galaxy one. They claim that this structure occurs because the intracluster medium (ICM) experiences ram pressure but dark matter (DM) and galaxies do not. Although the above-mentioned naive ram pressure-stripping scenario seems to be correct, such characteristic off-sets have never been reported in previous simulations. In this paper, we show the first results that successfully reproduce such characteristic structures in the simulations, and discuss their implications using a simple analytical model [4].

2 The Simulations

We use the Roe TVD scheme to solve the hydrodynamical equations. Gravitational forces are calculated by the PM method with the standard FFT technique. We consider mergers of two virialized subclusters with an NFW density profile [5] in the ΛCDM universe for DM. DM masses of the larger and smaller subclusters are $1.00 \times 10^{15} M_\odot$ and $6.25 \times 10^{13} M_\odot$, respectively.

The initial density profiles of the ICM are assumed to be those of a beta-model. The gas mass fraction is set to be 0.1 inside the virial radius of each subcluster.

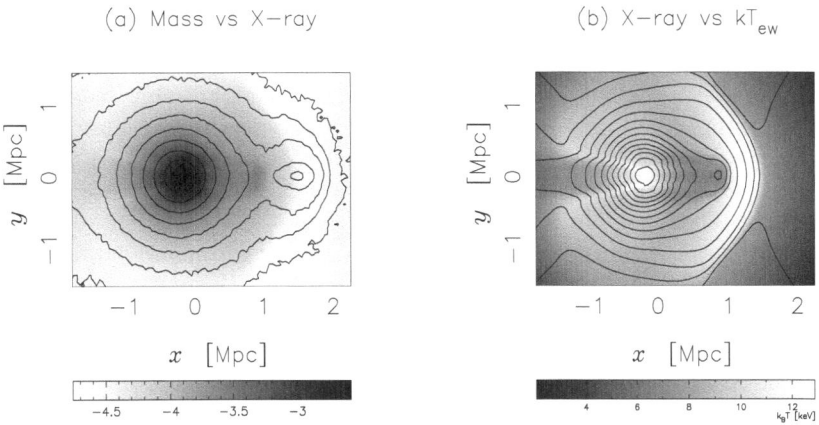

Fig. 1. (a)Projected total surface mass density (contours) overlaid with X-ray surface brightness distribution (grayscale) at a time of 0.67 Gyr after the core passage. (b)X-ray surface brightness distribution (contours) overlaid with emissivity-weighted temperature distribution (grayscale) at the same epoch.

Figure 1(a) shows the X-ray surface brightness distribution (grayscale) and projected total surface mass density (contours) at a time of 0.67 Gyr after the passage of the subcluster through the core of the larger one. A clear off-set of the mass density peak from the X-ray peak is seen for the smaller subcluster remnant. This is clearly because the ICM in the smaller subcluster is lagged by the ram pressure. Figure 1(b) shows the X-ray surface brightness distribution (contours) overlaid with the emissivity-weighted temperature distribution (grayscale) at the same epoch. A weak jump in the X-ray surface brightness distribution at $x \simeq 1.5$ (near the smaller mass peak) is a bow shock. A more prominent jump in the X-ray brightness just in front of the smaller X-ray peak is a contact discontinuity, which will be recognized as a cold front in actual X-ray observations. As for the overall ICM and DM structures of 1E 0657-56 around the smaller western X-ray and mass peak, our results agree qualitatively with the observations.

3 Discussion on the Ram Pressure-Stripping Conditions

Let us discuss the ram pressure-stripping conditions in merger of two clusters with an NFW DM density profile. If the gravity on the subcluster's ICM is weaker than the ram pressure force in unit volume, the ICM will be stripped from the substructure potential. This stripping condition becomes (see [4] for detailed calculations.)

$$F(\alpha : M_1) \equiv \alpha^{2/3-w}\frac{1+\alpha^{1/3}}{1+\alpha} - \frac{3A}{2g(\alpha M_1)c(\alpha M_1)} < 0, \qquad (1)$$

where α, c, g, A, and M_1 are the mass ratio, concentration parameter, ratio of the mass inside the scale radius to the virial mass, fudge factor of an order of unity, and mass of the larger cluster, respectively. Figure 2 shows the function $F(\alpha : M_1)$ for $M_1 = 1.0 \times 10^{15} M_\odot$ in the ΛCDM universe. The solid, dashed, and dot-dashed lines represent the cases of $A = 0.6, 0.4,$ and 0.2, respectively. In any case, α is less than ~ 0.1 when $F(\alpha) < 0$. It is interesting that this criteria is close to the mass ratio of our simulations where the clear off-set appears. Obviously, such an off-set does not appear at all if the ram pressure-stripping does not work effectively. On the other hand, if the ram pressure overwhelms the gravity, we will see a mass peak associated with no X-ray peak. Clear off-sets in the simulation and 1E 0657-56 suggest that the parameter α is close to the above-mentioned critical value.

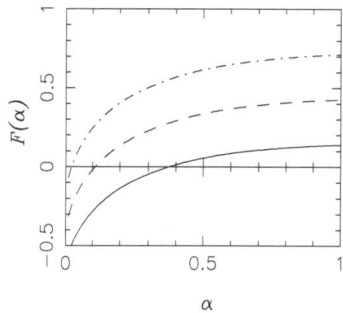

Fig. 2. Function $F(\alpha : M_1)$ defined by equation (1) for $M_1 = 1.0 \times 10^{15} M_\odot$ in the ΛCDM universe. The solid, dashed, and dot-dashed lines represent the cases of $A = 0.6, 0.4,$ and 0.2, respectively.

References

1. D. Clowe, A. Gonzalez, M. Markevitch, 2004, ApJ, 604, 596
2. M. Markevitch, A.H. Gonzalez, L. David et al., 2002, ApJ, 567, L27
3. R. Barrena, A. Biviano, M. Ramella et al., 2002, A&A, 386, 816
4. M. Takizawa, PASJ, in press, astro-ph/0608693
5. J.F. Navarro, C.S. Frenk, S.D.M White, 1997, ApJ, 490, 493

Cosmological Structure Formation Shocks and Cosmic Rays in Hydrodynamical Simulations

C. Pfrommer[1], V. Springel[2], T.A. Enßlin[2], and M. Jubelgas[2]

[1] Canadian Institute for Theoretical Astrophysics, University of Toronto, 60 St. George Street, Toronto, Ontario, M5S 3H8, Canada
pfrommer@cita.utoronto.ca
[2] Max-Planck-Institut für Astrophysik, Karl-Schwarzschild-Straße 1, Postfach 1317, 85741 Garching, Germany

Summary. Cosmological shock waves during structure formation not only play a decisive role for the thermalization of gas in virializing structures but also for the acceleration of relativistic cosmic rays (CRs) through diffusive shock acceleration. We discuss a novel numerical treatment of the physics of cosmic rays in combination with a formalism for identifying and measuring the shock strength on-the-fly during a smoothed particle hydrodynamics simulation. In our methodology, the non-thermal CR population is treated self-consistently in order to assess its dynamical impact on the thermal gas as well as other implications on cosmological observables. Using this formalism, we study the history of the thermalization process in high-resolution hydrodynamic simulations of the Lambda cold dark matter model. Collapsed cosmological structures are surrounded by shocks with high Mach numbers up to 1000, but they play only a minor role in the energy balance of thermalization. However, this finding has important consequences for our understanding of the spatial distribution of CRs in the large-scale structure. In high resolution simulations of galaxy clusters, we find a low contribution of the averaged CR pressure, due to the small acceleration efficiency of lower Mach numbers of flow shocks inside halos and the softer adiabatic index of CRs. These effects disfavour CRs when a composite of thermal gas and CRs is adiabatically compressed. However, within cool core regions, the CR pressure reaches equipartition with the thermal pressure leading, to a lower effective adiabatic index and thus to an enhanced compressibility of the central intracluster medium. This effect increases the central density and pressure of the cluster, and thus the resulting X-ray emission and the central Sunyaev-Zel'dovich flux decrement. The integrated Sunyaev-Zel'dovich effect, however, is only slightly changed.

1 Motivation

Cosmological shock waves form abundantly in the course of structure formation, both due to infalling cosmic plasma which accretes onto filaments, sheets and halos, as well as due to supersonic flows associated with merging substructures. Additionally, shock waves in the interstellar and intracluster media can be powered by non-gravitational energy sources, e.g. as a result of supernova explosions. Cosmologically, shocks are important in several respects for the thermal gas as well as for CR populations. (1) Shock waves dissipate gravitational energy associated with hierarchical clustering into thermal energy of the gas contained in dark matter halos, thus supplying the intra-halo medium with entropy and thermal pressure support: where and when is the gas heated to its present temperatures, and which shocks are mainly responsible for it? (2) Shocks also occur around moderately overdense filaments, heating the intragalactic medium. Sheets and filaments are predicted to host a warm-hot intergalactic medium with temperatures in the range $10^5 \, \mathrm{K} < T < 10^7 \, \mathrm{K}$ whose evolution is primarily driven by shock heating from gravitational perturbations developing into mildly nonlinear, non-equilibrium structures. Thus, the shock-dissipated energy traces the large scale structure and contains information about its dynamical history. (3) Besides thermalization, collisionless shocks are also able to accelerate ions through diffusive shock acceleration. These energetic ions are reflected at magnetic irregularities through magnetic resonances between the gyro-motion and waves in the magnetized plasma and are able to gain energy in moving back and forth through the shock front: what are the cosmological implications of such a CR component, and does this influence the cosmic thermal history? (4) Simulating realistic CR distributions within galaxy clusters will provide detailed predictions for the expected radio synchrotron and γ-ray emission. What are the observational signatures of this radiation that is predicted to be observed with the upcoming new generation of γ-ray instruments and radio telescopes?

To date it is unknown how much pressure support is provided by CRs to the thermal plasma of clusters of galaxies. A substantial CR pressure contribution might have a major impact on the properties of the intracluster medium (ICM) and potentially modify thermal cluster observables such as the X-ray emission and the Sunyaev-Zel'dovich (SZ) effect. In contrast, CR protons play a decisive role within the interstellar medium our own Galaxy. CRs and magnetic fields each contribute roughly as much energy and pressure to the galactic ISM as the thermal gas does. CRs trace past energetic events such as supernovae, and they reveal the underlying structure of the baryonic matter distribution through their interactions. CRs behave quite differently compared to the thermal gas. Their equation of state is softer, they are able to propagate over macroscopic distances, and their energy loss time-scales are typically larger than the thermal ones. Therefore, CR populations provide an important reservoir for the energy from supernova explosions or structure formation shock waves, and thereby help to maintain dynamical feedback for periods longer than thermal gas physics alone would permit.

2 Structure formation shock waves and cosmic rays

We have developed a formalism that is able to measure the shock strength instantaneously during an smoothed particle hydrodynamics (SPH) simulation [1]. The method is applicable both to non-relativistic gas, and to plasmas

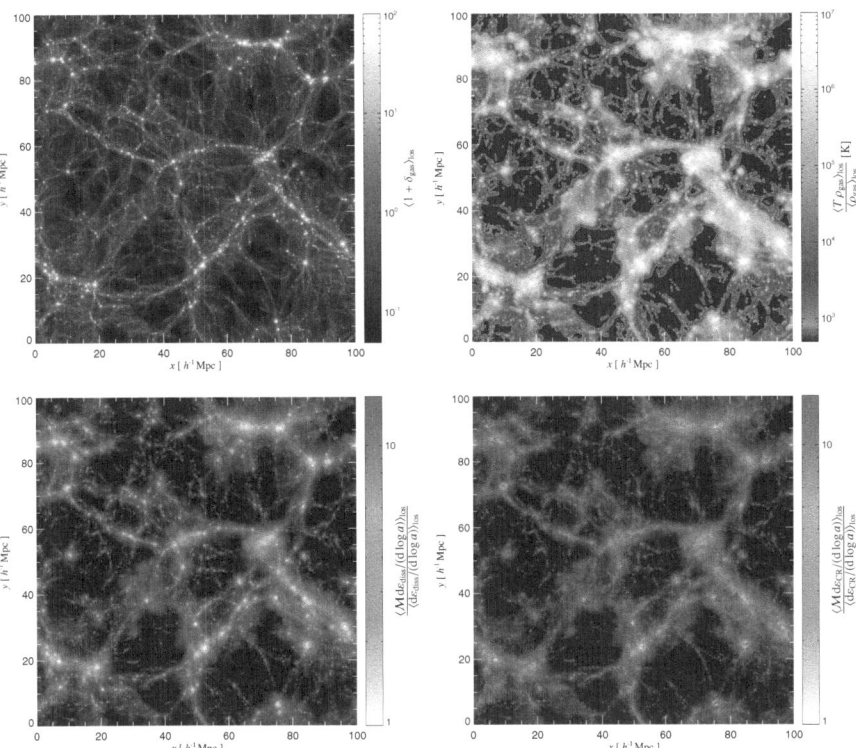

Fig. 1. Visualization of a non-radiative cosmological simulation at redshift $z = 0$ where the cosmic ray (CR) energy injection was only computed while the effect of the CR pressure on the dynamical evolution was not taken into account. The *top panels* show the overdensity of the gas and the mass weighted temperature of the simulation. The *bottom panels* show a visualization of the strength of structure formation shocks. The colour hue of the map on the left-hand side encodes the spatial Mach number distribution weighted by the rate of energy dissipation at the shocks. The map on the right-hand side shows the Mach number distribution weighted by the rate of CR energy injection above the momentum threshold of hadronic CR p-p interactions. The brightness of each pixel is determined by the respective weights, i.e. by the energy production density. Most of the energy is dissipated in weak shocks which are situated in the internal regions of groups or clusters, while collapsed cosmological structures are surrounded by strong external shocks (shown in blue). Since strong shocks are more efficient in accelerating CRs, the CR injection rate is more extended than the dissipation rate of thermal energy.

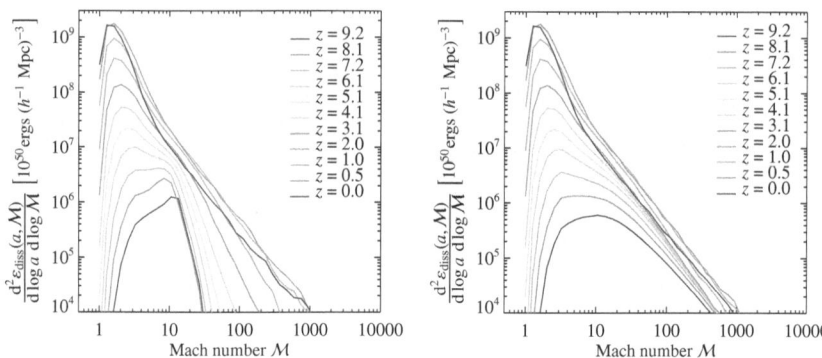

Fig. 2. Influence of reionisation (at redshift $z = 10$) on the Mach number statistics of non-radiative cosmological simulations. The figure on the *left-hand side* shows the differential Mach number distribution $\mathrm{d}^2\varepsilon_{\mathrm{diss}}(a, \mathcal{M})/(\mathrm{d}\log a\,\mathrm{d}\log\mathcal{M})$ for our simulation with reionisation while the figure on the *right-hand side* shows this distribution for the simulation without reionisation. Strong shocks are effectively suppressed due to an increase of the sound velocity after reionisation.

composed of CRs and thermal gas. We apply our methods to study the properties of structure formation shocks in high-resolution hydrodynamic simulations of the Lambda cold dark matter (ΛCDM) model using an extended version of the distributed-memory parallel TreeSPH code GADGET-2 [2] which includes self-consistent CR physics ([3], [4]). Fig. 1 shows the spatial distribution of structure formation shocks in comparison to the density and temperature distribution while Fig. 2 shows the cosmological Mach number distribution at different redshifts.[3]

The main results are as follows. (1) Most of the energy is dissipated in weak shocks internal to collapsed structures while collapsed cosmological structures are surrounded by external shocks with much higher Mach numbers, up to $\mathcal{M} \sim 1000$. Although these external shocks play a major role locally, they contribute only a small fraction to the global energy balance of thermalization. (2) More energy per logarithmic scale factor and volume is dissipated at later times while the mean Mach number decreases with time. This is because of the higher pre-shock gas densities within non-linear structures, and the significant increase of the mean shock speed as the characteristic halo mass grows with cosmic time. (3) A reionisation epoch at $z_{\mathrm{reion}} = 10$ suppresses efficiently strong shocks at $z < z_{\mathrm{reion}}$ due to the associated increase of the sound speed after reionisation. (4) Strong accretion shocks efficiently inject CRs at the cluster boundary. This implies that the dynamical importance of shock-injected CRs is comparatively large in the

---

[3] Note that we corrected a missing factor 10 in the normalization of Fig. 6 in [1].

low-density, peripheral halo regions, but is less important for the weaker flow shocks occurring in central high-density regions of halos.

3 Cosmic rays in hydrodynamic cluster simulations

To study the impact of CRs on cluster scales, we performed cosmological high-resolution hydrodynamic simulations of a sample of galaxy clusters spanning a large range in mass and dynamical states, with and without CR physics. These clusters have originally been selected from a low-resolution dark-matter-only simulation of a flat ΛCDM model and then re-simulated using the 'zoomed initial conditions' technique. We account for CR acceleration at structure formation shocks and consider CR loss processes such as their thermalization by Coulomb interactions and catastrophic losses by hadronic interactions with ambient gas protons (see [5] for details). Within clusters, the relative CR pressure $X_{CR} = P_{CR}/(P_{CR} + P_{th})$ declines towards a low central value of $X_{CR} \simeq 10^{-4}$ in non-radiative simulations due to a combination of the following effects: CR acceleration is more efficient at the peripheral strong accretion shocks compared to weak central flow shocks, adiabatic compression of a composite of CRs and thermal gas disfavours the CR pressure relative to the thermal pressure due to the softer equation of state of CRs, and CR loss processes are more important at the dense centres. Interestingly, X_{CR} reaches high values at the centre of the parent halo and each galactic substructure in our radiative simulation due to the fast thermal cooling of gas which diminishes thermal pressure support relative to that in CRs. This additional CR pressure support has important consequences for the thermal gas distribution at cluster centres and alters the resulting X-ray emission and the SZ effect significantly (cf. Fig.).

4 Conclusions

We studied the properties of cosmological shock waves using a technique that allows us to identify and measure the shock strength on-the-fly during an SPH simulation. Invoking a model for CR acceleration in shock waves, we have carried out the first hydrodynamical simulations that follows the CR physics self-consistently. These simulations show that it is crucial to consider the dynamical back-reaction of a non-thermal cosmic ray (CR) component in order to describe the intracluster medium reliably. The X-ray luminosity from galaxy clusters is boosted predominantly in low-mass cool core clusters due to the large CR pressure contribution in the centre that leads to a higher compressibility. The integrated Sunyaev-Zel'dovich effect is only slightly changed while the central SZ flux decrement is also increased. These CR-induced modifications can imprint themselves in changes of cluster scaling relations or modify their intrinsic scatter and thus change the effective mass threshold

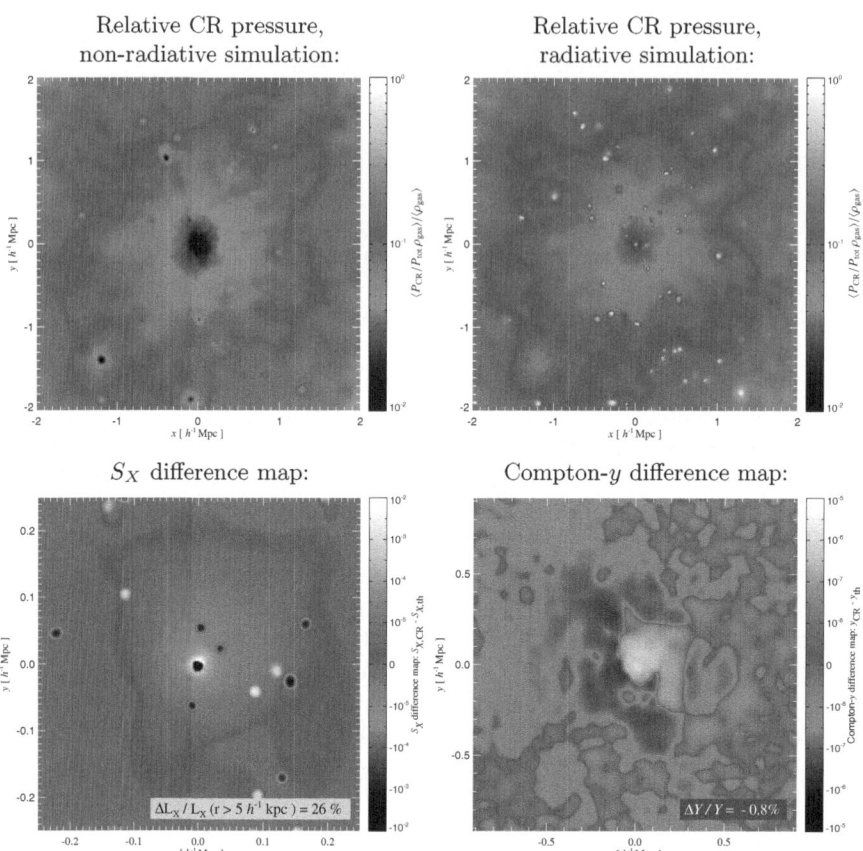

Fig. 3. The top panels show a visualization of the pressure contained in CRs relative to the total pressure $X_{CR} = P_{CR}/(P_{CR} + P_{th})$ in a zoomed simulation of an individual galaxy cluster with mass $M = 10^{14} h^{-1} M_\odot$. The map on the *left-hand side* shows a non-radiative simulation with CRs accelerated at structure formation shock waves while the map on the *right-hand side* is from a simulation with dissipative gas physics including cooling, star formation, supernova feedback, and structure formation CRs. The lower panels show the CR-induced difference of the X-ray surface brightness S_X (*left-hand side*) and the Compton-y parameter (*right-hand side*) in a radiative simulation with structure formation CRs compared to the corresponding reference simulation without CRs. The relative difference of the integrated X-ray surface brightness/Compton-y parameter is given in the inlay. Within cool core regions, the CR pressure reaches equipartition with the thermal pressure, an effect that increases the compressibility of the central intracluster medium and thus the central density and pressure of the gas. This boosts the X-ray luminosity of the cluster and the central Sunyaev-Zel'dovich decrement while the integrated Sunyaev-Zel'dovich effect remains largely unaffected.

of X-ray or SZ surveys. Neglecting such a CR component in reference simulations can introduce biases in the determination of cosmological parameters.

References

1. C. Pfrommer, V. Springel, T.A. Enßlin and M. Jubelgas, 2006, MNRAS, 367,113
2. V. Springel, 2005, MNRAS, 364, 1105
3. T.A. Enßlin, C. Pfrommer, V. Springel and M. Jubelgas, 2006, astro-ph/0603484
4. M. Jubelgas, V. Springel, T.A. Enßlin and C. Pfrommer, 2006, astro-ph/0603485
5. C. Pfrommer, T.A. Enßlin, V. Springel, M. Jubelgas, and K. Dolag, 2006, in prep.

Determination of Cosmological Parameters using XMM-Newton Observations of the HIFLUGCS Cluster Sample

O. Nenestyan and T.H. Reiprich

Argelander-Institute for Astronomy, Bonn, Germany,
oxana@astro.uni-bonn.de, reiprich@astro.uni-bonn.de

1 Introduction

Determining the past, present and future of our Universe was and remains the major challenge in cosmology. Several independent methods exist to achieve this goal, each of them having its positive features but also drawbacks: e.g., optical observations of distant supernovae, microwave measurements of CMB fluctuations and X-ray observations of galaxy clusters. In this contribution we have presented our approach in trying to contribute to the solution of this problem using XMM-Newton observations of galaxy clusters.

2 Method

Our studies are based on XMM-Newton observations of a highly complete and flux-limited sample of the brightest galaxy clusters (*HIFLUGCS*), which comprises 64 objects ($\bar{z} = 0.05$, 3 clusters with $z > 0.1$). 60 of them have been observed, the remaining 4 with low fluxes have been proposed. The sample construction was done using *RASS*, which assures the homogeneity of the cluster sample. To ensure a high completeness of the sample, necessary for an accurate determination of the cluster mass function, the region of ± 20 deg from the Galactic plane, and also the areas corresponding to Magellanic Clouds and Virgo galaxy cluster were excluded.

Using the best sample existing up to now, in terms of homogeneity, completeness, high quality X-ray observations and size, the main goal of our work is to determine the cosmological parameters (mean matter density of the Universe, Ω_m, and the amplitude of the initial density fluctuations power spectrum, σ_8), and to set a precise local baseline for studying the cluster evolution. An accurate determination of these parameters requires a highly complete cluster sample, and very acurate mass determinations of the galaxy clusters. Because of a tight correlation between X-ray luminosity and total gravitational mass ([1], [2]), X-ray flux-limited samples of galaxy clusters are very appropriate for our purpose. Under the assumption that the intracluster gas is in hydrostatic equilibrium:

$$\frac{dP_{\mathrm{gas}}(r)}{dr} = \frac{-\rho_{\mathrm{gas}}(r)GM_{\mathrm{tot}}(<r)}{r^2}, \tag{1}$$

and considering the ideal gas equation as the equation of state:

$$P_{\mathrm{gas}}(r) = \frac{k_{\mathrm{B}}}{\mu m_p}\rho_{\mathrm{gas}}T_{\mathrm{gas}}, \tag{2}$$

(where P_{gas} represents the gas pressure, G the gravitational constant, M_{tot} the total gravitational mass) the total gravitational masses of the clusters can be determined:

$$M_{\mathrm{tot}}(<r) = \frac{-k_{\mathrm{B}}T_{\mathrm{gas}}(r)r^2}{\mu m_p G}\left(\frac{1}{\rho_{\mathrm{gas}}(r)}\frac{d\rho_{\mathrm{gas}}(r)}{dr} + \frac{1}{T_{\mathrm{gas}}(r)}\frac{dT_{\mathrm{gas}}(r)}{dr}\right). \tag{3}$$

The XMM-Newton observations allow us to determine $\rho_{\mathrm{gas}}(r)$ and $T_{\mathrm{gas}}(r)$. Using these masses we will construct a cluster mass function, and by comparison to a theoretical one, for instance predicted by Press-Schechter formalism :

$$\frac{dn(M)}{dM} = \sqrt{\frac{2}{\pi}}\frac{\rho_0}{M}\frac{\delta_c(z)}{\sigma(M)^2}\left|\frac{d\sigma(M)}{dM}\right|\exp\left(-\frac{\delta_c(z)^2}{2\sigma(M)^2}\right), \tag{4}$$

(or, e.g., the Jenkins mass function [3],...) tight constraints on Ω_{m} and σ_8 can be obtained.

3 Current status

X-ray data reduction entails many steps (point source subtraction, filtering the events files from high flare periods and creation of the GTI's, creating images, spectra, response files, radial temperature and surface brightness profiles, etc), and we developed automatic procedures for most of them.

At present the major concern of our work is to find a reliable method for the background subtraction. Because nearby galaxy clusters fill the entire field of view, it is impossible for us to use the double background subtraction method to correct for the cosmic X-ray background (CXB). We are currently testing different methods to correct for the CXB using the highest redshift clusters of our sample: A2204 and A2163, for which we can apply the double background subtraction method. Some preliminary results are shown in Fig.1 where the two symbols correspond to the temperature profiles obtained by using the double background subtraction method (triangles) respectively using a simple thermal model for the CXB (circles). We also experimented with a version of the Snowden & Kuntz method [4] for one of the REFLEX clusters, A1689. The preliminary results shown in Fig.2 look promising, but still more work is required to constrain the temperature profiles reliably.

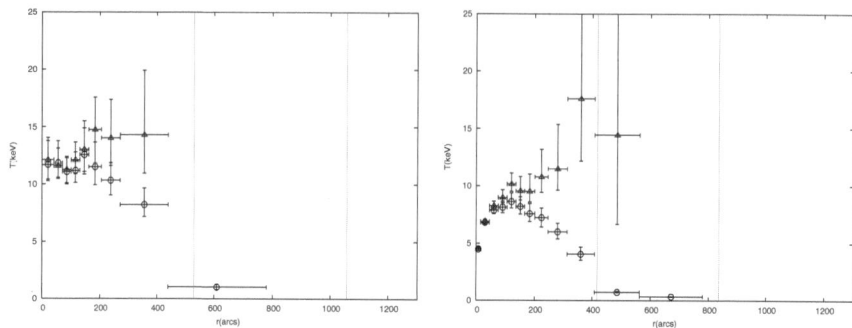

Fig. 1. Preliminary temperature profiles of A2163 and A2204 using the double background subtraction method (triangles) and simple modelling of the CXB (circles). The vertical lines indicate the virial radius (determined from the Rosat data) and half of the virial radius.

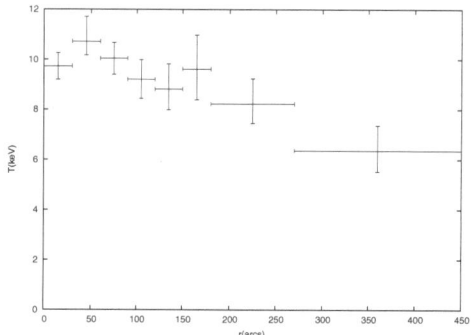

Fig. 2. Preliminary temperature profile of A1689 obtained by applying a version of the Snowden & Kuntz method.

References

1. Reiprich, T. & Böhringer, H. 2002, ApJ 567, 716
2. Reiprich, T. 2006, A&A, 453, L39
3. Jenkins, A., Frenk, C.S., White S.D.M., Kolberg, J.M., Cole, S., Evrard, A.E., Couchman, H.M.P., Yoshida, N. 2001, MNRAS, 321, 372
4. Snowden, S.L., Kuntz, K.D., Cookbook for analysis procedures for XMM-Newton EPIC MOS observations of extended objects and the diffuse background, 2006

Complex Physics in Cluster Cores: Showstopper for the Use of Clusters for Cosmology?

T.H. Reiprich and D. S. Hudson

Argelander-Institut für Astronomie, Universität Bonn, Auf dem Hügel 71, 53121 Bonn, Germany `thomas@reiprich.net`, `dhudson@astro.uni-bonn.de`

1 Introduction

Galaxy clusters can be rather messy objects, e.g. [7]. Why should one use them to help solve pressing cosmological problems, especially about the nature of dark matter and dark energy? Are there not cleaner probes for this purpose?

First of all, the possible implications of dark energy, e.g., a modification of the fundamental gravity law or an introduction of a fifth force, are too far-reaching that we could afford to rely on just one single method: several independent observational methods are necessary if our picture of the universe is to be changed dramatically. Secondly, measurements of the primary anisotropies in the cosmic microwave background are not sensitive to any evolution of the equation of state of dark energy. Thirdly, it would appear that we may be able to simulate relevant physical processes in galaxy clusters actually *more* realistically than, e.g., in galaxies or supernovae. That is, clusters may indeed be *relatively* simple and clean probes. Fourthly, with purely geometric tests, e.g., using supernovae as standard candles, we cannot differentiate between, e.g., quintessence and a possible breakdown of general relativity. This can, however, be achieved with tests based on structure growth, e.g., the evolution of the galaxy cluster mass function.

Moreover, clusters are unique cosmological probes in the sense that there are many, more or less independent methods to constrain cosmological parameters with clusters and basically all wavelengths can be used to study clusters. Tests include, e.g., cluster baryons (fraction and its *apparent* evolution), power spectrum (normalization, shape, and baryonic wiggles), mergers (frequency and its evolution), and mass function (normalization, shape, and evolution). Wavelengths to find and study clusters include, e.g., optical/infrared (galaxies, lensing), radio (Sunyaev–Zeldovich-effect, halos and relics, wide and narrow angle tailed galaxies), γ-rays (especially with future instruments like GLAST), and X-rays.

Finally, after a phase of skepticism, renewed trust in clusters seems to spread. Skepticism was in part caused by low values of $\sigma_8 \sim 0.7$ (for $\Omega_m = 0.3$) indicated early from cluster studies [1, 6, 9, 12], which seemed to be at variance with σ_8 values obtained from other probes, including the 1st year

WMAP data [10]. This has changed since the release of the 3rd year WMAP data, which now confirms the low σ_8 values [11]. Furthermore, the new best fit results from WMAP indirectly suggest that the intrinsic scatter and bias of cluster scaling relations like the X-ray luminosity–gravitational mass (L_X–M_{tot}) relation may be smaller than previously thought [8].

For future determinations of the evolution of the cluster mass function with the new generation of X-ray surveys (e.g., eROSITA is expected to detect about 100 000 clusters), primarily only X-ray luminosites will be available (gas temperatures only for a small subset of clusters). Therefore, we concentrate in this contribution on effects of cluster physics on the L_X–M_{tot} relation. And since this is a cooling flow conference, we concentrate on the influence of cool cores on this relation.

2 Cool cores and the luminosity–mass relation

As mentioned above, indirectly the WMAP 3rd year data require no large bias or intrinsic scatter in the L_X–M_{tot} relation. However, we would rather like to determine the intrinsic scatter directly from the data. This can be difficult because the measured scatter is a combination of statistical, systematic, and intrinsic scatter. So, a detailed understanding of all relevant systematic effects is required for a reliable determination of the intrinsic scatter. We are confident that the high quality cluster samples and state of the art data now available from Chandra, XMM-Newton, and Suzaku will be sufficient for a good estimate. We are currently working on this using the *HIFLUGCS* clusters. The preliminary results we show in this contribution are very closely related to other work that has been done recently with older data [5, 2].

HIFLUGCS contains the 64 X-ray brightest clusters in the sky excluding ± 20 deg around the Galactic plane and some small regions around the Magellanic clouds and the Virgo cluster. It is a complete X-ray flux-limited sample selected from deep surveys based on the ROSAT All-Sky Survey [6] (RB02). It is currently the best available sample in terms of homogeneous selection, size, completeness, representativeness, and full Chandra and (almost) XMM-Newton coverage. We are currently analyzing >120 Chandra and >100 XMM-Newton observations with a total exposure time approaching 7 Ms (see Fig. 1, and Hudson & Reiprich, these proceedings, and Nenestyan & Reiprich, these proceedings).

We are currently studying several methods to classify clusters as cool core (CC) and non-cool core (NCC) clusters with Chandra, including the slopes of the inner temperature and density profiles, central cooling times (the time the gas needs to cool below X-ray emitting temperatures), and central entropies. There is a large but not complete overlap between the results of these methods. Here we use a special "central" entropy to select CC (low entropy; i.e., high density and low temperature) and NCC (high entropy) clusters (see Hudson & Reiprich, these proceedings).

Fig. 1. Chandra observations for nine exemplary clusters in *HIFLUGCS*. All available observations and all usable CCDs are analyzed in order to maximize signal-to-noise ratio and field-of-view.

Now let us check if the two populations, CC and NCC clusters, behave differently in the L_X–M_{tot} diagram. For the nearby clusters in *HIFLUGCS*, L_X is best determined with ROSAT data due to its large field-of-view and low background. Gravitational masses have not, yet, been determined with Chandra or XMM-Newton for all *HIFLUGCS* clusters so we simply use the old masses determined from the ROSAT gas density profiles and overall (primarily ASCA) gas temperatures (RB02). Figure 2 (left) shows that the L_X–M_{tot} relation for CC clusters has a factor of 2.5 higher normalization (at $5 \times 10^{14} M_\odot$) than the relation for NCC clusters – the CC clusters segregate out to the high L_X (or low M_{tot}) side (see also Chen et al., these proceedings). Also, the CC clusters seem to exhibit smaller scatter around their best fit relation than the NCC clusters. This may be at variance with the results of O'Hara et al. [5] who found a *larger* scatter for CC clusters in the L_X–T_{gas} relation. Furthermore, it appears that all low mass clusters and groups in the sample have a cool core.

The factor 2.5 offset between the two best fit relations may indicate significant intrinsic scatter; i.e., since CC clusters have higher central densities and since X-ray emissivity is proportional to density squared, CC clusters may have significantly higher L_X for given M_{tot} compared to NCC clusters, if the central regions of CC clusters account for a very significant fraction of the total cluster luminosities.

On the other hand, systematic effects can play a role as well. If, e.g., cool cores bias overall cluster temperature estimates low compared to their virial temperatures then the estimated masses will be biased low, too. A possible

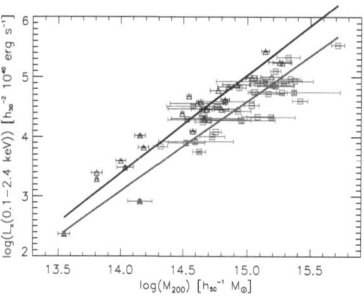 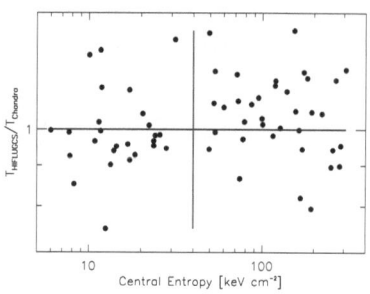

Fig. 2. *Left:* L_X vs. M_{tot} for the *HIFLUGCS* clusters (RB02). Blue triangles represent CC clusters and red squares NCC clusters as classified through the central entropy determined with Chandra. The best fit (bisector) relations show that CC clusters have a higher normalization. *Right:* Ratio of original *HIFLUGCS* and new (preliminary) Chandra temperatures vs. central entropy (CC clusters are to the left, NCC clusters to the right). Overall, there is quite good agreement between the temperature estimates; however, most of the CC clusters have a ratio below 1 while most of the NCC clusters have a ratio larger than 1.

mass bias is enhanced compared to a temperature bias because $M_{tot} \propto T_{gas}^{1.5}$ and the offset to be accounted for in M_{tot} direction is smaller then the offset in L_X direction because $M_{tot} \propto L_X^{1.4}$. So, even relatively small T_{gas} biases can have a significant effect on the L_X offset in the L_X–M_{tot} relation. Also other systematic differences between CC and NCC clusters with the potential of biasing simple mass estimates might be important, e.g., a difference between the steepening of the surface brightness profiles in the very outer CC/NCC cluster parts (e.g., Burns et al., these proceedings).

Many of the temperature estimates we used for the original mass determination in RB02 were, one way or another, "corrected" for cooling flows. So, we actually do not expect a very large bias. With the new preliminary Chandra temperature profiles for all *HIFLUGCS* clusters available (Hudson & Reiprich, these proceedings) it is straightforward to exclude thoroughly any cool core emission for overall temperature estimates. Figure 2 (right) shows the ratio of the original temperature estimates and the preliminary Chandra overall T_{gas} determinations as a function of central entropy. Clusters to the left in this diagram are CC clusters, those to the right NCC clusters. While in general there is very good agreement between the temperatures, one notes that most of the CC clusters have a ratio below 1 while most of the NCC clusters have a ratio larger than 1, indicating that indeed a small temperature bias is present. However, currently the magnitude of this effect alone does not seem large enough to account for all of the observed L_X offset. Soon the Chandra analysis will be completed (including the mass determination). We will then be able to derive very tight and robust limits on the intrinsic scatter in the L_X–M_{tot} relation.

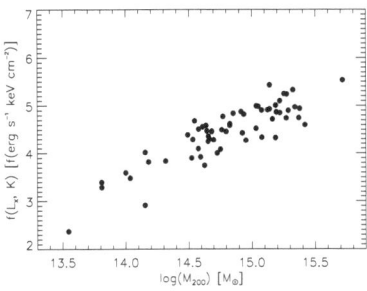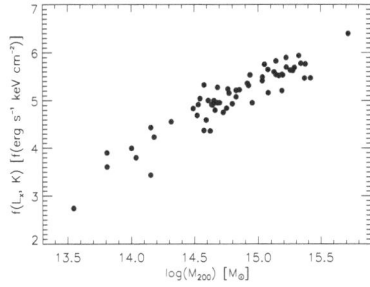

Fig. 3. *Left:* L_X vs. M_{tot} for the *HIFLUGCS* clusters (RB02). *Right:* Same as left but all luminosities multiplied by the central entropy to the power of 0.361, resulting in a reduction of scatter.

Having a continuous measure for the "strength" of a cool core one can try to include it as a scaling parameter in the L_X–M_{tot} relation; e.g., O'Hara et al. [5] used the central surface brightness for this purpose. Here we play with the central entropy. Figure 3 shows the L_X–M_{tot} relation again (left) and then, on the right, the same relation but all L_X values multiplied with the central entropy, K^α, and $\alpha = 0.361$ chosen such that scatter is minimized. And, indeed, such a scaling does reduce the scatter. Again, we will work this out in more detail once we are completely done with the Chandra analysis. The specific choice of using central entropy to reduce scatter will possibly only be of limited practical value because if the data are good enough to determine the central entropy then M_{tot} is probably better determined directly from the density and temperature profile than from the L_X–M_{tot} relation.

3 Summary

Soon we should be able to quantify robustly the intrinsic scatter in scaling relations directly from cluster data, eliminating the need to estimate it indirectly by comparison to other cosmological probes. Even if it turns out that cool cores cause a relatively large intrinsic scatter, it is straightforward to correct for the resulting effects in cosmological tests. So, cool cores do not appear to be a showstopper for using clusters for precision cosmology.

Something else that will be required in the near future from the X-ray cluster community is a coordinated effort to perform detailed consistency checks, similarly to what the weak lensing and simulation communities have already done [3, 4]. We are trying to do a first simple step in this direction by analyzing the *HIFLUGCS* sample independently with Chandra and XMM-Newton but a larger scale effort involving several more groups and also simulations is necessary to convince the general cosmology community that cluster systematics are sufficiently under control.

References

1. S. Borgani, et al., 2001, ApJ, 561, 13
2. Y. Chen, T. H. Reiprich, H. Böhringer, Y. Ikebe, and Y.-Y. Zhang, 2006, A&A, submitted.
3. C. S. Frenk, et al., 1999, ApJ, 525, 554
4. C. Heymans, 2006, MNRAS, 368, 1323
5. T. B. O'Hara, J. J. Mohr, J. J. Bialek, and A. E. Evrard, 2006, ApJ, 639, 64
6. T. H. Reiprich and H. Böhringer, 2002, ApJ, 567, 716
7. T. H. Reiprich, C. L. Sarazin, J. C. Kempner, and E. Tittley, 2004, ApJ, 608, 179
8. T. H. Reiprich, 2006, A&A, 453, L39
9. U. Seljak, 2001, MNRAS, 337, 769
10. D. N. Spergel, et al., 2003, ApJS, 148, 175
11. D. N. Spergel, et al., 2006, ApJ, subm. (astro-ph/0603449)
12. P. T. P. Viana, R. C. Nichol, and A. R. Liddle, 2002, ApJ, 569, L75

Part IX

Suzaku Observations

New Results with the XIS Onboard Suzaku

N. Ota[1], K. Matsushita[2], and the Suzaku cluster analysis team

[1] Cosmic Radiation Laboratory, RIKEN (Institute of Physical and Chemical Research), 2-1 Hirowasa, Wako, Saitama 351-0198, Japan ota@crab.riken.jp
[2] Department of Physics, Tokyo University of Science, 1-3 Kagurazaka, Shinjuku-ku, Tokyo 162-8601, Japan

Summary. We present new results from the Suzaku XIS observations of two nearby clusters of galaxies, the Centaurus cluster and the Fornax cluster. For the Centaurus cluster, we investigated the Doppler shift of the iron K lines to constrain the bulk velocity of the intracluster medium (ICM) in its core region. We found that there is no significant velocity gradient within the calibration uncertainty: the 90% upper limit on the line-of-sight velocity difference is $|\Delta v| < 1400$ km s^{-1}, providing a tighter constraint than the previous observations. Regarding the Fornax cluster, we studied metal abundances of the ICM out to a large radius. K-lines of O and Mg were clearly resolved and the abundance profiles of O, Mg, Si, S and Fe were derived to high accuracy. The central $r < 4'$ region shows the Fe, Si, and S abundances to be ~ 1 solar, while the Fe and Si abundances drop to about 0.5 solar in the outer region. O/Fe and Mg/Fe are about 0.4–0.5 and 0.7 in unit of the solar ratio, respectively, and do not show any significant radial gradient. Implications of the results on each cluster are briefly discussed.

1 Introduction

The 5th Japanese X-ray satellite "Suzaku" has been launched successfully in 2005 [6]. The satellite carries four X-ray telescopes (XRT; [9]) for the X-ray Imaging Spectrometer (XIS; [3]) and non-imaging hard X-ray detector (HXD; [10]). The XIS consists of four X-ray sensitive CCD cameras: three front-illuminated (FI) CCDs and one back-illuminated (BI) CCD. It also has a relatively wide ($18' \times 18'$) field of view. The effective area of the XRT+XIS system is about 1000 cm^2 in the 1–6 keV band and 550 cm^2 at 8 keV. The background level is the lowest among three large X-ray observatories currently in operation (Chandra, XMM-Newton, and Suzaku). Therefore the improved performance of the XIS in the $\sim 0.3 - 10$ keV band will offer unique opportunities for cluster studies. In §2 and §3, we show results from the Suzaku observations of the Centaurus cluster and the Fornax cluster, both of which have been carried out in the initial performance verification phase by the Suzaku Science Working Group.

2 Gas bulk motions in the Centaurus cluster

2.1 Objectives

Clusters of galaxies are thought to grow into larger systems through complex interactions between smaller systems. Numerical simulations predict the existence of ICM bulk flows with a velocity of ~ 1000 km s^{-1}, lasting several Gyr after each merging event [7]. If the ICM has a significant bulk velocity compared to its thermal velocity, the associated non-thermal pressure would endanger the assumption of hydrostatic equilibrium in cluster mass estimation. Therefore, the presence/absence of ICM bulk motion has a great impact on cluster studies. X-ray spectroscopy of 6.7 keV He–like iron K-lines provides the most sensitive way of investigating bulk flows. Since a line-of-sight velocity of 1000 km s^{-1} translates to an energy shift of 22 eV, we need not only sufficient photon statistics and good energy resolution but also precise instrumental gain calibrations. We thus performed a detailed analysis of the XIS data to constrain the bulk velocity in the Centaurus cluster [8]. The XIS has an excellent sensitivity at the iron K-line energies and the current accuracy of the in-orbit gain calibration is already high enough to derive a meaningful constraint.

2.2 Observation

The Centaurus cluster ($z = 0.0104$) is a nearby X-ray bright cluster hosting the cD galaxy, NGC 4696. We observed the cluster with the XIS in three pointings in 2005 December: one centered on the cD galaxy and the other two offset by $\pm 8'$ in declination (Figure 1). The net exposures are 29.4, 33.4, and 31.6 ksec for the central, southern, and northern fields, respectively. As detailed in [8], based on a careful examination of the position dependence of the instrumental gain, the 1σ systematic error on the energy scale is estimated to be ± 10 eV in terms of the iron-K line energy.

2.3 Data analysis and results

To study the velocity structure of the ICM on arc-minute scales, we divided the central $18' \times 18'$ field into 8×8 cells (Figure 1). We co-added the spectra from the three FI-chips, and merged the three pointings where they overlap. Excluding the calibration source regions, 52 cells in total were used in the spectral fitting. Figure 2 shows an example of the 5–10 keV XIS spectrum. We observe a very intense He-like Fe-Kα line, a weaker H-like Fe-Kα line, and a blend between He-like Ni-Kα and He-like Fe-Kβ lines.

The spectra were fitted with the APEC model with metal abundances reset to zero for a continuum emission and three gaussian lines, which resulted in acceptable fits in almost all the cells. The line centroids were constrained with 1σ statistical errors of 3–4 eV for the inner 2×2 cells and 4–10 eV for the surrounding cells. As shown in Figure 3, the measured redshifts are consistent, in almost all the 52 cells, with the optical value within the 90% calibration errors. Thus the 90% upper limit on the velocity difference is $|\Delta v| < 1400 \text{ km s}^{-1}$ between any pair of regions within the field of view.

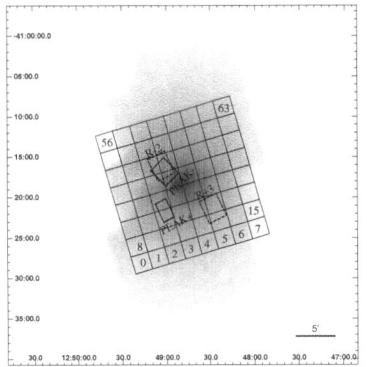

Fig. 1. XIS image of the Centaurus cluster. The central 8×8 cells were used for the study of small-scale velocity structure. The four rectangular regions were used for comparison with the Chandra results.

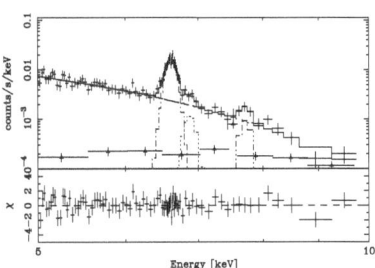

Fig. 2. Background-subtracted XIS0+2+3 spectrum of the Centaurus cluster (the crosses) fitted with the continuum (the dashed line) plus three gaussian models. The background spectrum taken from blank-sky observations is also shown (the crosses with triangle).

2.4 Discussion

From the XIS spectral analysis, no evidence of significant spatial gradients in the Fe-K line energies was found in the central 240 kpc region. This result apparently excludes the large ($\sim 2400 \text{ km s}^{-1}$) velocity difference previously claimed in [1]. To investigate this issue, we analyzed the same rectangular regions as studied in [1] and also considered the effect of smearing due to a wide response of the XRT. However, we found that the present result cannot be easily reconciled with the Chandra observations (see [8] for details).

The obtained upper limit gives a tighter constraint on the ICM bulk motion than suggested from previous reports. This implies that the line-of-sight bulk velocity does not greatly exceed the sound velocity in the ICM. As an elementary exercise, we evaluate the upper limit on rotational motion assuming that the gas is rigidly rotating at a typical circular velocity of $\sigma_r \sim |\Delta v|/2$ and derive its quantitative effect on cluster mass estimation. If we assume spherical symmetry and include the centrifugal force, the total mass should be higher than the hydrostatic mass by a factor of $[1 + \mu m_p \sigma_r^2 (kT)^{-1}]$. Therefore with the upper limit on the velocity difference, the non-thermal pressure support due to hypothetical rotational motion will not alter the hydrostatic equilibrium by more than a factor of ~ 3.

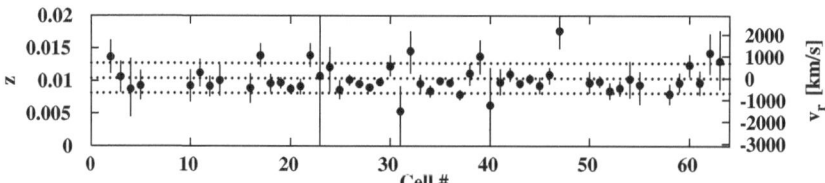

Fig. 3. Results of the velocity measurements in the 52 cells. The left and right axes show the redshift (converted from the measured He-like Fe-Kα line centroid energy using the rest-frame value of 6.677 keV) and radial velocity of the ICM, respectively. The optical redshift of 0.0104 and the range of the present calibrational error at the 90% confidence level, ±0.002 in redshift or ±700 km s^{-1} in the line-of-sight velocity, are indicated with the horizontal dotted lines.

3 Abundance distributions in the Fornax cluster

3.1 Objectives

The metal abundances in the ICM and the hot X-ray emitting interstellar medium (ISM) in early-type galaxies provide important clues to understand the metal enrichment history. The metals in the ICM are mainly synthesized by supernovae (SN) in early type galaxies. Fe and Si are synthesized in both SN Ia and SN II, while O and Mg are not synthesized to any great extent in SN Ia. Thus the O and Mg abundances prove unambiguous information about the formation history of massive stars. XMM-Newton have measured O and Mg abundances, but reliable results are obtained only for the central portion of very bright clusters with cD galaxies. We have conducted Suzaku observations of the Fornax cluster, to measure the abundance profiles accurately [5]. The XIS has a good energy resolution at the O line energy and low background, and thus provides better sensitivity than the XMM-Newton EMOS cameras.

3.2 Observation

The Fornax cluster is a poor cluster at the distant of 19 Mpc, whose X-ray temperature is 1.3–1.5 keV. We observed this cluster in two pointings: one for the central field centered about 2′ southeast of the cD galaxy NGC 1399 and the other for the north field. The two observations were carried out in September 2005 and January 2006, and the net exposures after data screening are 68 ks and 85 ks, respectively. Two X-ray peaks corresponding to NGC 1399 and the elliptical galaxy NGC 1404 are clearly detected.

3.3 Analysis and results

The XIS spectra were accumulated from concentric rings centered on NGC 1399 (Figure 4), where NGC 1404 was masked out. The background was subtracted by using the night-earth data as the non X-ray background and modeling the Cosmic X-ray Background and the Galactic emission. Then by fitting the background-subtracted spectra with a single- or two-temperature vAPEC model diminished by the Galactic absorption, we derived the ICM temperature and the metal abundances. Since there is a decrease of the low-energy efficiency due to some contamination material on the filter of the XIS, we allowed the contaminant thickness to vary in the fit.

The fitted ICM temperature is almost constant at 1.3 keV for $4' < r < 16'$, and it drops to 1 keV outside $16'$. Within $r < 4'$ the two-temperature model gave a significantly better fit, where the metal abundances of the two components were tied to each other. The best-fit temperatures of the two components are obtained to be 0.8 and 1.4 keV. As shown in Figure 5, the Fe abundance is 1 solar and decreases to about 0.5 solar and stays at this level outside $6'$. Si and S are found to show almost the same values as Fe. On the other hand, the O abundance is a factor of 2–3 lower than that of Fe, and O/Fe and Mg/Fe are about 0.4–0.5 and 0.7 in units of the solar ratio, respectively. In addition, we obtained similar results for NGC 1404. Figure 6 summarizes the abundance ratios of various elements.

3.4 Discussion

The Suzaku observations of the Fornax cluster revealed the abundance profiles in the ICM. In NGC 1404, the abundances of O, Ne, Mg and Fe in ISM are also derived. In elliptical galaxies, O and Mg abundances in the hot gas should be equal to those in mass losing stars, since these elements are not synthesized to any great extent by SN Ia. In contrast, most of the Fe around NGC 1399, NGC 1404 and the ICM should have been synthesized by SN Ia. Adopting the new solar abundance model in [4], O/Fe and Si/Fe in the gas are $\sim$ the solar ratio and consistent with a mixture of SN Ia ejecta of the W7 model and metal-poor Galactic stars. The O mass-to-light ratio and Fe mass-to-light ratio within 130 kpc are $2 \times 10^{-3} M_\odot/L_\odot$ and $4 \times 10^{-4} M_\odot/L_\odot$, respectively. These values are more than an order of magnitude lower than those for richer clusters, indicating that most of the metals may be outside of the region we observed. The metal distribution of the ICM may be used as a tracer of the history of the cluster.

4 Summary

Based on the high-sensitivity spectroscopy with the Suzaku XIS, we studied the ICM bulk motion in the Centaurus cluster and the abundance distributions in the Fornax cluster. At the same time, these two observations demonstrate the capability of the XIS, which is expected to provide a new, powerful tool to study cluster physics.

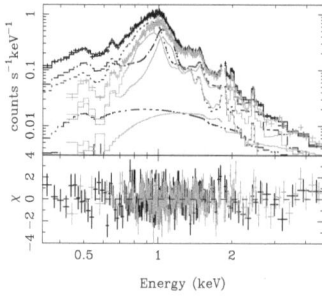

Fig. 4. XIS0 (gray) and XIS1 (black) spectra of the Fornax cluster ($r < 2'$).

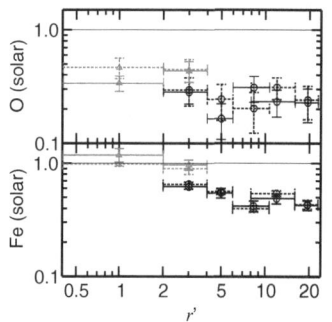

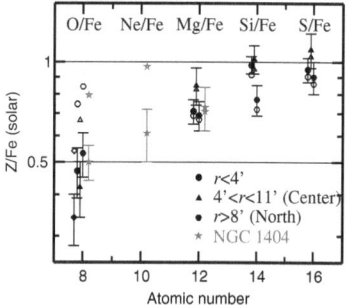

Fig. 5. O and Fe abundance profiles of the Fornax cluster based on the single temperature model (black open circles) and the two temperature model (gray open triangles) derived from the XIS1 (solid lines) and XIS0,2,3 (dotted lines).

Fig. 6. Abundance ratios of various elements compared with Fe for three radial regions and NGC 1404. The closed and open symbols are the abundance ratios using the solar abundance by [2] and [4], respectively.

References

1. R. A. Dupke, J. N. Bregman, 2006, ApJ, 639, 781
2. U. Feldman, 1992, Physica Scripta, 46, 202
3. K. Koyama et al., PASJ, in press
4. K. Lodders, 2003, ApJ, 591, 1220
5. K. Matsushita et al., PASJ, in press (astro-ph/0609065)
6. K. Mitsuda et al., ApJ, in press
7. M. L. Norman, G. L. Bryan, 1999, LNP, 530, 106
8. N. Ota et al., PASJ, in press (astro-ph/0609560)
9. P. Serlemitsos et al., PASJ, in press
10. T. Takahashi et al., PASJ, in press

Suzaku Observations of A2218

Y. Takei[1], T. Ohashi[2], J. P. Henry[3], K. Mitsuda[1], R. Fujimoto[1], T. Tamura[1], N. Y. Yamasaki[1], K. Hayashida[4], N. Tawa[4], K. Matsushita[5], M. W. Bautz[6], J. P. Hughes[7], G. M. Madejski[8], R. L. Kelley[9] and K. A. Arnaud[9]

[1] ISAS/JAXA, 3-1-1 Yoshinodai, Sagamihara, Kanagawa 229-8510, Japan
  takei@astro.isas.jaxa.jp
[2] Tokyo Metropolitan University, 1-1 Minami-Osawa, Hachioji, Tokyo 192-0397, Japan
[3] University of Hawai'i, 2680 Woodlawn Drive, Honolulu, HI 96822, USA
[4] Osaka University, Toyonaka 560-0043, Japan
[5] Tokyo University of Science, 1-3 Kagurazaka, Shinjuku, Tokyo 162-8601, Japan
[6] Massachusetts Institute of Technology, 70 Vassar Street, Building 37, Cambridge, MA 02139, USA
[7] Rutgers University, Piscataway, NJ, 08854-8019, USA
[8] Stanford Linear Accelerator Center, 2575 Sand Hill Road, Menlo Park, CA, 94025, USA
[9] NASA Goddard Space Flight Center, Code 662, Greenbelt, MD 20771, USA

1 Introduction

A large (30–50%) fraction of baryons is thought to reside in the warm-hot intergalactic medium (WHIM), the intergalactic medium with $T = 10^{5-7}$ K and $n_H = 10^{-6}$–10^{-4} cm^{-3} [1, 3, 2]. The WHIM may be detected in X-ray spectra via emission or absorption lines from highly ionized elements, such as O and Ne. Although there are several reports of likely detection of redshifted OVII and/or OVIII lines probably due to the WHIM [12, 6, 4], some of them are controversial. Hence, further sensitive study of other systems is desirable. Note that the observation of the WHIM is also important for understanding the origin of the cluster soft excess.

We report our observation of a cluster of galaxies A2218 ($z = 0.1756$ [14]), with the XIS [7] onboard *Suzaku* [11]. An important characteristic of the XIS is the absence of a large low-energy tail in the pulse-height distribution function, even in the very soft energy ($E \lesssim 0.5$ keV) band. The fairly large redshift of A2218 would also help us observe a significant energy shift of emission lines with the XIS instrument.

2 Observations and data reduction

We carried out four observations: two on the cluster, ($16^h35^m54^s$, $66°13'00"$) in J2000.0, and two in offset regions, ($16^h17^m48^s$, $65°27'36"$) as Offset-A, and ($16^h39^m31^s$ $66°13'31"$) as Offset-B. The offset observations were performed to measure the foreground Galactic contribution. The positions of Offset-A and Offset-B pointing are about two and one degree from the cluster.

We extracted spectra from an annulus region between 3′ and 8′ from the cluster center, corresponding to 540 and 1430 kpc at the source, respectively. We generated the ancillary response files (ARFs) using the arf builder 'xissimarfgen' [5] with ae_xi[0123]_contami_20060525.fits contamination tables. Response matrix files (RMFs) ae_xi[0123]_20060213(c).rmf were used. In the analysis shown below, the spectra and ARFs for two A2218 observations were added. Further, the spectra and response of the three FI sensors were combined. The summed FI and the BI spectra were fitted simultaneously.

3 Spectral analysis and results

After internal background was subtracted, each offset-pointing spectrum was fitted with a sum of the local hot bubble (LHB), Milky-Way halo (MWH) and the Cosmic X-ray background (CXB) models. The model fairly represented the spectra. We then fitted the A2218 spectra with a model including the thin thermal plasma (ICM) of $z = 0.1756$ and the background emission determined from the offset-region spectra. The spectra were well fitted with this simple model, suggesting that no obvious soft excess is present. The temperature and abundance are generally consistent with the values previously determined using *XMM-Newton* [13] or *Chandra* [9], and with the different background models (Offset-A and -B).

We performed another spectral fit by adding two Gaussian emission lines to the model, in order to test for the existence of redshifted O lines. The energies of the lines were fixed to 488.22 eV (O VII resonance) and 555.99 eV (O VIII). The intrinsic widths were fixed to zero. The obtained surface brightness I of the lines is small and consistent with zero. We then constrained the upper limits for the O VII and O VIII I. Since evaluation of the systematic uncertainties is crucial for this purpose, we took into account uncertainties in the degradation of the energy resolution of the XIS, in the thickness of the contamination layer on the OBFs, in the O abundance in the ICM, and in the spatial variation of the Galactic emission. Then, the conservative upper limits are derived by assuming no O in the ICM and employing the ARFs with 20% thicker contaminant and 10% fainter Galactic model for the Offset-B observation; the values are 1.1×10^{-7} photons cm^{-2} s^{-1} arcmin^{-2} and 3.0×10^{-7} photons cm^{-2} s^{-1} arcmin^{-2} for O VII and O VIII lines, respectively.

Fig. 1 compares the O line surface brightness I for our results and other observations reported by [6] and [4]. The left and right panels are for the O VII and O VIII cases, respectively. The intensity of the O lines of Galactic emission measured by [10], [8] and our Offset-A are also shown. the upper limit of O VII line intensity in the A2218 vicinity obtained in this work is about six times lower than the Galactic level. The upper limit for O VIII is also lower than the level reported as a positive detection in other works.

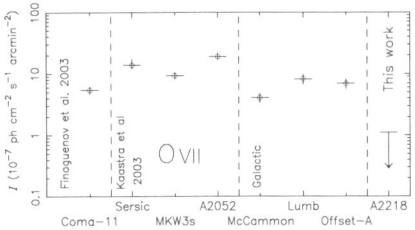 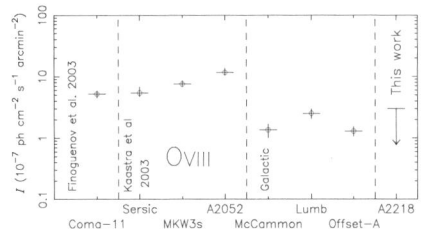

Fig. 1. Comparison of O VII (**left**) and O VIII (**right**) surface brightness. From left to right, those in the Coma-11 field [4], Sérsic 159–03, MKW 3s and A2052 [6], Galactic emission of [10], [8], and our Offset-A observation, and the upper limits in A2218 outskirts (this work) are plotted.

4 Summary

We used *Suzaku* XIS observations to constrain the surface brightness of O emission lines around A2218. After considering systematic uncertainties, we obtained upper limits for the O VII and O VIII lines that are significantly lower than the previously reported ones around other clusters. Our tight constraints demonstrate the sensitivity of the XIS for redshifted O lines.

References

1. Cen, R., & Ostriker, J. P. 1999, ApJ, 514, 1
2. Chen, X., Weinberg, D. H., Katz, N., & Davé, R. 2003, ApJ, 594, 42
3. Davé, R., Cen, R., Ostriker, J. P., Bryan, G. L., Hernquist, L., Katz, N., Weinberg, D. H., Norman, M. L., & O'Shea, B. 2001, ApJ, 552, 473
4. Finoguenov, A., Briel, U. G., & Henry, J. P. 2003, A&A, 410, 777
5. Ishisaki et al., 2006, PASJ, 58, in press
6. Kaastra, J. S., Lieu, R., Tamura, T., Paerels, F. B. S., & den Herder, J. W. 2003, A&A, 397, 445
7. Koyama et al., 2006, PASJ, 58, in press
8. Lumb, D. H., Warwick, R. S., Page, M., & De Luca, A. 2002, A&A, 389, 93
9. Machacek, M. E., Bautz, M. W., Canizares, C., & Garmire, G. P. 2002, ApJ, 567, 188
10. McCammon, D., et al. 2002, ApJ, 576, 188
11. Mitsuda et al., 2006, PASJ, 58, in press
12. Nicastro, F., Mathur, S., Elvis, M., Drake, J., Fiore, F., Fang, T., Fruscione, A., Krongold, Y., Marshall, H., & Williams, R. 2005, ApJ, 629, 700
13. Pratt, G. W., Böhringer, H., & Finoguenov, A. 2005, A&A, 433, 777
14. Struble, M. F., & Rood, H. J. 1999, ApJS, 125, 35

A Suzaku Observation of the Cluster of Galaxies A1060

K. Sato[1], N.Y. Yamasaki[2], M. Ishida[1], Y. Ishisaki[1], T. Ohashi[1], T. Kitaguchi[3], M. Kawaharada[3], M. Kokubun[3], K. Makishima[3], N. Ota[4], K. Nakazawa[2], T. Tamura[2], K. Matsushita[5], N. Kawano[6], Y. Fukazawa[6] and J.P. Hughes[7]

[1] Department of Physics, Tokyo Metropolitan University, 1-1 Minami-Osawa, Hachioji, Tokyo 192-0397 ksato@phys.metro-u.ac.jp
[2] Institute of Space and Astronautical Science (ISAS), Japan Aerospace Exploration Agency,
3-1-1 Yoshinodai, Sagamihara, Kanagawa 229-8510
[3] Department of Physics, University of Tokyo, 7-3-1 Hongo, Bunkyo, Tokyo 113-0033
[4] Cosmic Radiation Laboratory, The Institute of Physical and Chemical Research (RIKEN),
2-1 Hirosawa, Wako, Saitama 351-0198
[5] Department of Physics, Tokyo University of Science, 1-3 Kagurazaka, Shinjuku-ku, Tokyo 162-8601
[6] Department of Physical Science, School of Science, Hiroshima University, 1-3-1 Kagamiyama, Higashi-Hiroshima, Hiroshima 739-8526
[7] Department of Physics and Astronomy, Rutgers University, Piscataway, NJ 08854-8019, USA

Summary. We carried out observations of the central and $20'$ east offset regions of the cluster of galaxies Abell 1060 with *Suzaku*. Spatially resolved X-ray spectral analysis has revealed temperature and abundance profiles of Abell 1060 out to $27' \simeq 380\ h_{70}^{-1}$ kpc, which corresponded to $\sim 0.3\ r_{\mathrm{vir}}$. The temperature decrease of the intra cluster medium from 3.4 keV at the center to 2.2 keV in the outskirt region are clearly observed. Si, S and Fe abundances also decrease by more than 50% from the center to the outer parts, while Mg shows a fairly constant abundance of ~ 0.7 solar within $r < 17'$. O shows a lower abundance of ~ 0.3 solar in the central region ($r < 6'$), and indicates a similar feature with Mg, however it is sensitive to the estimated contribution of the Galactic components at the outer annuli ($r > 13'$). Results on temperature and abundances of Si, S, and Fe are consistent with those derived by *XMM-Newton* at $r < 13'$. The formation and metal enrichment of the cluster are discussed based on implications of our results.

1 Introduction

Abell 1060 (hereafter A 1060) is a nearby cluster of galaxies ($z = 0.0114$) characterized by a smooth and symmetric distribution of the intracluster medium (ICM), and has no cD-galaxy at the center. [5] discovered $\sim 30\%$ of the temperature drop from the central region to $r \sim 13'$ with *XMM-Newton*, while it was considered as being flat on the basis of the ASCA observation [4].

Suzaku carried out two pointing observations for A 1060 in November 2005, the central region and 20′ east offset region, with exposure of 40.5 and 55.5 ks, respectively. Owing to the low-background nature of the Suzaku XIS, we are able to measure the temperature and abundance profiles to a much larger radius than XMM-Newton. We use $H_0 = 70$ km s^{-1} Mpc^{-1}, $\Omega_\Lambda = 1 - \Omega_M = 0.73$ in this paper. At a redshift of $z = 0.0114$, 1′ corresponds to 14 kpc, the virial radius, r_{180}, is 1.35 Mpc with the average temperature of 3.3 keV [5]. Throughout the paper we adopt the Galactic hydrogen column density of $N_H = 4.9 \times 10^{20}$ cm^{-2} [2] in the direction of A 1060.

2 Spectral Analysis and Results

We extracted spectra from seven annular regions of 0–2′, 2–4′, 4–6′, 6–9′, 9–13′, 13–17′ and 17–27′ from the cluster center. The spectrum of the 17–27′ region is shown in figure 1:left. The ionized Mg, Si, S, Fe lines are clearly seen in each ring. The O$_{VII}$ and O$_{VIII}$ lines are prominent in the outer rings, however, most of the O$_{VII}$ emission is supposed to come from the local Galactic emission. We used the vAPEC model as the Intracluser medium (ICM) of the cluster, and the APEC with 1 solar abundance as Galactic emission. The spectra were well-fitted by the $apec_1 + apec_2 + phabs \times vapec$ model.

The temperature observed with Suzaku dropped from 3.3 keV in the central region to 2.2 keV in the outermost region of $\sim 0.3\ r_{180}$. This is consistent with the previous results for inner region with XMM-Newton [5]. As is discussed in [5] the temperature of A1060 dropped faster than the averaged $T/\langle T \rangle$ curve by [6] obtained from 30 nearby clusters with ASCA observations.

In this paper, we obtained luminosity weighted abundance profiles of Fe, Si, S, Mg, Ne and O in the ICM of A 1060 within a radius of 27′ $\simeq$ 380 kpc. Abundances of Si, S, and Fe decrease from 0.7, 0.8 and 0.5 solar in the central region, to 0.3, 0.2 and 0.2 solar, respectively, in the outskirts of the cluster. As pointed out in [3], the abundance gradient over a few hundred kpc scale in clusters of galaxies follows the mass ratio between the galaxies and the ICM gas, because heavy ions released from galaxies without kinetic energy diffuse slowly in the intra-cluster space. [1] showed that non-cooling flow clusters after major merger events do not exhibit that kind of abundance gradient due to disruption of the central regions. In contrast, Mg and O abundances show a flatter distribution. The resultant Mg abundance is fairly constant, $\sim$0.7 solar, throughout the observed region within 240 kpc from the cluster center. Oxygen also exhibits a flat distribution or a slight increase from $r \simeq 10′$ toward the outer region.

In order to check the difference of the abundance profiles, we show the abundance ratio of O, Mg, Si compared to Fe from the cluster center to the outermost region in figure 1:right. Obviously, the profile among Si/Fe, O/Fe, and Mg/Fe ratios differ in the gradients. Si/Fe ratio is consistent with a constant, but O/Fe is not. As this difference can not be explained by ion mass, it should be caused by the period and process of the O and Fe enrichment of the ICM. The O and other SN II origin metal enrichment is considered to occur before the last merger mixing. It does not deny the possibilities of very flat metal distribution from the start, such as starburst-driven galactic wind at the early phase of the cluster evolution (eg. [8]), or metal injection to the primordial gas by population III stars. In these cases, metals can exist outside the cluster potential. Future quantitative studies about metal abundance of outer ($\sim r_{\rm vir}$) region of ICM or IGM outside clusters with enough energy resolution to distinguish between the local and cluster emission will be required. For further details, please see [7].

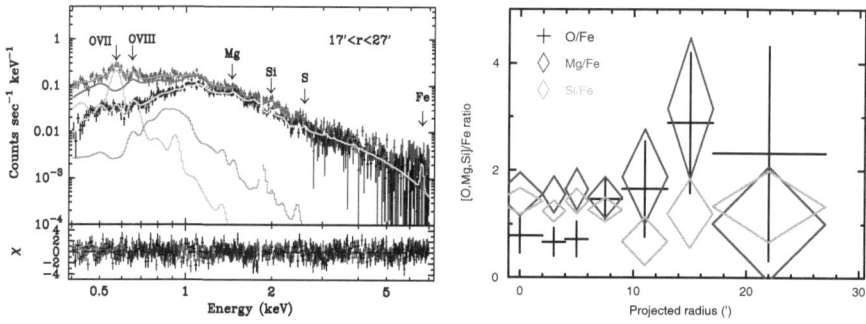

Fig. 1. *Left*: The left panel shows the observed spectra of the outermost region, and they are plotted by upper and lower crosses for BI and FI, respectively. The *apec* components for the BI spectra are indicated by the lines. The energies of several prominent lines are also indicated in the panels. The lower panels show the fit residuals in unit of σ. *Right*: The right panel show the abundance ratio of O (black crrosses), Mg (dark-gray diamonds) and Si (light-gray diamonds) compared to Fe.

References

1. De Grandi, S., & Molendi, S., 2001 ApJ, 551, 153
2. Dickey, J. M., & Lockman, F. J. 1990, ARA&A, 28, 215
3. Ezawa, H. et al. 1997, ApJL, 490, 33
4. Furusho, T., Yamasaki, N. Y., Ohashi, T. S. R., Kagei, T., Ishisaki, Y., Kikuchi, K., Ezawa, H., & Ikebe, Y. 2001, PASJ, 53, 421
5. Hayakawa, A., Hoshino, A., Ishida, M., Furusho, T., Yamasaki, N. Y., & Ohashi, T. 2006, PASJ, 58, 743
6. Markevitch, M., et al. 1998, ApJ, 503, 77
7. Sato, K., et al. 2006, in preparation
8. Strickland, D.K., & Stevens, I.R., 2000, MNRAS, 314,511

Initial Results from Suzaku Hard X-ray Observations of Abell 3376

K. Nakazawa[1], N. Kawano[2], Y. Fukazawa[2], T. Kitaguchi[3] and the Suzaku-team

[1] Institute of Space and Astronautical Science, JAXA, Sagamihara, Kanagawa 229-8510, Japan, nakazawa@astro.isas.jaxa.jp
[2] Hiroshima University, Higashi-Hiroshima, Hiroshima 739-8526, Japan, kawano@hirax6.hepl.hiroshima-u.ac.jp and fukazawa@hirax6.hepl.hiroshima-u.ac.jp
[3] The University of Tokyo, Hongo, Bunkyo-ku, Tokyo 113-0033, Japan, kitaguti@amalthea.phys.s.u-tokyo.ac.jp

1 Introduction

Clusters of galaxies are known to host large amount of accelerated (relativistic) particles in their vicinity. The strongest evidence is the Mpc-scale radio halos and relics, which are interpreted as a synchrotron emission from GeV electrons. These electrons scatter the cosmic microwave background (CMB) up to the 10–100 keV energy band via Inverse-Compton (IC) process. Since the CMB density is well known, hard X-ray observations thus can be a powerful tool to directly measure the total number of GeV electrons. The lack of sensitivity in hard X-rays, however, make it very difficult to examine. Although radio halos and relics are detected in more than 50 clusters so far (e.g. [2]), non-thermal hard X-rays are suggested in only a handful of objects, including the Coma cluster in particular (e.g. [3]), using the hard X-ray instruments on-board, e.g., Beppo-SAX and RXTE.

Suzaku satellite [6] is the fifth of a series of Japanese X-ray observatories, launched last summer into a low earth orbit with an inclination of 31 degree. With the combination of the X-ray CCD cameras (XIS: [5]) and the hard X-ray detector (HXD: [9][4]), Suzaku can observe wide energy band from 0.3 keV up to 600 keV. Both the XIS and the HXD is characterized by its low and stable background. In particular, the HXD obtained the lowest background level ever achieved in the energy band from ~ 10 keV to ~ 100 keV. Thus, Suzaku can provide a new observational evidence for non-thermal X-rays in clusters of galaxies. In this paper, we present the results from initial analysis of the hard X-ray data from a cluster of galaxies, Abell 3376, observed with Suzaku in its PV-phase.

2 Merits of the HXD

The HXD is characterized by its deep well-type active shield to anti-out any coincident events in order to obtain lowest background [9][4]. The main de-

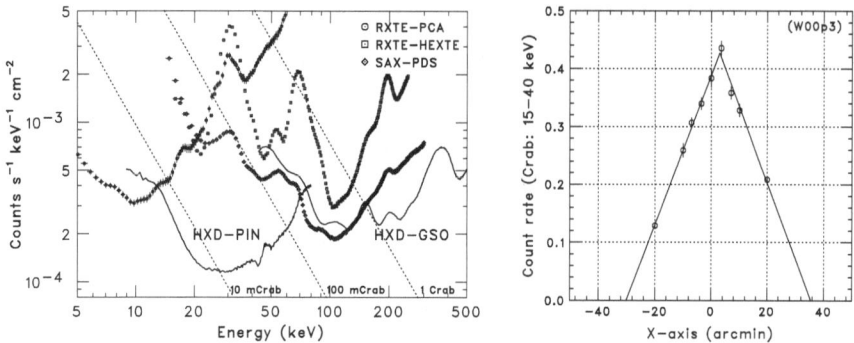

Fig. 1. *left* : A comparison of the in-orbit detector background of PIN/GSO with those of RXTE-PCA, RXTE-HEXTE, and BeppoSAX-PDS. Dotted lines indicate 1 Crab, 100 mCrab, and 10 mCrab intensities. *right* : An angular response of single fine-collimator along the satellite X-axis, obtained from offset observations on the Crab nebula

tector consists of 2 mm thick Si PIN diodes and 5 mm thick GSO scintillator crystals, each covering 10-70 keV and 40-600 keV, respectively. The background spectra observed in orbit is very low (< 10 mCrab), especially in the PIN energy band. Thanks to its very low background, the HXD can achieve 3–5 mCrab sensitivity in the 10-40 keV band with relatively short observation of ~ 40 ks. Another key characteristics of the HXD is its narrow field of view (FOV), 35′ in full-width-at-half-maximum (FWHM). With this narrow beam size, not only the Cosmic X-ray background (CXB) flux and source contamination probability are minimized, but also a moderate spatial resolution for near-by clusters can be obtained.

3 Observation and data reduction

Considering these merits, we observed as the first trial a cluster with relatively low hot gas temperature ($kT < 6$ keV) and a structure with angular scale similar to the HXD FOV. Due to the calibration accuracy including the response and the background modeling, the HXD sensitivity around ~ 100 keV was expected to require time to reach a certain level. Thus, we did not select the Coma cluster in the PV-phase, since its strong thermal emission dominates the spectra up to ~ 80 keV. After considering several candidates, Abell 3376 is selected. It hosts a pair of Mpc-scale radio relics[1] and among the clusters with highest significance of hard X-rays with PDS observations [7].

Two observations were performed at October-November 2005; one pointed at the cluster center and the east-relic region, and another one pointed at the west relic. In the following sections, we only mention about the HXD data

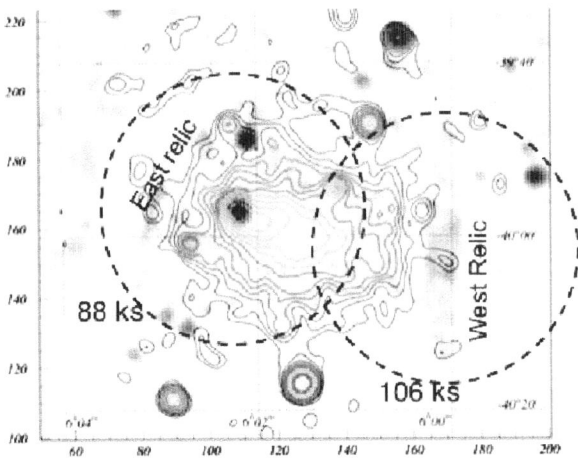

Fig. 2. ROSAT PSPC contour image overlaid on the NVSS gray scale image. Two Suzaku pointing positions are indicated by dashed circles, roughly corresponding to the region in which the effective area decreases to 50% of that of the FOV center.

analysis in detail, since it is the main part of the current study. When analyzed in the standard manner, i.e., cut-off-rigidity higher than 8 GV, elevation from the earth rim larger than 5 deg, and time-after-SAA larger than 500 sec, the former observation resulted in 88 ks of good exposure in the HXD, while the latter in 106 ks. The final internal version of the calibration information (referred to as "rev0.7" in the early Suzaku papers) is used in this paper.

4 Preliminary results on the Hard X-ray signals from Abell 3376

By generating a spectra and subtracting the background model provided by the HXD team, we obtained the hard X-ray spectra from this object. The signals thus detected were about 5–10% of the detector background. Therefore, the hard X-ray flux is smaller than 1 mCrab. With this small flux, it is not sufficient to only consider the statistical errors, but we also need to estimate the systematic errors in the current response and background modeling of the HXD. From the very preliminary study briefly presented in the appendix, we found that with current background modeling and uncertainties, analyzing only the data obtained in orbits with no SAA passage ("SAA-free analysis") provides smaller errors, sacrificing a large loss of exposure.

Spectra obtained with the "SAA-free" method are presented in Fig.3. The center/east observation resulted in a 31 ks good exposure, while that of the west relic in 21 ks. With this spectra, the count rate of both observations were generally consistent with the expected CXB. Thus, from the systematic

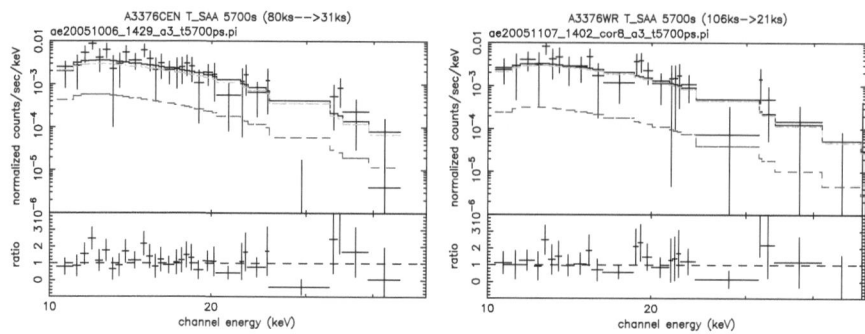

Fig. 3. Center/east relic (*left*) and west relic (*right*) spectra after "SAA-free" screening. Lower dashed lines represents the best-fit $\Gamma = 2$ power law model, while the upper dashed lines the CXB model.

error analysis presented in the appendix, the preliminary upper limit flux in the 20-80 keV band is evaluated to be $\sim 7 \times 10^{-12}$ erg s^{-1} cm^{-2} when the hard X-rays are modeled with $\Gamma = 2.0$ fixed power-law component. Since the systematic error is evaluated using only 27 observations, we conservatively quote 1×10^{-11} erg s^{-1} cm^{-2} as a current $\sim 90\%$ upper-limits for both observations.

Note that this is a very rough trial and there is a big room to improve both the selection criteria to obtain the highest sensitive results and the background modeling accuracy itself. Currently, with much improved background modeling, we are re-analyzing the data in detail. The final upper-limit quoted above is about $\sim 7\%$ of the detector background, while our goal is to achieve $\sim 3\%$ accuracy or better. Thus, we think there is a room even for detecting the excess hard X-rays in future, through intense studies.

5 Conclusion and future prospects

By preliminary analysis of the Suzaku-HXD observations of the Abell 3376 cluster, we did not detect significant excess hard X-rays from this cluster. Considering the systematics included in the current analysis, we quote a tentative $\sim 90\%$ upper limit of 1×10^{-11} erg s^{-1} cm^{-2} in the 20-80 keV band. Our upper limit is consistent with the SAX-PDS analysis [7], but with smaller best-fit flux. Future (i.e. on-going) analysis includes the better background modeling and screening, and the search for non-thermal emission from the XIS data obtained within the west relic region.

Appendix: A rough estimate of background systematics

The simplest way to evaluate the systematic error associated with the HXD data analysis for dim objects is to analyze dozens of such objects with the same manner. Since we are discussing about a steady source, here we only present results on the data averaged over a long exposure. From pure statistical point of view, 3–5 mCrab sensitivity can be obtained with only ∼ 40 ks exposure, i.e. about 1-day long observation, thanks to the low background of HXD-PIN. As a rough estimate, then we generated 1-day averaged background subtracted count distribution of 27 periods ("days") using actual data pointing at apparently dark objects. For simplicity, we concentrate on only the 20-30 keV band here.

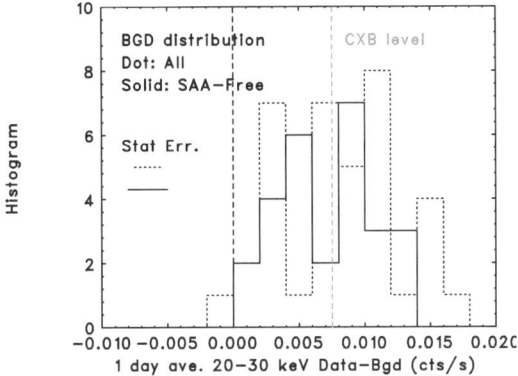

Fig. 4. Distribution of the residual counts of 27 observations of dim-sources. Only the results of the 20-30 keV band is presented. Dotted line presents the results from original screening analysis, while the solid line from the "SAA-free" analysis. See text for detail.

The distribution of the original screening is presented as a dotted histogram in Fig.4. The CXB level was calculated to be ∼ 5% of the total background in this energy band. In the plot, we clearly see offsets on average consistent with the estimated CXB, while the distribution is slightly larger. Since the statistical error is much smaller, the data is dominated by systematic errors. Carefully looking at the data, we noticed slight degradation of background reproducibility right after the SAA passage, which is not surprising.

Accordingly, as a very rough trial, we analyzed the data only obtained in the orbit without SAA passage ("SAA-free analysis"). For simplicity, we set the screening criteria as time-after-SAA larger than 5700 sec, the orbital period. This selection resulted in reducing the original exposure to a half. Although the statistical error increases by a factor of ∼ 1.5, the actual residual significantly decreased to 5% bottom-to-peak of detector background, using this selection method (see the solid line in Fig.4).

References

1. J. Bagchi et al., 2005, In *29th International Cosmic Ray Conference Pune*, pp101–106
2. L. Feretti, A. Burigana and T.A. Ensslin: New Ast. Rev., 2004, 48, 1137
3. R. Fusco-Femiano, et al., 1999, ApJ, 513, L21
4. M. Kokubun et al., 2006, PASJ, 58, in press
5. K. Koyama et al., 2006, PASJ, 58, in press
6. K. Mitsuda et al., 2006, PASJ, 58, in press
7. J. Nevalainen, T. Oosterbroek, M. Bonamente, and S. Colafrancesco, 2004, ApJ, 608, 166
8. M. Rosseti and S. Molendi, 2004, A&A, 414, L41
9. T. Takahashi et al., 2006, PASJ, 58, in press

Part X

Heating vs. Cooling in Galaxy Formation

A New Chemodynamical Tool to Study the Evolution of Galaxies in the Local Universe: a Quick and Accurate Numerical Technique to Compute the Gas Cooling Rate for any Chemical Composition

N. Champavert and H. Wozniak

Université Lyon 1, CRAL, Observatoire de Lyon, 9 avenue Charles André,
Saint-Genis Laval cedex, F-69561, France
champavert@obs.univ-lyon1.fr

Summary. We have developed a quick and accurate numerical tool to compute gas cooling irrespective of its chemical composition.

1 Introduction

Metal abundances in galaxies vary from place to place and with time. These variations reflect the star formation history within galaxies because the chemical enrichment results from synthesis of heavy elements by successive generations of stars. Gas radiative cooling is very sensitive to the ISM chemical composition. Cooling influences both the gas dynamics and the star formation rate. Therefore we need to follow self-consistently dynamical and chemical evolution. We have developed a method to compute in a very short CPU time accurate cooling functions dependent on the exact abundances of elements. It will be used in high resolution chemodynamical simulations of galaxies.

2 Description of the method

It is well known that cooling rates depend on gas chemical composition. A gas with solar composition has a cooling rate greater than without metals by more than an order of magnitude for some temperatures (Fig. 1). Cooling rates are generally interpolated between rates computed for different metallicities with solar abundance ratios (see [1] for instance) or with solar abundance ratios and enhanced abundances for some elements (see [2] for instance). The gas abundances are obviously not always solar because the gas ejected from stars by winds and SNe has a non-solar composition. Hence we need to take into account the real abundances of each element in the gas to build more realistic cooling functions whichever the chemical composition. However, computing

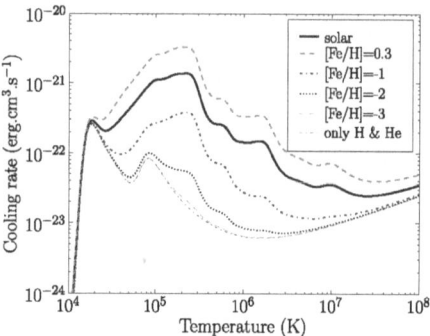

Fig. 1. Cooling rates as a function of temperature for different metallicities computed with Mappings III. The abundances used in calculations are taken from [2]

cooling rates with Mappings III for each gas particle at each timestep of a chemodynamical simulation is a very CPU time consuming task. We thus have developped a recipe to reconstruct the cooling curve on the fly.

We consider the ISM as an optically thin gas in collisional ionization equilibrium. Calculations are performed in the temperature range 10^4 K to 10^8 K using Mappings III, successor of Mappings II whose cooling computations details are described in [2]. All the abundances of elements are relative to hydrogen. The hydrogen density is $1.0 \, \text{cm}^{-3}$, independent of temperature. We drop the temperature dependence of cooling rates ($\Lambda(T)$) for the sake of legibility.

Many cooling processes depend on the density of electrons and on the abundances of elements. A simple hypothesis is to suppose that cooling rates are proportional to these two quantities. Thus, we can try to reconstruct the total cooling curve Λ_{tot} with a linear combination of individual curves for each element. We first need to compute these individual curves. We obtain Λ_H with a purely hydrogen gas. For all the heavier elements used for calculations in Mappings III (He, C, N, O, Ne, Na, Mg, Al, Si, S, Cl, Ar, Ca, Fe, Ni), we have calculated cooling rates for mixtures of hydrogen and one element X_i at its solar abundance $\Lambda_{X_i,H}$. Individual cooling curves Λ_{X_i} have then been obtained substracting Λ_H from each Λ_{H,X_i}: $\Lambda_{X_i} = \Lambda_{H,X_i} - \Lambda_H$. For a given chemical composition, we then compute the quantity α_{X_i} for all the elements previously listed, where α_{X_i} is the abundance of element X_i normalized by its solar abundance: $\alpha_{X_i} = n_{X_i}/n_{X_{i\odot}}$. For helium, our hypothesis is too simple. Its cooling rate cannot be considered proportional to its abundance with a good approximation as there are too strong non-linearities. We choose to linearly interpolate in a grid of precomputed curves for different abundances to obtain the helium cooling curve $\Lambda_{He,\alpha_{He}}$. As hydrogen and helium are the two most abundant elements, we assume that they are the only sources of electrons. We indeed consider that the contribution of the metals are negligible because of their much lower abundances in the ISM. By construction,

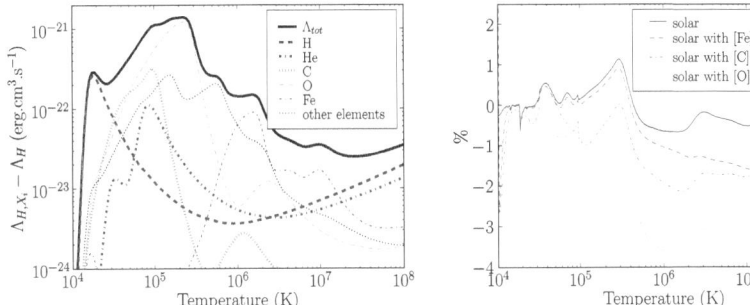

Fig. 2. (left panel) Cooling rates for several elements at solar abundances. The curve for other elements is the sum of the contributions of N, Ne, Na, Mg, Al, Si, S, Cl, Ar, Ca and Ni. The solid broad line Λ_{tot} is the cooling curve for solar abundances reconstructed using our recipe. **(right panel)** Percentage of relative error made when reconstructing cooling functions with our recipe for different compositions

our individual cooling functions for metals take into account the electrons of the element and hydrogen. Thus we only need to add the contribution of electrons provided by the ionization of helium $ne\,(He)$. We add the terms for hydrogen, helium and metals to build Λ_{tot} so that we finally obtain the following approximation: $\Lambda_{tot} \approx \Lambda_H + \Lambda_{He,\alpha_{He}} + (1 + ne\,(He))\sum_{metals}\alpha_{X_i}\Lambda_{X_i}$

To check the accuracy of our tool, we have compared cooling curves computed with Mappings III and the ones reconstructed with our recipe for various chemical compositions. We choose a solar composition to test a well-known chemical composition. Carbon, oxygen and iron are three of the most important coolants (Fig. 2, left panel). Furthermore, they are three of the most abundant elements in stellar winds and supernovae ejecta. Thus, we increase by a factor of 10 the abundance of these three important coolants. This factor is somewhat arbitrary and was only chosen to test the accuracy of our recipe in the case of a very high-enhanced gas. In all cases, the relative errors remain below a few percent (Fig. 2, right panel). The errors due to the computation of cooling rates are thus comparable to the others coming from, for instance, the hydrodynamical scheme.

References

1. Böhringer, H. and Hensler, G., 1989, A&A, 215, 147
2. Sutherland, R.S. and Dopita, M. A., 1993, ApJS, 88, 253

A Heating Model for the Millennium Gas Run

L. Gazzola and F. R. Pearce

School of Physics and Astronomy, University of Nottingham, NG7 2RD, UK
ppxlg@nottingham.ac.uk

Summary. The comparison between observations of galaxy clusters thermo-dynamical properties and theoretical predictions suggests that non-gravitational heating needs to be added into the models. We implement an internally self-consistent heating scheme into Gadget-2 for the third (and fourth) run of the Millennium gas project (Pearce et al. in preparation), a set of four hydrodynamical cosmological simulations with $N = 2 \times (5 \times 10^8)$ particles and with the same volume ($L = 500h^{-1}\mathrm{Mpc}$) and structures as the N-body Millennium Simulation (Springel et al. 2005). Our aim is to reproduce the observed thermo-dynamical properties of galaxy clusters.

1 Model

The large dynamical range that characterises cosmological simulations and the fact that the physics of heating mechanisms like AGN feedback, galactic winds and conduction has typical scales much smaller that those describing galaxy clusters (few kpc versus Mpc), make their implementation challenging. In addition the Millennium gas runs do not have a high enough resolution to properly model these phenomena ($m_{gas} = 3.12 \times 10^9 h^{-1} M_\odot$, softening $25h^{-1}\mathrm{kpc}$) and therefore we need to adopt a relatively simple heating scheme. Our model seeks to improve on the simple preheating scheme that was implemented in the second run of the Millennium gas project.

We implement a self-regulated star formation plus feedback procedure by selecting gas particles to convert into stars by imposing a temperature plus over-density threshold (in units of critical density). We choose an over-density rather that a physical density threshold. We then inject an energy E_{inj} to the neighbours of the new star, weighted in distance by the standard SPH smoothing kernel.

We choose a temperature threshold of $10^5 K$ in both models and density threshold, ρ_{thr}, to be:

- $\rho_{thr} = 200$: mimics the energy deposited by supernovae in galaxy cluster outskirts, low mass objects and at early times
- $\rho_{thr} = 2500$: mimics the effect of star formation in the center of galaxy clusters

We then tune E_{inj} in order to reproduce the observed luminosity-temperature relation of galaxy clusters at $z = 0$.

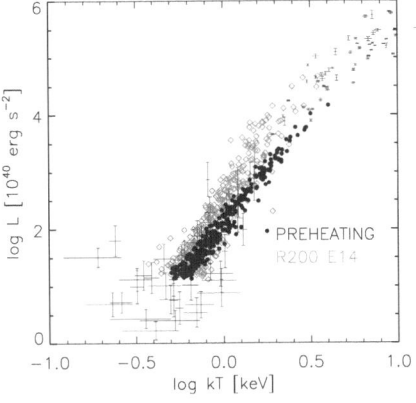 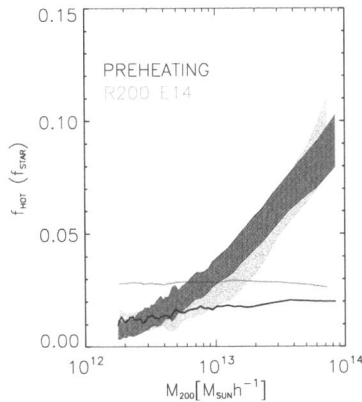

Fig. 1. Left panel: bolometric luminosity versus emission weighted temperature for groups and clusters in a test box of size $L = 125h^{-1}\mathrm{Mpc}$. Black dots are for the preheating run, gray diamonds for $\rho_{thr} = 200$. Observations from Ponman et al. 1996; Helsdon et al. 2000 and Novicki et al. 2002. Right panel: hot gas (shaded area) and stellar fraction (lines) for the preheating run (black lines) and $\rho_{thr} = 200$ (gray lines).

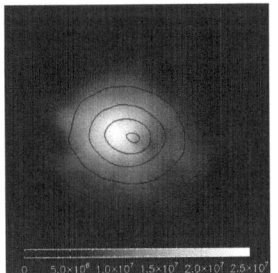 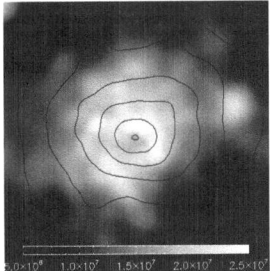 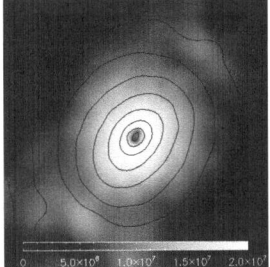

Fig. 2. Emission weighted temperature maps for three clusters. The panel size is four times the virial radius. Superimposed surface brightness contours.

2 Results

The amount of energy that we inject, E_{inj}, is our free parameter and it will be different for the two runs, which vary the density of gas into which the energy is injected. From a test run with $L = 250h^{-1}\mathrm{Mpc}$ we found that choosing $E_{inj} = 14keV$ for $\rho_{thr} = 200$ we manage to reproduce the observed $L - T$ relation for group size objects as well as clusters in the range of $1 - 6keV$. Comparing this run with the preheating model (similar to the model proposed by Kay et al. 2004), we find that the two runs have a slightly different normalisation but still in agreement with observations (Figure 1). We also notice that the $L - T$ relation has too little scatter in the preheating run, compared to the observations, while the current model appears far more reasonable. This difference is at least partially due to the difference in the temperature structure: in the preheating run the gas is brought to such a high temperature at early times that no cool gas is found at redshift zero and the haloes are characterised by a very smooth gas distribution. On the contrary cool cores can form in our current model and the clusters often show multiple-structure in temperature maps and offsets between the emission peak and the temperature peak.

In Figure 2 we show the hot gas fraction and the star fraction as a function of cluster mass. Tuning the preheating run and $\rho_{thr} = 200$ to match the observed $L - T$ relation at $z = 0$, results in a much reduced baryon fraction relative to the non-cooling case, with $f_{bar} \leq 0.13$ while the universal fraction is 0.18 for this cosmology. It is interesting to notice that we obtain similar trends despite the two very different heating mechanisms: one is an external heating (preheating run) and the other one an internal heating. Once the Millennium volume will be available we can use the simulations to test how powerful $f_{hot}(z)$ -hot gas fraction- and $f_{bar}(z)$ -baryonic fraction- are as cosmological probes.

The full $500h^{-1}$Mpc millennium run will also be suitable for time evolution studies. The major drawback of the preheating scheme is that the gas is brought to such a high adiabat at z=4, when the preheating occurs, that no gas can condense any more and star formation is quenched. With this alternative model we get a more realistic star formation history and time evolution.

3 Conclusions

We present an alternative to a simple preheating model by introducing a self-regulated star formation plus feedback scheme. The improvements of this model relative to the preheating one are: an increased scatter in the luminosity-temperature relation, the existence of cool core clusters and a more realistic time evolution. On the other hand the amount of energy required in order to match the observational data with this scheme is quite high, especially if we employ a high density threshold that has been tuned to match the observed $L - T$ at $z = 0$. We are currently exploring higher and lower resolution simulations to constrain the systematic effects introduced by our model.

References

1. Helsdon, S. F., & Ponman, T. J. 2000a, MNRAS, 315, 356
2. Kay S. T., Thomas P. A., Jenkins A., Pearce F. R., 2004, MNRAS, 355, 1091
3. Novicki, M. C., Sornig, M., & Henry, J. P. 2002, AJ, 124, 2413
4. Ponman, T. J., Bourner, P. D. J., Ebeling, H., & Bohringer, H. 1996, MNRAS, 283, 690
5. Springel V., White S. D. M., Jenkins A., et al., 2005b, Nature, 435, 629
6. Pearce, F. R, Gazzola, L., Kay, S., Thomas, P. A. , in preparation

Powerful Radio Galaxies at z=2−3: Signposts of AGN Feedback in the Early Universe

N.P.H. Nesvadba and M.D. Lehnert

GEPI - Bâtiment des Communs, Observatoire de Paris - Section de Meudon, 5, Place Jules Janssen, 92195 Meudon Cedex, France
nicole.nesvadba@obspm.fr, matthew.lehnert@obspm.fr

1 Introduction

Recent models of galaxy evolution postulate a phase of vigorous AGN feedback to explain some of the outstanding mysteries in the evolution of massive galaxies (e.g. [16, 7]). Such a phase appear neccesary to explain, for example, the slope of the upper end of the galaxy mass function, which is steeper than expected from the mass distribution of dark-matter halos [1]. Moreover, AGN feedback in the early evolution of massive galaxies also provides an attractive solution to the "hierarchy problem": Massive galaxies at low redshift have old stellar populations and high metallicities indicating vigorous starbursts at high redshift while being devoid of cold gas and recent star formation. To do this, AGN feedback must have effectively removed significant amount of gas (10^{9-10} M$_\odot$) from the massive galaxies near the end of their formation, thus suppressing further star formation. To unbind such large gas masses, kinetic energies of about the binding energy of the host galaxy must be injected into the ISM (10^{60-61} ergs). The observed [α/Fe] relative overabundance in massive low redshift galaxies sets a tight upper limit to the length of the starburst of $<$ few $\times 10^8$ yrs [19], indicating nearly instantaneous gas removal.

Starburst-driven winds is likely not a good candidate to explain these characteristics: They are predominately energy driven and are likely not powerful enough to unbind significant gas masses from the deepest gravitational potential wells. Momentum-driven AGN winds do not suffer these limitations – they are likely sufficiently energetic to influence the most massive galaxies because of their supermassive black holes, helping to explain why the [α/Fe] relative overabundance and the "hierarchy problem" scale with galaxy mass.

Powerful radio galaxies at high redshift are an ideal sample to directly search for the signposts of AGN feedback in the early Universe. HzRGs have large molecular gas and dust masses [14], on-going star-formation of up to $>$ 1000 M$_\odot$ yr^{-1} [8], and particularly large stellar masses (few $\times 10^{11-12}$ M$_\odot$, [20]) for their redshifts. They reside in galaxy overdensities indicating strongly biased environments [21], and their redshifts are near the redshift of the peak of the co-moving number density of AGN [15]. All these arguments make them the likely progenitors of the most massive galaxies at low redshift seen during an epoch of rapid growth. Thus, they are ideal targets to study the impact of AGN feedback on the early evolution of massive galaxies.

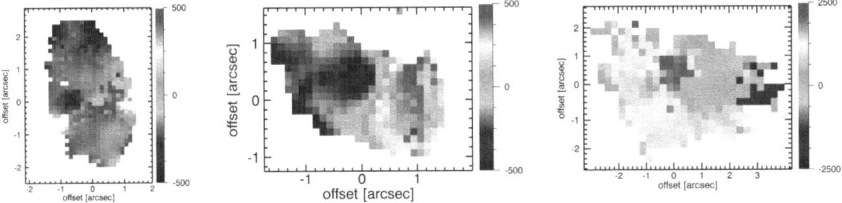

Fig. 1. (left to right:) Velocity maps of TXS0828+192 , MRC0316-257 , and MRC1138-262 . Color bars show relative velocities in km s^{-1}.

2 Near-Infrared Imaging Spectroscopy of High-Redshift Radio Galaxies

The extended nebulosities of HzRGs often have extreme kinematics with relative velocities and FWHMs >1000 km s^{-1} that appear related to the radio jet, as is well known from a series of longslit studies aligned with the axis of the radio jet [12, 18, 22]. However, given the complex morphologies and velocity fields of HzRGs, one-dimensional spectra are insufficient for investigating the impact of the radio jet on the overall dynamics of the ISM.

Using the near-infrared integral-field spectrograph SINFONI on the VLT, we have observed a small sample of 6 HzRGs at redshifts $z = 2.2 - 3.5$. SINFONI is an image slicer with an 8" ×8" field of view at a pixel scale of 0.25"×0.25". Our data have a spatial resolution of ∼ 0.4"−0.6", and a spectral resolution of R∼ 2000−4000 in the J, H, and K bands [5]. Typical integration times are 2−3 hours per band and source. See [13] for a pilot study of MRC1138-262 at z=2.16.

Four galaxies in our sample have extended radio lobes (R>50 kpc), two galaxies are compact, with radii < 10 kpc in the radio. At redshifts z∼ 2−3, the bright optical emission lines fall into the near-infrared atmospheric windows. We measured the gas kinematics from fitting the cores of emission lines of [OIII]λ5007 and Hα, depending on the redshift and line widths. For galaxies at z≥2.6, Hα falls outside the K band, and for FWHMs ≥700 km s^{-1}, it blends with [NII]λ6583. In both cases Hα is not a robust tracer of the kinematics. Comparison of the kinematics measured with both lines (and Hβ) indicates that both [OIII] and Hα yield similar results.

The velocity fields of the galaxies with extended radio emission are shown in Fig. 1. In all cases, the kinematics are not indicative of large-scale, gravitationally-driven, ordered motion. All galaxies have nebulosities with ∼

20-30 kpc radii that are aligned with the radio jet axis. The velocity fields are reminiscent of inflating "bubbles", with uniform internal motion, but large relative velocities between these regions ($\sim 1000-2000$ km s^{-1}). In all cases but MRC1138-262, there seem to be two roughly conical outflows of ionized gas [13]. MRC1138-262 has $3-4$ individual regions and a more spherical morphology, but a similar spatial extent, and equally extreme kinematics. The different appearance of MRC1138-262 might be simply due to orientation or differences in the local environments of the host galaxies as the jet pushed outwards.

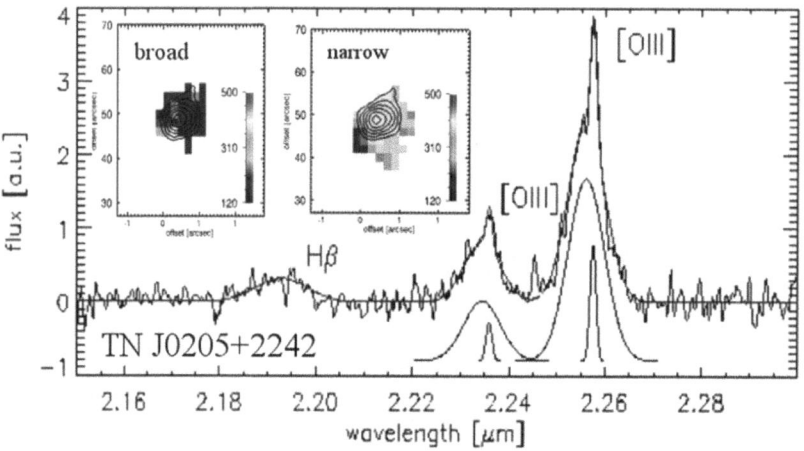

Fig. 2. Integrated K band spectrum of TN J0205+2242 . Insets show the velocity fields of the narrow and broad component. Note that the spatial resolution of the AO assisted data set is 0.4"×0.4" (~3×3 pixels).

The two galaxies with compact radio sources have complex emission line profiles, and compact emission line morphologies. Fig. 2 shows the integrated spectrum of TNJ0205+2242.

3 The Driver and Outflow Energy

Several arguments indicate that the AGN is the most plausible driver of the outflow, and interacts with the AGN through interactions with the radio jet (see also Fig. 3). These can be summarized as: (1) The nebulosities in all extended and compact targets are confined within the radio lobes and approximately aligned with the jet axis; (2) The kinematics of the emis-

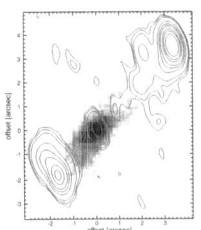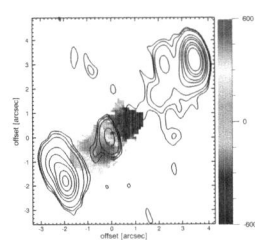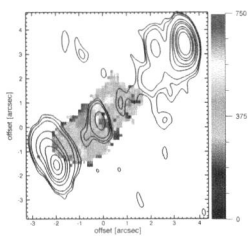

Fig. 3. Relationship between optical and radio emission in MRC0406-244 .

sion line regions are consistent with conical outflows with relative velocities Δv=1000−2000 km s^{-1}; (3) Low-surface brightness radio emission in
MRC0406-244 appears to be related to velocity gradients within each bubble,
and apparently coincides with a "hole" in the line image of this galaxy; (4)
Velocities and spatial extent of the nebulosities indicate dynamical timescales
of a few$\times 10^7$ yrs for the gas, similar to estimated AGN and jet lifetimes [11].

Assuming case B recombination, and neglecting extinction, the observed
Hα luminosity implies ionized hydrogen masses of at least few$\times 10^9$ M$_\odot$ with
low filling factors, $\mathcal{O}(10^{-6})$. These characteristics arise naturally when a sufficiently intense, expanding wind sweeps up, accelerates, and ionizes dense
clouds in the host galaxy's ISM. Following, e.g., [9], we estimate that a total
of $\sim 10^{60-61}$ ergs must be injected into the ISM of the host during a dynamical timescale to explain the observed kinematics in our sample, which closely
follows the standard mechanism assumed for starburst-driven winds (see [13]
for details).

Does the radio jet provide enough energy to plausibly drive the outflow?
Kinetic luminosities of radio jets are notoriously difficult to estimate. Various
methods have been proposed [3, 23, 4], which yield kinetic luminosities that
differ within factors of a few for a given radio luminosity. For our sample,
they imply kinetic luminosities of $0.4 - 2 \times 10^{46}$ erg s^{-1}, on order of what
is required to power the observed outflow. These estimates suggest that the
radio jet might well be the dominant or only driver of the outflow, if the
coupling efficiency between the radio jet and the ISM is $\mathcal{O}(0.1)$. (See also
[13].)

4 Consequences for Galaxy Evolution and Cosmological Impact

The observed outflows can only be "smoking guns" of the ISM clearing phase
of the AGN (§1), if they can unbind 10^{9-10} M$_\odot$ in gas within the tight
time constraints set by the observed [α/Fe] enhancement observed at low
redshift (few$\times 10^8$ yrs). In §3, we estimated kinetic energies of $\mathcal{O}(10^{60-61})$
ergs within the nebulosities, which corresponds to the binding energy of a

massive galaxy. From the dynamical timescales and observed hydrogen gas masses [1] in the extended sources follows a strict lower limit to the mass loss rates, $\dot{M} \geq 100$ $M_\odot$ yr^{-1}, corresponding to $\sim 10^{10}$ $M_\odot$ in a few$\times 10^8$ yrs, and sufficient to fulfill the criteria given by the models. [7] estimate gas infall rates in massive halos at $z \sim 3$ of ~ 200 $M_\odot$ yr^{-1}; simple energy estimates suggest that the energy injection rates through the radio jet are sufficient to shut down accretion over cosmologically significant timescales, and thus to suppress subsequent galaxy growth ([2] find a similar result at low redshift). This is a fundamental difference to starburst-driven winds, which may not be able to remove the gas from the dark matter halos of massive galaxies, and prevent it from subsequently "raining down" onto the galaxy.

To gauge the relative impact of AGN–driven winds compared to starburst-driven winds (see also [13]), we use the co-moving number density of high-redshift quasars of [15], $\Phi_{QSO} = 5 \times 10^{-7}$ Mpc^{-3}, correct for a duty cycle $f \sim 300$ to account for the finite 10^7 yrs lifetime of the radio jets, and adopt a canonical total energy injection of $\mathcal{O}(10^{60})$ ergs. If all luminous AGN go through exactly one radio-loud "clearing phase", this implies that AGN feedback will release an overall energy density $\geq 4.5 \times 10^{56}$ erg s^{-1} Mpc^{-3} into the IGM, which is similar to the energy release of starburst-driven winds [10]. Estimating the mass and in particular the metal release, we find that $\sim 0.6 - 14 \times 10^6 M_\odot$ Mpc^{-3} of mass are expelled into the IGM, and $\sim 1 - 30 \times 10^4$ $Z/Z_\odot$ $M_\odot$ in metals. Compared to the $\sim 1.3 \times 10^6$ $M_\odot$ of metals released through starburst-driven winds at high redshift [6], this suggests that AGN winds are likely not the dominant mode of metal enrichment through cosmic time, they may be however a significant contributor.

References

1. Benson, A. J., Bower, R. G., Frenk, C. S., Lacey, C. G., Baugh, C. M., Cole, S., 2003, ApJ, 599, 38
2. Best, P. N., Kaiser, C. R., Heckman, T. M., Kauffmann, G., 2006, MNRAS, 368, L67
3. Bicknell, G. V., 1995, ApJS, 101, 29
4. Bîrzan, L., Rafferty, D. A., McNamara, B. R., Wise, M. W., Nulsen, P. E. J., 2004, ApJ, 607, 800
5. Bonnet, H. et al., 2004, the Messenger, 117, 17
6. Bouché, N., Lehnert, M. D., Péroux, C., 2006, MNRAS, 367, L16
7. Croton, D. J., Springel, V., White, S. D. M., De Lucia, G., Frenk, C. S., Gao, L., Jenkins, A., Kauffmann, G., Navarro, J. F., Yoshida, N., 2006, MNRAS, 365, 11
8. Dey, A., van Breugel, W., Vacca, W. D., Antonucci, R., 1997, ApJ, 490, 698
9. Dyson, J. E., Williams, D. A., 1980, Physics of the interstellar medium, (New York, Halsted Press)

---

[1] Note that the mass loss estimate only includes ionized hydrogen; intrinsic gas masses, e.g. due to other phases of the ISM, will be significantly higher.

10. Lehnert, M. D., Heckman, T. M., 1996, ApJ, 472, 546
11. Martini, P., 2004, in Coevolution of Black Holes, Galaxies, ed. L.C. Ho (Cambridge University Press)
12. McCarthy, P. J., Baum, S. A., Spinrad, H., 1996, ApJS, 106, 281
13. Nesvadba, N. P. H., Lehnert, M. D., Eisenhauer, F., Gilbert, A., Tecza, M., Abuter, R., 2006, ApJ, 650, 693
14. Papadopoulos, P. P., Röttgering, H. J. A., van der Werf, P. P., Guilloteau, S., Omont, A., van Breugel, W. J. M., Tilanus, R. P. J., 2000, ApJ, 528, 626
15. Pei, Y. C., 1995, ApJ, 438, 623
16. Silk, J., Rees, M. J., 1998, A&A, 331, L1
17. , 2005, MNRAS, 361, 776
18. Tadhunter, C. N., 1991, MNRAS, 251, 46
19. Thomas, D., Maraston, C., Bender, R., Mendes de Oliveira, C. , 2005, ApJ, 621, 673
20. van Breugel, W. J. M., Stanford, S. A., Spinrad, H., Stern, D., Graham, J. R., 1998, ApJ, 502, 614
21. Venemans, B. P., 2006, AN, 327, 196
22. Villar-Martín, M., Binette, L., Fosbury, R. A. E. , 1999, AAP, 346, 7
23. Wan, L., Daly, R. A., Guerra, E. J., 2000, ApJ, 544, 671

2dF–SDSS LRG and QSO (2SLAQ) Survey: Evolution of the Most Massive Galaxies

R. C. Nichol[1], R. Cannon[2], I. Roseboom[3] and David Wake[4], for the 2SLAQ Collaboration

[1] ICG, University of Portsmouth, Portsmouth, PO1 2EG, UK
[2] Anglo-Australian Observatory, PO Box 296, Epping, NSW 1710, Australia
[3] Dept. of Physics, University of Queensland, QLD 4072, Australia
[4] Dept. of Physics, Durham University, South Road, Durham DH1 3LE, UK

Abstract: The 2dF–SDSS LRG and QSO (2SLAQ) survey is a new survey of distant Luminous Red Galaxies (LRGs) and faint quasars selected from the Sloan Digital Sky Survey (SDSS) multi–color photometric data and spectroscopically observed using the 2dF instrument on the Anglo-Australian Telescope (AAT). In total, the 2SLAQ survey has measured over 11000 LRG redshifts, covering 180deg^2 of SDSS imaging data, from 87 allocated nights of AAT time. Over 90% of these galaxies are within the range $0.45 < z < 0.7$ and have luminosities consistent with $\geq 3L^\star$. When combined with the lower redshift SDSS LRGs, the evolution in the luminosity function of these LRGs is fully consistent with that expected from a simple passive (luminosity) evolution model. This observation suggests that at least half of the LRGs seen at $z \simeq 0.2$ must already have more than half their stellar mass in place by $z \simeq 0.6$, i.e., our observations are inconsistent with a majority of LRGs experiencing a major merger in the last 6 Gyrs. However, some "frosting" (i.e., minor mergers) has taken place with $\sim 5\%$ of LRGs showing some evidence of recent and/or on–going star–formation, but it only contributes $\sim 1\%$ of their stellar mass.

1 2SLAQ Survey

The 2dF–SDSS LRG and QSO (2SLAQ) survey is a collaboration of over 70 UK, US and Australian astronomers focused on studying the evolution with redshift of Luminous Red Galaxies (LRGs) and faint quasars. The details of the 2SLAQ survey can be found in a series of recent papers, including Cannon et al. (2006) and Richards et al. (2005). In this paper, we focus on two recent works on the evolution of LRGs with redshift (Wake et al. 2006; Roseboom et al. 2006) from the joint SDSS and 2SLAQ surveys.

2 Evolution of the Luminosity Function

As seen in Figure 1, the SDSS and 2SLAQ luminosity functions (LFs) brighter than $M_{0.2_r} = -22.6$ are in excellent agreement when the passive evolution cor-

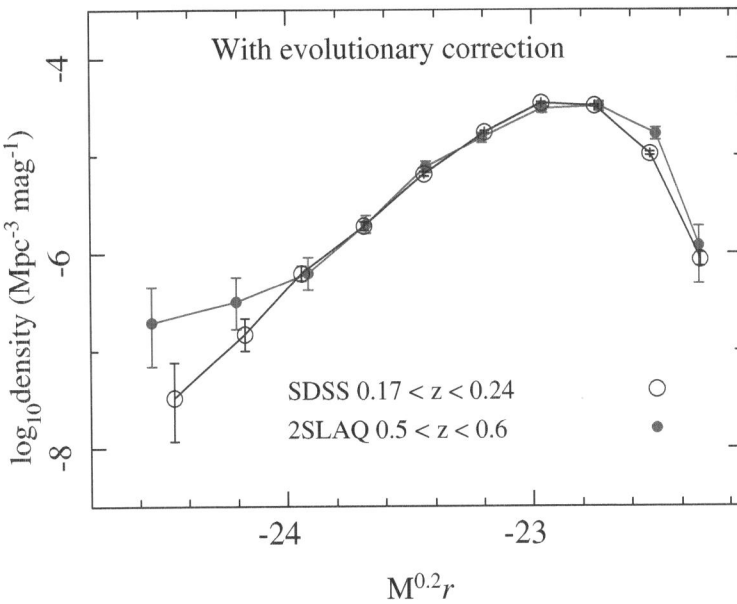

Fig. 1. The $M_{0.2_r}$ (absolute magnitude at $z = 0.2$) luminosity function after passive evolution corrections for both the SDSS (open data points) and 2SLAQ (solid data points) LRG samples have been made. The points are plotted with their one sigma error bars as described in Wake et al. (2006)

rections are included. The agreement of these luminosity functions is further confirmed by calculating the integrated number and luminosity density of LRGs brighter than $M_{0.2_r} = -22.5$, which agrees to better than 10% out to $z = 0.6$. Throughout the analysis, the same simple passive evolution model was used for predicting and correcting the colours and luminosities of LRGs as a function of redshift, and this agreement demonstrates the lack of any extra evolution, beyond the passive fading of old stars, out to $z \simeq 0.6$.

It may appear that our lack of extra evolution beyond passive (out to $z \sim 0.6$) is in conflict with recent results from the COMBO-17 (C17) and DEEP2 (Bell et al. 2004; Faber et al. 2005). These smaller–area, but deeper (in magnitude limit and redshift), surveys find evidence for a change in the density of red galaxies out to $z \sim 1$ beyond that expected from passive fading of the stellar populations. For example, Faber et al. (2005) report a quadrupling of ϕ^* for red galaxies since $z = 1$, although this result is strongest in their highest redshift bin, where they admit their data are weakest. A direct comparison with these deeper surveys is difficult because of the differences in colour selections used for the surveys, as well as the relative luminosity ranges probed by the different surveys, i.e., the 2SLAQ survey is designed

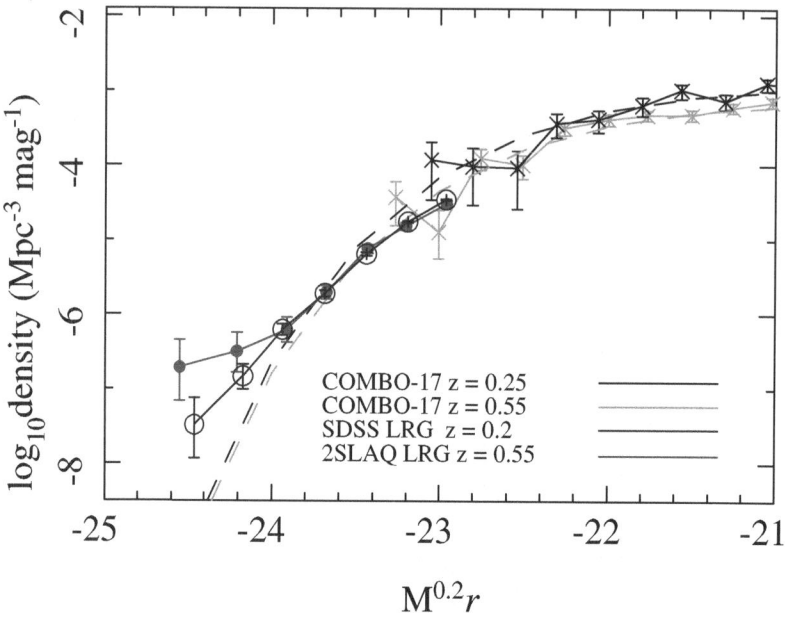

Fig. 2. The $M_{0.2_r}$ luminosity function with passive evolution corrections for the SDSS (blue open data points), 2SLAQ (red solid data points) LRG samples, and the COMBO-17 red galaxies at z = 0.25 (black open stars) and z = 0.55 (green solid stars). The dashed lines show the Schechter function fit to the COMBO-17 points. The points are plotted with their one sigma errors.

to probe galaxies brighter than a few L^*, while the DEEP2 and C17 surveys effectively probe galaxies below L^* at $z \sim 0.6$ (due to their smaller areal coverage and fainter magnitude limits).

However, to facilitate such a comparison, we show in Figure 2, the LFs from Figure 1, and the C17 red galaxy LFs for the same redshift range and K+e corrected to $M_{0.2_r}$. We only plot our LFs to $M_{0.2_r} < -22.9$ as we do not include all the red galaxies fainter than this due to the SDSS LRG selection criteria. Figure 2 demonstrates that when one restricts the data to the same redshift range, there is excellent qualitative agreement between the 2SLAQ and C17 luminosity functions. We are unable to make a quantitative comparison due to the difficulty in exactly matching the selection criteria of the two surveys. Taken together, the surveys shown in Figure 2 extend the evidence for no evolution in the LF of LRGs to $M_{0.2_r} < -21$, which is close to L^* in the LF. Figure 2 also demonstrates that these two surveys are probing different luminosity regimes at $z < 0.6$ as there is at most only 0.5 magnitudes of overlap in their LFs in which the C17 survey is becoming seriously affected by small number statistics due to its smaller areal coverage.

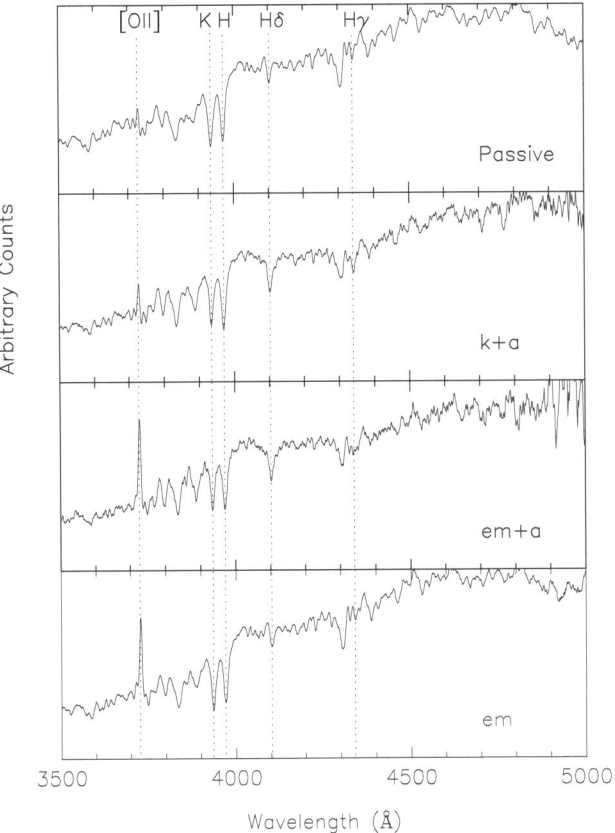

Fig. 3. The combined spectrum for the four spectral classifications considered in Roseboom et al. (2006) i.e. passive (top), "k+a", emission ([OII]) plus absorption (Hδ), and then just emission ([OII]) on the bottom. Note the increase in the absorption strength of the higher order Balmer lines (Hγ,Hϵ, etc) in the middle two combined spectra.

The luminosity functions given in Figures 1 and 2 place tight constraints on models of massive galaxy formation and evolution. Our results appear to favour little, or no, density evolution, i.e., there are already enough LRGs per unit volume at $z \simeq 0.6$ to account for the density of LRGs measured at $z \simeq 0.2$. Using a simple model for "dry (major) mergers", we find that the 2SLAQ and SDSS LFs are consistent with each other without any need for merging. At the 3σ level, we can exclude merger rates of $> 50\%$, i.e., more than half the LRGs at $z = 0.2$ are already well-assembled, with more than half their stellar mass in place, by $z \simeq 0.6$. Our limit is barely consistent with the predictions in Figure 5 of De Lucia et al. (2006), where they show that $\sim 50\%$ of $z = 0$ massive ellipticals have accreted 50% of their stellar mass since $z \simeq 0.8$.

3 Minor Mergers

Our simple model for "dry (major) mergers" above does not constrain the rate of minor mergers involving LRGs. For example, Roseboom et al. (2006) determined the recent star formation histories of the SDSS and 2SLAQ LRGs based on the Hδ and [OII] lines (see Figure 3 for examples). While the majority (>80 per cent) of LRGs show the spectral properties of an old, passively evolving, stellar population, a significant number of LRGs show evidence for recent and/or ongoing star formation in the form of "k+a" (2.7 per cent), emission+absorption (1.2 per cent) or just [OII] emission LRGs (8.6 per cent). Therefore, $\sim$ 5% of LRGs have evidence for recent star–formation (k+a's and the emission+absorption), while it is unclear what percentage of the [OII] emitters for due to on–going star–formation or have an AGN. Furthermore, the [OII] emitters could be interlopers in the sample, scattered over the color boundaries.

By dividing the sample into 2 redshift subsamples from $0.45 < z < 0.55$ and $0.55 < z < 0.65$, and comparing to a $z \sim 0.15$ sample selected from SDSS, it is observed that the fraction of "k+a" LRGs increases with redshift as $(1 + z)^{2.8 \pm 0.7}$. Spectral synthesis models suggest that these LRGs could originate from passive LRGs which have undergone a starburst involving only $\sim$ 1% of their total stellar mass. Therefore, the $\simeq$ 5% of LRGs that show evidence of recent and/or on–going star–formation (k+a and emission+absorption galaxies) are probably produced by minor mergers with gas–rich dwarf galaxies and represent a "frosting" of new stars on top of the majority older stellar population.

Acknowledgement. RCN thanks Hans Boehringer for still including this paper in the proceedings even though he missed the conference due to the August 10th 2006 terrorist alert at Heathrow Airport. We all thank the 2SLAQ team for their hard work and allowing us to present their work here. We again thank the AAO staff for all their assistance during the collection of the 2SLAQ data.

References

1. Bell, E. F., et al. 2004, ApJ, **608**, 752
2. Cannon, R., et al. 2006, MNRAS, **372**, 425
3. De Lucia, L., et al. 2006, MNRAS, **366**, 499
4. Faber, S., et al., 2005, astro-ph/0506044
5. Richards, G. T., et al. 2005, MNRAS, **360**, 839
6. Roseboom, I. G., et al. 2006, MNRAS, **373**, 349
7. Wake, D. A., et al. 2006, MNRAS, **372**, 537

Part XI

Conference Summary

Heating vs. Cooling 2006: Conference Summary

R. Bower

Institute for Computational Cosmology, Durham University, UK
r.g.bower@durham.ac.uk

1 Introduction

This is an exciting time. We are getting a good grip on the galaxy formation problem [25, 7, 3], and a lot of that development has come from the progress being made on clusters cooling flows. This conference comes immediately after the Cosmic Frontiers meeting in Durham [6]. That meeting focussed largely on the issue of galaxy formation. The juxtaposition of the meetings emphasises the fundamental interconnection of galaxy formation and galaxy clusters.

To progress in both areas, there are three key problems that need to be understood. 1) The break in the luminosity function of galaxies. We are only just beginning to understand the process that creates the break at the bright end of the galaxy luminosity function. The solution should explain why the milky-way is a typical large galaxy, and why the spectrum of galaxy masses does not extend to much higher masses. 2) The "hierarchy problem". In a CDM universe, structure forms from the smallest masses up, so that the largest structures have formed most recently. In contrast, the largest galaxies have some of the oldest stellar populations. They are "red and dead". 3) The "cooling flow problem" that we are here to discuss. The optimists amongst us hope that by solving the last of these problems, we will solve the other two.

Cluster formation, and galaxy formation are intimately connected. And, at this meeting, I'll argue that the clusters are the more important. After all, more than 85% of the baryonic mass is in the form of the hot intra-cluster medium (ICM) and less than 15% is in the form of stars and cold gas. The ratios may be similar in the universe as a whole, but, because of the high temperature of the ICM in groups and clusters it is only here that we can make a complete census of all the forms of baryonic matter. Thus clusters make ideal astrophysical laboratories where we can develop our understanding and then perhaps apply it to smaller haloes. Perhaps in future meetings, we'll also be able to use clusters as calorimeters, helping us understand the total energy budget of the universe.

I started writing this summary before the meeting, but I have to admit that I've had to throw a lot away. I'm pleased to see that some of the things I thought would be contentious no longer are. It seems that people are generally happy to accept that cooling flows don't flow, and that AGN do heat.

So it seems that I'm allowed to describe cool-core clusters as cooling flows again. Although there's relatively little mass dropping out of the flow, we realise that there's still energy flowing to replace (somehow) the energy radiated. There might even be mass flowing around in the central region. However, something that's not quite clear to me after listening to all these talks: what property of a cluster makes it count as a "cooling flow" or a "cool-core" cluster: is it a central temperature decrement? a mass deposition rate (even if this is just a notional concept)? a central cooling time or entropy? or even a central density slope? This is something we all need to agree on. I vote for the third option.

In the rest of this summary, I briefly try to pull together and tease-out some of the ideas that I've seen presented. Inevitably I'll miss out many excellent and exciting strands.

2 Is cooling self-regulated?

Many talks have investigated the energy budget in various clusters, measuring the energy that is being pumped into the ICM from the size and pressures of either cavities in the X-ray gas, or radio synchrotron maps. The cavities seem to be almost everywhere they are needed (eg., Fabian suggested that they could be seen in more than 70% of cool-core clusters [9]) and that the energy available could comfortably reheat the radiated energy, especially if we might be catching strong variations in the power of the central source.

The central cooling times of cool-core clusters are short. If there's enough energy, and if we can identify a mechanism to tightly couple the energy being injected to the cooling rate, we have all the ingredients of a stable feedback loop. We could then expect that the energy from the central source would adjust the entropy profile of the cluster until the average heating and cooling rates balance.

Churazov's talk [5] illustrated how this can be formalised. He suggested that, if the jet was powered by Bondi accretion from the hot gas, the heating rate and cooling luminosity would both depend on the central entropy but with very different powers (the cooling luminosity would have a much steeper dependence). In this case, its simple to imagine that the system will evolve towards a stable equilibrium, and we can expect cooling and heat to balance when we average over a period of time. This all sounds very optimistic, but we have yet to show that the system is sufficiently tightly coupled for the feedback loop to be stable. There are many different timescales in the problem [13]. For example, the cooling time, the sound crossing time (the time that the system takes to react to energy input), the star formation timescale, the timescale for cooling material to reach the accretion disk, the timescale for material to flow into a central black hole and launch a jet. The potential for interaction between these processes were nicely illustrated in [7, 3]. The ingredients are all there, but it is far from clear that they lock together to form a stable system.

A particular fly in the ointment is that accretion rates that are too high might lead to dramatic drop in jet power, since jets are believed driven by geometrically thick, optically thin disks [5, 18] with relatively low accretion rates. At higher, accretion rates, the geometrically thick disk would collapse and the system might get trapped in a run away state with high mass in-flow rates and low jet powers. It seems, however, that this rarely happens in the local universe.

3 How might self-regulation work?

Self-regulation is an appealing idea, but we are only just beginning to understand how it might work. Everyone's favorite mechanism seems to be heat input from jets fired from a central black hole. If we accept that jets from the central black hole could provide enough power to offset the energy being radiated by the cluster, the next question that we need to address is how the jet energy gets dispersed into the cooling ICM. The mass of material being injected seems to be a small fraction of the mass cooling flow rate. Thus we seem to need to mix the injected energy into the ICM in order to make the system work. At present its not clear how this happens, and trying to smoothly heat the large volume of the cooling region with a heat source as small as a black hole seems problematic [9]. However, it's not clear to me that the cluster really is in a steady state, and perhaps this is part of the solution, rather like the manner in which a small radiator heats can effectively heat a large room.

So how might the AGN actually heating the ICM? Several ideas are vying for this privilege. Good candidates include: weak shocks generated as the bubble is inflated [20]; the buoyancy of the bubble and the ICM motions that it generates [5]; uplift of material from the densest regions [15]; and mixing the bubble energy with the surrounding ICM. There isn't currently a consensus over these ideas. If you want my personal opinion, it's the third option that dominates. The idea of circulation flows was introduced by Matthews [15]. In a recent preprint, McCarthy [17] shows that a many of observed properties of cluster can be simply and naturally solved by combining circulation flows with a moderate level of cluster pre-heating.

There are many challenges that these candidate mechanisms need to over come. Two, in particular, occupied people's attention at the meeting: bubble morphology and metallicity gradients. There was also the mysterious "isothermal shock" in Perseus [9].

While observations of the cavities seem to be naturally explained by hot plasma bubbles, this morphology is a challenge for simulations. Hydrodynamical simulations do not appear to produce the right solution: the bubbles are rapidly shreded by Rayleigh-Taylor instabilities as they rise through the ICM. At best, the simulations produce turbulent smoke rings. It seems very likely that we need to include weak magnetic fields in the simulations in order to suppress the shredding process and preserve the bubble morphology out to larger radius in the cluster [11]. It was encouraging to see that work on magneto-hydrodynamic simulations is gaining pace, even though the existing simulations are still rather simple [8]. Indeed, it's not yet clear that the magnetic fields actually make the bubble shapes a better match to observations.

At the meeting many seemed convinced that circulation flows couldn't work (eg. how are the metallicity gradients preserved? the energy might be carried to too great a radius; why does the process result in a universal density profile? - but see [17]). One possibility is that everything is much more complicated and cannot be described by hydrodynamics alone. Perhaps the heat input is mediated by things that most simulators would rather not consider (but see [21]). For example, cosmic rays and relativistic particles might provide a viable heating scheme. Viscosity might also play an important role in distributing the energy input from the AGN. Such "Funny Fizzics" might be the solution, but, to be honest, I'd rather it wasn't! If it is cosmological simulations of galaxy formation are going to get a whole lot more difficult, and it may never be possible to simulate cosmologically significant volume in sufficient detail.

It was interesting that some possible heating mechanisms didn't get many votes: conduction seemed to be out of fashion at the meeting (it's too unstable); galaxy motions received little consideration too (why? - that's less clear to me).

4 Where does the AGN power come from?

Although AGN seem to be the likely power-house, we are still far from a complete solution. The cooling flow is happening on scales of 10 kpc and larger. We are far from being able to connect this to the fueling of the black hole. The jet power is much generated close to the last stable orbit ($\sim 10\,\mathrm{au}$); however, the Bondi radius (where the gravity of the black hole begins to dominate over the pressure of the ICM) is much closer to being observable at around $\sim 10\,\mathrm{pc}$. As a starting point, Bondi accretion onto the central black hole is a good guess. Indeed, Allen's results [1] suggest a remarkably good connection between the Bondi rate in elliptical galaxies and the power needed to offset the cooling radiation (if the BH is rapidly spinning [18]). However, it is far from clear just how relevant the Bondi rate really is: the cluster gas has significant angular moment and thus it seems unlikely that the ICM can accrete directly onto the BH with out first passing through a large-scale disk. It's unclear why the fueling rate of this disk should be closely related to the power of the AGN. More over, Allen's results apply to elliptical galaxies (which are close enough the the gas temperature and density at the Bondi radius can be inferred with minimal extrapolation). In more massive systems, the Bondi rate must be inferred by extrapolation. It is worrying that making a careful attempt at this suggests that the energy budget falls short of what is required in high-mass clusters [23].

Another fundamental question that we need to address is why the black holes in the centres of clusters launch high efficiency jets while other black holes appear to radiate most of their accretion energy as powerful quasars. There's at least a sketchy picture emerging, in which the power of the jet is related to the scale height of the magnetic field threading the disk [5, 14, 18]. Thus the jet efficiency may be connected to the advection dominated and radiatively efficient solutions for the accretion disks, as well as to the spin of the black hole.

Self-regulation is an appealing idea, but a cascade of processes (each within their own timescales) is required to make it work. Until we can clearly demonstrate how they are linked together, it's only optimism that suggests we have the right solution.

5 What about the rest of the clusters?

The meeting was clearly focussed on "cool core" (aka. cooling flow) clusters. It is important to remember that this is less than 50% of the cluster population, and to think about how the science results we've discussed so far generalise to the rest of the population. Do their entropy profiles differ at large radius? (no) Do they have cavities, suggestive of past episodes of AGN activity? (not clear yet). What happened to make these systems different? Burns [4] suggested that it was early mergers, another possibility is that it is the variation in the amount of entropy injection by AGN early in the formation history of the cluster (ie., pre-heating, [17]). Clearly this is an important issue, and we need to view the cluster population as a whole. One of the problems is that many cluster studies are biased to the most "interesting" systems. For example, many studies preferentially target "cool core" systems because of their high surface brightness. The XMM large programme [22] will make a significant improvement to the staus quo.

Considering the diversity of clusters brings out two related issues. A number of talks targeted lower mass haloes (see [10] for an overview). It is crucial to understand how and why these systems differ from their high-mass counterparts. The observations suggest that heating becomes even more important, apparently destroying the system and pushing baryons out beyond the observable X-ray region [2]. This is fascinating.

Conversely, in larger clusters, we see convergence in the scaled profiles at large radius. Fantastic new data is emerging that probes entropy profiles out to large radius [22, 24]. These measurements can be compared directly to numerical simulations of clusters [12, 19]. Although these produce far too higher a fraction of stars, they show that the measurements give reliable constraints on the baryon fraction with the virial radius. It also possible to show that the energy available for the AGN is unlikey to expel a significant fraction of the baryons from massive system (hence the convergence that I alluded to above). This has important implications since the measured baryon fraction in clusters are incompatible with latest results for the CMB anisotropy and polarisation from WMAP [16]. The discrepancy also has important ramifications for the power spectrum amplitude (σ_8). I believe we will see these values revised downwards (for Ω_b, upwards for σ_8) in the near future.

6 The connection to galaxy formation

The three problems of galaxy formation that I high-lighted at the beginning all stem from the "cooling flow problem" that we are here to discuss. What's impressive, they are simply different aspects of the same problem. It's now possible to implement a guess at these schemes in semi-analytic models of galaxy formation [25, 3]. Applying a simple physical model for AGN feedback not only solves the "cooling flow problem", it also explains why the the "hierarchy problem" that has dogged galaxy formation theories. By providing a mechanism that makes a sharp cut in the galaxy luminosity function the parameters of the model naturally adjust to produce a population of early massive galaxies similar to that observed. In [3], we called this the "broken hierarchy" of galaxy formation because of the clean way in which AGN heating breaks galaxy formation away from the hierarchical growth of cold dark matter haloes. It solves many other problems too.

In my summary, I've tried to stress that the problem is far from solved. While the paradigm seems capable of explaining the observed universe, the devil lies in the detail, and the model must be fleshed out to actually explain how the AGN heat the ICM, and how the AGN is fuelled. Only then will we be able to demonstrate that the process contains the right scales, and that the feedback loop is sufficiently tightly coupled. Personally I find it amazing that something as small as a black hole might determine the entire future of the luminous universe.

Acknowledgement. I would like to thank the organisers of the workshop for a particularly interactive and stimulating meeting. In particular, I am indebted to Ian McCarthy and Mark Voit for trying to help me understand how it must all hang together. I would also like to thank PPARC for the support of a senior research fellowship.

References

1. Allen S., this conference
2. Arnaud M., this conference

3. Bower R. G., Benson A. J., Malbon R., Helly J. C., Frenk C. S., Baugh C. M., Cole S., Lacey C. G., 2006, MNRAS, 370, 645

4. Burns J.O., this volume

5. Churazov E., this conference

6. Cosmic Frontiers, ed. Shanks, et al., 2006, ASP Conf. Series, in press.

7. Croton D. J., Springel V., White S. D. M., De Lucia G., Frenk C. S., Gao L.,. Jenkins A., Kauffmann G., Navarro J. F.,

8. De Young D.S., this volume Yoshida Y., 2006, MNRAS, 367, 864

9. Fabian A., this volume

10. Jones C., this volume

11. Kaiser C. R., Pavlovski G., Pope E. C. D., Fangohr H., 2005, MNRAS, 359, 493

12. Kravtsov A. V., this conference

13. McNamara B. R., this volume

14. Meier D. L., 2001, ApJ, 548, L9

15. Matthews W. G., Brighenti F., Buote D. A., 2004, ApJ, 615, 662

16. McCarthy I. G., Bower R. G., Balogh M. L., 2006, astro-ph/0609314

17. McCarthy I. G., Babul A., Bower R. G., BAlogh M. L., 2007, MNRAS submitted.

18. Nemmen R. S., Bower R. G., Babul A., Storchi-Bergmann T., 2006,astro-ph/0612392

19. Nagai D., Kravtsov A. V., Vikhlinin A., this volume

20. Nulsen P. E. J., this volume

21. Pfrommer C., this volume

22. Pratt G.W., this conference

23. Rafferty D. S., this volume

24. Vikhlinin A., this volume

25. White S. D. M., this conference

Part XII

Colour Plates

Perseus pressure map with radio overlay. See also page 70

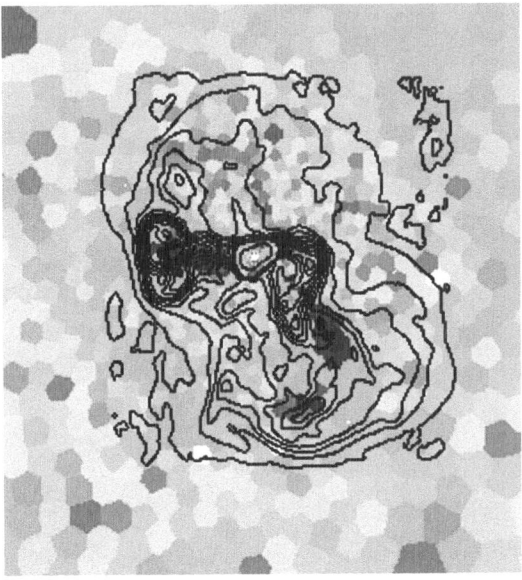

Entropy deviations from a smooth radially symmetric model overlaid with radio contours (90 cm). Radio map kindly provided by F. Owen. The E and SW radio lobes coincide well with the regions of low entropy in the X-ray arms. Also, the edge of the large radiolobe to the north roughly coincides with a NW edge in the entropy map. See also page 88

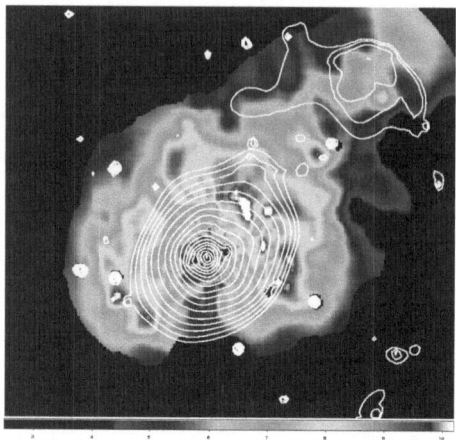

Temperature map of the Cygnus cluster. See also page 103

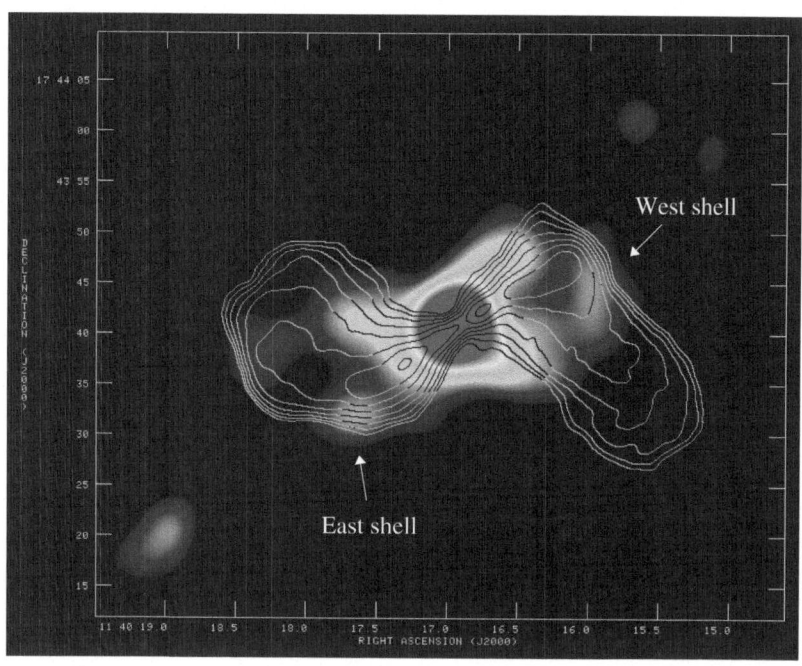

Gaussian smoothed 0.5 - 2 keV image with 20 cm radio contours. See also page 98

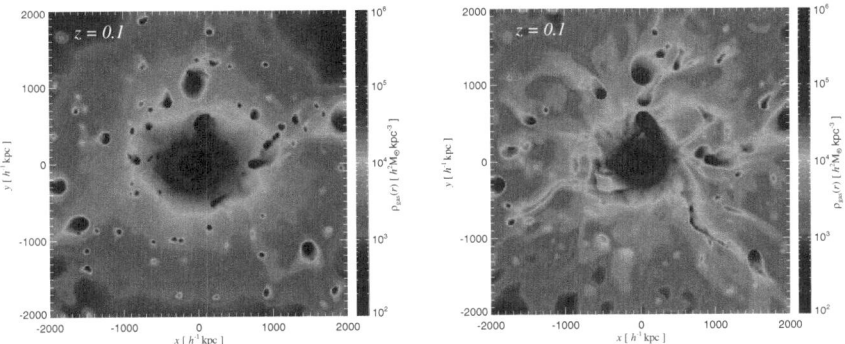

Projected gas density maps of a galaxy cluster simulation at redshift $z = 0.1$, as indicated in the upper-left corner of the panels. The left-hand panel shows the gas density distribution in the case of a non-radiative run, while the right-hand panel gives the gas density distribution when Braginskii shear viscosity is "switched-on", using a suppression factor of 0.3. See also page 239

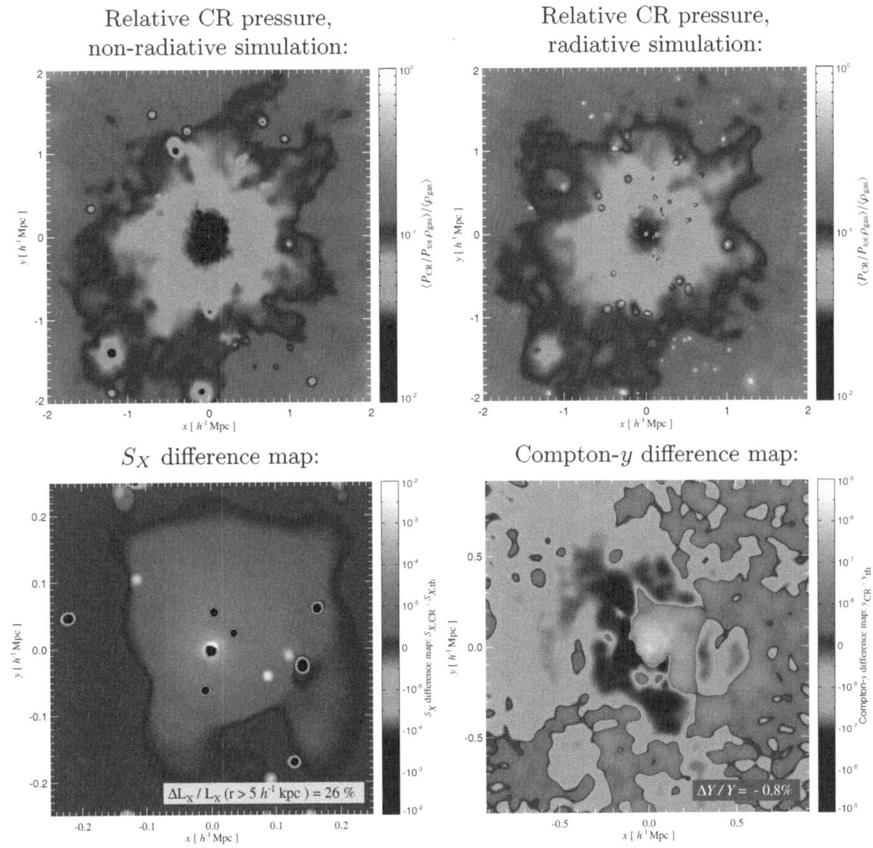

Relative CR pressure, non-radiative simulation:

Relative CR pressure, radiative simulation:

S_X difference map:

Compton-y difference map:

"Top panels: Relative cosmic ray pressure in a $10^{14}\,M_\odot$ cluster in non-radiative simulations (left) and simulations with dissipative gas physics (right). Bottom panels: Cosmic ray-induced difference of X-ray surface brightness (left) and Compton y-parameter (right) compared to corresponding simulation without cosmic rays. See also page 376."

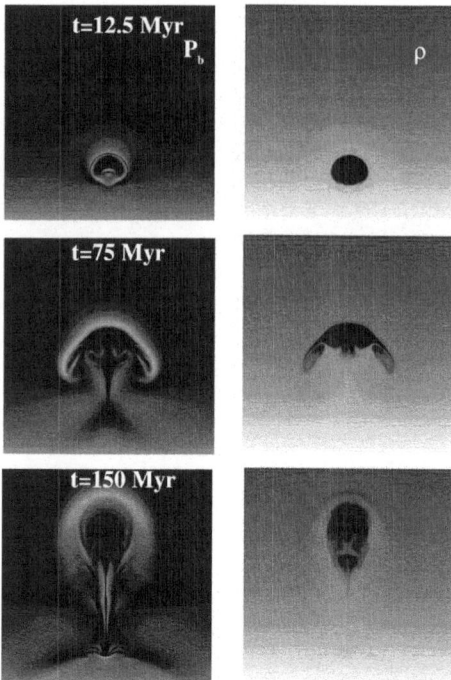

Time evolution of a radio source bubble in a 2D MHD simulation for $\beta = 120$. Magnetic pressure is on the left, and density is shown on the right. (From Jones & De Young 2005.) See also page 206

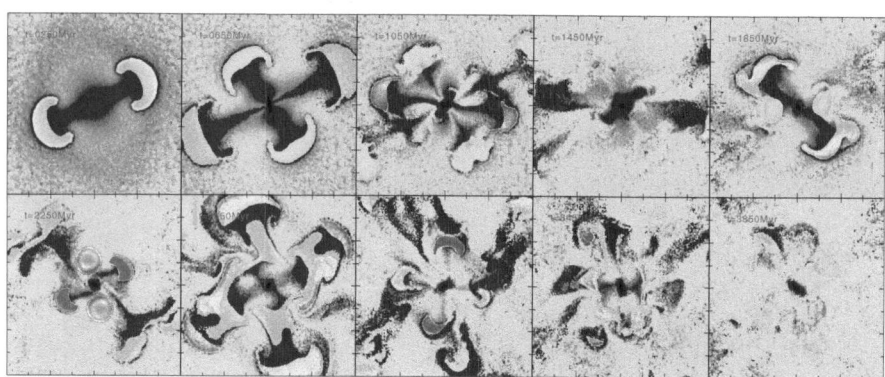

Global evolution of the cluster temperature during 4 Gyr. The box dimension is $400 \times 400\,\text{kpc}$. See also page 331

Object Index

ESO ASTROPHYSICS SYMPOSIA
European Southern Observatory

Series Editor: Bruno Leibundgut